W9-AQW-268

A THEORY OF DETERMINISM

The Mind, Neuroscience, and Life-Hopes

TED HONDERICH

CLARENDON PRESS · OXFORD
1988

Oxford University Press, Walton Street, Oxford OX2 6DP
Oxford New York Toronto
Delhi Bombay Calcutta Madras Karachi
Petaling Jaya Singapore Hong Kong Tokyo
Nairobi Dar es Salaam Cape Town
Melbourne Auckland
and associated companies in
Beirut Berlin Ibadan Nicosia

Oxford is a trade mark of Oxford University Press

Published in the United States
by Oxford University Press, New York

British Library Cataloguing in Publication Data
Honderich, Ted
A theory of determinism: the mind,
neuroscience and life-hopes.
1. Neurophysiological determinism & free
will – Philosophical perspectives
I. Title
123
ISBN 0-19-824469-X

Library of Congress Cataloging in Publication Data
Honderich, Ted.
A theory of determinism, the mind, neuroscience, and life-hopes/
Ted Honderich.
Bibliography: p. Includes index.
1. Neuropsychology. 2. Determinism (Philosophy) 3. Quantum theory. I. Title
QP360.H665 1988 612'.8--dc19 87-34926
ISBN 0-19-824469-X

Set by Pentacor Limited,
High Wycombe, Bucks
Printed in Great Britain
by Biddles Limited,
Guildford & King's Lynn

A THEORY OF DETERMINISM

To Bee, Janet, John, Kiaran,
Pauline, and Ruth

Acknowledgements

The following, to whom I am most grateful, commented on some or all of the last draft of this book: Prof. Sir Alfred Ayer, Prof. Colin Blakemore, Dr Malcolm Budd, Dr Kerry Greer, Dr W. D. Hart, Ms Cynthia Macdonald, Mr Derek Parfit, Prof. Christopher Peacocke, Ms Nicola Lacey, Mr Derek Matravers, Mr Wayne Norman, Prof. Anthony O'Hear, Prof. Timothy Sprigge, Prof. Euan Squires, Dr Anthony Sudbery, Ms Patricia Walsh, Prof. J. Z. Young, and an anonymous reader for the Oxford University Press.

The following, to whom I am also very grateful, commented on a previous draft, or on writings which issued directly in the book: Prof. Donald Davidson, Prof. Daniel Dennett, Prof. Sir John Eccles, Prof. Paul Fatt, Dr Gertrude Falk, Dr Vivette Glover, Prof. Alastair Hannay, Prof. Benjamin Libet, Mr David Lloyd-Thomas, Mr John Mackie, Mr Peter Morriss, Mr Nicholas Nathan, Dr Carlos Nino, Prof. Hilary Putnam, Ms Janet Radcliffe Richards, Mr Richard Rawles, Prof. David Sanford, Prof. Amartya Sen, Mr Peter Smith, Prof. Stephen Stich, Prof. Patrick Wall, Prof. V. M. Weil, Dr. Edgar Wilson.

I have also benefited from questions, objections, and stoicism on the part of postgraduates and undergraduates at University College London—particularly in a seminar on the manuscript in 1986—and stalwarts who turned up to guest lectures and meetings of philosophy societies in British, American, and Canadian universities.

I also thank, although we have had our differences, my teachers, past colleagues, and colleagues at University College London, including, first of all Prof. Jerry Cohen for our many spirited encounters, and also Prof. Ayer, Prof. Stuart Hampshire, Dr Peter Downing, Dr John Watling, Prof. Bernard Williams and Prof. Richard Wollheim.

Mr John Allen and Mr John Spiers, librarians at University College London, were most helpful.

Typing was done with forbearance by Ms Catherine Backhouse, Ms Dinah Perry, and Ms Wendy Robbins.

Contents

List of Figures

Introduction

Is there a clear and explicit, consistent, complete, and in general a *conceptually satisfactory* theory of our ongoing lives, above all our deciding and acting, in terms of causal or other necessary connections? Such connections, called nomic or lawlike connections to distinguish them from logically necessary connections rooted in words and other symbols, are pervasive in the natural world, and fundamental to it. They are connections not between words but in reality. They are the cement of the universe.

The same large question about our own ongoing lives, above all our deciding and acting, can be put in several different ways.

Is there a conceptually satisfactory theory of ourselves which locates us within the natural world rather than apart from it? Again, is there a conceptually satisfactory theory of our lives which *eschews* explanations, if they can be called such, which are intrinsically mysterious? Such mysterious explanations include within themselves a certain proposition, on reflection a striking one. It is that *every* fact about a person, including every fact about brain and Central Nervous System, and character and personality, and thought and feeling, might have been exactly as it was before and at the moment when the person understood something, or hoped, or decided, or acted, and nevertheless the understanding, hoping, deciding, or acting might never have occurred. That was a possibility in reality, not merely something that can be thought without contradiction. Is there, to put the question in a last and the most familiar philosophical way, a conceptually satisfactory *determinist* theory of our existence?

In the second half of this century, and perhaps before, it has been replied, and perhaps more often thought, that there is not. This book gives the answer that there is. It does so by providing one. This is the main endeavour of the first of its three parts.

However well supported or confirmed, what is ordinarily called a theory remains at least in some way tentative, not advanced or taken as a settled truth, or at any rate not something on a level with *simple* truths. A theory is not a report of observations, but a deeper account of them. Our relation to the atomic theory, or the theories of evolution or gravity or of reciprocal synapses, is not our relation to the simple truths that things have weight or hearts, or that they fall, or change together. Still, the fundamental fact for us about any theory, however conceptually satisfactory, is the fact of how it stands to reality. What reason can we have for accepting it?

Hence the second question addressed in this book, in its second part, is whether the theory of determinism expounded in it is well supported. There is reason to think so. If it cannot be said to be proved true, it cannot be said to be proved false either. Certainly I believe the determinism of this book, at any rate the main lines of it, or, to speak a bit more carefully, I take it to have the strength, say, of the theory of evolution. It is, in my feeling about it, no speculation. To concede that belief in the theory goes somewhat beyond the evidence and is consistent with some hesitation and diffidence is perhaps not to all greatly more than that the belief in question, like a great deal e indeed everything of interest, is not a matter of sense-experienc simple logical necessity.

The third and last question, quite as large, is that of the *con quences* of this determinism. What human consequences are there of any theory of our existence which is determinist in the strong way of the theory of this book? Many theories of our existence, including Freud's, whatever else is to be said of them, can be, and often are, tal in the strong deterministic way. The same question of conseque arises, as it turns out, about certain widely accepted outlo theories which are less deterministic, or indeed partly indeterr and wholly consistent with a common interpretation of Theory. It arises, for example, about what is someti naturalism, or the scientific vision of ourselves, which ne micro-indeterminism with macro-determinism. (D 984; Searle, 1984, Lect. 6) Philosophers have given more o this question of consequences than to the first two, thos ceptual adequacy and of the truth of determinism. Expressed ly, and as generally, it is the question of what we are to make of r lives if or since determinism is true.

It has given rise to two traditions, each seeming to be as strong now, in 1988, as it has been in past centuries. The answer given in one tradition is that if determinism is true, we are not free. Determinism and freedom are incompatible, irreconcilable. Almost always, in this tradition, out of the conviction that we *are* free, or out of the deep desire to be so, it is also concluded that determinism is not true. The answer to the question of the consequences of determinism given in the second tradition is that the truth of determinism does not or would not touch our freedom. We are free despite its truth, as many say, or, as some say, because of it. Determinism and freedom are compatible, reconcilable.

My greatest hope for this book, and my greatest claim for it, is that it resolves the problem of the consequences of determinism. At least it gives the fundamentals or outline of the resolution. This is the endeavour of its third part. It resolves the problem not by succeeding where there must be no hope of succeeding, where whole declamatory lecture-halls of philosophers, as audacious or arrogant as I, and certainly more able, have failed. That is, it does not succeed by establishing that one or the other of the two traditions is correct. It is part of the resolution of the problem that both are mistaken. Their fundamental propositions are false.

Part One—To return to the first question, that of the existence of a conceptually satisfactory determinist theory of our ongoing lives, the view that there was none was taken near the mid-century by two of the most acute of philosophers, J. L. Austin (1956, p. 131) and P. F. Strawson (1962, p. 187). Neither of them, it is worth saying, was moved to his conclusion by the sorts of *feeling*—one is tempted to say *mere* feeling—including spirituality and residual religiousness, noticeable enough among opponents of determinism. Neither was dedicated to the preservation of the human mystery. They have not remained alone in their scepticism. David Pears, despite being a remover of barriers in the way of a determinist theory, takes us to lack a theory that works. (1973, p. 110) A. J. Ayer is of a similar mind. (1984c, p. 3) Such opponents of determinism as Stuart Hampshire and Anthony Kenny, although concerned to refute determinism, also have doubts as to its conceptual adequacy. (Hampshire, 1965, 1972a; Kenny, 1975, 1978; cf. Watson, 1982, p. 9; Trusted, 1984, p. 138; W. H. Davis, 1971, p. vii)

It would be difficult for sceptics to imagine and clarify, let alone defend, a general line of argument or a principle of inquiry which would issue in the conclusion that a conceptually satisfactory determinist theory is *impossible*. Indeed, the scepticism of Austin, Strawson, and others is properly to be taken as a lesser if a substantial thing, a claim as to just the *absence*, and the likelihood of a continued absence, of such a determinist theory.

It is one thing to declare or more likely to presuppose that Free Will, the Faculty of the Will, the Self-Conscious Mind, or what I may seem to speak of in speaking of my self, are pieces of metaphysical excess, or even nonsense, or that our choices and decisions are quite clearly effects, or that our behaviour is lawlike, or that the brain is a machine or the biological instantiation of certain formal systems. It is another thing to set out a determinist theory which is explicit, complete, and at a proper level of specificity. To do the latter thing is to do no less than set out a philosophy of mind and action. To do so is necessarily to deal with three problems: the relation between neural and mental events, the antecedents of those events, and the nature of action. Each of these can reasonably be taken to generate further conceptual problems. There is also a prior problem, yet more general than the three mentioned, which is the nature of necessary or nomic connection.

As it seems to me, then, the given scepticism as to the likelihood of a conceptually adequate determinism of the given kind has had as its best ground a true perception of the *largeness* of the enterprise, and a true awareness that it has not been succeeded in, or even much attempted. Determinism as an account of our existence, when it has not been presupposed, has been talked of rhetorically, briskly defined in formal ways, remarked on as the basis of one or another model of the mind, had its truth affirmed or denied on the basis of premisses distant from the subject-matter, and had its human consequences assiduously contemplated and disputed. It has not been much *expounded*.

The first chapter of this book is given over to the prior problem in any satisfactory exposition of determinism, that of causal and other nomic connections. It gives an independent account of these. The second, third, and fourth chapters comprise a philosophy of mind and action. The second has to do with simultaneous mental and neural events, the psychoneural relation. It presents a first part of a new theory of that relation, the Union Theory, which supplants Identity Theories of mind and brain. The third chapter concerns the explanation of mental events, their antecedents. It examines the conceptual viability of Free Will—indeterminist thoughts and theories of mental events—and also completes the Union Theory. The fourth chapter gives an account of the nature of actions and their explanation. The three hypotheses advanced in these chapters are, with the prior account of nomic connection, the principal parts of the determinist theory advanced. The theory is, in certain reasonable senses, complete and specific.

It is distinct, certainly, from other determinisms. It is distinct from, although consonant with, determinist theories of absolute generality, having to do with all that exists. Newton and LaPlace come here, and, to the discomfiture of many contemporary physicists, Einstein. My interest is not in the whole universe—or how determinism works in physics (Earman, 1986)—but ourselves. A part of what will be said is of relevance to LaPlacean or universal determinism, but that is not our subject. To set aside other things which *are* of similar limited scope but none the less irrelevant, the theory of this book is distinct from deterministic Freudianism and similar congeries of ideas. To linger a moment with the example of Freud, it has sometimes been thought by philosophers, as it was by Hampshire, that psychoanalysis is 'the new positive science of human conduct', and that it is to be taken as providing the principal challenge to human freedom, that above all with which freedom must be shown to be compatible. (1959, p. 255) My own persistent attitude in this matter is close enough to one once tartly expressed by Bernard Williams. 'To say that a view of human freedom is compatible with scientific advance because it is compatible with developments in psychoanalysis is much like saying that a material is uninflammable because it doesn't burst into flames when one shines an electric torch on it.' (1960, pp. 41–2) Whatever else of a more tolerant sort is to be said of Freudianism (MacIntyre, 1971b), it is not the theory of this book and plays no part at all in it.

Nor is the theory to be identified with various determinisms other than the universal kind and the Freudian. It is not a *weak* determinism that involves something less or looser than causal and nomic connection properly conceived. It is different in kind from any view that involves so-called causally sufficient conditions, or so-called complete causes, that merely probabilify and do not necessitate their effects. Again, it is of course not a *partial* determinism, which limits its causes to certain levels or items, perhaps *events* somehow conceived, and allows that when *persons* or *agents* perform their acts of volition or whatever, this is not a matter of ordinary cause and effect.

The theory is not, to use again a certain description in a way now common, a model of the mind, or the brain, or both. It is not, that is, a middle-level speculation as to the organization or function of either mind or brain or both—where many such speculations at this moment of history are computer-based conceptions issuing in characterizations of the brain as a semantic engine, an information-processing system, and so on. The theory does not conflict with such models of how mind or brain work, but rather is logically prior to them. It is best regarded, perhaps, in so far as its relation to such models is concerned, as establishing certain fundamental constraints on their formulation—constraints given by the three hypotheses and associated propositions. The theory is of the nature of mind and brain, including the fundamental principles of their operation. Finally, with respect to various overviews or taxonomies of determinisms, it is clear that an attempt to put the determinism of this book into the three or four categories of determinism supplied by one philosopher (Popper, 1982b), would do great damage either to it or the categories. Nor does it fit easily into other taxonomies. (Ayers, 1968; Trusted, 1984)

Reassuringly, the theory is not a world apart from everything. It has inexplicit historical antecedents, above all in Hume's splendid Sections 1 and 2, Part 3, Book 2, of *A Treatise of Human Nature*. The theory in its several parts approximates to a number of views in the contemporary philosophy of mind, views on brain and mind, or action, or on lesser subjects, these being at least tacitly determinist in nature. It is identical, I think, with no set of these views—and not identical either, incidentally, with my own previous writings on the subject. I have changed my mind about important issues. It is approximate, too, to certain ruling presuppositions in contemporary neuroscience.

Part Two—To return to the second question, that of the truth of the theory, it will be clear already that its three hypotheses are properly if briefly described as claims of fact, to be settled in part by empirical evidence. They differ somewhat among themselves in this respect, but none purports to be merely a logically necessary or conceptual truth, in no need of empirical confirmation. Despite philosophy's proper preoccupation with the unempirical, it would be wholly wrong to suppose that such factual claims have not in fact been part of philosophy, or are not such now. No philosophical journal is without them, or what immediately presupposes them. It is sometimes said or implied that such claims ought not to be within philosophy, roughly on the ground that questions of fact are properly the business of scientists. There are excellent reasons for dissenting, and thus for supposing that the ongoing history of philosophy does not involve a kind of general trespass on science. The main reason, to restrict ourselves to claims about the relation between mental and neural events, and the explanation of those events, and the explanation of actions, is that such claims do, despite their factual nature, raise conceptual issues. They raise issues fundamental to philosophy and to be resolved only by methods and ways of reflection that are most developed within philosophy. To say so is not to embrace just the anti-scientism, arguable as it can seem, that supposes there actually are two ways of understanding, the subjective and the objective. (Cf. T. Nagel, 1986, p. 9; Davidson, 1980e, p. 216.) Science is such as to give it a right to judgement on the claims in question. So too is philosophy.

Something else quite simple is more important than the possibility that the determinist theory of this book would be more properly advanced by a philosopher or a scientist. It is that any adequate treatment of the fundamental question must be subject to both philosophical and scientific constraints. It must aspire to both philosophical and scientific virtues. That is not to say that any adequate treatment of the question must in part be a piece of science, but it goes some small way in that direction. There are writings on determinism and freedom that must be allowed philosophical respectability, and which very notably lack the other respectability, as there are writings whose neurophysiology, psychology, computer science, or Quantum Theory must be taken as authoritative, but whose departures into vagueness and looseness, indeed flights of fancy, are philosophically unsatisfying, or breath-taking, or close to absurd.

The fifth chapter of this book is largely given over to a survey of directly relevant neuroscience. It gives a certain summation as of the year 1988 of the sciences that fall under that heading, a summation for philosophical and like readers. In so doing, it gives an assessment of the body of empirical evidence that is most important to this book's theory, and to any similar determinism, and also most important to opposition to them. There is no other evidence so important, in any other science, any body of knowledge, any set of propositions whatever. There is no evidence so important to the questions of the mind–brain relation, the explanation of mental and neural events, and the explanation and nature of actions. My confidence that the theory advanced here is well confirmed is not wholly, but it is in the main, owed to this evidence.

It will come as an irritation to some readers, and a relief to others, that despite what has just been said, this book is no piece by a handmaiden of science. If it is no piece of anti-scientism in the sense mentioned above, it is certainly no piece of scientism, which is to say dismissive of all that is not science. (Cf. P. F. Strawson, 1985, p. 67; Mackie, 1976, p. 169.) Nor does it have in it the odd dream of philosophy's being succeeded by 'neurophilosophy'. (P. S. Churchland, 1986) Also, I have found it impossible to believe that the evidence for determinism which is provided by neuroscience is overborne by the familiar, obscure, and too often merely magisterial considerations of a philosophical kind drawn from something else commonly accorded pride of place, which is physics, and in particular Quantum Theory. The fifth chapter, if it finds support for the theory of the book in neuroscience, also argues against the idea, certainly not uncommon among scientists, that any determinism is refuted by Quantum Theory. One of my several reasons is that the subject-matters of the first part of this book—the psychoneural relation, the psycho-behavioural relation, and so on—are under the very eye of neuroscience. They are not near to being under the eye of physics. Neuroscience is in fact touched not at all, or barely, by indeterminist ideas drawn from Quantum Theory. It is also true that the incorporation of an interpretation of Quantum Theory into indeterminist pictures of the mind produces something very approximate to incoherence and contradiction.

Finally, with respect to the truth of determinism, philosophers as distinct from scientists have advanced a considerable number of what they take to be refutations. Some of these are in fact considered in the four chapters (1–4) which are primarily concerned to expound and clarify. This is so since many objections are of a mixed character. The sixth chapter is concerned with a set of related objections of roughly an epistemological kind.

Part Three—To return to the question of the consequences of determinism and of the near-determinism noticed earlier, the question most considered by philosophers, at least since Thomas Hobbes's *Of Liberty and Necessity* of 1646, it has almost always been considered in a certain way. The question has been asked whether, if determinism is true, we are free in such a way that we are morally responsible for our actions. One tradition answers no, the other yes. To that question has occasionally been attached another closely related moral issue, that of the justification of punishment.

To proceed in this way is not to come into close touch with the true subject of the consequences of determinism. It is to leave out a lot of that subject, most of it. P. F. Strawson's 'Freedom and Resentment' (1962) rightly directed philosophers away from moral responsibility (although with the official aim of returning to it) to another part of the general question. It has to do not with feelings of moral approval and disapproval with respect to someone's decision or action, which is in fact the subject-matter of the question of moral responsibility, but with what may be called *personal* feelings, of which resentment and gratitude are two. However, there are more parts to the question of consequences, and they require attention as much. Indeed, in my view, one of these can be said to demand attention more than either the given moral disapproval and approval or personal feelings. It is the question of how determinism stands to what can be called our *life-hopes*. It is with respect to this above all that a theory of determinism appears to be most challenging. It can be, as it has been to many, including myself, a black thing.

The seventh chapter of the book begins a consideration of life-hopes, personal feelings, and moral disapproval and approval, and also additional consequences of determinism. One, which has hitherto been misplaced and isolated in philosophical discussions, and thus been misconceived, has to do with knowledge. The fifth and sixth matters considered in an initial way in the chapter have to do with morality, but not moral responsibility. They are the matter of right action and principles, and the matter of the general moral standing of persons.

It is my contention that a broad fact established in the seventh chapter, having to do with our common attitudes and responses in the matter of determinism and life-hopes, determinism and personal feelings, and so on, is fully sufficient to provide a refutation of the propositions fundamental to the two traditions of Incompatibilism and Compatibilism. Those propositions are elicited by means of a brief historical survey carried forward in the eighth chapter. It is indeed the conclusion of the eighth chapter that both traditions consist in provable error.

The ninth chapter completes the inquiry into determinism and life-hopes, personal feelings, knowledge, moral responsibility, actions and principles, and the general moral standing of agents. It gives my resolution of the problem of the consequences of determinism. It is also a resolution, as already indicated, of the problem of consequences raised as much by widely held near-determinist theories of the mind which assert or contemplate indeterminism at the micro-level, by way of Quantum Theory, but none the less accept or tend to accept macro-determinism, a determinism which certainly includes what are commonly called neural events. This principal conclusion of the book may not be right, as I am confident it is, but very certainly it is not more of the same. It is not one more attempt to have one of those two so overridden and wearied nags, Incompatibilist or Compatibilist, plod at last into the winner's enclosure. They are both put out to pasture.

The tenth chapter does not carry the fundamental argument onward, but applies it in connection with a final consequence of determinism, its consequence for an array of institutions, practices, and habits in our societies. Among these but only one of seven, although it is given most attention, is punishment. If there has been an insufficient awareness of the range of consequences of determinism, there has also been an insufficient awareness of the extent of what can properly be regarded as this last one of those consequences, having to do with the array of institutions, practices, and habits. Reflection here finally issues in a consideration of determinism and politics.

It would be satisfactory to conclude this introduction by saying that my aim in what follows has been no more than to advance truth, or, even, to advance truth and do myself some good by doing so. Whatever can be said along those lines, something needs to be added. Philosophy in general does seem one of the intellectual endeavours which most gives rise to kinds of assertiveness, combativeness, and dismissiveness. The latter traits, as it seems, are pervasive, not a lot less noticeable when they are officially disavowed. (Nozick, 1981) Philosophers cannot but seem to be *advocates*—learned or not so learned counsel. I do not mean that they do not believe in their briefs, but rather that they certainly press them. The truth may be owed to the subject-matters of philosophy, which are such that the facts in them do not in the end speak for themselves, or speak still more uncertainly than elsewhere. Whatever the explanation of this truth about philosophers and philosophy, I cannot claim to have risen far enough above it.

PART 1

Necessary Connections and Three Hypotheses

1

Necessary Connections

1.1 THE NATURE OF CAUSES AND CAUSAL CIRCUMSTANCES

Suppose we believe that the snow is what is muffling the sound of the traffic, or that flipping the switch made the windscreen wipers start to work, or that it is the position of the car's heater that accounts for the driver's left knee being warm. Each of these three is an ordinary belief, well-founded or not, that one thing caused or is causing another. But *what* is it that we believe when we believe that one thing caused another, that the second was the effect of the first? This much-handled question is not part of the high life of philosophy. It would be agreeable to pass it by, as have many inquiries into determinism pertaining to our decisions and actions.

It would be agreeable to pass the question by, but unwise. One large reason is that it is now often enough claimed that we take effects to be events that might not have occurred, given all things exactly as they were beforehand. Such claims about our ordinary beliefs are commonly depended on in asserting Free Will, which is to say indeterminist accounts of decision and action. They have been given a confidence and liveliness through the influence of a common interpretation of Quantum Theory.

A second and smaller reason for attending to causation is that too dramatic conceptions of it, such as those which connect it with certain images or ideas of power, or fate or plan, or compulsion, or logical connection, distort one's responses to determinism.

A third is that there is among scientists a kind of general scepticism of talk of causation, and in particular a suspicion of general conceptions of it used by philosophers. Certainly any determinism requires such a thing. The scepticism needs dealing with not only to satisfy scientific readers, but because the theory of determinism in this book has its most important empirical support in science.

A fourth and perhaps the largest reason for attending to causation is that we do well to avoid the general conceptual uncertainty that must be part of an inquiry which leaves undefined any of its fundamental ideas which are open to definition. Causation, it can be supposed, is less clearly conceived by us than other basic facts. The question 'What do our beliefs about causation come to?', it can be supposed, has no clear and plain answer. Simple truth, which must be the highest aim of any real inquiry, has explicitness as a condition. We can, to my mind, have good hope of both in the present inquiry.

To answer the given question about our ordinary causal beliefs we must also answer another, more fundamental, and in fact more important to our inquiry. We take an effect to result not only from a *cause*, but from a set of things which includes the cause. That the windscreen wipers started to work can properly be said to have been caused by a set of things including the state of the wipers' mountings and the smooth surface of the windscreen as well as the switch's being flipped. What do we believe of the connections between such a set, which we may call a *causal circumstance*, and the effect?

In much of our reflection on this question, we shall also be dealing with a third, which will be as important, that of the nature of certain other necessary or nomic connections, which hold between what may be called *nomic correlates*. These are pairs of things in a fundamental way like causal circumstances and effects. Two of many examples of nomic correlates, although the matter is in several ways complex, are provided by the interdependent variation of the pressure of a gas and its volume and temperature, according to the Boyle–Charles law for ideal gases, and the orbits of the two stars which make up a double star, which are held close together by mutual gravitational attraction.

Strictly speaking, in what ontological category of things are causes and other conditions, the things which comprise causal circumstances? To that first question, there is the answer that necessarily they are spatio-temporal—things in space and time. This is fundamental, and will be of importance later, in connection with Quantum Theory, but does not take us very far forward. It seems very likely that causes and other conditions are all of one and the same fundamental sort, despite differences that can be made between them. What sort of things are effects? There must be a strong presumption that at bottom they are of this same nature, despite an occasional argument to the contrary (Vendler, 1962), since effects *are* also causes.

We often refer to causes as *events*, and no doubt it is possible, given one or another idea of an event, to call all of them events. (Davidson, 1980c, d; Kim, 1973) However, it is as natural to say that it was the *fact* that the switch was flipped that caused the windscreen wipers to start to work. (Pollock, 1976, p. 157) Also, at least some of what we call causes may seem to be ordinary *things* or indeed *stuff*, such as the snow. These usages have given rise to philosophical views as to the nature, strictly speaking, of causes and effects. Various objections to these views—like much else about nomic connection—must go unconsidered here. (Cf. Honderich, 1989)

If we think of natural and explicit answers to the question of what caused something, what we come to are items like the weight of a thing, or its colour, hardness, shape, nearness, transparency, weakness, speed, pitch, temperature, chemical constitution, positive charge, protein content, molecular structure, or spin. It is not the age or colour of the bottle of wine that is flattening the napkin, but its weight. It is not the weight of the cup that makes it reflect the light as it does, but rather its glaze. It is the position of the switch—on rather than off—that made the windscreen wipers start to work, as it is the position of the heater and the density of the snow that cause the other mentioned effects. The most natural and explicit answer to the question of what caused something, then, is simply a property or relation of an ordinary thing.

What needs to be resisted immediately, however is the idea that what is in question is a *general* property or relation, something had by many things. It is not the *general* property or *universal* of weighing two pounds, whatever it is, that is flattening the napkin. The general property will exist just as it did—at least on several views of general properties—if the weight of this bottle of wine changes and the napkin is not flattened in the same way. Yet more decisively, it is not *all* of the general property, including that part or whatever that is involved in the soup tureen, which also weighs two pounds, that is flattening the napkin. We come to the idea, then, that what is flattening the napkin is *this bottle's weight*, an *individual or particular property* of this bottle of wine and of nothing else whatsoever. We come to a spatio-temporal individual or particular, but not one which is an ordinary thing.

Evidently such individuals can be said to exist, despite the fact that they lack the independence of ordinary things—they cannot exist on their own. Evidently, secondly, they can be individuated. Both points are implicitly granted by philosophers who have been concerned with general properties. (Armstrong, 1978a, b; Loux, 1976; P. F. Strawson, 1979) Those philosophers who have been most Platonic or 'realistic' about general properties necessarily speak of their instances. It is not *all* of the bottle of wine, or any individual property of it other than its own property of weighing two pounds, that is an instance of the general property of weighing two pounds. Individual properties have sometimes been made somehow dependent on general properties by these philosophers, but that is nothing to the point.

Individual properties and relations have been with us since Aristotle, and have had the support of such diverse philosophers as Peirce and Stout, but they have not always got a good name. (Armstrong, 1978a, pp. 77–87) This is so since, rightly or wrongly, they have been put to grander use than any that is contemplated here. They have been used to try to solve the problem of general properties or universals. It needs to be granted, certainly, that in speaking of a thing's properties one is not always speaking of individual properties, and that the proponents of more traditional solutions to the problem of universals, unsuccessful though they have been in their own proposals, have made trouble for the solution in terms of individual properties.

All this is distinct from, and does not endanger, the truths that there exist individual properties and relations, and that they can be individuated. To repeat, in the words of another, 'Some philosophers, no doubt, have made too much of the category of particularized qualities. But we need not therefore deny that we acknowledge them.' (P. F. Strawson, 1959, p. 169; cf. K. Campbell, 1981)

Causes and effects, then, are certain spatio-temporal items, individual properties and relations, as distinct from a good deal else. As will also turn out to be the case, given the definition of the *physical* to which we shall come, they are all of them physical. (2.2.3; cf. Earman, 1986, p. 249) What is not physical is not a cause or effect either.

As for what may be called, although not very properly, the conflict with our ordinary ways of speaking, in which events, facts, and things are mentioned as causes or conditions, and also other items (Ayer, 1972, p. 133), the principal explanation is that we are in accord with the rooted practice of taking the whole for the part. It is perfectly natural to say that the candle is alight or the house is on fire when strictly speaking only a part of it is. Individual properties are parts of events or in a related relation to events, however events are conceived. The same is true of facts as typically conceived, and of ordinary things. In what follows, I shall almost always speak in the ordinary way of causes and conditions as things in a generic sense, events, facts, ordinary things, and stuff. All that will be said could be put, although often less comfortably, in terms of individual properties or, occasionally, sets of them.

1.2 CONNECTIONS STATED BY DEPENDENT CONDITIONALS

The three sample causal beliefs with which we began have a number of features. We believe in each case there actually occurred two things, the cause and the effect. What will or would cause something is not a cause, but only a possible cause or potential cause, until or unless the effect actually occurs. Certainly for cause and effect we do need *two* things, things that in the most fundamental sense are two in number. We have an intuitive grip on numerical non-identity which is wholly sufficient for us to judge proposed criteria of it. It will suffice to say here, making use of one familiar criterion, that two things can be causally related—cause and effect—except in two cases. They cannot be so related if (i) they share all truths: if any truth about either is a truth about the other. Hence, two things cannot be causally related as wholes if (ii) a part of one is such that the part shares all truths with the other whole individual. (Cf. Swain, 1980.)

We do not believe that the connection between cause and effect is logical or conceptual, despite the fact that an event or thing may be defined in terms of its effect. (Ayer, 1972, p. 6 f., 1980, pp. 60–63) We do none the less believe that certain connections hold between cause and effect. One is, if we make a certain common assumption, and taking the first example, that *if the switch had not been flipped, then the windscreen wipers would not have started to work.* To state the same single fact of connection in the world by way of a different conditional statement, the contrapositive of the first, *if the wipers did start to work, then the switch was flipped.* Given that we do indeed believe that the flipping and the starting each occurred, it is also true, and would certainly be more natural to say, that *since* the wipers started to work, the switch was flipped. However, we are now interested in the connection by itself. It is expressed without superfluity by the conditional with 'if' rather than the conditional with 'since'.

We do *not* believe, certainly, that if the situation had been different, it would still be true that if the switch had not been flipped the wipers would not have started to work. We know it would have been possible to have left the switch alone but—given a bit of forward planning and some wire—to have completed an *ad hoc* electrical circuit so that the wipers *did* start to work, just as and when they did. In that different situation, to speak in the other way, using the contrapositive, it would not have been true that if the wipers started to work, the switch was flipped.

Take the flipping, a particular or token event which occurred only in a particular place at a particular time, to be *f*, and the starting to be *s*. (The *type* of event, of which *f* was an instance, is of course something else. It is no part of the present story. Speaking more precisely, *f* is an *individual* property or relation, not a general one.) Take the variable *x* to cover any other particular event or condition, or set of events or conditions, which did not occur, such as the completing of an *ad hoc* electrical circuit. The belief that *f* caused *s*, then, is in part that if *f* had not occurred, *s* would not have, that if *s* occurred so did *f*. The belief is *not* in part what is certainly different, that even if *x* occurred, if *f* had not occurred, *s* would not have—or that even if *x* had occurred, if *s* occurred, so did *f*.

It is then evident that the statement, 'If *f* didn't occur, neither did *s*', like its partner, depends for its truth on the fact that certain events did not occur. However, the conditional statement wears its meaning on its face, as much as any statement does. It does not mean greatly more than it says, as a good many philosophers have seemed to think, and as surveys of their theories indicate. (Mackie, 1973a) It is about *f* and *s*. It is not about *x* and nor is it about any other event or condition that *did* occur, and, to speak vaguely, is part of a full story of why the wipers started. That the situation was as it was, and hence that certain events did not and certain events did occur, is indeed much of our *reason* for asserting the conditional statement but not part of the statement. I am not talking about the absence of an *ad hoc* electrical connection or about the presence of other things, either specifically or under some general description or by some general means, when I make the conditional statement.

There is more to be said of these conditional statements and their difference from others, and hence of the connection they state. For the moment let us merely have some abbreviations. We may say, in abbreviation of the conditional that if the cause had not occurred neither would the effect, that the cause was *required for* the effect. In place of saying that if the effect occurred so did the cause, we may say that the occurrence of the effect *required* that of the cause.

A fourth example of a causal belief has to do with someone's being taught to drive by a prudential instructor, whose car has two brake pedals, each moving the same single connecting rod. It happens, extraordinarily, that both instructor and pupil depress their pedals simultaneously, with the same force. The rod moves, and the car stops. It would have stopped in exactly the same way if only one of the two had stepped on his pedal. In this situation of overdetermination, what caused the car to stop? We *can* say that the instructor's braking caused it, and it is natural to say that it was a cause of the stopping. However, the very same is true of the pupil's action. Each action was a cause of the stopping, then, and so we have two causes. Here we do not make the common assumption mentioned but not specified in connection with the first three examples, which is that an effect had a single cause. The important point about the example, though, is that neither the instructor's action nor the pupil's action was *required for* the stopping. The stopping did not *require* either, since the other also occurred.

Still, something of a related kind is true. The instructor's braking would have been required for the stopping if the pupil had not braked. That is, *if the pupil had not braked, then if the instructor hadn't, the car wouldn't have stopped.* So with the pupil's braking. Again, the stopping would have required the instructor's braking if the pupil had not braked, and it would have required the pupil's braking if the instructor had not braked. We may say of each braking that it was *alternatively required for* the effect, and that the effect *alternatively required* each braking.

Causes are therefore either required *or* alternatively required for their effects. They need not be required. To draw an explicit conclusion about what we have so far of causes and effects, if a particular event *c* caused some event *e*, then one of these two criteria was satisfied:

(1) If *c* had not happened, then *e* would not have happened. Stated differently, if *e* did happen then so did *c*.

(1a) Event *c* and one or more others occurred. If the other or others had not occurred, then if *c* had also not occurred, *e* wouldn't have. If the other event or events had not occurred, then if *e* had, so would *c*.

We have come to this conclusion quickly. (Cf. Mackie, 1974, p. 47.) It may well be that events can be causes in virtue of satisfying criteria about their being required or alternatively required for effects which are different from the mentioned two. For a complete account we should perhaps have to add further criteria. We shall not do so, since a complete account is not essential to our coming concerns.

If causes are required or alternatively required for their effects, they evidently are not the only things that are such. The other conditions of an effect, by which I again mean particulars or tokens rather than types—it would again be more accurate to speak of individual properties and relations—are also required or alternatively required for the effects. Also, effects require or alternatively require other conditions as well as causes. It is fairly clear what sorts of conditions are singled out from their fellows as causes, although giving a systematic account is not easy.

In many cases we speak of a given condition as cause and it is the one action or piece of behaviour involved, something to which responsibility attaches. It is a person's or animal's contribution to the effect. In a second class of cases we refer to a condition which is unusual or abnormal. The soup is staying hot because the soup-plate is heavy Derbyshire stoneware and was well heated—the soup's being ordinarily hot to start with, also a condition of its staying as hot as it does, is not mentioned as cause. This criterion of unusualness is in fact a criterion having to do with a particular situation. The presence of oxygen, not usually cited as the cause of a fire, *is* put forward as having caused a fire in a laboratory or a space satellite for the reason that oxygen is usually and by intention excluded from some apparatus or chamber. In a third class of cases we may take the latest condition, the one closest in time to the effect, as the cause. Fourthly, we may single out what is ordinarily called an event or change, partly because of its brief duration, and thereby distinguished from what is called a standing condition—say the movement of a lever as against its tensile strength. Fifthly, although the fact is often overlooked by philosophers, we commonly talk of causes in situations where we are attempting to explain some event and have not yet done so. In these situations we commonly know or think we know of many of the conditions which must occur for the occurrence of the event. Some further condition or number of conditions is needed. This condition or set of conditions, as yet unidentified, is given the name of cause.

These five groups of cases are perhaps the most important. Sometimes they are described differently and more detail provided. (Berofsky, 1971, pp. 58 f.; Mackie, 1974, pp. 34 f.; Hart and Honoré, 1959) The principal point for us in this neighbourhood is a clear one. It is *not* the case that we name one condition as cause because it is different from others in any respect having to do with the relations of which know. It is not *more* required for the occurrence of the effect. Nor is it different in terms of another relation between cause and effect to which we shall come, or any relation between causal circumstance and effect.

There is a further cause–effect or condition–effect relation to be noted, but it will be best to do so after noting one between causal circumstance and effect, implied already. The conception of a causal circumstance, although the name may be new, is of a familiar kind. The idea of some set of things including a cause, all of it somehow involved in the occurrence of the effect, is explicit or implicit in our ordinary thinking about the world. In saying that an event *c* caused an event *e*, or that *e* was the effect of *c*, we typically have in mind but do not say that a set of things including *c*, but not necessarily all occurring at the same time, was *required for e*. The idea is of *some* set of things which was required for the occurrence of the effect, which is not to say that the set includes *all* the things prior to the effect that were required for it. Evidently there is a problem, to which we shall return, of the general definition of a causal circumstance.

In the three examples given at the beginning, having to do with the sound of the traffic, the windscreen wipers, and the warm left knee, a cause and each condition are naturally taken to be required for the effect, as distinct from alternatively required. Hence, indeed, there is the further conclusion that the whole circumstance was required for the effect, that the effect required the circumstance. Partly because of this relation it is possible to speak, as we have, of the circumstance as well as the cause as causing the effect or being the cause of the effect.

There are situations, it seems, where it is reasonable to say that there are two or more causal circumstances for a single effect. We have such a situation of overdetermination of a kind in the fourth example, that of the car with two brake pedals. There is the circumstance which includes the instructor's braking, and the one which includes the pupil's braking. The two circumstances have some conditions in common, evidently, but by the general definition of a causal circumstance we shall come to adopt (1.5), there are two circumstances. Not all of either circumstance was required for the effect, given that the other whole circumstance existed. Each, however, was alternatively required. This is not to say that if the situation had been different then it would still be true that if all of either had not existed, then if the other had not, the effect would not have occurred. Something else might have stopped the car.

Let us then specify two criteria about the requiredness of causal circumstances to effects. If a circumstance *cc* was in fact a causal circumstance for an event *e*, then one of the following criteria was satisfied.

(2) If *cc* had not existed, *e* would not have happened. Since *e* happened, *cc* existed.

(2a) Circumstance *cc* and one or more others existed. If each of the others had not been complete, then if *cc* had not been, *e* would not have happened. If each of the other circumstances had not been complete, then if *e* occurred, *cc* was complete.

The first criterion can be satisfied, of course, without its also being true that even if *x* (any other event or condition or set of them) had also occurred, then if we had not got *cc* we would not have got *e*. Similarly, the second criterion can be satisfied despite what would follow with respect to the several circumstances if some *x* had also existed. In what follows, in so far as the requiredness of causal circumstances to effects is concerned, it will be enough to keep in mind only the great majority, which are of the *ordinary* kind. Each ordinary causal circumstances is simply required for the effect, the effect requiring it.

There is one general caveat and a smaller qualification to be added to the conclusion we now have (2, 2a) about causal circumstances and effects. The caveat also applies to the earlier conclusion about causes and effects (1, 1a). The caveat pertains too to what follows in this chapter, further conclusions about causal relations. The conclusions are not to be taken as true of *every* pair of things of which the second is called an effect. Consider those effects, as they are sometimes called, which are choices or decisions of persons, or actions following on choices or decisions. Consider someone deciding not to try to go to the theatre. This may be said to be the effect of his thought that probably there are no decent seats left. We are unlikely to have in mind a causal circumstance in this case. What is important, however, is that if the question of whether there was one is raised, we may be uncertain about the answer. It is arguable, with *some* hope of success, that we can speak of a choice or a decision as an effect and not be committed to there having been a causal circumstance which was required for it, and speak of an action as an effect and need not be committed to there having been a causal circumstance that was—roughly speaking—prior in time to the agent's initial neurophysiological activity. This touches on the fundamental subject of this book. What is being proposed at the moment, however, is no more than a certain view of our conception of causation.

The caveat, to state it simply, is that it is to be taken as a view of only our conception of what can be called *standard effects*: all those where the effects are *not* decisions, choices, like mental acts of persons, or ensuing actions. It may or may not be that decisions and the like are in fact standard effects. Whatever the truth, it is to be allowed that in speaking of them as effects in the ordinary way, to the extent we do, we may make use of a conception other than the one we have been examining. It is to be allowed then, although something more will be said of the matter (1.6), that we have two conceptions.

One further and less important qualification is needed. We have been conceiving and speaking of causal circumstances as *sets* of conditions, and will continue to speak in this way. This accords with our ordinary causal beliefs, as is apparent. However, it is not to be taken as part of the definition of a causal circumstance that it must include more than one element—more than one condition, event, individual property or whatever. We rarely if ever have in mind a circumstance described as a single event or whatever, but the possibility is not to be excluded by our definition. Thus all references to sets of conditions are to be understood as allowing for the possibility of one-member sets.

Evidently, there is more to our belief that the flipping caused the wipers to start to work than has so far been specified. If we do have that belief, we do not suppose that the cause might have happened, given things as they were, without the effect. Given things as they were, the effect had to happen. But what has been said so far, that the flipping and the circumstance including it were *required for* the wipers' starting to work, is consistent with the cause existing without that effect. A first thing's being required for a second is consistent with the possibility of the first thing's happening and the second not happening.

We need to add at least that if the flipping caused the wipers to start to work, then *if the switch was flipped the wipers started to work*. Otherwise put, *if they hadn't started to work, the switch would not have been flipped*. More generally, if *c* caused *e*, then

(3) If *c* occurred, so did *e*. If *e* had not occurred, neither would *c* have occurred.

Our beliefs here, that a cause requires its effect, that an effect is required for its cause, are not to be confused with others. We do not believe, although *c* caused *e*, that if *c* occurred then whatever the situation had been, *e* would still have occurred. We do not believe that whatever the situation had been, if the switch was flipped, the wipers would have started. They would not have done so if they had been loose on their mountings. The truth of the conditional is dependent on other things. Secondly, however, we must also avoid the mistake of thinking that the conditional 'If the switch was flipped, the wipers started' means more than it says: it is about the flipping and the starting. So much for the reminders. Finally here, do we need some criterion additional to (3) and having to do with situations in which there is more than one cause and hence more than one causal circumstance? We do not. No matter how many causes and causal circumstances existed for an effect, it was true of each cause that if it occurred, so did the effect.

Even some like-minded philosophers (Mackie, 1974, Ch. 2; Sanford, 1985) have been ready enough to take it that if *c* caused *e* in an ordinary situation then it is true, as we have it in (1), that if *c* hadn't occurred, neither would *e*, but they have omitted or denied what we now have in (3), that it is also true that if or since *c* occurred, so did *e*. What leads these philosophers to their omission or denial is just the fact that *c* might have happened without *e*. Indeed it might have—in a different situation where some or all of the rest of the causal circumstance for *e* was missing. But this dependency of *e* on other than *c* has a perfect counterpart with respect to the first conditional. The truth that if *c* hadn't occurred then neither would *e* is equally dependent on other things. The absence of *c*, if accompanied by an alternative causal circumstance for *e*, would not have been followed by the absence of *e*. There is no reason here, and I think no reason at all, to doubt that we take it of a cause and its effect that if or since the cause occurred then so did the effect. Something more will be said of the matter, however. (p.37)

Our fourth connection, as may be anticipated, is that a causal circumstance, like a cause, requires the effect, and the effect is required for the causal circumstance. We believe, of the causal circumstance for the starting-to-work of the wipers, that since it happened, so did the effect—and that if the effect were missing, the causal circumstance would be missing or incomplete. If a circumstance *cc* was a causal circumstance for *e*, then

(4) If *cc* occurred, so did *e*. If *e* had not occurred, neither would *cc* have occurred.

That is not dependent on other things in just the way that (3) is dependent on other things. It is dependent, however, on another and more important connection, a fifth, between a causal circumstance and its effect.

1.3 'IF A, EVEN IF X, THEN STILL B'

The fifth connection is different. To approach it informally, suppose that we spend time and arrive at a thorough understanding of the ordinary operation of this Citroen's windscreen wipers. We come to believe that whenever ten specified types of conditions obtain, including a flipping of the switch, there is the effect of the wipers' starting to work. What do we believe if on a certain occasion we have taken it that there exist conditions of exactly the ten types, including a flipping, but the wipers do not start? We may suppose that we are mistaken on this occasion in taking it that all the ten types of conditions do obtain. We may suppose, differently, that our prior thorough understanding of the ordinary operation of the wipers was not thorough enough. That is, there is an additional type of condition which obtains when the wipers start—not ten in all but eleven. More likely, we may suppose that we have not arrived at an exact specification of one or more of the ten. What is needed is not exactly a particular type of condition specified before the present occasion, but a slightly different one, which did indeed obtain previously when the wipers started.

What is common to these and related responses is that if we take the starting of the wipers to be an effect we believe at least that there is *some* type of circumstance which is *uniformly* connected with the wipers' starting. Whenever a circumstance of this kind exists, there also occurs a starting-to-work of the wipers. Certainly we do take any standard effect to be an instance of such a *uniform connection*. However, is this *all* that there is in the world, along these lines, to the connections between an effect and its causal circumstance?

Hume gave one of philosophy's most famous answers, an answer whose strength is owed to its great clarity and simplicity, when he said yes. (1888 (1739), pp. 73 ff.) To give the answer is to refuse to go far beyond what we already have, or have as implied, in connection (4). If the answer leads very naturally to the truth, it is nevertheless mistaken, as is shown by the philosophically familiar but evergreen fact that certain items constitute an instance of such a uniform connection or constant conjunction, but the second is *not* the effect of the first. Although each causal circumstance and effect, likewise, is an instance of such a uniform connection, that is not its unique nature.

Consider a particular day and the night that follows. The example is of course that of Hume's early critic Thomas Reid (1969 (1788)), and has many counterparts, some of them being members of runs of total coincidences. Let us have in mind, only slightly less imprecisely, a period of light in London and thereabouts, that one I now call yesterday, and a following period of darkness, last night. We could of course give precise spatio-temporal specifications. It seems that if any two things whatever satisfy the Humean requirement, yesterday and last night do—all days are followed by nights. But yesterday and last night, however they are related, are not related as causal circumstance and effect. Yesterday did not cause last night. More must be true of any different pair of things which in fact are causal circumstance and effect.

There have been many attempts to save the Humean account, or some development of it. (Ayer, 1940, pp. 179 f.; Hempel and Oppenheim, 1953, pp. 337 ff.; Nagel, 1979, pp. 64 f.; Berofsky, 1971, pp. 203 f.; cf. Earman, 1986, Ch.5; Honderich, 1989) They cannot but strike one as unsuccessful, partly because *ad hoc*. The Humean view has persisted, among all those disinclined to mystery in connection with causation, not because of these defences, but for want of a satisfactory alternative. The alternatives have for the most part consisted in elusive doctrines of 'natural necessity', causal 'power', 'agency' or some kind of 'logical connection' and in inexplicit declarations of the reality of causal necessitation. Now there are superior alternatives, not at all of the unsatisfactory kinds. One of them, to my mind, gives an unfanciful, clear, and true view of the relation between causal circumstance and effect. This alternative view, a member of a small family of related although differing views, follows on naturally enough from a consideration of Hume's.

Why do we not take yesterday as the causal circumstance for last night? What do we take to be the difference between yesterday and last night, on the one hand, and, on the other, another instance of constant conjunction, the one comprising the true causal circumstance and last night? It is that we might, in other than merely a logical sense, have got yesterday but not last night. *Certain other events or conditions might have occurred such that we got yesterday but not last night.* One would have been the creation of a new light source, about as great as the sun. It is thus false that if certain other things had happened, although we got yesterday, we would still have got last night.

Compare what we take to be the true causal circumstance for last night, which we may label *the solar conditions*. They included, roughly speaking, the earth's London face being away from the sun for a time, the absence of any light source like that of the sun, in the right place at the right time, and conditions having to do with the behaviour of light. It is *not* true that certain other events or conditions might have occurred such that we got the solar conditions but not last night. If or since we had the solar conditions, even if certain other things had happened, we would still have got last night. This is indeed what distinguishes causal circumstances and effects (and nomic correlates) from other instances of constant conjunction. This, if it needs to be made more precise, is what we need to concentrate on, as some others have before, although without coming to the conclusion we shall reach. (Mill, 1961 (1843), Bk. 3, Ch. 5, s. 6; Ayer, 1963, 1963a, pp. 231–4; Goodman, 1954; Hospers, 1956; Downing, 1958–9, 1959, 1970; Honderich, 1981b, pp. 421 f., 1982a, pp. 302–3)

Let us take the variable *x* to cover or range over certain conceivable events or conditions or whatever—in fact individual properties or sets of them—which in fact did not occur. They are, we can as well say, certain conceivable changes in the universe, ways in which the universe might have been different. Let us contemplate, first, that they include all such changes save for the absence of *cc* or of *e*. They include all such changes save for logical excluders of *cc* or *e*. We can now contemplate that the relevant *cc–e* connection, when we suppose that *cc* caused *e*, can be stated in this way: If *cc* occurred, then even if *x* had occurred, *e* would still have occurred. That is on the right lines, of course, since certainly we do not suppose that *e* would have occurred even if either *cc* or *e* itself did not. However, it will not do. We regularly take it that a causal circumstance is linked by way of a causal chain or sequence to its effect. Without attempting a characterization of such sequences, let us suppose that a link *k* occurred in a causal sequence connecting *cc* with *e*. Clearly we do not believe that since *cc* occurred, *e* would still have as it did, *even if k had been missing*. We need to restrict *x* a bit more in order to express what we want.

What we come to is this. If *cc* was a causal circumstance for *e*, then

(5) If *cc* occurred, then even if there also occurred any change *x* logically consistent with *cc* and *e*, it was never the less the case that *e* occurred—or, *cc* began and *e* ended a sequence of things such that

it was true of each one and its immediate successor that if the first occurred, even if there also occurred any change *x* logically consistent with both, then the second also occurred. To speak differently, if *e* had not occurred, then even if there had also occurred any change *x* logically consistent with the absences of *e* and of *cc*, and consistent with the absences of links between *cc* and *e*, it would also have been the case that *cc* did not occur.

This fifth causal relation, like several to come, is stated by what we can call *independent nomic conditional statements*, or simply *independent conditionals*. Their truth, in brief, is independent of what else is true. Expressed one way, as we generally shall, they are of the form *If a occurred, then even if any events or conditions logically consistent with a and b had also occurred, in place of those which did, b would still have occurred*. Or, as we can as well say, *Even if any events or conditions logically consistent with a and b had occurred, rather than those which did, if a occurred then b did*. Or again, independent nomic conditionals come to this: *Given the rest of the world as it was, or given that it was different in any way we can conceive it as being, without logically excluding a and b, then if a happened so did b*.

Independent conditional statements are thus different in kind from those *dependent nomic conditional statements* or simply *dependent conditionals*, which state the first four of our causal relations. Dependent nomic conditionals are certain of the statements of the form *If a occurred, then b occurred*. Their truth, in brief, is dependent on what else is true.

By way of abbreviation of what is stated by the independent conditionals in (5), *cc* can be said to have *necessitated e*, and *e* can be said to have been *necessary to cc*. We can also, in abbreviation, speak of an event as *necessitated* without identifying or indeed knowing its causal circumstance. Here a necessitated event is of course to be understood as an event which does stand in the given relation to some or other antecedent. Like remarks might have been made elsewhere—with respect to a *required* event, for example. What we have in (5) might be improved in a number of ways so as to deal with questions and indeed objections, and thereby complicated and indeed greatly complicated. In particular the contrapositive formulation might be considered further. What we have, further, might be expressed in several different logical notations. We might consider problems (e.g. Wiggins, 1973) and proposed solutions (e.g. Thorp, 1980, pp. 16–26) which arise in connection with notations. What we have will suffice as it stands. It does indeed distinguish yesterday and last night from the other instance of constant conjunction, the solar conditions and last night. The solar conditions but not yesterday count as causal circumstance for last night. There is no peculiarity, incidentally, about this particular very grand causal circumstance and effect. Reflection on smaller examples of causal circumstance and effect, such as those with which we began, is quite as capable of illustrating this fifth causal connection.

The given connection between causal circumstance and effect is in fact the principal instance of what can be called *fundamental nomic connection* or *fundamental necessary connection*. Such connection is what is stated by independent nomic conditionals and, of course, holds between any two things when it is true that if or since the first occurred, then even if any change logically consistent with either had also occurred, the second would still have occurred. Fundamental nomic connection, as will be made clearer, is just that—fundamental. It is the stuff of or the basis of all the relations specified so far or still to be specified between cause and effect, causal circumstance and effect, and nomic correlates.

There are two more causal connections to be noted. As we saw earlier, (2) an ordinary causal circumstance is *required for* its effect. If, say, the ten conditions including the flipping of the switch had not occurred or existed, the wipers would not have started. This is a truth dependent on the situation as it was—there was no *ad hoc* electrical circuit and so on. There are related connections, however, which have the independence of (5) the connection just noted. One is bound up with the fact that we do indeed suppose that there is some *set* of types of circumstances, each type related in the same way to startings-to-work of the wipers. We believe that if the wipers did start, even if certain changes had taken place in the situation, there would have occurred an instance of one or another member of this set of types of circumstances. Either the switch was flipped and other conditions existed, *or* an *ad hoc* electrical circuit was completed and other conditions existed, *or*. . . . More generally, suppose again that *cc* was a causal circumstance for *e*. Suppose also that any one of *cc′* or *cc″* or. . ., if it existed, would also have been a causal circumstance for *e*, and, we might add, would not have been part of a causal sequence including *cc*. Then

(6) If none of *cc* or *cc′* or *cc″* or . . . existed, even if there occurred any change *x* consistent with that and with *e*'s absence, *e* would still not have happened. If *e* happened, at least one of *cc* or *cc′* or *cc″* or . . . existed, even if there also occurred any change *x* consistent with both.

By way of abbreviation, one or another of a set of possible

circumstances was *necessary to e*, and *e necessitated* the occurrence of one or another of the set.

Our last relation follows on from this. In terms of the example it has to do with the fact that if in the situation there existed only the one circumstance for the starting of the wipers, then, even if certain other events or conditions had occurred or existed, the wipers would not have started. More generally, if *cc* was a causal circumstance for *e*, and with the other terms defined as with (5), we have this:

> (7) If none of *cc′* or *cc″* or. . . existed, then if *cc* had not existed, even if any change *x* had occurred logically consistent with that and with *e*'s absence, *e* would not have happened. If none of *cc′* or *cc″* or. . . existed, then if *e* happened, *cc* would still have existed even if any change *x* had occurred consistent with *e* and *cc*.

By way of abbreviation, for what it is worth, we can say that *cc* was *dependently necessary to e*, that *e* was such as to *dependently necessitate cc*.

Our principal conclusion about causation is now at least in distant view. Causation and other nomicity consists in no less than, and not greatly more than, *a web of connections* between things or events, at bottom individual properties. What are these connections? They are those asserted by the two kinds of conditional statements. Causation is not, as some suppose, anything less than these connections—which conclusion will be defended in what follows. Nor is it greatly more. There is thus a clear and plain answer, if one which requires complication, to the question of what causation and other nomicity comes to.

The web may be open to other styles of description. Any of these must give it as having a certain structure. Each of the connections stated by independent nomic conditionals gives rise to others. For example, suppose *cc* necessitated *e*, *e* thus being necessary to *cc*, and that *cc* consisted in *c* and *c′*. It follows that if *e* had not occurred, and *c′* had occurred, then *c* did not occur. We shall not pursue these matters further here, but they will be noticed again in connection with the nature of conditional statements and the subject of causation and science. (1.4, 1.6)

It is worth emphasizing what has already been said or implied, that all seven of the connections at which we have looked, and the further subordinate ones at which we shall not look, are indeed *objective* connections, connections in reality. They are entirely independent of minds, theory, conceptual schemes, the statements which state them, and so on. There are philosophers, some of them inclined to Kant's doctrine that we *impose* the category of causation on reality, some of them freer spirits, who think or at any rate say differently of causation and of nomic connection generally—in a phrase, that it is part of the mental order. One of these philosophers presses on forward, with agreeable audacity, to characterize the view I have expounded as Idealist or even Scholastic. (Putnam, 1983) That is, the view expounded is seen as one which 'mentalizes' the natural world by intruding the mental order—nomic or necessary connection—into it. The view, on the contrary, is precisely one of *Causal Realism* rather than *Causal Idealism*. (Kim, 1981) It is exactly unlike any theory which *does* somehow locate nomic connection in the mental order, whether or not it then relocates it elsewhere, and thus *is* properly labelled Idealist.

The point stands in connection with another. Those familiar with philosophical writing on causation, or touching on causation, will have noticed that our analysis so far of it has taken the terms 'necessary connection', 'nomic connection', and 'lawlike connection' as synonymous, but has made little reference to *laws*. The analysis may appear to be unlike those which, to speak quickly and only of one central matter, describe something like a causal circumstance and an effect as two items which fall under a law, and then proceed to attempt to give an account of what a law is—a true proposition of a certain character. (Hempel and Oppenheim, 1953; Hempel, 1965) These different analyses may appear to describe a connection in reality by way of what can be called our linguistic response to it, or the character of our belief about it. In fact, our analysis and these seemingly different analyses are basically alike. Both characterize connections in reality and both give an account of the character of our beliefs about them. It could not be otherwise.

Our analysis, in a way more direct, specifies necessary connections, but in so doing does provide an account of the nature of laws. It does so by actually giving their form or structure. The most fundamental kind of them are independent nomic conditional statements, general rather than particular. Laws of the fundamental kind thus *are* general propositions to the effect that if something is the case, then no matter an alteration in certain logically consistent concomitants, something else is also the case.

The alternative procedure, although its focus is different, is indeed basically alike, as it must be. Here, one starts with a connection in the world, and appears to describe it by way of our characterization of it, the character of our belief about it. To do the latter thing, however, if the procedure can have any hope of success, *is* to describe the connection in the world. If it were not to do this, it would be no more than the futility of changing or avoiding the subject.

1.4 THE ANALYSIS OF CONDITIONAL STATEMENTS

What we mainly have in answer so far, about causes and causal circumstances, is that they stand in seven connections—the last three of which are also fundamental to what will be said of nomic correlates. All are connections stated by either dependent nomic conditionals or independent nomic conditionals. What we need now, to have a better grasp of these connections, is a better grasp of the two kinds of statements. To understand them more fully is to see more clearly what we believe about the real connections, connections in extra-linguistic reality. The subject of dependent nomic conditionals has for long been a disputed one, and part of larger disputed subjects, those of larger categories of 'if' statements and of 'if' statements generally.

Dependent conditionals can initially be *identified*, as they have been here, as typified by the 'if' statements we accept in connection with our standard causal beliefs—'If the switch hadn't been flipped, the wipers wouldn't have started', 'Since the switch was flipped, the wipers started', and the like. The idea, of course is not to *elucidate* dependent conditionals by relating them to causal statements and the like, but to do just the opposite. It will be convenient, by the way, to abbreviate the conditional 'If the switch was flipped, the wipers started' not merely to 'If *f* occurred, then *s* occurred', as we have already, but to 'If *F* then *S*'. So too with all other conditionals: letters in lower case for *events*, *conditions*, and the like, the same letters in upper case for the *statements* that the event occurred or the condition existed. The custom will in fact be followed generally hereafter, with subjects other than that of conditionals.

Dependent nomic conditionals are readily distinguished from a number of other sorts of 'if' statements. First, they are not logically or conceptually necessary, as is 'If she has children, she is somebody's mother.' That they are not such statements is in accord with the fact, rightly insisted upon by Hume, as already noted, that causes cannot be said to be in a certain logical connection with their effects: the fact that it is not contradictory, however mistaken it may be, to assert that a causal circumstance for an event existed but that the event did not occur.

Dependent conditionals, secondly, are not to be identified with the 'material conditionals' of truth-functional logic, which rarely if ever turn up in ordinary language. The 'material conditional' is customarily written as $P \supset 3\ Q$. It is only misleadingly expressed as *If P then Q*, as is now widely accepted. (Bradley and Swartz, 1983, pp. 226–9; Anderson and Belnap, 1962) The material conditional $P \supset 3\ Q$ is by solely in virtue of *P* and *Q* both being true, or both false, or *P* being false and *Q* being true. It is *not* true in virtue of *any* further relation between *P* and *Q*. It is false only when *P* is true and *Q* false, and false solely in virtue of those truth-values of its parts. Despite ingenious if strenuous attempts (Grice, 1975; Ayer, 1972) to present 'if' statements in general as being material conditionals at bottom, it is evident enough that our *If F then S* is *not* true solely in virtue of the antecedent and consequent being both true or both false, or false and true respectively. (Mackie, 1973a)

Thirdly, dependent conditionals are unlike a very considerable and mixed assortment of 'if' statements. (i) 'If she feels so strongly, she'll decide against it.' (ii) 'If he is reasonable and understands the facts, he'll send the letter.' (iii) 'If you want them, there are biscuits on the sideboard.' (iv) 'I could have if I chose to.' (v) 'If I'm awake the sun will rise and if I'm not awake the sun will rise.' (vi) 'Since you moved your arm that way, you waved.' (vii) 'The offer was made and accepted, so there's a contract.' (viii) 'If that was painted in the eighteenth century I'm a Dutchman.' (ix) 'If you were Julius Caesar, you wouldn't be alive.' These are in various ways different, as reflection will show, and raise different questions. What is common to all of them and to others, as it is to the first two sorts of 'if' statements, is that none states the kind of connection of one thing with another which is expressed by any dependent conditional.

This general distinction, clear enough despite our not having an analytic account of it, is in part brought into sharper focus in a somewhat unexpected way. There is a difference, although an uncertain one, between some statements of the form *If P then Q* and others of the form *If P, Q*. (W. A. Davis, 1983a) Suppose that someone has unkindly disconnected the wiring between the switch and the wipers. It makes sense to say, and in a certain situation it will be true, that (1) if the switch is not flipped, the wipers will not start. One can say quite as truly in this way, of course, that (2) if the switch *is* flipped, the wipers will not start. (The case is then like (v) above.) But is it true that (3) if the switch is not flipped, *then* the wipers will not start? On the contrary, it seems false. This is so since this third statement asserts the existence of a connection between two things (no flipping and no starting), and *ex hypothesi* no such connection exists. The first statement, like the second, can naturally be taken as not asserting such a connection, and hence can be true. The third statement is a dependent nomic conditional, while the first, whatever else is to be said of it, is not. The point is instructive, but it would certainly be mistaken to suppose that all ordinary conditionals are of the form *If P then Q* and all other 'if' statements of the form *If P, Q*.

Dependent nomic conditionals can also be characterized in terms of their logical properties in a narrow sense. Let us notice two of these. The seven connections surveyed above (1.2, 1.3) were stated by both a conditional and its contrapositive. Dependent nomic conditionals, as can be anticipated, in fact have the logical feature that they do simply entail their contrapositives. *If not-F then not-S* entails *If S then F*, and the latter entails the former. *If F then S* entails *If not-S then not-F*, and here too the latter entails the former. That there is this mutual entailment with respect to the two members of each pair is, or is intimately connected with, the proposition that the two conditionals state the same fact of connection between two things in the world. The feature of entailing their contrapositives distinguishes dependent nomic conditionals from certain other 'if' statements. Some of these are exemplified by (iii) and (iv) above. From 'If you want them, there are biscuits' it does not follow that if there are none, you don't want any, and from 'I could have if I chose to' it does not follow that if I didn't choose to do the thing, I wasn't able to do it.

Dependent nomic conditionals also have the logical feature that they are transitive. That is, *If P then Q* and *If Q then R*, where these are such conditionals, entail *If P then R*. It has sometimes been said that certain other 'if' statements are not transitive—for example, 'If J. Edgar Hoover had been born a Russian, he would have been a Communist', 'If he had been a Communist, he would have been a traitor', and 'If he had been born a Russian, he would have been a traitor'. (Lewis, 1973, p. 33; Stalnaker, 1975, p. 173) It is said that this proves the failure of transitivity—the three statements are unexceptionable and the third does *not* follow from the first two. There is the objection, however, whatever else is to be said, that the third statement fails to follow from the first two only because of an ambiguity—and more precisely because the consequent of the first conditional is in fact not identical with the antecedent of the second. We do not actually have in this supposed counter-example to transitivity what we must have, statements of the forms *If P then Q*, *If Q then R*, and *If P then R*. (Mackie, 1980) Certainly, whatever is to be said of transitivity elsewhere, dependent nomic conditionals *are* transitive. They are thus perfectly suited to the analysis of our beliefs about causal chains or sequences. Certainly from the facts that *r* caused *s*, and *s* caused *t*, it follows that *r* caused *t*.

Are dependent conditionals to be characterized more generally in terms of two categories to which philosophers have given much attention, those of subjunctive and counterfactual statements? This seems often to have been assumed. To have a new example, consider the statement that (*A*) since it is raining, the balcony is wet. It is an indicative conditional, a conditional in the indicative mood. Consider also the statement that (*B*) if it were raining, the balcony would be wet. It is subjunctive. Is only one of these, perhaps the second, a dependent conditional in the sense we have in mind? No, both statements, although they are different in mood, are such conditionals. (*A*) is part of what is stated by stating that rain is making the balcony wet, or causing the balcony to be wet. (*B*) is part of what is stated by stating that rain would make the balcony wet. The distinction between our dependent conditionals and others is thus not a difference between the indicative and the subjunctive mood.

It is as clear that another difference between (*A*) and (*B*) is no more relevant. (*B*) is counterfactual: it implies the falsehood of its antecedent. (*A*), called by some a factual conditional, implies that its antecedent is true. The difference is not the distinction between the class of dependent nomic conditionals and other 'if' statements. *Both* (*A*) and (*B*), to repeat, are dependent conditionals. So is what is sometimes called an open conditional: If it is raining, the balcony is wet. It carries no implication as to the truth or falsehood of its antecedent.

Dependent conditionals have often not been distinguished by philosophers from one or another larger category of 'if' statements. Partly because of this fact, dependent conditionals have been taken as problematic. The principal problem about them has generally been said to be that of their *meaning* or *semantics*. The problem is to define the meaning of conditionals, to say what it means to say that if kangaroos had no tails they would topple over, to say exactly what conditionals mean. (Goodman, 1965, p. 17, p. 23, cf. p. 14; Ayer, 1972, pp. 120 f., cf. p 118; Lewis, 1973, p. 1; Mackie, 1973, p. 64) As a look at the philosophy of language and its analyses of 'meaning' or its uses of 'semantics' quickly shows, much more would need to be done to give us a well-defined problem, but let us not linger. The vague expression of it is sufficient for our purposes. Let us rather glance at two of what are presented as solutions to the problem, the *metalinguistic* and the *possible-worlds* proposals. By doing so we shall become clearer about the problems of nomic conditionals, and hence their solutions. We shall also avoid a doubt about what will be maintained here.

The metalinguistic proposal (Goodman, 1965), so named because, at any rate in the first instance, it presents conditionals as being about other linguistic entities, is along the following lines. What is it to say that if (*R*) it is raining, then (*W*) the balcony is wet? Roughly, it is to say that the statement (*R*), and (*C*) true statements of certain conditions, and (*L*) a true *lawlike* statement, together entail the statement (*W*). The proposal, as is allowed by its proposer, faces serious problems, notably that of explaining the nature of a *lawlike* statement. It is nonetheless advanced as being on the right lines.

The possible-worlds proposal can most easily be stated briefly in terms of a dependent conditional that is counterfactual. To say that if it were raining the balcony would be wet is to say this: among possible worlds where it is raining, the one which overall is most like our actual world is also one in which the balcony is wet. (Lewis, 1973; cf. Stalnaker, 1975) Or, to interpret the idea in a way less ontologically extravagant, a way which does not seem to commit us to a plurality of somehow existing worlds, what the conditional means is this: if our actual world were different in that it were raining, and differences overall were in a sense the smallest possible, the balcony would be wet. One source of this theory, to continue in terms of the example, is the truth that if it were raining, more other things would be different than that the balcony was wet. For a start, there would be a cause of the rain, and further effects of it—a wet garden and so on. This prevents us from supposing that the conditional in question, to speak in the ontologically extravagant way, comes to this: in the possible world where it is raining, but everything else is the same as in this world save that the balcony is wet, the balcony is indeed wet.

What we must then do, it is supposed, is to turn our attention to a primitive idea of over-all similarity between possible worlds. This has to do both with what are called states of affairs, which we may take ultimately to be a matter of individual properties, and also what are called laws. To note a possibility to which we shall return in a moment, it is allowed that a possible world *w'* might be more like our actual world than a possible world *w''* even though the laws of our world are to some extent suspended or do not exist in *w'* and are intact in *w''*.

A bit more will be said of particular features of the metalinguistic and possible-worlds proposals, but let us first consider something common to both of them and indeed to other proposals. All of these, to repeat, although there is some uncertainty and inconsistency, are presented as answers to the question of the *meaning* or *semantics* of certain 'if' statements, certainly including dependent nomic conditionals. To think about this even for a moment is to see that something is amiss.

Does 'If it's raining the balcony is wet' *mean*, in however large a tolerable sense, something about other conditions—say the absence of a canopy over the balcony and so on? As was maintained earlier (1.2), surely not. The unsatisfactory conclusion that the conditional is about so much, or rather the unsatisfactory conclusion that the conditional is about a further *statement* about so much, follows from the metalinguistic view. At any rate there follows some such unsatisfactory conclusion pertaining to other conditions somehow described.

Again, does 'If it were raining the balcony would be wet' *mean* something about *other* ways that the world would be different, over and above the balcony's being wet, if there were the difference that it was raining? Is the given conditional in part about a cause of rain, or the wet lawn? It is a remarkable idea, not made better by bravely labelling the conditional enthymematic. The unsatisfactory conclusion, or a related one, follows from the possible-worlds view.

The views are *more* plausible when taken as answers, or at any rate materials for answers, to a question quite different from the question of meaning. They are *more* plausible when taken as responses to a question about dependent conditionals which in fact has more claim to be regarded as the principal one. It can be called the *logical problem*, and briefly expressed it is this: in general, what are the *premisses* or *grounds* or *bases* for dependent conditionals? It is not the question of what in general we *say* when we assert such conditionals, or what they are *about*, but the question of what *reasons* we have for saying what we do. (This is the question that is fundamental with *every* sort of 'if' statement.) It is our reasons for asserting a dependent conditional which bring in a good deal more than what is brought in by the conditional itself. That there was no canopy may be part of why I say that if it's raining the balcony is wet, but it is not part of what I say.

The metalinguistic view remains in several ways odd and indeed unsuccessful when regarded in the more plausible way. It may be said to be on the right lines, but at best it provides materials for an answer to the logical question, materials which it does not combine properly. Further, so to speak, one of the materials is indeed inadequate. If we are seeking an explanation of the grounds of dependent conditionals, and one of these is given as a *lawlike* statement, we do indeed require an explanation of the nature of such a statement. As for the possible-worlds view, of which a great deal might be said, it too seems to involve an unanalysed notion of law, although the matter is more obscure here.

Let us notice only a clear objection for which the way has been prepared. What we have as premiss for the dependent conditional that if it were raining the balcony would be wet is roughly this: in that possible world most like our own in which it is raining, the balcony is wet. But it is specifically allowed that that world might *lack our laws*, including a law which pertains to the rain and the wet balcony. In that world, to be brief, it could be an accident or mere coincidence that the rain was accompanied by the balcony's being wet. *That* could not be our reason for asserting the given conditional, whatever else is. (Cf. L. J. Cohen, 1980; Pollock, 1976; Swain, 1978.) Whatever the strengths and interests of possible-worlds conceptions in several inquiries, notably formal semantics, we do not here have an acceptable answer to our question.

On what basis *can* we assert the dependent nomic conditional that if (*R*) it is raining, then (*W*) the balcony is wet? The short answer is that we assert it on the basis of two things, an *independent* nomic conditional, and (*C*) a belief about certain conditions, which is a belief that the antecedent of the independent conditional is in a certain part true. Again, we assert it since we accept (i) an independent nomic conditional roughly to the effect that in the world as it is, and within certain large limits as it might be, if it is raining and certain other things are the case, then the balcony is wet, and we also accept (ii) that those other things *are* the case. It follows that if it is raining then the balcony is wet.

To be more explicit, it is simplest to take the particular formulation of the independent conditional just suggested, and anticipated earlier (1.3), in place of *If R and C, even given any X consistent with R and C and W, then still W*. That is, let us have this: *Given the world as it is, or given any changes in it logically consistent with R and C and W, then if R and C then W*. From these two things it follows—as from *If A, then if B then C*, and *A*, it follows that *If B then C*—that *If R and C, then W*. From this in turn, together with *C*, there follows the dependent conditional *If R then W*. To repeat, let us have the statement (*Y*) describing the actual events and conditions accompanying *r* and *c* in the world as it is, and the disjunctive statement (*X*) to the effect that the world is in one way or another otherwise, logically consistent with *R* and *C*, and *W*. Then our premisses and conclusion are as follows.

If *Y* or *X*, then if *R* and *C* then *W*.
Y.

....

If *R* and *C* then *W*.
C.

....

If *R* then *W*.

This answer to the logical question about dependent nomic conditionals is reassuringly persuasive. Certainly it involves no unexplained notion of a lawlike statement. That is not to say that it involves no notion of a lawlike statement, or, to speak of reality rather than our language for it, no notion of lawlike connection. It is unthinkable that any arguable account of dependent conditionals could be without a notion of lawlike connection, and hence of law or lawlike statement. As can properly be said, the answer just given to the logical question rests essentially on an *explained* notion of lawlike connection. 'Lawlike connection' is simply another term for what was earlier (1.3) called *fundamental nomic connection* or *fundamental necessary connection*—and for connections related to it. Fundamental nomic connection *is* the connection stated by independent nomic conditional statements. To rest an answer to the logical question about dependent nomic conditionals on independent nomic conditionals *is* to answer the question in terms of explained or analysed lawlike connection. The plain answer also has other virtues (Honderich, 1982a), but they need not be sung here.

As for the meaning of dependent conditionals, it is possible and perhaps necessary to say of them, as it is commonly said of 'if' statements of various kinds, that they are to be taken as primitive, in the sense of not being open to analytical definition or reductive analysis. (Certainly one only gets something synonymous, at best, and no *analysis*, by rendering 'If *P* then *Q*' as 'On the assumption that *P* is true, so is *Q*' or 'In a possible world where *P* is true, so is *Q*'.) Dependent conditionals are thus to be regarded in the way of the primitive conception or conceptions at the base of any logical system. That is not to say, however, that their meaning cannot be characterized. It has been here, in what has been said already. Their meaning is such that they are to be distinguished from various other 'if' statements, that they have certain logical properties, and that they are entailed by independent nomic conditionals together with further premisses in a way derived from the antecedents of the latter conditionals.

To turn now to independent nomic conditionals, they can be identified initially, as they have been, as typified by 'if' statements we accept in connection with our beliefs as to causal circumstances and effects. They can, as we know, take the form illustrated by this 'if' statement of our current example: *If R and C, even if X, then still W.* Their meaning is evidently quite other than that of dependent conditionals, since they are in part and in a way *general*. Each such conditional asserts, with respect to all events or conditions of a certain class, that the occurrence of any or any set of them, or indeed all of them, would none the less leave it true that if the conditional's antecedent is true, so too is its consequent. By antecedent, in terms of the example, I of course mean only *R and C*. In virtue of this fact of generality with respect to independent conditionals they are not tied to a particular situation, as are dependent conditionals. Their truth is not dependent on a particular situation. They can be expressed formally in several ways, making use of the resources and notations of different logical systems, but are perspicuously expressed in just the forms we have. We can, as with dependent conditionals, distinguish them from other 'if' statements, specify their logical properties—including contraposition and transitivity—and give their logical relations, notably their relations to dependent conditionals.

On what is such an independent conditional as *If R and C, even if X, still W* based? The answer, in brief, is the method of empirical inquiry, at its best the method of science. There can be no doubt whatever about the validity of this method, and no doubt either that its description has been and remains a matter of controversy, or of several controversies. One of these, perhaps the most general and fundamental, has to do with the problem of induction. What is the explanation of the rational justification we evidently have when we reason in certain ways from certain premisses to particular or general conclusions about the world? What is the explanation of why I am right to conclude, as I am, that if it is raining and certain other things are true, the balcony is wet—and, in brief, that it would be wet no matter what else were true? The explanation will include, certainly, past situations both like and unlike the present one—like, in that they included events and conditions of the same type as *r* and *c*; unlike, in that they included events and conditions of other types than those accompanying *r* and *c*. The explanation will also include what is related to this and is absolutely fundamental to scientific method, which is the experimental procedure of testing and establishing connections by the 'varying of circumstances', which is essentially the discovery of what is relevant and what is irrelevant to a given event. (Mill, 1961 (1843), p. 249; Keynes, 1952, p. 393; Carnap, 1962, p. 230; Honderich, 1989)

To say this much of the method of empirical inquiry, above all the method of science, is of course to say little more than nothing. Anything like an adequate account of the method of empirical inquiry is out of the question here. One separate point is clear enough, however. It hardly needs remarking that the experimental procedure of varying the circumstances *fits exactly* the account of fundamental nomic connection which we have. It fits that account better than it fits others, including a probabilistic account of which a bit more will be said. (Skyrms, 1980, p. 16) That is a further if subsidiary argument for the account.

One thing remains to be noticed. It is now clearer than before (1.3) how fundamental nomic connection, the connection stated by independent conditionals, is either the stuff or the basis of all the seven causal connections. It is the stuff, so to speak, of the last three—(5), (6), and (7). It is the basis of the first four—(1, 1a), (2, 2a), (3), and (4). It is the basis in the sense that each of the dependent conditionals rests on some independent conditional and a further premiss related to the antecedent of the independent conditional. Consider the dependent conditional (3) *If c occurred, so did e.* Consider also *If cc occurred then, even if there also occurred any change x logically consistent with cc and e, it was also the case that e occurred*—which is the first part of the independent conditional in (5). Circumstance *cc*, we take it, consisted in *c* and also in *c'*, *c"*, ... As with the example lately considered, it is evident that (3) is entailed by the given part of (5) together with a statement of the occurrence of *c'*, *c"*, ...

It would be rash to make the conclusions of this chapter depend absolutely on exactly the account of certain 'if' statements that has now been given or intimated. These statements, as already remarked, make up a controverted subject. (Sanford, 1988) It is complete with competing predilections, schools, logics, methods, and terminologies—and indeed competing conceptions of the subject, by which I mean conceptions of just what 'if' statements are properly treated together. What I hope to have shown, which is consistent with a certain tentativeness about what has been said, and with incompleteness, is that we do have a grasp of both dependent and independent conditionals, which grasp can be clarified and which gives to us an explicit understanding of the seven causal connections that were set out. It is not as if conditionals of the two sorts were near to being sufficiently problematic or obscure as to make it unprofitable to use them in elucidation of causes and effects, causal circumstances and effects, and—to look forward—nomic correlates.

1.5 CAUSAL VERSUS OTHER NOMIC CONNECTIONS

We take causal circumstances and causes to have a nature lacked by effects. This nature presumably explains the truth that if *a* is a causal circumstance or cause of *b*, then *b* cannot be such of *a*. We ordinarily say of causal circumstances and causes that they *make their effects happen*, but we do not say, and will deny, that effects make either of the two causal items happen. The philosophical variations on this usage are many. The causal items are said to be active, to be productive, to be geneses, to have potency or efficacy. Some philosophical writings on causation consist in good part in a somewhat numinous insistence on the distinctive nature of the causal items as against their effects—causes, for example, are declared to be 'powerful particulars' or 'forceful objects at work'. (Harré and Madden, 1975)

We also say of causal circumstances and causes that they *explain* their effects, in a sense in which effects do not explain the causal items. Here, there is less possibility of philosophical variation, but this second characterization of the nature of the causal items is perhaps as important as the first. Finally, we take it that effects *depend on* the causal items, and that the latter do not in this way depend on the former. It is perhaps a good deal less than certain that this third characterization is conceptually distinct from the first two. I shall suppose it comes to much the same.

In our inquiry into causation so far, we have not attended specifically to this fact of difference or asymmetry between causal items and their effects—the fact of *causal priority* as it is sometimes called. We do indeed have it that a causal circumstance *necessitated* its effect. But to assert that is by definition to assert no more than a certain independent conditional—roughly, that since the circumstance existed, even if most other things had been different, the effect would still have occurred. We also have it that effects do no more than *dependently necessitate* their causal circumstances. That is to say, roughly, that if the effect occurred, and no other causal circumstance for it but one existed, that one would still have existed even if most other things had been different. It is not obvious, although it may be true, that the ideas that a causal circumstance made its effect happen, and explained it, and that the effect depended on the circumstance, somehow come to no more than these independent conditional claims. Philosophers have sometimes denied that the asymmetry between causal items and their effects is a matter of connections stated by 'if' statements. These, they feel, are not enough.

Certainly we cannot rest with the three ordinary ideas we have of the distinction between the causal items and their effects. The first idea is of a metaphorical and anthropomorphic kind, and the second and third also call out for analysis, if only for the reason that there are other non-causal pairs of things such that the first explains the second and the second depends on the first. The obvious example is that of the premiss and conclusion of a deductive argument. It is not *that* kind of explanation and dependence that is in question with causation. What kind it is needs to be explained.

If we cannot rest with the ordinary ideas, we can no more rest with their philosophical variants. It is all very well to insist that causes have or are *powers* or whatever, but we need to know what is to be understood by that. They do not give commands and they are not premisses from which many or important conclusions follow. Nor does it seem likely that the idea of causal power is not open to analysis, or, what comes to much the same, that it is somehow to be acquired without noticeable effort by thinking on what is common to such verbs as 'push' and 'pull', as has sometimes been supposed.

One persistent analytical account of causal priority does seize upon an indubitable truth, that causal items stand to their effects as our means to our ends, while no effect is our means to its cause or causal circumstance. When it is true that an effect—the wine bottle's being open—is my means, it is not such as an effect but as a cause of something else, which other thing is not a means to it. We do indeed manipulate and control our surroundings, in so far as we can, by way of things as causal rather than as effects. However, there is the immediate objection that not *all* causal items are the means of someone. No earthquake is, and in fact relatively few causal items in the natural world are such. The attempt has been made, inevitably, to extend the idea of a means to cover all causal items. (von Wright, 1971; Mellor, 1986) This stratagem is not reassuring, for several reasons, but there is a more fundamental objection which applies even to those causal items which really are our means.

It is that the fact that a causal item is a means is not a fact about *it*, but a fact about *us*. The fact that a cause of the wine's being cool, say refrigeration, is my means—this is the fact that (i) I can bring about that cause, and (ii) it is a cause of what I desire. This thought, that the given cause of the wine's being cool is not in or of itself a means, is reinforced by the truth, among others, that the given cause is precisely *not* a means to my idiosyncratic drinking companion, who likes his *Haut Poitou* uncooled. But the asymmetry of causal items and their effects is, of their very nature, a fact about them, a fact which would persist in a world devoid of desires, and, as might be added, a world devoid of our capability of bringing things about.

Is it possible to explain the asymmetry by way of a *clear* idea of power—or capacity, ability, or disposition? Well, we can give a certain clear sense to saying that the hot coffee is able or has a power to dissolve the cube of sugar. What it is in general for a thing *a* to have the power to produce *b* is for it to be true that an individual property or properties of *a*, together with other things, will constitute a causal circumstance for *b*. Anything that is a cause, then, is in this clear sense a power, a power to produce an effect. (Cf. Ayers, 1968.) There is a related secondary sense of the term 'power' and like terms, where the power is the *class* of differing individual properties or property-sets, each of which is nevertheless alike in entering into some causal circumstance for one effect. Or, better, a power of this kind is the *class of types* of such properties or property-sets. In this sense, hot coffee can be said to share a power to dissolve sugar with steam, certain chemicals, and so on. The secondary sense is clearly dependent on the primary.

We need to reflect, however, on what has just been said: in brief, that for something to be or to have a power in the primary sense is for it to enter into a possible causal circumstance. Given our account of causal circumstances, that is fundamentally to say, in line with the independent conditional (5) set out in Section 1.3 and mentioned at the beginning of this section, that for *a* to have the power to produce *b* is for roughly this to be true: *If A & C, even if X, still B*—where *C* asserts the existence of other conditions or events. The difficulty is that a like conditional (derived from the independent conditional (6) also set out in 1.3) may well be true of *b*. That is, it will be true, if there is no other causal circumstance for *b* on hand, and no alternative for *a*, that *If B & C, even if X, still A*. To speak informally, in terms of the example, the hot coffee together with other things guaranteed dissolved sugar, but it may also be true that the dissolved sugar, together with (different) other things, guaranteed the hot coffee. The upshot of this is that in this sense of 'power'—as of many like terms—it is at least arguable not only that a cause has a power to produce its effect, but *also* that an effect has a power to produce its cause. Here we have no adequate difference between cause and effect. It *is* true, somehow, that a cause has a power in a sense that an effect does not, but we have not got that sense.

There is the further grave difficulty about the idea in hand, as a little reflection will show, that in the given sense no *causal circumstance*, as distinct from *cause*, has the given power. Leaving aside several other good attempts to explain the difference between causes and causal circumstances and their effects, and also what can be said of great obstacles in the way of these attempts (Mackie, 1974, Ch. 7, 1979; Ayer, 1984a; Sanford, 1976, 1985; Papineau, 1985b; Honderich, 1986), let us return to and concentrate on our ordinary convictions about the difference.

What *do* we have in mind in taking it that a causal circumstance *makes an effect happen*? A good answer is that we regard the causal circumstance as leaving no room for any other eventuality than the effect. The causal circumstance settles that but one of certain possibilities becomes actual. Most plainly, the causal circumstance *fixes or secures the occurrence of just one thing*, as distinct from fixing the occurrence of that thing or a second or a third or. . . . What do we have in mind in taking a causal circumstance to *explain* an effect in the given sense? There are the same good answers. It is for the circumstance to leave room but for one eventuality, for it to settle things. It is not for the circumstance to give rise to something or other, but for it to give rise to *just the effect*.

These several glosses of the characterization of a causal circumstance as making its effect happen and explaining it, glosses which are surely very natural, lead us to a firm conclusion, one that may be anticipated. If it is not obvious, as remarked before, it surely *is* true that the nature of causal circumstances and causes, as against effects, is, at least in good part, explained by what we have already—it is explained in good part by the fact that causal circumstances necessitate effects, and effects merely necessitate one causal circumstance or another. That is, to simplify the independent conditional (5) a bit, a causal circumstance is such that if it happens, then *just its effect* does. But, to simplify (6), an effect is such that if it happens, then all that is true is that *one or another* of a set of causal circumstances has existed.

The clear distinction made by these two conditionals gives a clear sense to talk, mainly by philosophers, of causal circumstances having a power lacked by effects, and so on. To revert to what is fundamental, the distinction made by the two conditionals gives *some* clear sense to our saying that causal circumstances explain effects, and make them happen, and not the other way on. More needs to be said about our conviction, but here we have something. We have in a causal circumstance by itself a *complete* answer to the question of why an effect occurred. We do not have, in just an effect, such an answer to the question of why a causal circumstance occurred. That we do not have such an answer, it can be argued, is the fact that what follows, from the occurrence of the effect, is only that that circumstance *or* another occurred.

I have latterly been speaking only of causal circumstances, and not causes. What has been said can be extended to them. That is, in brief, it is reasonable to suppose that their nature, as distinct from that of effects, is to be explained by their membership of causal circumstances. What has been said, however, seems not enough. It is, I think, one of two parts of an adequate account of causal asymmetry. The additional part, which does not have to do with the connections stated by independent conditionals, is perhaps particularly necessary in connection with our conviction about the explanatoriness of the causal items.

Both that conviction, and the conviction that the causal items make their effects happen, can also be glossed as convictions that the causal items *bring into existence* their effects. Given this, it is impossible to avoid the idea that another part of the difference between the causal items and their effects is that the causal items exist at a time when their effects do not. They exist *before* their effects. If all causal circumstances and causes precede their effects in time, it seems we have in that temporal consideration a second basis for the asymmetry we are considering. *Do* all of what we take to be causal circumstances and causes precede their effects? Here there is a large philosophical dispute, and we shall be told by some that the answer is no. Is there not a causal circumstance, including the weight of the driver, for that simultaneous effect which is the flattening of the seat cushion?

One thing that can be said in opposition to the simultaneity idea is that if we persist in thinking precisely of *causation*, of one thing *causing* another, as distinct from any related kind of connection, we are inclined to try to substitute successions for simultaneities. We are inclined to think of connections between earlier and later events rather than connections between simultaneous events. The flattening of the seat cushion at *this* instant is owed to the driver's weight at a prior instant. The last instant of the flattening of the cushion, we are inclined to think, will be simultaneous with the beginning of a causal circumstance for the cushion's being other than flattened. In this inclination to take the causal items as prior to their effects, incidentally, we have the support of a good deal of science, indeed a strong scientific tradition, having to do with the principle of retarded action. (Bunge, 1959, p. 62f.)

In our ordinary thinking about causation and time there evidently is uncertainty, as is not the case elsewhere. In some respects there seems not much room for argument about our conception of standard effects and their causal circumstances. There is surely no doubt that we take them to involve the necessity relations and the relations of requiredness. In connection with time, our conception is not settled. This fact is consonant, to say the least, with the long philosophical dispute about causation and time, including the idea that causes might not only be simultaneous with their effects but might come *after* their effects. (Dummett, 1954) If we are subject to uncertainty, there is room for decision, as distinct from discovery. The definition we shall adopt here, in the tradition of Hume and many others, is one that takes causal circumstances and hence causes to *precede their effects in time.* It allows us the conclusion that has just been contemplated: that the difference between causal items and their effects has its basis not only in the consideration that causal circumstances fix uniquely the occurrence of their effects, but also in the consideration that causal circumstances precede their effects. The definition is adopted, of course, not merely for the reason that it gives us a further explanation of the difference between causal items and their effects. It has an independent recommendation, although one that needs more argument that has been supplied here.

Given the account we have of the difference, we can now proceed quickly to a final characterization of a causal circumstance. We have everything in hand save one consideration, having to do with non-redundancy. The difference between causal circumstances and their effects, further, is what distinguishes causal circumstances and effects from other things also in nomic connection, which is to say nomic correlates. We can also proceed quickly to a characterization of these.

We began with the idea that a causal circumstance consists in a set of conditions including a cause or causes—more precisely a set of individual properties—each being (1, 1a) required or alternatively required for the effect. From this it followed (2, 2a) that the circumstance too is required or alternatively required for the effect. Further (3) each condition requires the occurrence of the effect, and so too (4) does the circumstance as a whole. As for the fundamental necessity-relations, (5) the causal circumstance necessitates the effect, and hence the effect is necessary to the circumstance. (6) Also, the circumstance is one of a set of circumstances necessary to the effect. Further (7) the causal circumstance is dependently necessary to the effect, which is to say that the effect dependently necessitates the circumstance. We have it too (8) that the circumstance (whose constituents need not be simultaneous) is prior in time to the effect. It is in virtue of (5), (7), and (8) that we can truly say that the circumstance *makes happen* and *explains* the effect, and not the other way on.

The property of a circumstance that (5) it necessitates its effect, although we did not pause to consider the matter, is in a way essential to a final element in a definition of a causal circumstance, more particularly a specification of what is included in such a circumstance. To return again to the beginning, a circumstance (1) consists of items required for the effect. That is not to say that a circumstance includes *all* such conditions of the effect. Which ones then? The answer is that (9) a causal circumstance is to be taken to include *no more conditions than are needed to necessitate an effect*. That is, it includes *just* a set of conditions such that if the set existed, so did the effect, and still would have even if certain other conditions or events had also existed. Nothing is redundant. To take a causal circumstance as having no redundancy is obviously to exclude things wholly irrelevant to the effect. With respect to the circumstance for the starting-to-work of the windscreen wipers, the car's radio being on is likely to be irrelevant.

Other things are not irrelevant in the given sense, since they *are* required conditions of the effect—but they are *not* part of the causal circumstance. A causal circumstance, in accordance with the non-redundancy criterion, does not include a particular condition and also a causal circumstance for that condition, or any part of one. In specifying a circumstance for the working of the wipers, we may include the switch's being flipped, but if we do, we cannot also include the muscle movements which gave rise to the switch's being flipped. Certainly we do not need to try to go into the whole causal history of an event in order to specify something—one of the many sets of things—that had the property of making the occurrence of the event necessary. In general, if *c* is in a causal circumstance *cc*, and *c* is the effect of *cc*′, then *cc*′ cannot be part of *cc* and neither can any part of *cc*′. Equally, a causal circumstance *cc* for an event *e* does not include any other effect than *e*, perhaps an effect in a causal sequence connecting *cc* with *e*. To speak loosely, a causal circumstance does not include two or more links of any one causal line running through it from past to future.

We shall have no need of a fuller definition of a causal circumstance than the informal one we now have—given in the nine propositions above. What will be of greater value is a partial characterization which mentions only those features (9, 5, 8, 7, in order of appearance in the partial characterization) which will be of most importance in what follows immediately. Thus, if *cc* was the only causal circumstance for an event *e*, then

> *cc* was no more than a set of conditions or events which necessitated *e*, and preceded *e*, and was dependently necessitated by *e*.

As already noted, and as will be of some importance later in this inquiry, the temporal feature does not require that the constituents of the causal circumstance be simultaneous. The point will be of particular relevance in connection with *causal sequences* or *causal chains*, to be considered later in the most relevant context. (3.1)

What is bound to come to mind at this point is that there are pairs of things distinct from, but fundamentally like, a causal circumstance and its effect. Certainly the thought must occur to anyone acquainted with almost any part of science, or even a small selection of scientific laws. Such pairs of things enter into what is variously described as interaction, reciprocal causation, functional interdependence, functional relation, concomitant variation, and so on. Such pairs of things are like a causal circumstance and its effect in that they stand in fundamental nomic or necessary connection (1.3), which is to say some connection stated by an independent nomic conditional statement. They differ from causal circumstance and effect in that they lack either or both of the features that give rise to the difference between causal circumstance and effect—the priority of the causal circumstance. Thus (i) neither may precede the other, or (ii) they are not such that one necessitates the other while the other merely dependently necessitates it, or (iii) they may lack both features. What is true, rather, is (i) that they are simultaneous, or (ii) each necessitates the other, or (iii) they are simultaneous *and* each necessitates the other. Such pairs are by definition *nomic correlates*.

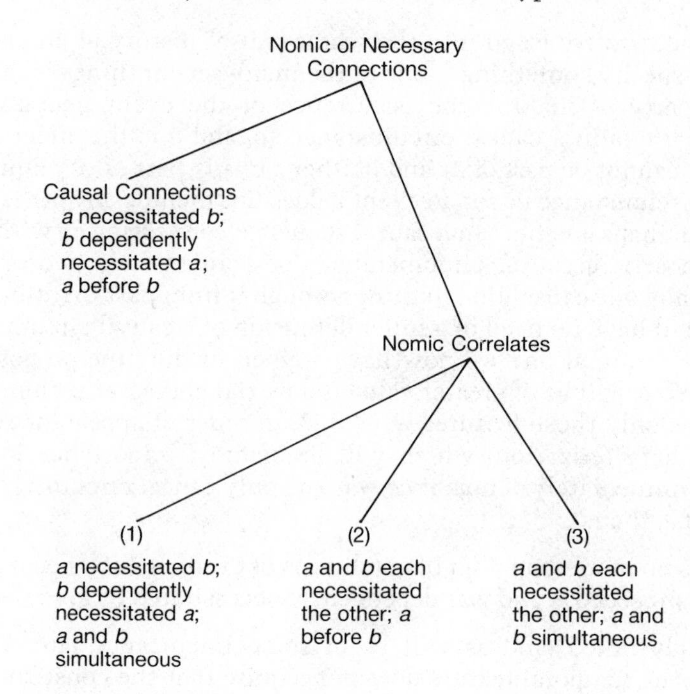

FIG. 1. Nomic or Necessary Connections

As in Figure 1, then, we have four principal categories of nomic or necessary connection, each involving some fundamental nomic connection but differing in some respect from each other category. Each of the three categories of nomic correlates is open to further description, as will be anticipated. For example, with respect to the first category of nomic correlates, the category which will be of most importance to us, there is the truth that if *a* necessitates *b*, then *b* is necessary to *a*. Also, if *b* dependently necessitated *a*, then *a* was dependently necessary to *b*.

Certain objections to these conceptions, objections having to do with science, will be considered later. Let me say now only that there is no established conception or usage for 'nomic correlate' or other more or less equivalent terms used in talk of interaction, functional interdependence, concomitant variation, and so on. No doubt, with respect to the first and second categories of nomic correlates, it is possible to call the second member an *effect* without seeming to misuse language, including this or that scientific language. It is also possible, also without such misuse, *not* to call the second member an effect, but to call it a nomic correlate instead, as we have. It is on the way to being as reasonable as refusing to say about the see-saw that one end's going down is the effect of the other end's going up. It is reassuring to be able to say that the inquiry to come does not depend on our decision. We shall be concerned with the first category of nomic correlates, and refer to them as nomic correlates, but it is their nature rather than their name that is important.

It will be noticed that in all of this the term 'correlate' is being used in a way consistent with our previous uses of 'event', 'cause', 'condition', 'causal circumstance', and 'effect': not for a type of thing, but for a thing itself, more particularly an individual property or set of such properties rather than a type. To use the term 'event' for a token rather than a type is ordinary. So with 'cause' and 'effect'. 'Condition' and 'causal circumstance' can also be used in this way. The ordinary use of the term 'correlate', where it is ordinarily used, is perhaps for a type of individual rather than an individual. We are departing from this usage, and must put up with some inconvenience, in order to have consistency and hence clarity in several contexts. Nothing substantial hangs on the decision.

One last and related point. We have found causation to consist, at bottom, in connections between particulars. We have analysed the *particular* statements which state these connections. Some of these, analysed by way of independent nomic conditionals, are in a way general. (1.4) They are distinct, however, from causal *generalizations*, which will be of some importance in what follows. Much might be said of the relation between particular and general causal statements, but here a little must suffice. If we maintain that *cc* was a causal circumstance for *e*, we are maintaining the conditionals of which we know, notably that if *cc* occurred, then even if certain things of a general class had also occurred, *e* would still have occurred. This is evidently in a way general. By asserting it, however, we are also committing ourselves to a *general* conditional proposition of a standard kind. We cannot make the causal claim about *cc* and allow that other circumstances *identical* with *cc* have different upshots. If we take the starting of the wipers to be an effect of a given circumstance, we must accept that there is a *type* of circumstance which is connected with startings of wipers generally. What is true, to express the matter simply, if *cc* was a causal circumstance for *e*, is this:

> If any circumstance of the type of *cc* occurs, even if certain other events or conditions also occur, so does an event of the type of *e*. If the event does not occur, neither does the circumstance.

We are committed to other such generalizations, less simple, one of them related to the fact that a circumstance is dependently-necessary to the occurrence of an effect. We are similarly committed in connection with nomic correlates.

1.6 THE VIEW DEFENDED

1.6.1 Causation and Science

What has been implied in this inquiry, and what it is natural to think, is that a conception of causal and other nomic connections is fundamental to, and pervasive in, our view of reality. If that is so, it constitutes *a* general reason for the ordinary and perhaps irresistible belief that reality *is* in whole or in part a matter of causal and other nomic connections. That ordinary belief in turn gives some support, to say no more, to theories of determinism pertaining to our choices and actions: theories whose burden is that *we* in our lives are a matter of such connections. The strength of the general reason for the ordinary belief depends partly on the resolution of large philosophical issues. These, having to do with whether we take our concepts *from* or *to* reality, will not be debated here.

Still, I do not wish in what follows to forego the support of what must seem to be the good sense of our *natural empiricism*, which is independent of philosophy. (It gets definition of a kind by way of the firm rhetorical question: 'If we do not get our conception of causal and other nomic connection from reality, where *do* we get it?') That is one reason for looking at a certain objection, that our conception of such connection is *not* fundamental to, or pervasive in, a certain body of knowledge and speculation, one which must have some pride of place in any informed view of reality. The body of knowledge and speculation is *science*.

There is also a larger reason for considering the objection about the place of causal and other nomic connections in science. It is that the particular theory of determinism in the chapters to come will depend considerably on the claim that the part of science most relevant to it, neuroscience, does indeed establish certain causal and other nomic connections. This could hardly be so if a conception of such connections was no part of neuroscience.

The objection in one audacious form is that *science has or should have nothing to do with causation*. Russell took this view at one point (1917a), giving a mixed bag of reasons. Most of them, in retrospect, are poor. (Mackie, 1974, Ch. 6) We will do better to consider the objection that *science does not in fact involve or make use of a conception of causal and other nomic connection*. What prompts it is the truth that many of the general propositions fundamental to science, most notably many of those dignified with the name of *laws*, are not of the form of ordinary causal generalizations or ordinary nomic but non-causal generalizations. That is, they are not of the form exemplified at the end of the last section when it was said that if *cc* was a causal circumstance for *e*, then there is the true generalization that all circumstances of the type of *cc*, even if certain other events or conditions occur, are followed by events of the same type as *e*.

There are several differences. The fundamental one is that as a result of the mathematization of physical science, what importantly occurs in it are equations or formulae of several kinds, propositions to the effect that one set of magnitudes is related in a certain way to another set or sets. More particularly, they assert in several ways that values of one variable or parameter are a certain function of values of other variables or parameters. The theory of mechanics, for example, consists in equations which specify that certain traits of bodies in motion are dependent on other physical properties. Is it then the case that no group of these fundamental quantitative propositions of science can be interpreted as stating causal connections? On the contrary, some can, and have often been called and regarded as *causal laws*. The subject is a large and complex one, but the essential point is simple, and easily illustrated.

The point is that an equation can be interpreted as stating this: given each of a number of possible extents to which specified things possess certain different general properties, it is then the case, despite accompanying conditions and events, that another specified thing would still possess another general property to one of a number of possible extents. That is, the equation can be *causally* interpreted, by way of a conditional of the sort we know. It is different in being numerical and in its generality. It is general in two ways. It has to do with *general* properties, certainly, but also it has to do with *ranges* of them.

To return to one of our initial examples, consider the belief that it is the position of the car's heater that accounts for the driver's left knee being warm. In believing this we believe that some causal circumstance necessitated the temperature of his knee. We could by measurement give a numerical specification of each of the parts of the causal circumstance—the operation of the heater, its relative location, and so on—and also such a specification of the effect. Or rather, we could replace our imprecise proposition with a precise one, numerically expressed. A further step, or rather a great deal of research, would be greatly more significant in terms of the question we are considering. This would bring us to something akin to the fundamental propositions of science. We could replace what we have with a proposition relating variation in the operation of the heater, variation in its relative position, and so on, with variation in the temperature of the driver's knee. This generalization would of course be efficiently stated only by way of a certain interpreted equation. The equation would involve both time and a many–one relation, the marks of nomic connection of the causal kind.

The moral of this simple reflection is that fundamental propositions of science can be and often are causal, despite generality and mathematical form. It is difficult to see why it has ever been supposed otherwise—that generality and number in themselves somehow fight with causation sensibly conceived. The simple reflection also illustrates something else, that the fundamental general causal propositions of science rest on particular causal propositions, of a kind with which we are all familiar.

None of this, of course, is to be confused with the idea that *all* of the causal laws of science, or even many, are in the given way analogues in form of the ordinary belief that a particular causal circumstance necessitated a particular effect. That is, roughly, they do not assert that variations in a fully specified type of causal circumstance are followed by variations in a type of effect. There are many alternatives. Some equations, asserted in a certain context or on certain assumptions, have to do with *parts* of causal circumstances. Others relate (i) what would ordinarily be called an effect and a part of a causal circumstance to (ii) another part of the circumstance. All of them have their foundation in the basic connections between causal circumstance and effect, above all that causal circumstances necessitate their effects.

The idea that science does not involve causal connection also faces other difficulties, one of which can be given briefly. It is possible, through concentration on parts and aspects of science, or by way of the refutable assumption that the word 'cause' does not turn up in it (Suppes, 1970, pp. 1–2; cf. Skyrms, 1980, p. 110 f.; Hesslow, 1981), to forget that science includes a *vast* number of propositions which are in various ways explicitly causal. For example, this is true of classifications and definitions, which typically involve causal conceptions. To anticipate something to come in this inquiry, an ordinary definition of the *neuron* is an example.

A special type of biological *cell*, being the unit of which the nervous systems of animals are composed. It consists of a *nucleus* surrounded by a *cytoplasm* from which thread-like fibres project. In most neurons impulses are received by numerous short fibres called *dendrites* and carried away from the cell by a single long fibre called an *axon*. Transfer of impulses from neuron to neuron takes place at junctions between axons and dendrites, which are called *synapses*. (Uvarov, Chapman, and Isaacs, 1979)

So much for the claim that science does not involve causation. The other part of the objection (p. 50), that it does not involve other nomic connection, is as mistaken. In fact, philosophers and others who have been sceptical about the place of causation in science have been inclined to put non-causal nomic connections, somehow conceived, in its place. (i) Connections of this kind are stated in, among others, laws which enter into the taxonomical classifications of 'natural kinds' or 'substances'. Such laws assert connection between determinate properties in every object or stuff that is of a certain kind—arsenic, benzopyrene, coal, and so on. (ii) Non-causal nomic connections are also stated, as remarked at the beginning of this chapter, in numerical laws such as the Boyle–Charles law for ideal gases, $pV = aT$, where p is the pressure of the gas, V its volume, T its absolute temperature, and a a constant pertaining to the mass and nature of the gas. (iii) Lastly, non-causal nomic connections are stated by numerical laws different in that they do specify how a magnitude varies with time. Galileo's law for freely falling bodies in a vacuum is an example. Such laws, although sometimes run together with causal laws, are best regarded as otherwise. This is so since the relation of dependence between the variables given in the law is wholly symmetrical: a *later* state of a system fixes an *earlier* state as much as an earlier fixes a later.

The necessary qualifications in the case of each of the three sorts of non-causal connection are many. To mention only the most general one, the non-causal connections asserted within science are of many diverse kinds, and they are usually in one way or another partial. They do not assert, simply, that x is a nomic correlate of y. It is usually the case that what is specified, against a certain background assumption, is a part of the nomic correlate of something else.

All of this has been set out in greater detail before now. (Feigl, 1953; Nagel, 1979; Mackie, 1974; cf. Bunge, 1959) A good deal of illustration of both causation and other nomic connection in science will be provided in due course (Ch. 5), when we consider neuroscience. What remains to be said here is that the objection—that science does not involve causal and other nomic connections—is sometimes a part of something more general and by its nature more difficult to deal with: scientific scepticism about philosophy and its categories.

If this scepticism has often been well-based, an all-encompassing and undiscriminating scepticism is absurd. (One occasionally encounters a burst of inanity of Nobel proportions, as in this: 'Our capacity for deceiving ourselves about the operation of the brain is almost limitless, mainly because what we can report is only a minute fraction of what goes on in our head. This is why much of philosophy has been barren for more than 2,000 years and is likely to remain so until philosophers learn to understand the language of information processing.' (Crick, 1979, p. 132) The language is not so hard to understand, and it is jejune to suppose that not knowing it has impeded philosophy, or that knowing it will help a lot.) It needs to be kept in mind, by both science and philosophy, that the concerns of the two are in large part different. A general account of nomic connection is an aim of philosophy but no more an aim of science, despite the fact that it rests on such connection, than, say, a general theory of prosody is an aim of poets. Nor is science concerned with just the kinds of generalization that make up a theory of determinism with respect to our lives. The latter have a scope suitable to the question in hand. Different concerns and procedures, and hence unfamiliarity, should not be allowed to stand in the way of understanding.

1.6.2 Specifying Causal Circumstances

A second and commodious objection to the view of nomic connection which has been set out was anticipated earlier (1.3), and may be shored up a bit by the fact lately mentioned, that science broadly speaking is not much engaged in arriving at complete descriptions of causal circumstances—causal circumstances as we have conceived them. Rather, as was mentioned, it is concerned in several ways with parts of such circumstances, often enough a single condition.

The objection is that neither in science nor elsewhere do we in fact have or use the conception of a causal circumstance that has been set out here. It is accepted that *some* conception of a set of conditions for an effect is used, but it is denied that this is the conception of a causal circumstance. *That*, as may be said in tones suitable to the objection, is the conception of something *so complete* as to necessitate *by itself* its effect. More precisely, it is the conception of something which has the nature assigned to it above all by the independent nomic conditional (5). Thus, in part, a causal circumstance is such that if it occurred, then even if there also occurred any change logically consistent with it and its effect, the effect occurred—or, it was the first and the effect was the last of a sequence of things such that the given connection obtained between each thing and its successor. To speak in the same right tones, a causal circumstance can be said to comprise *everything* needed so as in a way to guarantee its effect. It is such that *nothing* could get in the way of its operation save changes logically inconsistent with it or with its effect or with a causal sequence of which it and its effect are parts.

The objection, again, is that we do not in our standard causal thinking have or use such an idea. Certainly we have and use some lesser idea of a set of conditions and events for an effect, including what we call the cause, but we do not have this ambitious idea. The objection involves the feeling or perhaps the conviction that we can never *state* or *explicitly describe* such a thing as a causal circumstance is said to be, which is to say that we cannot do something like give a complete enumeration of its elements: we cannot give particular or individuating descriptions of all of them. (Cf. Russell, 1917a, p. 187; Lucas, 1962, pp. 57–9; Scriven, 1964, p. 409)

For illustration, consider a staple example in thought about causation, that of a house fire, caused by a short circuit. The view to which we have come is that the cause was part of a causal circumstance as conceived. But, it is maintained, we cannot state this circumstance. We may make an easy start with the short circuit and a bit more, but do we not then have to include in the circumstance the absence of a burst in a water pipe at the right moment? Do we not have to specify the particular condition which was the absence of a torrential rainstorm and an open window? Do we not have to include there having been no general power failure in South-east England, and the absence of an earth tremor that would have shifted whatever it was that first ignited, and the earth's not being destroyed by a nuclear explosion? And so on. To state the relevant causal circumstance, would we not have to do what we can hardly hope to do, which is to enumerate parts of much of the whole state of the universe at or during a time? In fact, it may be added, we do not take ourselves to be faced with such an overwhelming task in connection with condition-sets for an effect, and hence for this reason too those sets are wrongly conceived as causal circumstances.

The objection touches on large issues in the philosophy of language and in epistemology, but a relatively brief reply, distinguishing various lines of thought that enter into it, is certainly possible.

The first is that we do not have a certain general idea, since often we do not have certain specific ideas. *One* thing to be said of that is that it does not at all follow, from the fact that we cannot give particular descriptions of items that fall within a set, that we cannot satisfactorily conceive of and describe the set. (The measure of what is satisfactory will, as always, be fixed by a given practical or theoretical purpose.) No one can enumerate *the conditions and events which issued in the first sighting of Halley's Comet* or *the weights of man-made objects in the Northern Hemisphere in the nineteenth century*, but it would be bizarre to say that no clear ideas attach to those definite descriptions. It is as bizarre to suppose that I do not have a clear idea of a particular causal circumstance, say the one which gives rise to a query on the console of my computer, when I am unable to enumerate its elements. *Of course* the description I can give is general. But if the *general* were the *unclear*, we could with reason abandon *all* science and inquiry.

The objection in another part is that in order to *confirm* an hypothesis that an effect was in fact the effect of a causal circumstance of a certain type, we *do* need an enumeration of elements, and often we do not have an enumeration. We shall be unable to test an hypothesis unless we have a close specification of the circumstance-type, and often we lack it. That is true, but it does not issue in the conclusion that we do not ordinarily have and use the given conception of a causal circumstance, or that we could not use it in the confirmation of causal hypotheses. The fact of the matter is that we do not often have confirmed hypotheses of the given kind. In our day-to-day lives, including day-to-day scientific lives, we have little need of such confirmed hypotheses. The situation is not that we always have confirmed hypotheses, and sometimes or often lack an enumeration of the elements of a causal circumstance. *If* that were true, it would perhaps support the conclusion that a different conception of a condition-set for an effect enters into our beliefs and proceedings, although that is unclear. But it is not true.

In a third part, the objection must be that with respect to confirmed hypotheses as to condition-sets, it cannot be that those condition-sets are causal circumstances as conceived. Causal circumstances as conceived are unspecifiable. This central line of thought takes its force from overlooking an essential specification of a causal circumstance, one which derives directly from our ordinary beliefs.

Suppose we say that the causal circumstance for the house fire consisted in the short circuit, the location of inflammable material in contact with the relevant conductor, and the presence of oxygen. We could try harder and do better by specifying certain electrical facts, an ignition temperature for a material, and so on—in general by giving conditions for the particular chemical reaction of combustion that occurred—but there is no need. It is of course true that each of the three elements in the circumstance as described was itself an effect of something else, and, further, had a whole causal history. No general power failure in South-east England and no nuclear explosion were indeed part of the history of the short circuit. But a causal circumstance, as specified earlier (p.46), most certainly does *not* include all of the causal history of itself and hence of the effect. As specified, a causal circumstance includes no more than was needed to necessitate the effect, which is to say that it included just a set of conditions or events such that if the set existed, so did the effect, and still would have even if certain other conditions or events had also existed. The causal history of anything includes innumerable causal circumstances—none of them, however, includes any of the others. Thus the enterprise of confirming what was a causal circumstance for something else is not remotely like some hopeless speculation into the whole natural history of a thing, or even a stretch of it.

In a fourth part the objection takes some strength from the truth that, as in this case, our conceptions of particular causal circumstances, by certain possible comparisons, are inadequate. It is indeed true of the given description of the causal circumstance for the house fire that it could be much improved. But it also possible to think, with respect to any improvement, that *it* leaves room for further improvement. We may forever want confidence that we have come to the *ultimate facts* about some physical process. That is not to say we have no satisfactory conception of the particular causal circumstance. Or, if it is to say that, by way of some stringent idea of what is satisfactory, there is also another conclusion to be drawn. It will be along the lines that it is quite as true that we have no satisfactory conception of anything much—including, most piquantly, the elements which enter into any alternative conception of the condition-set for an effect. As elsewhere, a too zealous objector is hoist by his own petard.

Fifthly, to introduce something not so far mentioned, the objection may include the idea that if our causal thought *did* rest on the given conception of a causal circumstance, we should be able to do well at prediction—better than in fact we do. Causal circumstances in a way *guarantee* their effects. But it does not at all follow from the fact that causal circumstances in a sense guarantee their effects that we can predict those effects better than we do. To do so we would require knowledge which in fact we lack.

1.6.3 Condition-Sets and Backgrounds

The objection just considered is linked to another, or rather, to a positive proposal as to condition-sets and their effects. We may be told that what we understand of an event *e*, if it is taken as an effect, is that there existed a certain set of conditions—say *sc*—such that since it existed, *e* occurred, and *e* would still have occurred *so long as 'the usual background' or 'the usual environment' obtained*. It is therefore allowed, although *e* was an effect, that we might have got *sc* along with some extraordinary event or condition, and *not* got *e*. (Cf. Mackie, 1965, p. 250, 1974, Ch. 2; Lucas, 1962, pp. 57–9; Scriven, 1964, p. 409; Anscombe, 1971, pp. 1, 12, 23; Kenny, 1975, p. 110, p. 115.)

We are to understand that the idea of a usual background is *relative*: there are different usual backgrounds for different causal claims, or different groups of causal claims. With respect to the short circuit and the house fire, it is true that since a set of conditions obtained, there was the fire, and this would still have occurred so long as, say, there was not a flood at the right moment. The usual background does not include a flood. It is allowed, as it must be, that the idea of a usual environment is vague, and hence that our ordinary causal beliefs are vague.

A set of conditions of the given kind, one which is followed by the effect *if* the usual background obtains, is called by some philosophers a *sufficient condition* for the effect. Given what is said of the need for an ordinary background, the condition-set is of course *not* sufficient for an effect in an ordinary sense, where a sufficient condition is precisely one that necessitates the effect. The fact is recognized by some philosophers who take the given view of causation. They note in using the term 'sufficient condition' that they do not mean that the condition-set is *genuinely* sufficient for the occurrence of the effect. It is partly because 'sufficient condition' and 'causally sufficient condition' are so used, incidentally, that I have introduced the term 'causal circumstance'.

There is something more important, which is that the essential idea of the background view does not in itself conflict with our necessitation view. Those who propound the background view deny or doubt the necessitation view, of course, but they need not. We can in fact accept it along with our proposition that effects are necessitated events. We then have the composite view that an effect is necessitated by a causal circumstance and, furthermore, is related to a so-called sufficient condition in the way described, having to do with the usual background. It may be that there is something to be said for this composite view.

Can the background view be held *on its own* as an adequate account of causation? So it is supposed. However, there seems to be very strong reason indeed to think that there cannot be an acceptable account of effects as we conceive them which does not include necessitation.

Suppose a radical objector says of an effect, the smell made by the candles when they were snuffed, that there existed *no* circumstance such that no matter if certain other events or conditions occurred, the effect would still have occurred. He does not supplement but instead denies the view of causation to which we have come. The objector must be disinguished sharply from someone who merely attributes to us some particular idea of a particular necessitating circumstance for the effect, some particular idea of what was included in the circumstance for the candle smell, and believes that the circumstance included more. (The latter person merely maintains of a particular set of conditions that *it* was not the whole of the causal circumstance. *It* would not have been followed by the effect if some other event or condition had been missing.) *Our radical objector maintains about what we take to have been the causal circumstance that it might not have been followed by the smell, and moreover that this would not have resulted from the absence of some further condition.* He thereby makes what happens, the smell, into what can properly be called a matter of real chance. He maintains that *what we take to have been the causal circumstance might have existed as it did, and the rest of the universe might have been just as it was—and there might have been no smell.*

It is important to allow that he may be right about one thing. At any rate, we have no need to dispute one proposition. Things *might* have happened just as he says. Causation or some of it might have come to an end just when I set out to snuff the candles, or never existed. Despite what has been said (p. 50) about an inference from our conception of causation to a conclusion about the nature of reality, our present concern is that conception. We are not engaged in proving the universe to be nomic, or defying it to step out of line. We are examining one of our fundamental conceptions of it. It is a conception that might not be true of it, or all of it, and which might cease to be true of it or of all of it.

What our radical objector cannot maintain, if he sticks to his guns about the relation of the smell to what went before, is that the smell was an effect. He says that the circumstance might have existed as it did, the other events and conditions in the universe might have been as they were, and there might have been no smell. If this is so, the smell was not an effect. The connection between the circumstance and the smell was fundamentally unlike the connection which we take to hold between a causal circumstance and its effect. The connection between the circumstance and the smell, on his view, is precisely the kind of connection, or want of connection, that would be true of real chance or real random events.

It does not matter if our objector allows or insists, as no doubt he will, that it was in some consistent sense probable or highly probable that the smell would occur. It is also true, on his view, for all this probability, that it might not have. On his view the fact that it ceased to be a probability, and *did occur,* is something that might not have happened. Obviously no addition about background conditions, as understood above, will help.

There is another expression of this reply to the radical objector. It depends on the fact that standard effects are events which in the most fundamental sense can be *explained.* There is a theoretical possibility of explaining them. We are not here concerned with the elusive though connected fact that causal circumstances in a different sense explain their effects, that sense which concerned us earlier in connection with causal priority. To say that effects are events which in a fundamental or standard sense can be explained is to say this: *there is a possibility of our finding a proposition which fully answers the question of why they occurred*. Effects are such, on the view of causation to which we have come, since there is the possibility of finding those things which no matter their accompaniment would still have been followed by the effect. However, what we surely have on the objector's view is precisely that the smell of the candles was something that cannot possibly be explained. I do not mean, of course, that it was something that cannot be explained *yet*, but rather that it was something of which no explanation can conceivably be found. No fundamental or standard explanatory proposition is possible. This is so since in yet another sense of the term 'explanation', there was none. There was no such thing in the world. There was no thing in the world for an explanatory proposition to state. What we are supposing is that *everything* could have been exactly as it was except for the non-occurrence of the smell of the candles. Hence there is *nothing* that explains what in fact did occur.

It needs to be allowed that this reply is of a certain kind. The objector has it that what we taken to be effects need not be necessitated. The reply has been that this converts effects into what they are not: *chance* or *random* events, and *inexplicable* events. However, definition of these latter events perhaps must be that they are unnecessitated events. The reply, then, it can be said, does not make use of a feature of effects independent of their being necessitated events, which feature can be used to argue that they are necessitated events. The reply, however, consists in more than mere reiteration or insistence. It is, at bottom, that beyond doubt we use certain ways of speaking in connection with effects. They are *not random* and *not inexplicable*. It can be added that they are *not mysteries, not discontinuities, not small or large miracles*, and so on. It may be that these ways of speaking can be explicated at bottom only by way of the idea of necessitation. What is indubitable is that they are used of effects, that they cannot be used of unnecessitated events, and hence that effects as we understand them are not unnecessitated events. There is more to be said, but it will be better said in reply other related objections.

1.6.4 Causes as Merely Required for Effects

We have concluded that a standard effect—one that is not a decision, choice, or the like of a person, or a following action—is indeed taken to be an event necessitated by a causal circumstance. To speak of a standard effect *is* to speak of an event necessitated by a circumstance, and to imply the existence of such an effect by speaking of a cause *is* to imply the existence of a necessitated event. Necessitation, of course, is simply the fact of the connection stated by the independent nomic conditional (5). It needs particularly to be kept in mind, with the objection now to be considered, that it is indeed advanced as an objection to a given analysis of our causal beliefs, and hence is itself a proposal as to those beliefs. That is why it is now of interest to us. It is not unknown for a philosopher or scientist, lured by the song of the sirens, which is to say an interpretation of Quantum Mechanics, to drift off course to a proposal as to a *better* or *improved* conception of causation. That one conception is somehow better than a second, supposing that that can be made out, is of course not the proposition that it is *our* conception or one of them.

The objection is that it is enough, for something to be a cause, that it is *required for* the occurrence of another event. It is enough, for an event to be a standard effect, that something else was required for its existence. It is enough if the dependent conditional (1), or perhaps one of the variants (1a) or (1b), is true of the two things. Consider an *indeterministic* coin-in-the-slot machine. (Mackie, 1974, pp. 40–43; cf. Feynman, 1967, p. 147; Anscombe, 1971; Sorabji, 1980, Ch. 2; Mellor, 1986) A coin is put in and a chocolate bar comes out. If the coin had not been put in, the bar would not have come out—the first thing was required for the second. However, the bar might not have come out even though the coin was put in. This failure with this surprising machine would not have been at all like the failures of ordinary deterministic coin machines. We are not to suppose of this machine that when a chocolate bar does not come out there exists a causal circumstance which necessitates the failure. Nor, on the present occasion when the machine does produce a bar, is there a circumstance that necessitates that. Given that the coin was put in, it was then a matter of what can be called real chance that the bar came out. But, it is said, we *would* say of the bar's coming out that it was an effect, and in particular the effect of putting in the coin. Therefore no more is needed for an event to be an effect, at least in so far as relations of conditionality are concerned, than that the event was such that another one was required for it.

This opinion can seem persuasive for a time, but we cannot persist in it. Let us not lose hold on the essential proposition that the machine is indeterministic. It contains a mechanism, perhaps a Quantum Mechanical mechanism or what is as fairly called a mystery-mechanism, such that *everything* might have gone on just the same up to some instant, let us say the instant when the bar appeared, and it might have happened instead at that instant that no bar appeared. It is not that something, anything whatever, would have been different before that instant if the bar had not appeared. The coin would have been put in, and so on. There was no necessitating circumstance whatever for the appearance. There would have been none whatever for the non-appearance if that had happened.

What may lead one towards supposing, despite this, that the bar's coming out was an effect of putting in the coin, is a certain fact. It is that the coin was indeed a perfectly ordinary cause of *something*, which thing was not a thousand miles away from the bar's coming out. The coin caused it to be *possible that the bar come out*. If that sounds a touch mysterious itself, it only sounds so. What we can say instead, unmysteriously, is that putting in the coin, like every other putting-in of a coin, activated the indeterministic mechanism. But the fact that the coin caused the mechanism to operate is *no* reason for saying that the coin caused the bar to come out. What we will say, if we keep clearly in mind that *everything might have been the same up to the instant when the bar came out, and no bar might have come out* is that *nothing caused the bar to come out*. We will say that exactly what did not exist was a causal link between the operation of the mechanism, or anything else whatever, and the bar's coming out. The temptation to say the coin caused the bar to come out can be explained by seeing it for the ordinary cause that it was, and of what event it was the cause, and of what nearby event it was not the cause.

The objection as used by one philosopher (Anscombe, 1975) is embedded in further reflections. All of them, to my mind, are ambiguous. It is unclear whether the subject-matter is our beliefs as to causation, or the truth of those beliefs. That is, it is unclear whether what is being maintained is that our conception of effects is of unnecessitated events or that we should change it to that. One of the reflections is that we believe circumstances to be 'enough' to make their effects happen, which is consistent with their not happening. (Anscombe, 1975, p. 60) No argument is given. Nor is any suggestion made as to what we presumably also believe about a difference between a circumstance which was enough to make an effect happen, but didn't, and a circumstance which was enough to make its effect happen, and did. If we had the given conception of 'now-they-work-and-now-they-don't' circumstances, we presumably would also have an idea along these lines that might be reported. It would of course be precisely different from the idea that a complete causal circumstance produces its effect and an incomplete one doesn't. It would in fact be a pretty salient idea, since we would have a lot of use for it, on many unhappy occasions. It is my view, of course, that we have no such idea, and no need of it, since we do not take condition-sets for effects to be merely 'enough' for them in the given sense.

1.6.5 Non-Standard Causes

There is another possible argument, which can be dealt with quickly, for the conclusion that standard effects are no more than events for which other events are required. It relates to the very distinction between standard and other effects, and depends on what that distinction grants, that we sometimes speak of decisions and the like as effects without supposing that they issue from causal circumstances, which by definition are necessitating circumstances. (1.3) The possible argument consists in the premiss (1) that decisions and the like are taken to be effects but also to be no more than events which follow on conditions which are required for them, the further premiss (2) that it is to be presumed that we have a *single* conception of effects rather than several, and the conclusion (3) that *all* effects, including what have been called standard effects, are merely events preceded by conditions required for them.

Although I have earlier granted it to simplify matters (p. 21), premiss (1) is not nearly so firm as an objector might like. Certainly we often think and speak of decisions and the like as effects. 'In the end, what made him decide was the thought that time was running out.' 'Her choosing that wine was the effect of what you said about its edge.' In such cases, we do indeed seem to take decisions and choices as having required conditions. Do we also take them to be necessitated? Well, our usages *do* strongly suggest that this or that thought or feeling *completed* a set of thoughts and feelings, and the decision or choice resulted. The thought or feeling *tipped the balance, made the difference.* The decision or choice issued from a *completed* circumstance. However, a commitment to Free Will somehow understood, or the force of some image of Free Will, is likely to keep many of us from the further conclusion that the circumstance was exactly a *necessitating* circumstance. But if we do clear-mindedly and explicitly take decisions *not* to have necessitating circumstances, do we call them *effects* at all? I am inclined to doubt it, but let us not linger, and consider instead the second premiss of the argument.

It is that we have a single conception of effects, rather than several. On what is this based? In fact there is *no* established generalization about our concepts, or our fundamental concepts, on which it can be based. It would be exceedingly difficult even to formulate one. In all of *what* conceptual situations, so to speak, is it to be supposed that we have a single conception rather than several related ones? Certainly it is not to be supposed of *all* situations: for a start, we do not have a single conception of truth but rather at least two conceptions, those of analytic and synthetic truth. Putting aside the idea that there is an established generalization of the right kind, we might wonder if there is something related: a sound principle of inquiry, a presumption or guide to be followed for the most part. This would give us a weakened form of the argument we are considering. One response to this is, in part, a concession. Often, or at least sometimes, it is not a good idea to multiply senses. That, however, falls very short of something to be dignified as a principle of inquiry. A principle of inquiry cannot be something that may keep us from recognizing a clear and settled conception.

We do have a clear and settled conception of standard effects. We do not suppose that the wipers' starting to work was some sort of accident or mystery, of which the driver's flipping the switch was merely a required condition. We suppose the starting-to-work to have occurred necessarily. There is no need to say more of the argument. There is the sweet temptation to add, however, that there is a bad but better argument of the same form. (1) We take standard effects to be necessitated. (2) We have a single conception of effects. (3) We therefore take all effects to be necessitated.

1.6.6 *Probabilism*

The last sort of objection, perhaps more than the others, consists in certain alternative views. It is these views in particular that involve the ambiguity mentioned earlier. (p. 61) That is, they tend to move between a claim as to how we do in fact conceive of causation and a recommendation as to how we ought to conceive of it. The recommendation is partly based on the feeling or belief, owed to Quantum Theory, that universal determinism or the Law of Causation is false. The official business of these essentially philosophical rather than scientific views is and must be the first or conceptual matter, the analysis of our conception of causation. It is only so understood, certainly, that they are relevant to our present business. So understood, they share the idea that we, or many of us, take effects to be no more than *probable* events. (Suppes, 1970, 1984; Skyrms, 1980; Mellor, 1986) Some of these events are certain, or, to speak in terms of the Probability Calculus, have a probability of degree 1. More of them, however, have a lesser probability. According to most of these views, although they wisely do not parade the fact, effects may have a probability of 0.5 or indeed less. In what follows, I shall not reproduce the details of any of them, but proceed more generally and schematically.

It may simply be said that what is needed for two events to be cause and effect as we understand them is that the probability of the second, given the first, is higher than the probability of the second, given the absence of the first. Or, it may be said, two events are cause and effect if the conditional probability of the second—its probability conditional on the occurrence of the first—is greater than the unconditional probability of the second, which is a probability independent of whether the first occurs. Or, finally, two events are cause and effect if the probability of the second event, given the first event and certain accompanying conditions, is greater than the probability of the second event, given those conditions but not the first event.

As they have just been inexplicitly stated, each of these views is in fact itself better regarded as a *set* of possible views. The members of each set are distinguished one from another by what are fairly well known, the different theories of the nature of probability, all of them consistent with the Probability Calculus. There is no agreement as to how many of these there are, partly for the reason that there is no clear principle of counting for such theories, and no full agreement as to what, from some point of view, are the theories worth considering. (Mackie, 1973b; Weatherford, 1982; Benenson, 1984)

One thing said in support of the idea that we take effects to be probable events, and, more particularly, said against the analysis expounded in this chapter that we take them to be necessitated events, is relevant enough, although likely to be disdained by the high-minded. It is a *diagnosis* whose burden is that the supposedly mistaken analysis, of a kind which has persisted through centuries, is owed not to good reason but to a kind of desire. It now, persists, it may be said, because of a non-rational or irrational longing on the part of philosophers to have *order* in the universe, a longing that has replaced the lost comfort of a belief in an all-governing God. (Cf. Suppes, 1984, pp. 1–2, 17, 29, 51.) The proper reply to this is not that the truth of a view, strictly speaking, is independent of the motivations of those who hold it, although that is so. The proper reply, which accepts the fact that we may rightly have some suspicion of a view on the ground that its proponent wants it to be true, is that the satisfactions afforded by indeterminist as against determinist claims of one kind and another are *far greater*. The point is worth considering for a moment, partly because it is of relevance to greatly more than the present conceptual issue.

The first satisfaction of indeterminist claims, including those, like the present one, which are not within but are antecedent to the traditional area of dispute about Free Will, is of course that they leave, or promise to leave, room for a reassuring or enhancing view of ourselves and our situation. One proponent of effects as unnecessitated does not speak only for himself. 'I cannot say that I think of [the causal determinist's view] as having any plausibility. And I should certainly *hope* that it was false. For I believe that it is determinism that rules out moral responsibility and other things we believe in . . .' (Sorabji, 1980, p. 37) Another would-be defender of freewill is as explicit. 'Without freewill we seem diminished, merely the playthings of external causes. Our value seems undercut. . . . My concern is not only intense but directed. I want (to be able) to conclude that we *are* worthwhile and precious.' (Nozick, 1981, p. 2)

A second satisfaction of indeterminism is in some relation to the first. It is the wider satisfaction of a universe which is precisely *not* subject to order, but has in it uncertainty, novelty, and escape from the past, all somehow agreeably constrained by probability.

Thirdly, and perhaps most relevantly to our present concern, there is the more intellectual satisfaction to be had from extending the application or province of a favoured doctrine, theory, or calculus, perhaps something in which one has a proprietory interest. Here it is incidental that the doctrine or whatever conflicts with or does not fit in well with determinism. The thought must indeed come to mind with probability theory and the Probability Calculus, and their being imported into causation, despite what was said above about the logical consistency of probability theory and necessitation.

Fourthly, there is the satisfaction of a kind of orthodoxy, perhaps with respect to what is taken as somehow of the greatest significance or fundamentality. An interpretation of Quantum Theory comes to mind. For this and the previous three reasons, it is not difficult to conclude, at least tentatively, that more inclination is satisfied by indeterminist than by determinist claims.

The general subject of determinism and our inclinations, however, will get greatly more attention later in the third part of this inquiry. (Chs. 7–10) The idea now under consideration, that we take effects to be probable events, is announced or expounded but not much supported by argument. The explicit arguments of which I know are to my mind hopeless.

> . . . the everyday concept of causality is not sharply deterministic in character. All of us have said things like . . . 'His reckless driving is bound to lead to an accident'. What we mean . . . is that the probability of the person's having an accident is high, and his own manner of driving will be at least a partial cause of the accident. The phrase 'is bound to' means that the probability is high of his having an accident. The phrase 'lead to' conveys the causal relation

between the reckless driving and the predicted accident. (Suppes, 1970, p. 7; cf. Suppes, 1984, p. 54)

The claim or premiss reduces to this: that we often say and believe something is likely to cause something else. But that truth is wholly distinct from, and does not entail, the proposition that the connection we take to hold between cause and effect is one of probability. It is not too much to say that the quoted argument has the strength of this: We say it is probable that the spoon is under the napkin; the relation of 'being under' is therefore a relation of probability.

What gives what persuasiveness it has to the probabilistic idea about causation is neither such an argument for it nor the earlier diagnosis of the appeal of the opposed view about necessitation. It is that the probabilistic idea is in accord with a kind of empiricism. This is an empiricism related to kinds of positivism and to operationalism. It amounts to a determination or tendency to accept as conceptually respectable, and as a report of reality, only what is in accord with some part of scientific practice, or a fully specified calculus, or particular and explicit scientific laws, or particular kinds of quantification, or some principle in the philosophy of science. Since considerations of probability are to the fore in one part of actual scientific practice bearing on causation, and there is little attempt to go beyond them, and there is to hand the Probability Calculus, we are invited to take it that causation can come to no more than probability.

This empiricism, which is indeed preferable to paradigmatically unscientific excesses of speculation and the like, is nevertheless wholly inadequate, and generally accepted to be such. It can be no more tolerable than one of its most specific forms, behaviourism in psychology. (p. 73) That particular movement, essentially the attempt to reduce the facts of mentality to the measurable facts of bodily movement, has rightly collapsed. To come closer to our present concern, there is no more hope for the idea that a certain determinism is false because we do not aspire to *prove* it, but rather proceed in several alternative ways, including probabilistic ways, on the assumption of its truth.

Ordinary observation and participation in physical phenomena seem to argue decisively against nature being deterministic. Causal attention to the way that a breeze moves the leaves on a tree would seem to make it quite unrealistic to think of having a theory that could predict their motion. . . . To think that one could actually predict these phenomena, that there would ever be a possibility to do so, seems on the surface quite mistaken. It would indeed take something close to theological commitment to think otherwise. (Suppes, 1984, pp. 12–13)

It does not *begin* to follow, from our having no effective theory or practice for predicting breezes, that they are not necessitated phenomena. To return finally to our present conceptual concern, it does not follow either, from the fact that in parts of science there is a dependence on probability, that the subject-matter is not conceived in terms of necessitation. If the empiricism in question has the attraction of method and restraint, at least when compared with certain other things, it is not something in which anyone, including its proponents, can persist.

There is more to be said against the probabilistic analysis of our conception of causation that can or needs to be said here. (Papineau, 1985a) I shall content myself with little more than some comments.

(i) Proponents of probabilistic analyses have it that two events can be cause and effect if the presence of the first raised the probability of the second from 0 to only ·01, or of course less. That is, the second can be an effect of the first if the first left the second, to speak in ordinary terms, *very highly improbable*. The same will be true, presumably, of the causal condition-set for the effect. The thought may come to mind of stipulating, say, that a cause or condition-set made an effect more probable than not—that is, gave it a probability of 0.5+. The arbitrariness of this would be extreme. Further, it could not be supposed this is what we now believe. If it were, we should be subject to a kind of indecision about the causal connection between condition-sets and their effects which is in fact entirely missing from our deliberations.

(ii) On the given probabilistic analyses an event *e*, if it lowers the probability of an event *f*, cannot be cause of *f*. But an event which lowers the probability of a second event, on the informal or any other theory of probability, may in fact turn out to cause it. Some medicines have caused precisely what they were rightly judged likely to prevent. (Hesslow, 1981)

(iii) There are the greatest difficulties in fitting the particular theories of probability into any probabilistic idea of causation. The *subjective* theory, for example, when put to work on causation by a proponent, renders causation subjective. It is not such, and no one takes it to be such. The *chance* theory's obscurity and peculiarity (Mackie, 1973b) renders it, one cannot but feel, unlikely as a component of an analysis of our conception of causation. That conception, for all the problems it raises, cannot reasonably be said to be of the same character. At least certain of the theories of probability, notably the *frequency* theory, cannot possibly allow it to be meaningful to speak of the probability of *single* events, and hence cannot be used in analysis of *particular* causal beliefs, those which are most common. It is at least arguable, moreover, that *no* probability theory allows for the explanation of single events. What we know, rather, and roughly, is that a proportion of a certain number of events will be such-and-such. But our causal beliefs typically do pertain to and are explanations of single events.

(iv) The probabilistic analysis, in the last resort, is to be judged against the persuasiveness of the conception of causation developed in this inquiry, and of course of like conceptions, with which it conflicts as much—and also Humean conceptions. It is a contest that few will take it to win. On the probabilistic view, all of the conditional statements which we accept in connection with causation must be rewritten into conditionals such that the events mentioned in their consequents are events of a certain probability. Furthermore, in those cases where these events are allowed to have a probability of 1, that is no more than *logically consistent* with their being necessitated. The connection that is asserted to hold between a condition-set and its effect, when the effect has a probability of 1, is akin only to the insufficient one allowed by the Humean accounts, in terms of constant conjunction. Finally, consider a part of our belief as to causal priority, that *causes make their effects happen*. It is not too much to say that the probabilistic analyses must revise this into a belief that many effects are *not made to happen*. They are such that it is merely true that they *might happen* or *might have happened*.

1.7 SUMMARY

Causes, like the other conditions which enter into causal circumstances, are at bottom conceived by us to be certain spatio-temporal items, individual properties. (1.1) So too are effects, and the elements of the nomic correlates of three classes. However, we typically use the whole (usually an event) to speak of the part (the individual property). The facts of necessary connection as we conceive them, having to do with causes, causal circumstances, effects, and nomic correlates, are those stated by certain conditional statements. (1.2, 1.3) Our conception of necessary connection cannot be said to require any facts other than those stated by these conditionals. The conditionals are of two kinds, and the second, independent nomic conditionals, are fundamental. They are, in brief, of the form *If A, even if X, then still B*. (1.4) Certain facts stated by the latter conditionals, together with effects being taken as later in time, are all that is needed to explain the difference we find or make between causal circumstances and causes on the one hand, and, on the other, their effects. This difference distinguishes causal circumstances and their effects from the other instances of necessary connection, the nomic correlates. (1.5) All necessary connections are open to forms of mathematical expression, and typically are given it in science. Whatever may be true of the world, and, as follows from what has been said of necessitation, our conception of an effect is not a conception of an event that occurs if a usual background exists, or an event which follows on something required for it, but might not have occurred, or a merely probable event. (1.6)

2

Psychoneural Nomic Correlation

2.1 CONSCIOUSNESS—LOOKING ELSEWHERE

Each of the three hypotheses making up the theory of determinism to be expounded in the first part of this book has to do with mental events, which is to say events within consciousness. Thus we need, first, a conception of consciousness.

It is taken by some to be that which falls under statements of certain logico-linguistic kinds. As the story goes, these are statements, first, which have the distinction that their truth-values do not depend on the truth-values of statements they contain. This is so with the statement, about an episode of consciousness, that *Freddie thought that the play was to begin at eight*. Its truth, if it is true, does not depend on the truth-value of the contained statement that the play is to begin at eight. Compare the statement, containing two others, that *he is grey-eyed and his shoes were made in Italy*. A second group of statements has the distinction that their truth-values do not depend on whether there does exist a thing called for by a contained referring expression. This is so with the statement that *he hoped for a seat in the stalls for under £10*. Compare the statement that *he sat in a seat in the circle*. There is the statement-kind, thirdly, overlapping with the second, such that truth-value may be argued, to my mind unpersuasively, to depend on the inclusion of one particular referring expression, rather than another expression referring to the same thing. Take the statement that *he wanted to see the German play of which he had heard some praise*, and consider substituting the co-referring description 'the Left-wing play whose inept performance will make him wish he had stayed at home'. Compare the statement that *he was present for a play by Brecht*.

Attention is paid to this *logico-linguistic conception* of consciousness, in terms of independent and dependent truth-values, precisely because of its literalness and precision. Those virtues are in short supply in connection with conceptions of consciousness. Of course, in order to have something literal and precise in the desired sense, care must be taken to understand the given conception exactly as it is, without addition. It thus may need to be understood, incidentally, more strictly than its original proponent intended. (Chisolm, 1957) Consciousness is to be identified *only* in the way mentioned, by way of independent or dependent truth-values of statements. The conception does not have to do, above all, with the statement's having some otherwise specified content or subject-matter: that is, say, as being about thoughts, wants, or intentions, or conveying something about persons, minds, subjectivity, experience, or an inner world.

The sad fact, well enough known, is that the given conception of consciousness is open to serious counter-examples. (Heidelberger, 1966; Lycan, 1969; Davidson, 1980e) Not all of consciousness does fall under statements of the given kinds, and certain matters of non-consciousness *do*. Consider the statement that *I am in a jolly mood*, which clearly is not one of the chosen kinds. So with the statement that *I have an earache*. Consider the statement that *good weather made possible the early appearance of my snowdrops*. Its truth-value does not depend on whether there actually occurred the happy event referred to by the second referring expression. Such counter-examples have led some philosophers attracted to the logico-linguistic conception to take the audacious course which involves, in part, simply ignoring the very large part of consciousness which does not fall under the conception. (Davidson, 1980e) We shall certainly not do that. Other philosophers have struggled to save the conception in various ways, piling on the epicycles.

There is a further fundamental objection, generally overlooked. Even if it *were* true that all of consciousness and only that fell under statements of the three kinds, or related kinds, we would by means of this truth get only a wholly uninformative conception of consciousness. We would have nothing like an analysis or other understanding of it. It it true enough to say that we would have *no* conception of *it*. In particular, we would not know what it was about consciousness that gave rise to its falling under the given kinds of statement. If a large part of consciousness is ignored in the mentioned audacious way, and the logico-linguistic criterion is used to discriminate the remaining part, it gives us no analysis or understanding of that remaining part. (Cf. Kim, 1971.) If we set out to frame hypotheses about the chosen part of consciousness, we do so at our peril, flying in the dark. By way of a quite proper analogy, suppose there were a certain amazing truth, that all and only the members of one species of tree, say Turner's Oak, could be described in a given language by sentences of a certain deep structure, or simply a rare surface-grammatical sequence: indefinite article, a curious gerund, preposition of a certain sort, and so on. We could then define this splendid species of tree as that which falls under statements made by sentences of the given grammatical sequence in the given language. The definition would be wholly uninformative, and make error likely.

There are other conceptions of consciousness which also owe their existence to the pursuit of certain virtues. One is the familiar conception in terms of *behaviour*. It is very possible to sympathize in good part, if not entirely, with the psychologists and philosophers who were sceptical or uncertain of the worth of introspection as a source of knowledge, reluctant to attempt to deal with the unquantifiable, keen to be in accord with certain principles of scientific methodology, resistant to such free speculation as the Freudian kind, and who thus took the step of analysing ascriptions of consciousness into claims about no more than behaviour. It is now generally recognized that the attempt to *analyse* consciousness in terms of behaviour amounted to flying in the face of the facts. It was doomed to become, as it did, a kind of plodding in the face of the facts. (MacIntyre, 1971c; Mackenzie, 1977; P. M. Churchland, 1984; Flanagan, 1984)

Like the logico-linguistic criterion, although in a different way, behaviourism attempts to analyse consciousness by *looking elsewhere*. If it is claimed that reasons are needed for this verdict, which might be doubted, one has to do with the fact that we do not take all the causes of behaviour to be *other behaviour*. When we take a belief, desire, or intention to have caused an action, as we commonly do, we are not thereby explaining the action by referring to other actions, let alone mere movements. This would be true, evidently, if consciousness consisted in behaviour. Other objections to traditional behaviourism can be transferred from their use with other related conceptions, to which we now turn.

One of these, sometimes itself referred to as 'behaviourism', but better named *causalism*, takes consciousness to consist in no more than episodes which enter into certain causal relations. These episodes are causes of behaviour, or effects of stimuli, or both. (Smart, 1959; Armstrong, 1968, 1980; Lewis, 1966, 1972) What is distinctive about this view is not that it takes the episodes of consciousness to stand in such causal relations. That at least some of them do so is a platitude. Nor would the view serve any purpose, including the purpose for which it was devised, if it were no more than the idea that particular episodes of consciousness—say desiring—can be *identified* by their causal roles, where such identification does not give us their nature, or all of their nature.

What is distinctive is the idea that consciousness can be adequately described in terms of causal episodes. That is, the whole of its nature or reality is at least adequately given by this description. Furthermore, such a term as 'causal episode' is to be understood in a way which excludes certain assumptions or implications. A causal episode, in the given sense, is not in itself a physical process, or that particular kind of physical process which is a neural process—a process of the Central Nervous System of a person or other organism. Nor is it in itself what would ordinarily be called a mental process, say a process involving the Self or a process of which the person in question has a unique kind of direct awareness. Causalism, rather, gives a 'topic-neutral' conception of consciousness.

Causalism can be considered together with something else. It lies in the immediate background of another family of views of consciousness which, at the time I write, is propagated with the zeal which once went into behaviourism. This family of views, which derives from many of the same commitments as its predecessors, but owes a great deal to the development of the computer, has among others the rather unenlightening labels *functionalism*, *cognitive science*, *cognitive psychology*, and *artificial intelligence*—several of which labels obviously have other uses. (Putnam, 1975; Fodor, 1968; Dennett, 1978, 1984; cf. Boden, 1977) To proceed in terms of a kind of model of this family of views, or many of them, what we can call functionalism shares with causalism the idea that mental episodes are to be understood *relationally*, in terms of their relations to other things. Somewhat less clearly put, mental episodes are to be understood in terms of their roles or functions *vis-à-vis* other things.

Functionalism as we shall understand it differs from causalism, first, in not restricting the relata in question to two categories: external input or stimuli and external output or behaviour. This, certainly, is an improvement. It gives what is in a sense a richer account of a mental episode by including relations with other strictly mental episodes and facts—these too, of course, to be understood relationally. Very roughly, my wanting the window open a moment ago is to be understood not only in terms of the stuffiness of the room (the stimulus), and my subsequently opening the window (behaviour), but also in terms of various beliefs, attitudes, and the like, including certain ordinary causal beliefs about open windows and perhaps attitudes having to do with propriety and the neighbours.

Functionalism, as we shall understand it, also differs from causalism more fundamentally: in its conception of the nature of all the relations in question. They are what can be called *logical, conceptual, or formal* relations rather than causal relations. They are typified by the relations whose extended statement consists in a machine table, or flow chart, or computer programme. A table, chart, or programme, in this sense, is not a tape or a disc but a sequence of propositions, which is to say abstract objects. (1.1) To speak differently, causalism takes mental episodes to be occupants of causal roles, but functionalism takes them to be logical or Turing Machine states, or logical or computational processes. In the words of two proponents of the view, it is 'the emerging view of the mind as software or programme—as an abstract sort of thing whose identity is independent of any particular physical embodiment'. (Dennett and Hofstadter, 1982, p. 7) It is indeed a consequence of such a conception that consciousness is not necessarily an attribute only of persons and other organisms. Consciousness is not tied to neural structures and events. It is accepted by functionalists that if any computers, or indeed any artifacts or entities whatever, can be said to pass through the requisite states or processes, they must therefore be conscious. Passing through the states or processes *is* being conscious. Causalism taken by itself has a related consequence, which has been obscured by the addition to it of what are really independent ideas, notably the arbitrary idea that conscious episodes having been conceived causally are then to be identified with neural structures and events.

There is no shortage of objections to both causalism and functionalism. Here is a quick list.

(i) If I am anaesthetized I do not feel the prick of the needle. Suppose, however, that I simulate or even by coincidence pass through whatever sequence of states is taken to be sufficient for the conscious episode of feeling the prick of the needle. It follows, on the causalist or functionalist view in question, that I do feel the prick of the needle. The conclusion, it seems, is intolerable.

(ii) Suppose that my private visual experience is strikingly atypical in that systematically I see green where others see red. That is, I am caused, perhaps by a deformation of my visual cortex, to have the visual experience which others describe as seeing something green in colour when the thing in question is what gives rise to their seeing it as red. This experience of mine, further, thanks to my training, stands in satisfactory relations to other things. I *stop* at the traffic light when I have a visual experience which others would describe as seeing a *green* light. I, of course, describe it as red. It follows from causalism and functionalism, seemingly absurdly, that our private visual experience is identical.

(iii) Suppose I do not understand Chinese, but, solely on the basis of the shapes of the letters, and rules connecting them merely as shapes, do the right things. For example, I pass through a sequence of states identical to one passed through by a native Chinese speaker in being presented with a question in Chinese and giving the answer in Chinese. It follows, on causalist and functionalist views, that I am understanding Chinese, which again seems absurd.

(iv) Differently, it seems difficult to accept that consciousness is tolerably conceived when it is so conceived that it follows that *anything* that can be regarded as passing through certain sequences of causal or logical states is conscious. As suggested by one formidable opponent of the views in question (also the author of the objection having to do with understanding Chinese) we do not so understand consciousness that we can believe that it must be possessed by a set of water pipes, paper clips, and old beer cans, provided that they instantiate a certain programme. (Searle, 1980, 1981, and 1984, Lect. 2) That such a thing could instantiate such a programme, as already remarked, necessarily is allowed by the proponents of causalism and functionalism.

(v) To add something unfamiliar, functionalism will need to explain how it avoids the absurd conclusion that consciousness is, so to speak, not merely independent of biology, but *entirely* independent of *all* instantiations of programmes and the like, which is to say *entirely* independent of persons, computers, and *all* other spatio-temporal things. A sequence of logical or abstract states, in whatever way it exists, exists whether or not instantiated. It does not come into being only when instantiated. Functionalism, having fled the traditional mysteries of the mind, appears to rarify consciousness, to speak quickly, into a Platonic universe, out of space and time. In the enterprise of seeking to understand consciousness as something more manageable and decently scientific than what it calls ghostly stuff, it is understood as *yet less* than ghostly stuff.

The objections have been very lightly sketched, and, inevitably, there is the possibility of attempting replies. (Abelson *et al.*, 1980; P.M. Churchland, 1984; Flanagan, 1984) I have little doubt, however, that the objections will elicit or reinforce a fundamental conviction in readers, a conviction as good as universal. It is, in sum, that consciousness is not adequately conceived in the given causal and logical ways. Like the logico-linguistic and behaviourist enterprises, causalism and functionalism attempt to define or analyse consciousness by looking elsewhere. There is no doubt that our conceptions of various kinds of conscious episodes do indeed include relational components. That is not to say that we take such episodes to *consist* in relational facts. (Cf. T. Nagel, 1986, pp. 7–8; McGinn, 1982, p. 7.) They allow for, and require, a certain *realism*. (Cf. Dummett, 1978)

2.2 MENTAL REALISM

It is sometimes supposed, by those who do allow that consciousness is a reality not caught by any of the doctrines we have considered, that none the less nothing enlightening can be said of it. It is a kind of *given*, something of which we do have direct acquaintance, but it is impossible to give any analysis of it. It is sometimes supposed, more extremely, that direct acquaintance or introspection is so uncertain or fallible that not merely no analysis but nothing of value can be got by means of it. We are in possession of no concept of consciousness at all. The conclusion may be supported by psychological evidence of introspective error (Lackner & Garrett, 1973), which evidence seems incapable of supporting any such general conclusion. The conclusion may derive instead from a general scepticism about all our perceptual capabilities, internal and external, and also a certain amount of fast philosophy. An example of the latter is the argument that since we are mistaken about secondary qualities—the real world does not contain red, sweet, and hot things but only certain electromagnetic, stereochemical, and micromechanical properties—so we can have no faith in our direct awareness of our conscious events, states, and processes. (P. M. Churchland, 1984, p. 29) The conclusion that we can get nothing of value, no conception, by direct awareness is yet more excessive than the conclusion that we can get no analysis of consciousness. The best reply to both conclusions consists in providing what it is maintained we cannot have.

Several properly ambitious philosophers attempt to bring consciousness into view by speaking, with respect to a conscious thing, of what it is like to be that thing. (T. Nagel, 1979a, 1986, Ch. 1; Sprigge, 1971, 1982, 1983; cf. Farrell, 1950) Consciousness, we are invited to say, is a way it is like to be, or, perhaps better, a way of being. The consciousness of a bat is what it is like to be a bat, that way of being. This perception strikes one as promising, but the impression may be evanescent. It is difficult to avoid the further thought that the given characterization of consciousness is elliptical, and that when it is filled in, as it must be, we are no further ahead, but have our *definiendum* turning up in the *definiens*. It seems clear, that is, that we are to understand that a way of being is something possessed by or entered into, say, by us, but not by sticks and stones. But then 'a way of being', when filled in, becomes 'a way of being conscious'. Consciousness, we find ourselves saying, is a way of being conscious. There is a second consideration. If the understanding of 'a way of being' as elliptical is resisted, as perhaps it can be, and the term 'a way of being' taken to be serviceable as it is, there is another difficulty. To speak of my way of being, or what it is like to be me, seems to be to speak of what *distinguishes* my conscious life from the conscious lives of others. That is, it is not to characterize consciousness generally. But that is precisely our aim.

What is offered may owe something to Brentano's familiar characterization of consciousness. (Beyond doubt that characterization lies behind the logico-linguistic criterion considered earlier.) Consciousness is said by him to be, fundamentally, *activity which has reference to a content*, or *activity which is directed upon an object*. (1973 (1874))

A *content* is not a spatio-temporal state of affairs, such as the play's beginning in the Olivier Theatre at eight. If someone had the right thought, that the play did begin there at eight, then their thought did have a content that was true, but that content was not identical with the actual state of affairs. The content was precisely the same sort of thing as the play's having begun on Saturn with Helen of Troy in the audience, which never happened. The *objects* in question, similarly, do not exist in the spatio-temporal way of ordinary things. Even in the cases where what we want are things which exist in the ordinary way, these things are not to be identified with the mentioned objects. What lies behind these conceptions of content and object is the true idea that the nature of consciousness *itself* may not be different at all between the times when we think truly and when we think falsely, or when we want what exists and when we want what does not exist.

As for the mentioned *activity*, an attempt must be made to understand it in a way consonant with what has been said of contents and objects, that they are not states of affairs or ordinary things, and also with a further fact, that there are various modes of consciousness. With respect to the first point, none of the activity—no instances—can be *identified* with ordinary perception, plain seeing, which consists in activity in a plainer sense, having to do with ordinary things. The activity posited by Brentano, secondly, is common not only to seeing and other perception, thinking in the sense of deliberating or judging, and also remembering, intending, and deciding, all of which are ordinarily thought to be in different ways *active*, but also pain and other sensations, dreaming, experiencing an emotion, and being subject to a mood, all of which are thought to be in different ways passive. As for the ideas of *having reference to* and *being directed upon*, they too must be given sense consonant with the nature of the given contents and objects and the various modes of consciousness.

It has often been objected to this account that it does not fit all of consciousness, that some of consciousness refutes it. (K. Campbell, 1970, p. 23) It is allowed that it is enlightening or tolerable to speak of hoping, perceiving, thinking, and a good deal else as having objects or contents in the given sense. What, however, of states of feeling, perhaps being depressed or feeling good about nothing in particular? What of sensations, perhaps the warmth one feels while sitting in a sunny window? Surely these cannot be said to have objects or contents in the given sense.

Anyone who takes this line confidently must surely understand more about the mentioned objects and contents than was conveyed above. He must understand something more precise than Brentano explains. It is true, certainly, that feeling good for no particular reason, and the sensation of warmth, do not involve objects or contents that can be said to be *determinate* in a certain sense. There is a great difference between the feeling and the sensation and, on the other hand, having the thought that one's father wrote a pamphlet about the consistency of Christianity and Communism. Still, we are not debarred from attempting to conceive of an indeterminate object or content. Nor are we debarred from using 'content' rather than 'object' where the former is more natural.

A good deal of philosophical writing about what is called intentionality suggests that it is likely that this line of objection, that objects or contents are sometimes missing, has another wholly different root, in a certain confusion. Brentano evidently has in mind a relation between something not explicitly mentioned, that which is active, and, on the other hand, contents and objects. This is the relation of 'reference' or 'direction'. There is another matter. *Some* of the contents and objects, we can take it, are distinguished from the rest by being *representative*. That is, they involve a second and wholly different relation, a semantic or intentional relation between themselves and whatever they represent. It is precisely true of hoping, perceiving, thinking, and a good deal else that it can be conceived in terms of such representative objects or contents. It is precisely untrue of certain states of feeling and sensations. However, the absence of an object or content *of a representative kind* must obviously not be confused with the absence of an object or content. Again, the absence of the representing relation is not the absence of the more fundamental relation.

My present aim, in fact, is not to rely on Brentano's conception of consciousness, but to use it to introduce my own somewhat related one. It, like his, has the recommendation of being a realist conception, which is to say one which does not withdraw from its subject-matter. Realism about the mental, to speak freely, seems to me essential, and the opposed looking-elsewhere, whatever its roots, a kind of fastidiousness or want of nerve which is futile. (Cf. Searle, 1984, Lect. 1.) My conception is related, also, in that both conceptions derive in a direct way from our pre-theoretical, first-person grasp of consciousness, as cannot be said of any of behaviourism, causalism, or functionalism, nor really of the logico-linguistic criterion, derived though it is from Brentano. My alternative account, a minimal one, seeks among other things to give fewer hostages than Brentano's to philosophical fortune, or anyway to philosophical doubt.

To think of any of one's conscious episodes in the moment after it has happened, is to think of a certain duality, one which has nothing to do with dualistic doctrines of mind and body. Think of feeling a sensation in one's knee—or noticing a cup, having the usual inattentive visual experience of a room, feeling good or depressed about nothing specific, being struck by the fact of a recent death, wanting to go to bed, deciding not to, momentarily intending to watch the news on television, picturing a face, thinking a question or a sentence, writing one, having a dream. Certainly there is great diversity here, and causal and other relations enter into our various conceptions of these things. It is also indubitable, and a fact of which philosophers and psychologists of different inclinations have made different things, that to think of any episode of experience is to think of *two elements*, two elements *within* the experience.

One of these elements turns up in each of one's conscious episodes, or is of a kind such that each of one's conscious episodes contains an instance of the kind. That is, there is that which is common, or of one kind, in each of the experiences of feeling a sensation, being depressed, thinking a question, taking a decision, and so on. This is not to assert, certainly, that there exists some entity, *outside* an experience or episode, which experiences or possesses it. The second element in each of one's conscious episodes is almost always different—different from its counterparts in all other episodes. The first element can be referred to as the *subject*, the second as the *content*.

The terms are of no great importance, and do little more than mark what is the incontrovertible fact of the duality within conscious episodes as we recall them. They mark our perception of that fact, and are not to be taken as carrying theory with them—not even so much theory, perhaps, as is suggested by William James in speaking of a 'dualism' of 'mind knowing and thing known'. (1890, p. 214) The subject–object relation is not to be characterized as the subject's *attending to* the content, or anything of the kind. To characterize it in this way would be, it seems, to introduce the *definiendum* into the *definiens*. Certainly, to repeat, the terms 'subject' and 'object' are not to be taken as carrying with them the Cartesian theory of a subject in the sense of a simple indivisible mental substance whose identity over time is primitive and irreducible. (McGinn, 1982, pp. 121–2) Even acute defences of Descartes' argument for an ego (B. Williams, 1978, pp. 95–100), prove to be ineffective (Parfit, 1984, pp. 225–6), but their failure does not affect the present view.

It is to be admitted that any particular conscious episode, in the moment of experiencing of it, seems to consist at least largely in but one thing—as, say, when the address '4 Keats Grove' comes to mind. One's experience is certainly in some sense not *of* two things. No doubt this is the truth on which Hume relied in denying the existence of the Self as commonly conceived. However, to think afterward of the having of the thought *is* to think of two things, both somehow integral to it. Having an idea of having had an idea is having an idea of two things. This is as much a truth about such episodes as unfocused depression. It is not as if one's depression could be captured in recollection by the idea of bare content. Nor is it possible to capture one's experience by the idea of a bare subject. That, surely, is inconceivable.

In any case, it is not quite true that the subject can be perceived, so to speak, only in recollection. Certainly the subject, sought from within a particular conscious episode, is peculiarly recessive. There is a difference, evidently, between the manner of the occurrence or contents in consciousness and the manner of the occurrence of the subject. Contents are at least to the fore. However, it seems impossible to deny that each of us also has a sense of a self, or a centre of awareness, somehow within and integral to our ongoing experience, and such as to give a kind of unity to it.

The duality is consistent with the mentioned diversity of contents. They are indeed of different characters. (i) In the case of visual and other perception, they are bound up with ordinary things—cups, rooms, and landscapes. *How* they are bound up is the philosophical problem of perception, into which we shall not enter. (ii) Contents, in the case of sensation, lack such a connection, but are in a special way qualitative and are somehow assigned a bodily location. These contents of sensation are, despite their difference from the contents of perception, no less distinguishable from that element in an experience which is the subject. (iii) So too are distinguishable the contents of moods, such as feeling good about nothing specific, despite their difference from the contents of both perception and sensation. Feeling good about nothing specific is not, so to speak, an unattached or free-floating phenomenon, unpossessed or unhad. (iv) Finally, and clearly, all of the contents so far mentioned are different in character from those which get definition by way of language—or rather, get definition only or more explicitly by way of language. Here, as in the example of thinking a question, we have what were earlier called representative contents, or the most important ones. They are of great importance in any full characterization of consciousness, and are rightly the subject of inquiry of several kinds, sometimes in the endeavour to give a limited characterization of the mind. (Searle, 1983, Harman, 1973, Fodor, 1979) None of this diversity in contents, to repeat, should distract us from perceiving the single duality within consciousness. That what is on one side is in fact a category of diverse contents does not take away from the fundamental duality.

Consciousness, then, to speak somewhat differently, consists in general in the relation of subject and content. This relation, further, is such that its terms *cannot* exist independently of one another. They can occur only in the given relation. Each term depends for its existence, that sort of existence it has, on being in relation with the other. There can be no subject without a content, and no contents not in relation to a subject. It is perhaps audacious, given the undeniable uncertainty in which we find ourselves in thinking of consciousness, but it may be possible to regard this mutual dependency as a matter of nomic correlation of one of the kinds specified earlier (1.5), although nomic connection of a uniquely primitive character.

Given the terms 'subject' and 'content', which lack close definitions, it would perhaps be yet more audacious to speak of logical or conceptual dependency. Still, it is tempting to think of a kind of logical connection. It seems that any attempt to conceive of a content of a mental event, separately from its inherent subject, is in a way futile. It appears to be futile in that inevitably one finds oneself conceiving something else. If I now consider an event of a moment ago, my idle contemplation of the cup on my table, and attempt to subtract from my present conception only a part of it—the subject within the event of a moment ago—and to hang on to the remainder, I am in fact left with something other than the content of the event. I am left with the cup—the idea of that ordinary physical object—or a proposition about the cup. These, however different, are independent existences. To succeed in thinking instead of the content, it seems, is necessarily also to think of something else, to think of that which exists *for* something else.

We shall leave the matter as we have it. Consciousness, to come to a kind of summation, consists in general in this *interdependent existence of subject and content.* If we have the content primarily in mind, and are willing to strain language somewhat, consciousness can be described as the *existence-for* of contents, or their *existence-to*. Contents exist *for* or *to* subjects. To go further, consciousness is *forness* or *to-ness*. We can as tolerably say, however, having the subject primarily in mind, that consciousness consists in the *existence-*

through of subjects. They exist only *through* contents. Consciousness is *existence-through*. Perhaps the straining of language is not worth the effort.

More might be attempted (Ayer, 1954a; Alston, 1976; Hannay, 1979; Wilkes, 1978) but we have, I submit, gone some way in analysing consciousness with the general conception of it as the interdependent existence of subject and content. Still, admissions cry out to be made. The conception is not literal and precise, in the sense in which, say, the logico-linguistic criterion is such. Nor does it satisfy all the demands mentioned above (p. 73) in connection with behaviourism, causalism, and functionalism. Not all of those, of course, as already noted, are reasonable—above all, any wholesale dismissal of what we get by direct awareness is entirely futile. As will be evident, it is my view that we have a choice between attempting to satisfy certain demands and failing to get hold of our subject-matter, or relaxing the demands and getting *some* hold on it. The latter course is not merely preferable but necessary.

2.2.1 Definition of Mental Events

Three important matters remain, of which the first has to do with the things we discriminate within consciousness—the parts, segments, or passages of consciousness, as distinct from subject and contents. I have spoken so far in an ordinary way of episodes of consciousness and of experiences, occasionally of modes and states of consciousness, and also of sensations, moods, perceptions, desires, thoughts, intending, deciding, and so on. Each of these things, whether taken as type or token, is itself a matter of subject and content. It is fairly common to speak of these various things as events, states, and processes of consciousness, but we do not have a single wholly general term for them. I shall continue to speak in these and the more ordinary ways of the things in question, but it will be at least economical to have a single wholly general term for all of them. Or rather, it will be economical to use one of these terms in a yet more general way, to cover all of these things and also things that do not fall naturally into any of the mentioned ordinary or semi-technical categories—or ones like them. This will be the term 'mental event'.

It will cover *anything whatever that occurs within consciousness*. If it is supposed that there is a want of definiteness about the idea of something's 'occurring within consciousness', we can as well speak of *anything whatever of which there is a logical possibility that an individual can discriminate it in his or her or its consciousness*. The discriminating of something within consciousness is to be understood as a case of remembering. It is fixing attention upon something afterwards, thereby assigning it, by way of its content, to a type. I cannot, it seems, both be aware of my hand and at the same time aware of that awareness. To attempt to perform the double act is instead to oscillate between the two. It is of course necessary that the alternative definition of a mental event not be of something which *in fact* can be recalled afterwards, but of something of which there is a logical possibility of recall. It is clear that we do not in fact have the capability of sharply recalling more than a very few of our mental events as defined, even immediately afterwards. This is so, to speak metaphorically, since almost all mental events, although not the ones in which we are principally interested, are elsewhere than at the focus of consciousness.

It will be as well to make explicit the third-person criteria of individuation for a mental event, criteria which are implicit in what has just been said. A mental event is individuated, first, as being within the consciousness of a given individual. Secondly, its individuation is a matter of content and/or temporal location. Two tokens of the same type of mental event, all of whose tokens are indistinguishable to the owner in terms of content, are distinguished by temporal location. In the case of many mental events, such as the elements of one's visual field, their discrimination is one with, or bound up with, the discrimination of ordinary things, their features and so on. The distinguishing of mental events by content is of course open to error, including error owed to memory and the sorts of error established by psychological research into introspection or direct awareness. It is perhaps not too hazardous to say, despite this, that our capability is such that there is not much else that we can do better. The perceptual distinctions we make are of the very stuff of our existence, and in some way the foundation of all else.

Given what has been said, a mental event can be simple or complex and of any duration. It may then be a momentary thought, a fleeting hope, the colouring of a feeling, the forming of an intention, a decision, a choice, a continuing sensation, an inactive or an active sensation (pp. 221 f.), a persistent mood. Given what has been said, further, a mental event is not necessarily what is sometimes called a whole mental state, which is to say all of a person's consciousness at or for a time, but typically is a part of a whole mental state, a part which itself has parts. Mental events, like most events, are composite—they contain or are constituted of other mental events. Certainly there is no need, as is sometimes supposed (Thorp, 1980, pp. 43–4), to consider whole mental states in connection with the relation between the mental and the neural.

Also, given what has been said, that a mental event is *wholly within consciousness*, it is also true that in speaking of *the mental life* of people, in a large sense of the term, we are speaking of *more* than mental events. In this sense of the term, my mental life contains my knowing things, and remembering things, and seeing things. To claim that I know that my postal code is NW3 2RT is evidently to claim or be committed to more than that I believe it. It is to claim or be committed to the fact external to me that my code *is* NW3 2RT. To say that I see the squirrel with the black nose is to be committed not only to the existence of a visual experience, but also to the existence of something else. Mental events are contained in rather than identical with what can be called *personal epistemic facts*. It is mental events that raise the questions with which we are concerned.

As in the case of the physical events considered in the last chapter in connection with causation and other nomic connection, mental events strictly speaking are to be regarded as individual properties or sets of such properties. That is not to say that necessarily they are in any sense physical, that a theory which *identifies* the mental with the physical, or some of the physical, is *forced upon us*, whatever our attitude to such a theory. It is to say, rather, that mental events have a kind of dependent existence. (1.1) We have in fact already come close to the view that mental events are at bottom individual properties, or rather sets of them, in allowing that the interdependent existence of subject and content may be a matter of nomic correlation.

Mental events, however language enters into their specification, as of course it does (McGinn, 1982, Ch. 4), are not individuated only by our descriptions, or at any rate our ordinary descriptions. We have in consciousness a finer mesh than that. Our awareness is finer-grained than our language. I can say truly on two occasions 'I'm getting alarmed' and yet the mental events not be of the same type. That is, there might have been some discernible difference in content between them. Any will do. They will, of course, despite this, have been 'of the same type' in some looser sense. It is likely to be, even, that two mental events are not of the same type on two occasions when I give a more explicit description of my experience, perhaps that I was thinking that my daughter is a quick thinker. Whether two or more mental events count as of the same type in our sense, to repeat, is a matter of whether there is a discriminable difference in content, any such difference, whether or not the events are included under the same more or less general description.

It is not to be supposed, either, that the same type of mental event cannot occur in the lives of two or more individuals. That is, it is a logical and conceptual possibility that two mental events, one in my conscious life and one in yours, be of the same type. Certainly there would be overwhelming or very great practical difficulty in the way of establishing that such a thing had happened. What we would need to do is establish about Brown's mental event and Green's mental event that if either man also had the other's experience, he could not distinguish in content between the two events.

2.2.2 Mental Dispositions

What has been said in this section concerns consciousness itself, and hence, as remarked above, does not concern all of our *mental life*. It does not concern all, either, of what can be called *the mental*. This is so, since the mental in a wide sense, as often conceived, includes not only the facts of consciousness but also dispositional facts. There is the dispositional fact that I believe my name is what it is, which is a fact about me when I am not thinking of my name. So too there are dispositional facts of desire, fear, intention and so on. For there to be such a fact about me is for it also to be true, roughly, that in certain circumstances I would consciously believe, desire, fear, intend, or whatever.

Determinism has to do in part with consciousness itself, most notably with the mental events of choosing, deciding, and the like, and with mental events which precede them. It also has to do with what else gives rise to all these events, and here the given dispositional facts are of large importance. They will get attention in connection with the second hypothesis of the theory of determinism of this book. In answer to the question of what sort of facts they are, the answer will be given that they are *neural* facts, facts about the Central Nervous Systems of ourselves and other species. The determinism to be considered here, then, like other determinisms, has to do with all of the realm of the mental as widely conceived. While it is true that we have until now been concerned with consciousness, and used the term 'mental events' in the ordinary way so as not to include all that may be included in *the mental*, the rest of the subject-matter of that domain is not thereby excluded from our coming reflections. That is not to say that we shall take dispositional mental facts or what is called the unconscious or the subconscious to be 'another realm of consciousness'—that we shall suppose there is consciousness and then inaccessible consciousness and then the brain, the latter involving dispositional mental facts. If there is a similarity between mental events and the given dispositional facts—that both enter into the explanation of behaviour—that is no sufficient reason for blurring the distinction between them, for etherealizing the dispositional facts.

2.2.3 The Mental as Physical

All three hypotheses of our determinism concern the physical in several ways, including that part of the physical which is the neural. Neural facts, pre-eminently facts about brains, will be the subject-matter of Chapter Five. How, in general, are physical events, facts and so on to be understood?

According to a long and dominant tradition, the physical is bound up with the spatial. To follow a recent and acute account in this tradition, the physical is what is spatial and also has certain further properties. (Quinton, 1973, pp. 46–53) More precisely, with respect to the spatiality, since we wish to exclude points, lines, and two-dimensional surfaces, we shall take it that the physical occupies a continuous three-dimensional region of space—that it is voluminous or geometrically solid. Hence what is physical has shape, however irregular, and also size and location. Further, in its geometrical solidity, what is physical is the unique occupant of its space. These primary qualities are insufficient by themselves for a physical thing, however, since they are possessed by a volume of empty space. Are we then to add only that what is physical must also be observable, that it must possess some secondary quality—colour, hardness or solidity, texture, smell, taste? To do so would exclude many very small particles whose existence is well established in physics, and also physical forces, such as magnetism. The obvious solution is to take into account that these stand in causal or other nomic connections with somehow observable space-occupants.

We thus come to the definition of the physical as (i) that which is an observable space-occupant or (ii) a space-occupant which is in causal or other nomic connection with an observable space-occupant. There are other views of the physical related in several ways to this. The physical is defined, for example, as that which is recognized by science, that which is 'an essential part of the coherent and adequate descriptive and explanatory account of the spatio-temporal-causal world'. (Feigl, 1958, p. 377) It may be defined, similarly, in terms of falling under scientific law, which is to say in terms of entering in nomic connections. (E. Nagel, 1979, pp. 146 ff.)

We shall take it, as anticipated earlier (p. 16), and in some agreement with these views, that all causes and effects are physical events, but we need not look further into the views. Given the conception we have, are mental events as we have conceived them excluded from being physical? They are not. That an event is characterized in terms of the interdependent existence of subject and content does not entail that it is not physical according to such a conception of the physical. We are not forced to the conclusion that mental events are somehow outside of the physical world. Nor, however, are we prevented from embracing that conclusion. It does not follow from the conception of an event in terms of interdependent subject and content that it *is* physical in the sense defined.

In fact, however, there are independent good reasons to suppose that mental events *are* physical in the given sense. For a start, there *is* a clear answer to the question of *where* thoughts and feelings take place, if not so fine an answer as we might like. They take place in and indeed are sometimes identified in part by a location, which is to say the location of a person or other organism. I may indeed remind my friend of a thought I had by referring to it as the one I had on first wakening in Venice. I can indeed mean the thought that occurred there, rather than mean the thought that occurred nowhere on an occasion when my location was Venice. Further, there seems no reason to suppose that mental events do not also *occupy* space—as do other events of which we cannot specify the minute space or the minute and myriad spaces which they occupy. (Thorp, 1980, pp. 60 f.) Finally, as we shall see more fully, and as generally is supposed, there are causal relations between mental events and observable space-occupants.

Certainly we may feel a resistance to the idea that mental events are in space. The resistance, to my mind, does not amount to what can properly be called an argument. It may amount to the irrelevant truth, that propositional or abstract objects, which enter into the contents of many mental events, are not in space. Whatever is claimed of these traditional philosophical mysteries, including the claim that they enter into mental events, they surely cannot be regarded as *parts* of them. Only if they were parts could they generate an argument for the non-spatiality of mental events. The resistance to taking mental events as spatial seems also to have to do with something less specific, pertaining to those of them which are perceptual and also those which are representative. We are tempted, perhaps, to associate them with what they are *of*, or what they represent, and hence to remove them from where they otherwise would seem to be, and hence to be lost for a location to which to assign them. However, there can be no ground for the initial association: my perception of the pen in my hand is not that pen, however much I 'enter into' the latter.

The idea that mental events are physical, however, is not the idea, whether true or false, that mental events are in fact *identical with* the particular kind of physical events with which they have often been and are still identified, which is to say neural events. These, as remarked, are events of the Central Nervous Systems of ourselves and other species. They are, by way of the summary description on which we shall depend until Chapter 5, *electrochemical events*. We now turn to the question of the relation of mental and neural events, and first to propositions to the effect that they are indeed identical.

2.3 IDENTITY THEORIES

What we have so far enables us to rule out certain purported answers to the question of how mind is related to brain, or, more precisely, how mental events are related to neural events. Any purported answer which does not begin from an adequate conception of consciousness and hence of mental events is in fact no apposite answer at all—good, bad, or indifferent. It does not get into touch with half of its subject-matter. Thus there is no apposite answer to be had which begins from the conception of mental events as merely behavioural events. Quite as certainly, any purported answer which begins from the logico-linguistic criterion strictly used, or causalism or functionalism, is no apposite answer, since it too in fact does not deal with half of the subject-matter. (Cf. T. Nagel, 1986, pp. 13–14.)

In particular, we have no answer to the question of the nature of the psychoneural relation in what can be called *Causalist Identity Theories*, often taken as the pre-eminent Identity Theories. In them it is said that certain events *causally conceived* are in fact *identical*, which is to say numerically identical, with neural events. (Smart, 1959; Armstrong, 1968, 1980; Lewis, 1966, 1972) The latter proposition of identity might be true—*all* that is required for its truth is that neural events cause certain things and are effects of others. But that proposition of identity can be true without its also being true that mental events adequately conceived, realistically conceived, are or could be identical with neural events.

Again, we have no account of the psychoneural relation when it is said, roughly, that certain formal sequences are *instantiated* or *realized* in sequences of neural events, as they may be in other sequences. (Putnam, 1975; Fodor, 1968; Dennett, 1978) The Functionalist proposition of instantiation or realization could be true without its being the case that mental events adequately conceived stand or stand only in the given relation to sequences of neural events.

What criteria, other than the criterion of sticking to the subject, must be satisfied by a satisfactory account of the psychoneural relation? A second is suggested by, although it does not depend on, certain ordinary beliefs. Have in mind, for a moment, all the neural sequences which took place in a man while he was wondering if the chair he was sitting on was level. Readers hesitant about the nature of neural facts can have in mind, more generally, all the *non-mental* events which took place during the time he was having the momentary experience which he would then have described in the mentioned way. Might the neural or all the non-mental sequences have been occurring, just as they were, and the man *not* have been wondering as he was? Might he have been thinking about something wholly different, say the age of his brother? Might he even have been wondering about what he was wondering, but in some slightly different way?

If the question is taken just for what it is, and no more, perhaps few of us will answer yes. What most of us are inclined to accept is that it was not a matter of coincidence or accident that some of the neural or non-mental processes were occurring as they were while the man was wondering as he was. Whatever the nature of the connection between the two things, there *was* a connection, rather than two things accidentally or coincidentally occurring together. The proposition is supported by a vast amount of entirely ordinary experience of connections between conscious life, one's own and that of others, and non-mental facts.

What can be called *the conviction of psychoneural intimacy* or perhaps *the axiom of psychoneural intimacy* is suggested by these ordinary beliefs, but is a good deal closer to many neuroscientific assumptions and propositions, as we shall see. (5.2) This second criterion of a satisfactory account of the psychoneural relation, again to express it quickly, is that mental and neural events are intimately related, that they are bound up together—but something more definite needs to be said.

Each type of mental event stands in a certain relation to types of neural events in a larger or smaller *locale* or collection of locales in the brain. That is, each specific type of mental event occurs simultaneously with a specific type or specific types of neural events or neural sequences in the locale or locales—call this co-occurrence. Further, this co-occurrence is in some way guaranteed, which is here to say that the events of the two kinds are *somehow or other* necessarily connected, in a direct way. The conviction or axiom is not that they are nomically connected, in one of the causal or other ways defined earlier. (1.5) It is that they are nomically connected or else in some other way necessarily connected. One alternative possibility might be said to be that they are logically or conceptually connected, as they would be if there were entailments between mental and neural statements—statements asserting the occurrence of mental or of neural events. Another alternative is that mental and neural events are somehow identical. There is in fact no hope for the first alternative possibility. In fact, then, the axiom of intimacy limits us to (i) psychoneural nomic connection or (ii) psychoneural identity or something like it. We need to choose between them.

The third criterion for a satisfactory account of the psychoneural relation is that of *mental indispensability*. It derives from a fact on which there is agreement by almost all philosophical parties. Philosophical disagreement is rare, and reluctant. The fact, I think, must be regarded as a pre-theoretical or pre-philosophical, something against which theory and philosophy are to be tested rather than a proposition derived from theory or philosophy. It is true, for example, in the situation that existed, that the mental event of my having seen the olives on the table was somehow indispensable to my wanting one and my reaching for one. The conviction or axiom of mental indispensability is not in itself the proposition that earlier mental events *cause* later events and actions, although such causation will indeed make for indispensability. That the conviction is not identical with the causal proposition follows from the fact that one can have the conviction without having a belief as to causal relations between mental events, or between mental events and actions. Typically, indeed, people have the conviction without also having an articulated view of the mind. Also, the conviction is shared by philosophers who espouse Free Will, and take at least some mental events to be in no causal or nomic connection with anything prior to or simultaneous with them.

The conviction or axiom of mental indispensability comes to this: earlier mental events are essential to any full explanations of certain later mental events and also actions. Earlier mental events are ineliminable parts of any full explanations of many mental events and also actions. Any full answer to the question of why each of many mental events occurred, or of why an action occurred, must mention earlier mental events. There is no full explanation of my reaching for an olive which does not mention my wanting one, no full explanation of my wanting one which does not mention my seeing them on the table. The conviction of mental indispensability defeats any true epiphenomenalism, which is to say any picture of the mind which makes mental events into wholly inefficacious side-effects or accompaniments of neural events. This nineteenth century picture (Huxley, 1893) is improved upon in certain ways by the several contemporary philosophers who feel themselves driven to it, but is not made tolerable. At the very least, as one allows, it is 'rather paradoxical'. (K. Campbell, 1970, p. 111)

There is more to be said of the convictions of psychoneural intimacy and mental indispensability—in part, about some neuroscientific doubts of the latter (Ch. 5)—and also of what will be called personal indispensability (3.1), and of pictures of the mind which go against them. Our present imperative business, in which we will take the two convictions as true, which requires no boldness, is that of one set of accounts of the psychoneural relation, and of how they stand to the convictions. Those accounts are Identity Theories other than those put aside a few pages back. Most of these implicitly *do* involve an adequate conception of consciousness and mental events. They assert that mental events, conceived with a proper realism, are numerically identical with neural events.

The clarifying and discriminating of these Identity Theories requires an understanding of what is to be meant by saying that one thing is identical with another—that a mental event *m* was identical with a neural event *n*. To put the question in a form that does not carry the paradoxical suggestion that two things might be one thing, what is to be meant by the claim that two descriptions or terms, say '*m*' and '*n*', pick out or designate one thing? Here and hereafter, by the way, the term 'neural event' and like terms are used without any suggestion of simplicity or unity. A neural event may be of many parts and so on.

It is a habit among some philosophers to engage in what is perhaps to be called *theorism* (P. S. Churchland, 1986, p. 258; E. Nagel, 1979), and hence to avoid giving an actual analysis of their principal claim, that a mental event is identical with a neural event. They do not attend directly to the relation of the phenomena. What is supplied, instead of a clarification of their principal claim, is a certain amount of analogy, sometimes a great deal (P. S. Churchland, 1986, Ch. 7) having to do with scientific theories and the reduction of one to another. The psychoneural identity-claim is to be understood on the analogy of the inter-theoretic reduction of heat to total molecular kinetic energy, lightning to electrical discharge, and so on. The analogies are in fact distant from our subject-matter, and in some respects hopeless. (Cf. Brandt & Kim, 1967, p. 515, Mackie, 1979, p. 21.) The principal thing to be said, however, is that whatever good or bad analogies exist, the given theories of the mind, if they are to achieve their ends, must *identify* neural and mental events. What is it to do exactly that?

One familiar and long-running understanding of any identity-claim is that the item which falls under one description is in just the same place, at just the same time, as the other. No two things, it may be thought, can be in precisely the same spatio-temporal position. This criterion of identity-claims seems unacceptable. There is no doubt it gives a logically necessary condition of identity, and no doubt that we make use of it much of the time. But it is at least uncertain that identity-claims can be understood in this way—that spatio-temporal coincidence gives logically necessary *and* sufficient conditions of identity. The spatio-temporal coincidence of what falls under two descriptions seems to be consistent with there being two things in question rather than one. As has been objected, two physical processes which are in the same place and time may indeed be two processes rather than one. (Davidson, 1980c) Consider the rotation and the warming-up of an iron ball.

The mentioned objector has in the past offered another account of identity for events: that what seemingly are two events are one if the two have the same causes and effects. (Cf. Steiner, 1986, p. 250.) Again it is to be allowed we make use of the criterion, but it must seem at least doubtful as the whole truth about identity. One reason is that it is at least a conceptual possibility that seemingly two things have the same causes and effects but are not spatio-temporally coincident. Thus they cannot be identical. Secondly, it is a conceptual possibility that there be an event without causes and also without effects. What would seem to be two such events *could not*, in terms of the given account of identity, really be two. But surely that must be left as conceivable.

The acceptable and most established understanding of identity-claims, quite as long-running as spatio-temporal coincidence, lies behind both our inclination towards that criterion, and our opposed willingness to say that what falls under one description may be spatio-temporally coincident with what falls under another description and yet there may be two things in question. In this acceptable understanding, to say that what falls under one description is identical with what falls under another is to say that *what falls under the first has true of it all and only the propositions that are true of what falls under the second*. What may seem to be two individual properties, or two collections of individual properties, is in fact one property or property-collection if any truth about the first is a truth about the second, and vice versa. What may seem to be two things is in fact one thing if the seemingly two things share all truths. This Leibnizian understanding of identity-claims, in terms of what is called the Indiscernibility of Identicals, although it certainly raises problems, is surely the correct one. (Ayer, 1954b, 1984b; Wiggins, 1980, Ch. 1; Munitz, 1971)

2.3.1 Dualistic Identity Theory With Neural Causation

The first realistic Identity Theory we shall consider combines the assertion that mental events and neural events are identical with the claim, as it is sometimes made, that it is *neural* events that have certain causal roles. Let us have the mental event m, my seeing the olives, and the simultaneous neural event n, and also the later mental event m', my wanting an olive, and the simultaneous neural event n'. The theory asserts that m was identical with n, and m' was identical with n'. Further, with respect to any causal relations, they hold between *n* and *n′*. The earlier event was cause or causal circumstance of the second.

That is to say of the earlier items, given our understanding of identity-claims, that that which falls under '*m*' shares all truths with that which falls under '*n*'. A fundamental and critical question arises, one which philosophers have not given the attention it requires, and whose answer has large consequences. The question, in our current terms, is this: What are the truths about that which falls under '*m*'? There are two possible answers to this question. To give one of them is to give further and essential definition to the Identity Theory in question. (Giving the other answer takes us to another theory, or rather several of them, to be considered below.) Let us say here that the truths about that which falls under '*m*' are of two kinds. There are truths, first, which pertain to consciousness or mentality. They have to do, that is, with the interdependent existence of a subject and content. There are truths, secondly, of a neural kind. That is, to use the summary description mentioned earlier, they are truths about the electrochemical. The truths which are shared by that which falls under the first term and that which falls under the second term are of these two kinds.

What this must come to is that *m* has both a mental part or property or character or whatever, and also a neural part, property, character, or whatever. The assertion of identity, thus, is just to the effect that there is a single entity, which entity has both a mental and a neural part or whatever. To claim *m* is identical with *n* is to say there is one thing with the two sorts of part or whatever. We shall say the same, of course, of *m′* and *n′*, having in consistency given the same answer to the question about the nature of the truths involved. What is to be added to this, as remarked above, is that any causality involved is a matter of the neural. What is to be added, as it is often expressed, is as follows: it is the first event *as neural* or *under its description as neural* that has a causal role. On the assumption that the first event caused the second, it is *m* (or as we can as well say, *n*) as neural or under its description as neural which caused *m′* (or as we can as well say, *n′*). What *this* comes to, as it must, is that it is the neural part of *m* and not the mental part of *m* which is a cause of the later event.

I suspect this Identity Theory has as many supporters as any, partly for the reason that it involves mental realism, which is in fact the most common view of consciousness. It is partly misleading to speak of it as an Identity Theory or a Monism, of course, although it has to do with a single entity, since it involves two categories of parts or properties. To label it, taking into account its causal part, it is the *Dualistic Identity Theory With Neural Causation*. Figure 2 models it. What we can refer

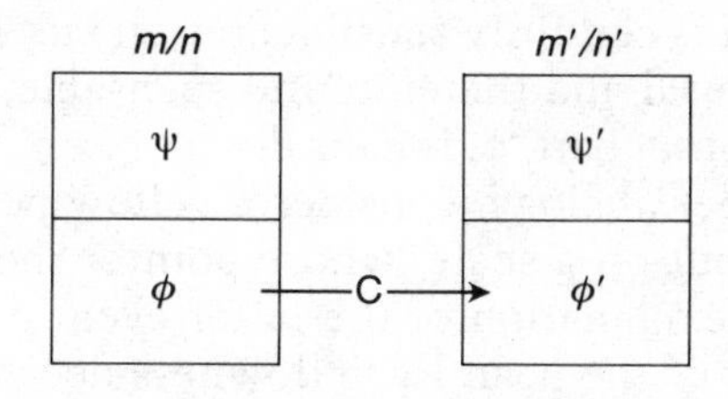

FIG. 2. Dualistic Identity Theory with Neural Causation

to either as *m* or *n* has the mental part or property ψ and the neural part or property φ. Here and hereafter, it is to be noted, 'φ' and the like will designate the *neural*. They will not be used more generally, as is common, for the *physical*. As for causal connection in the model, it is of course represented by —C→.

The Dualistic Identity Theory With Neural Causation may be taken, too quickly, to satisfy the conviction of psychoneural intimacy. This is so since it does speak of the mental and the neural in terms of one thing. Without further reflection here on whether the theory satisfies that conviction, let us consider whether the theory satisfies the conviction of mental indispensability. It attempts to do so by way of its identity-claim. To repeat, the theory gives a certain definition to the claim that *m* was identical with *n*. It is that there is one thing which has both mental and neural properties. Certainly we can say this—as readily as we can take what I now hold in my hand to be one thing, a pen with its new nib, rather than two things, pen and nib. Again, what is on the table can be regarded as one thing, rather than two things (a book and its jacket), or 403 things (jacket, binding, and 401 pages). In each case, it is a matter of choosing a classification of things, and thus a principle of counting. We may not come to discover that the world is different from what we thought by choosing a different classification, but we can indeed classify as we want, with lesser or greater profit.

It is essential to see that the given proposition of identity about *m* and *n* does not give us the conclusion that all the parts or properties of the single entity have the same standing. The fact that ψ and φ are parts of a single entity does not give us the conclusion that ψ's relation to other things is the same as φ's relation to those things, however we may speak. (Cf. 1.1.) By analogy, the fact that pen and nib are parts of one thing does not give us the conclusion that the pen stands to the depositing of the ink on paper in just the way that the nib stands to it. In particular, to come to the fundamental point, we do not have the conclusion that since ψ and φ are parts of one thing, and since φ is causal with respect to the later item, ψ is also causal. The given theory attempts to secure the indispensability of the mental through causal connection, which is certainly sensible, but in fact it fails to do so. We cannot get ψ as causal and therefore indispensable by speaking of one thing of which *another* part, ϕ, is causal.

By way of another analogy, consider a yellow pear, a half-pound in weight, which is put on a scale, whose pointer then goes to the half-pound mark. The explanation of the latter event does not necessarily include the colour of the pear. Its yellowness does not become part of the explanation of the pointer-event in virtue of being a property of a thing which also has the property of being a half-pound in weight.

We are driven to conclude that this Identity Theory, if it does have the recommendation of sticking to the subject of the psychoneural relation, and might be supposed to satisfy the requirement having to do with psychoneural intimacy, does not satisfy the requirement of mental indispensability. It is therefore untenable. Further familiar objections have been attempted against this and other Identity Theories, objections derived from a theory of identity statements as a species of necessary truths. (Kripke, 1971, 1980; cf. Ayer, 1984b) We can rest content with the objection from mental indispensability, which has what it is reasonable to call theory-independent strength.

2.3.2 A Variant of Anomalous Monism

There is a related view of the mind which merits special attention, the doctrine of Anomalous Monism. (Davidson, 1980e) Part of its interest is that it incorporates a particular argument for identifying mental and neural events. In setting out a *variant* of Anomalous Monism, I shall keep to the conception of the mental which we have, take the mental to be part of the physical, in general use the terminology with which we are familiar, and also pass by certain matters without discussion.

The view emerges from reflection on what seems to be a contradiction produced by three propositions. The first, called the principle of causal interaction between mental and otherwise physical events, is that there are causal connections between mental and otherwise physical events. It is the proposition that mental events cause events in brains and nervous systems, and also ordinary physical events external to individuals, such as the movement of olives, and that neural and other external physical events also cause mental events. The second proposition is the principle of the nomological character of causality: wherever there are causal connections between events, there are also nomic connections between them. This, certainly, can be taken as in accord with our own findings in the previous chapter. It is more plainly expressed as the truth that causal connection *is* a species of nomic connection. The third proposition, more adventurous, is in part that there are no psychoneural nomic connections. Indeed, there are no nomic connections between mental events and *any* other physical events. Arguments for this principle of the anomalism of the mental will be considered later. (2.7)

We escape the seeming contradiction in the three propositions and arrive at a certain view of the mind by way of a certain understanding of the second proposition, that causal connection brings in nomic connection. Given this understanding, we can hold all three propositions together. The understanding depends on the idea that a thing can be identified by one of its properties, and be said to cause a second thing, but it need not be in nomic connection with the second thing in virtue of the mentioned property. More particularly, a thing picked out by certain of its individual properties can truly be said to cause another, and be in nomic connection with it in virtue of other individual properties. When I put the yellow pear on the scale, it is not false to say that something yellow caused the scale's pointer to move to the half-pound mark. However, there is no nomic connection between yellowness and the pointer's so moving. There is no nomic connection between the thing in virtue of its being yellow and the pointer's moving to the half-pound mark.

It is not logically necessary, then, if a mental event causes another physical event, that it is in this nomic connection in virtue of its being a mental event. It is not logically necessary that the second thing is caused by the first in virtue of the mental or conscious part or property of the first. Furthermore, the third of our principles denies that there are any such nomic connections.

The upshot is clear enough. To state it in one way, the mental event must be identical with some other physical event, a neural event. There is one thing which is both a mental and a neural event. We can at this point ask the critical question posed earlier about the Dualistic Identity Theory With Neural Causation. What are the truths shared by the mental event and the neural event? The answer, as before, is that they are both mental and neural truths. We also have another result, essentially of the third principle. It is that there is no nomic connection between the mental part and the neural part of our single thing. We thus arrive at a particular Identity Theory. It differs from what we have lately considered, the Dualistic Identity Theory With Neural Causation, in one way. As shown in Figure 3, it does not contain the idea that it is the neural part of an entity, rather than its mental part, which causes subsequent items, notably later entities of the same sort. It also asserts (by '—— $\overline{\text{Nom}}$ ——' in Figure 3) what was left unasserted in the previous view, the absence of nomic connection between the mental and the neural part of our single thing.

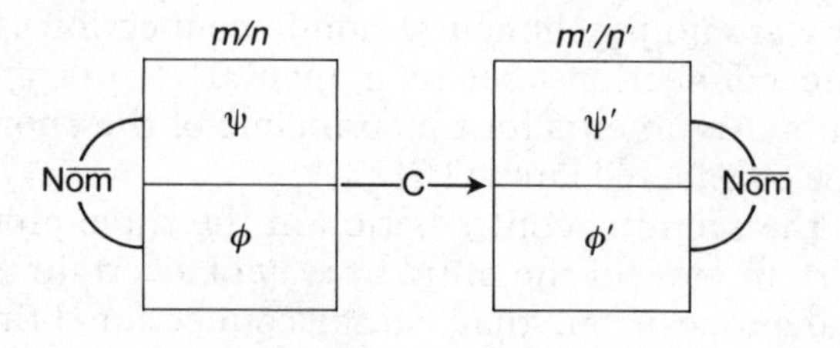

FIG. 3. Variant of Anomalous Monism

A question longs to be asked. Certainly we may speak of the whole and mean the part in talking of things in causal and nomic connection. Or, we may speak of something by way of certain of its properties, speak of it as causing something else, and it may not be in causal and nomic connection with the second thing in virtue of the mentioned properties. Suppose, however, that we speak more carefully, as we *must* in order to be clear, and specify just the individual properties of the first thing that are relevant to its causing the second thing. Does *that* causal connection bring in nomic connection between the very same items? *Evidently it does.* The only causal connections that there are, strictly speaking, *are* nomic connections. To express the point differently, if it is specifically the weight of the pear that made the pointer move to the half-pound mark, then the weight of the pear was in nomic connection with the pointer's movement.

Is there a difficulty in this idea that it is in virtue of certain of its properties rather than others that an event is the cause it is? Well, it may be said that the event of the pear's being put on the scale would not have been the event it was if the pear was not yellow. Thus there would be a barrier to saying that *that* event would have caused the pointer's movement to the half-pound mark if the pear had not been yellow. *That* event would not have occurred. Does that make the pear's being yellow causally relevant to the given effect?

It is clear that it does not. Certain conditional connections hold between the weight of the pear and the pointer's movement. They do not hold between the yellowness of the pear and the pointer's movement. The yellowness was logically necessary to the event's being the event it was, but not logically necessary to the event's being in a certain sense the cause it was. Certainly it may be said that the cause there was, in a loose sense of that phrase, would not have existed if the pear had not been yellow. That is consistent with the yellowness being causally irrelevant to the effect. That we can say, as we do, that the cause that there was would not have existed if the pear had not been yellow is owed to the fact of language mentioned above—roughly the fact that we take the whole for the part—and not to any fact of causal necessity about all properties of the pear. There is no such fact.

We can then rightly identify the causally relevant property or properties of a thing, and the resulting causal connection is in fact a nomic connection. To return to our variant of Anomalous Monism, we can and must ask what are the causally relevant properties of an entity of the kind we have, one that is both a mental event and a neural event. To avoid the question would be to fail to have a determinate picture of the mind. There are three possible answers: it is the mental part of the entity that is causally relevant, or the neural part, or both parts. Not much needs to be said of any of these three possibilities.

(i) If it is the mental part that causes another physical event, then we have nomic connection between the mental and the otherwise physical. That contradicts the third of the three principles with which we began. Causal connection of the specified kind gives us nomic connection, which connection is denied by the third principle.

(ii) If both parts of the entity are said to cause another physical event, then it remains true that the mental part of the entity causes the later physical event. We have just the same upshot of contradiction.

(iii) If we take the view that the neural part of the entity causes another physical event, there is a different embarrassment. It is of course exactly the one we have encountered with the preceding theory—the Dualistic Identity Theory With Neural Causation. We do not get the indispensability of the mental. That the neural part of the entity has a certain causal role does not give us the upshot that the mental part is an ineliminable part of any explanation of a later thing. In Figure 3, we do not have it that ψ is indispensable with respect to the later event.

We cannot but conclude that this account of the psychoneural relation fails. The initial three propositions, together with the fact of causally relevant properties and the fact that causal connection is nomic connection, issue either in contradiction or in a denial of mental indispensability. Essentially the same argument can be directed against the original doctrine of Anomalous Monism. (Honderich, 1982, 1983, 1984; Smith, 1982, 1984; cf. Stoutland, 1980; Macdonald & Macdonald, 1986; Melchert, 1986; Ayer, 1984c, pp. 3–4)

2.3.3 Two-Level Identity Theory

Another view of the mind presented as an Identity Theory also merits special attention, not only because of its emphasis on the reality of mental events. (Searle, 1983, Ch. 10, 1984, Lect. 1) Again, partly since we shall persist with exactly our own conception of mental events, what follows here is properly regarded as a variant of the original doctrine. One fundamental element of the Two-Level Identity Theory, to put it in one way, is that the events in which we are interested are open to two levels of description. At one level they are to be described as neural events, and at another level as mental events. Or, to speak differently, what we have are phenomena at different levels in the very same underlying stuff. A second element of the view is that the event at the neural level or micro-level causes the event at the mental level. What we may call the neural phenomenon causes the mental phenomenon. By way of analogy to these two elements of the view, consider the individual molecules and molecular behaviour in a bucket of water, and the liquid properties and behaviour of the water. The same stuff, at one level of description, has molecular properties and involves molecular behaviour, and at a higher level of description has liquid properties and involves liquid behaviour—it is wet, it flows, and so on, none of which is true of an individual molecule. Further, the molecular facts cause the liquid facts. A third element of the view is that an event as neural causes later such events, and also actions. My arm's going up is caused by an earlier series of neuron-firings. Indeed, perhaps, it can be said to be *caused entirely* by the firings. (Searle, 1983, p. 268) It will be thought, at this point, that the view can be diagrammed as in Figure 4.

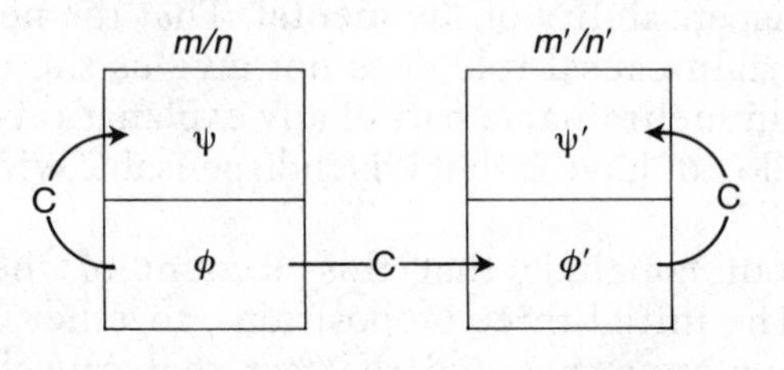

FIG. 4. Incomplete Model of Two-Level Identity Theory

If that were so, it would evidently offend against the axiom of mental indispensability. However, the first and third elements of the view can be argued to have consequences to which we have not attended. In virtue of the first element, according to which ψ and φ are the same event at two levels of description, or are phenomena at different levels in the very same underlying stuff, we *can* add that an event as mental causes a later such event as neural. Further, the first as mental also causes the later as mental, and the first as neural causes the later as mental. Thus, as the proper diagram in Figure 5 indicates, there is the possibility of arguing that we do well and truly cater for the axiom of the indispensability of the mental.

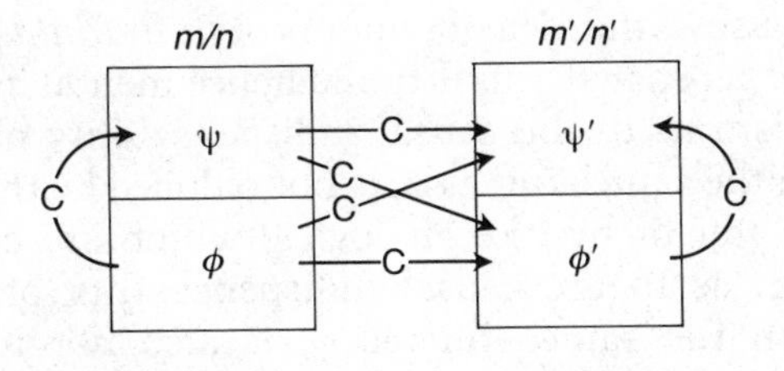

FIG. 5. Two-Level Identity Theory

This view evidently gives the same answer as its two predecessors to the critical question about what the shared truths are when a mental event is said to be identical with a neural event. The truths about that which falls under 'mental event' include both mental and neural truths. However, the present view is not well described, or at any rate not fully described, as conveying the idea of an entity of two parts or properties or characters. The essential idea conveyed by the claim of identity, that a mental event is a neural event, is that the mental event is *of the same stuff* as the neural event. Figures 4 and 5, then, must be read differently than their predecessors. For example, the relation of φ to the whole *m/n* rectangle is not of the order, or not simply of the order, of part to whole.

Exactly how, if it is, is mental indispensability achieved? It is not achieved by the second element of the view, that the event as neural causes that event as mental, or that the neural phenomenon causes the mental phenomenon. That is, in terms of Figure 5, mental indispensability is not achieved by way of the proposition that φ causes ψ. The causation *by* ψ of the later ψ′ and φ′, if considered only in terms of the causation of ψ by φ, would not give us the indispensability of ψ. This is so for the reason that φ by itself directly causes φ′, and, further, φ is at least part of a causal circumstance for φ′ which does *not* include ψ.

Mental indispensability, if it is achieved, is achieved by the first element of the view, which, again to put it as plainly as it can be put, is that ψ is of that stuff of which it is true that φ is also of it. (It would make little difference, perhaps, to speak of φ as *being* the stuff of ψ.) The essential question is therefore this: If stuff is causal, does it follow that what is realized in it is in the same way causal? Consider the pear again. Is the stuff of the pear causal with respect to the movement of the scale's pointer? It is hard to resist the answer yes. Is the *shape* of the pear, which is realized in the stuff, causal in the same way? The answer is certainly no. A reply to this argument would depend on a good deal of notably difficult metaphysical or ontological clarification, having to do with the essential ideas of stuff, levels, phenomena, and so on. This clarification is not provided. If, in clarification of the two levels, we make use of the idea of two sets of *individual properties*, it is clear we shall not get shared causality and hence mental indispensability.

One's scepticism as to the causal indispensability of ψ, despite its being realized in the same stuff as φ, is not reduced by the analogy with the bucket of water or by like analogies, which are certainly weak. One's scepticism as to the causal indispensability of ψ, despite its being realized in the same stuff as φ, is also not reduced by the persistent assertion that in ψ and φ we do not have 'two ontological categories', 'two mutually exclusive classes of things', 'mental things and physical things'. (Searle, 1983, p. 271) Against these unspecific assertions, which call out for further analysis, there is the ruinous fact of the second element of the view: that φ causes ψ. For any causal relation or connection, as is undeniable, *two* of something are required. Thus the monistic character of this view of the mind, to speak generally, is sharply reduced, or indeed reduced to a mere appearance, by the dualism brought into sharp focus by the given causal connection.

The Two-Level Identity Theory is a strong attempt to clarify the psychoneural relation. It is among several views closest to the one to which we shall come. None the less, it is difficult or impossible to escape the conclusion that like its predecessors it fails to secure mental indispensability and so is unacceptable.

2.3.4 Local Idealism

There are two Identity Theories which unquestionably deserve the name. They attend to our conviction of the indispensability of the mental in different ways. The first derives from giving a different answer to a question of which we know. (p. 94) We are told here as elsewhere that a mental event is identical with a neural event. This must mean that what falls under the term 'mental event' shares all truths with that which falls under 'neural event'. In answer to the question of what the truths are about that which falls under 'mental event', the answer given until now has been that they are both mental and neural truths. Suppose the answer given is instead that all of them are mental truths. They have to do exclusively with interdependent subject and content.

There is a disastrous upshot. What follows from this answer, together with the assertion of Leibnizian identity, is that the truths about that which falls under the term 'neural event' are wholly mental. What we took to be neural, to be a matter of electrochemical events, is transformed into wholly a matter of interdependent subject and content. The brain is, to speak quite accurately, mentalized. Neuronal events are assigned *only* the character of thoughts, desires, and the like, as we ordinarily understand them.

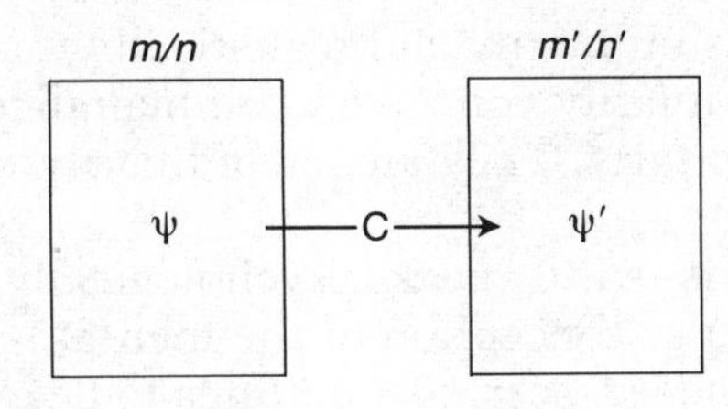

FIG. 6. Local Idealism

What we have is what can be called Local Idealism. In the tradition of philosophical Idealism, associated with Hegel and Berkeley, all that exists, including all of the physical, is taken to be of some mental character. In the place of this global Idealism, what we have in this Identity Theory is the proposition that a selected part of the neural world is turned mental. Neural events are turned mental. (Figure 6) It is a bizarre view, considerably more bizarre than Idealism itself, and cannot detain us. That it secures the indispensability of the mental, as indeed it does, cannot begin to give it a recommendation.

2.3.5 Eliminative Materialism

Mental events are identical with neural events, we are again told. If a question can be asked about the truths with respect to mental events, so a question can be asked about the truths with respect to neural events. One answer, that they are both neural and mental, would give us, if we added a proposition about neural causality, just the first view considered above, the Dualistic Identity Theory with Neural Causation. Suppose, however, as is more natural, that we answer that the truths about that which falls under the term 'neural event' are all of them electrochemical truths. (Figure 7)

Again there is a disastrous upshot. Given Leibnizian identity, we have the conclusion that mental events are wholly neural in character. What we have, plainly, is a denial of the existence of consciousness and

FIG. 7. Eliminative Materialism

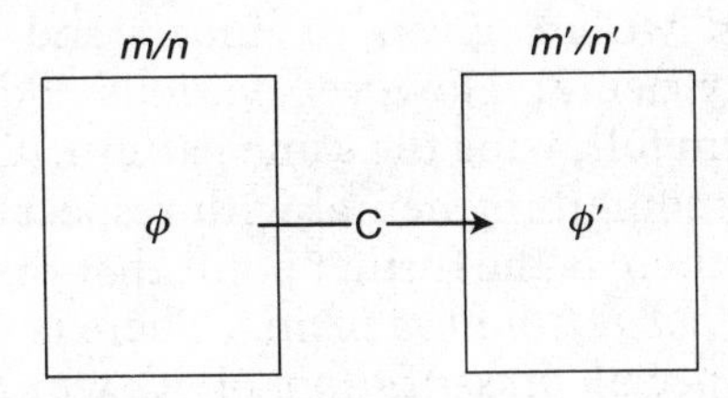

of mental events as ordinarily and realistically understood. Thoughts and feelings, as ordinarily conceived, are neuralized. They become a matter of neural activity. They are, as ordinarily understood, eliminated.

The proposition is, so to speak, psychologically less impossible to affirm, if *no* particular conception of the mental is employed in one's reflections. That, indeed, seems sometimes to be exactly the case. For example, in considering whether the mental and the neural are identical, which question may already be obscured by reflection on inter-theoretic reduction rather than the actual relation of identity, we may in effect be invited to operate with *the conception of the mental in 'folk psychology'*, which conception is in fact not supplied. We have *no* conception in the relevant sense, *no* account of the nature of the mental, when we are in effect directed to the conception of the mental in 'that rough-hewn set of concepts, generalizations and rules of thumb we all standardly use in explaining and predicting human behaviour'—preeminently the concepts of belief and desire. (P. S. Churchland, 1986, p. 229 f.) *Every* conception of the mental—Brentano to Functionalism—is a conception of that which includes beliefs and desires in a standard sense, and *none* is given relevant content by that fact.

Hardly any philosophers have actually *embraced* what we can call Eliminative Materialism, although some have been inclined to it and others have contemplated it. (P. S. Churchland, 1986; P. M. Churchland, 1981, 1984; Feyerabend, 1963; Rorty, 1970) Philosophers described as materialist have generally not gone so far as to exclude the existence of consciousness from the universe. Any such exclusion, it must surely seem, however much it accords with certain programmes and credos in the philosophy of science, is no less bizarre than the denial of neural reality in Local Idealism.

2.3.6 Semi-Eliminative Materialism

Let us finish by looking at what can be regarded as departures from Eliminative Materialism. They are in fact departures owed to the difficulty or impossibility of denying the existence of consciousness realistically conceived. It is said that neural events are identical with mental events, and we are given to understand that the truths in question are wholly neural. However, various additions are made at this point, all of them following the same pattern. (Cf. Wilson, 1980.)

There exists only something neural with respect to *m/n* in Figure 8, but, it may be said, there is the further point that the owner of *m/n*, the person in question, *has it from the inside*. There is the neural *m/n*, but there is also the special *presentation* of it, which is in some way

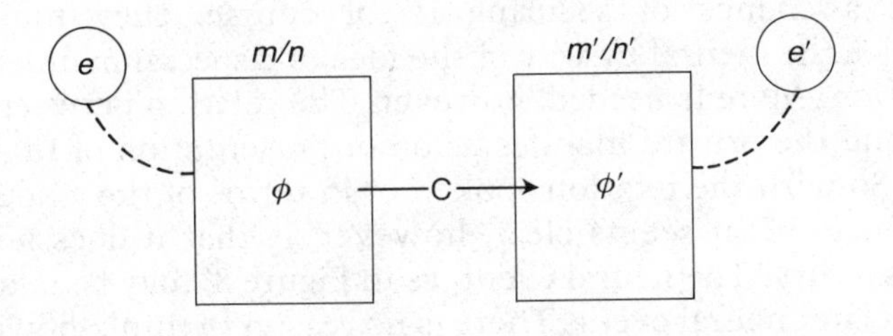

FIG. 8. Semi-Eliminative Materialism

private to the person in question. There is this special *manifestation* of the event. Or, it may be said, we must take into account a certain general distinction, between what is called the *evidence* and the *evidenced*. The event *m/n*, which is neural, is that which is evidenced by certain evidence. (Cf. P. M. Churchland, 1984, pp. 33–4.) Or, it may be said, finally, the assertion that *m* is neural is to have added to it a familiar doctrine owed to Frege. It is that certain fundamental terms in language have both reference and sense. That is, such a term picks out a given referent, and the term also has a sense. In the standard example, the terms 'the Morning Star' and 'the Evening Star' have the same referent, a certain heavenly body, but there is the difference between them that they have different sense. The term '*m*' has a referent which is wholly neural, but, we are to understand, there is also the matter of the term's sense.

The third addition is perhaps least arguable, since it simply is not true, as a fact of language, that such mental terms as 'feeling sad' or 'thinking about Oxford' or '*m*' are understood to have neural referents. However, putting aside any claim as to standard usage, we can recall that Frege defines the sense of a term as *the mode of presentation of the referent*. (1960 (1892), p. 57) This third addition could be taken as simply the claim that for certain neural events there is a special mode of presentation. The third addition is thus tantamount to one or both of the previous two.

There seems an inescapable contradiction about Identity Theories of this kind if they *do* add anything significant to Eliminative Materialism. An attempt is made to avoid it, certainly, but it is exceedingly difficult to escape the idea that the existence of consciousness is both denied and asserted. With the idea of evidence, as with those of a special manifestation or presentation (*e* in Figure 8), we appear to have consciousness reintroduced after its existence has been denied. The fact of mentality, not to be got rid of, is allowed to persist in a certain way. However, what can be said briefly is that these theories fail for the reason that they fail to secure the indispensability of the mental.

To have a chance of securing it, of course, they must indeed reintroduce the mental by way of the idea of a special manifestation or presentation. More is needed, however. The relation between a neural event *n* and the private manifestation or presentation of that event is obscure. So with the relation spoken of in terms of the evidenced and the evidence. What seems clear, however, is that it does not give us indispensability. The neural event, as in Figure 8, may be taken to be a cause of a later neural event. There is no reason to think, however, that the *evidence* or *manifestation* of the first event stands to the second event in the same relation as the first event itself stands to the second event. We have no reason to take it as part of any explanation of the second event and its manifestation. This is, in fact, a theory of the psychoneural relation which amounts to traditional epiphenomenalism.

This survey of attempts to identify the mental and the neural strongly suggests a general and I think hitherto unnoticed truth about all Identity Theories, and not only those surveyed. All Identity Theories must answer the critical question. Inevitably it must be answered in such a way as to produce either a kind of dualism or a true monism. In the former case, mental indispensability cannot be achieved by what is on hand. In the latter case, the upshot is Local Idealism or Eliminative Materialism, neither near to being tolerable. Identity Theories, in sum, face a defeating dilemma.

2.4 THE HYPOTHESIS OF PSYCHONEURAL NOMIC CORRELATION

Our progress so far has taken us through the subjects of nomic connection and in particular nomic correlation, the nature of consciousness or mentality, some of the criteria to be satisfied by a defensible account of the psychoneural relation, and the failure of the account of that relation given by Identity Theories. Thus we can now move quickly—to a conclusion as to part of what is to be put in place of Identity Theories. The conclusion is tentative, since it depends also on what is to come, a consideration of other pictures of the mind which include accounts of the psychoneural relation. Some (3.2) are akin to Identity Theories and some—the Indeterminist Theories (3.5–3.8)—wholly different. These are better considered in connection with the second hypothesis of the theory of determinism of this book.

The Hypothesis of Psychoneural Nomic Correlation, which as a result of our progress so far will need a good deal less clarifying than defending, is to the effect that each of our thoughts and feelings, indeed every item in our ongoing conscious experience, is in nomic connection of a certain kind with a simultaneous neural event. The hypothesis applies not only to wonderings, believings, inclinations, desires and moods, but also to guessings, judgings, resolvings, intendings, choosings and decidings. The first group of examples is of mental events that are in some sense passive and the second of mental events in some sense active. Whatever this distinction comes to, the Correlation Hypothesis makes no improbable distinction between them in terms of nomic connection, as used to be and sometimes still is attempted. (Johnson, 1940, Pt. 3, Ch. 7, s. 6; cf.Broad, 1925, p. 121 f.; Popper 1982b, pp. 155–6)

Something popular in one school of psychology (Kohler, 1966; Wertheimer, 1966) and sometimes still proposed, 'psychophysiological isomorphism', suggests this hypothesis: every type of neural event simultaneously necessitates and is necessitated by a single type of mental event. Hence, of course, each is necessary to the other. An hypothesis of this kind would make mental and neural events into nomic correlates of the third of the three kinds specified earlier: simultaneous mutual necessitators. (1.5) Such an hypothesis, if it is superior to Identity Theories, has a fairly familiar disadvantage. To state it quickly, there is at the very least a possibility that different types of neural events go with one and the same type of mental event. This was concluded early in the modern history of neurophysiology partly because of the surprising results of certain injuries to the human brain.

As we shall understand it, then, the Hypothesis of Psychoneural Nomic Correlation has to do with *sets* of neural events. The hypothesis is this:

> For each mental event of a given type there exists some simultaneous neural event of one of a certain set of types. The existence of the neural event necessitates the existence of the mental event, the mental event thus being necessary to the neural event. Any other neural event of any of the mentioned set of types will stand in the same relations to another mental event of the given type.

Another formulation is somewhat less abbreviated.

> Each mental event of a given type is such that there occurs simultaneously some neural event of one of a certain set of types. Since the neural event occurred so did the mental event, and would still even if any other events logically consistent with both events had also occurred. If the mental event had not occurred, neither would the neural event, even if any other events logically consistent with the absence of both had also occurred. Any other neural event of the mentioned set of types will stand in the same relations to another mental event of the given type.

This makes neural event and mental event into lawlike correlates of the first of the three types distinguished. (1.5) What we have, then, more fully, are neural and mental events that are simultaneous; the neural necessitating the mental, the mental thus being necessary to the neural; the mental dependently necessitating the neural, the neural thus being dependently necessary to the mental. (p. 75) We here have a clear and explicit account of the psychoneural relation, owed to our inquiry into nomic connection (Ch. 1) as distinct from accounts in terms of unexplained 'mapping' and the like. (Thorp, 1980, Ch. 3)

Consider, by way of example, my wondering a moment ago if the chair is level—consider, that is, the mental event which I would so describe. The hypothesis has the consequence that there also occurred a certain neural process which necessitated the mental event, and that the mental event was necessary to the process. Also, according to the hypothesis, should there ever occur another neural process of the same type, in me or anyone else, there would also occur a mental event of the given type. Should there ever occur another mental event of the given type—which is to say a mental event indistinguishable in content from the first—it would be accompanied by some neural event of a certain set of types but not necessarily the type of the first neural event. If in the future there never occurs a neural event of one of the set of types, neither will there ever occur another mental event of the given type. The psychoneural relation, then, to use a familiar terminology, is not a one–one or a many–many but rather a one–many relation.

The Correlation Hypothesis thus asserts neural and mental events to be in one of a certain family of relations well established in science. As remarked earlier (pp. 14, 47, 53), these relations are spoken of in terms of functional interdependence, concomitant variation, uniformities of co-existence, and so on, and given mathematical expression. (Cf. Mackie, 1974, Ch. 6; Bunge, 1959.) There would be little to be gained from attempting to find close analogies within physics, chemistry, or whatever. There is little point of thinking, exactly, of the gas laws or electromagnetism or the orbits of a double star. This is so partly since the problem of the *nature* or *explanation* of psychoneural nomic correlation—of which something more will be said below—remains insoluble.

It is essential to the argument to come not to confuse the Correlation Hypothesis with anything else. It is unlikely to be confused with one particular thesis of mental supervenience. Here it is asserted that identical neural events, if accompanied by mental events, are accompanied by identical mental events, and that there is no mental change without a neural change—but that these relations are *not* a matter of nomic connection but rather, to speak plainly, accidental connection. (Cf. Davidson, 1980e; Honderich, 1981a, p. 302, note 19.) More is to be said of this. (p. 141 ff.) The Correlation Hypothesis is more likely to be confused with one of several other theses of mental supervenience. These describe mental events as supervenient on neural events in the sense that mental events are *determined by* or *dependent on* neural events. One thesis given this general characterization is that identical neural events, as a matter of nomic connection somehow conceived, are accompanied by identical mental events. Another such thesis, somewhat stronger, is that for each mental event, as a matter of nomic connection somehow conceived, there exists a neural event such that any event of that type is accompanied by a mental event of the type of the given mental event. (Kim, 1979, 1982, 1984; P. F. Strawson, 1985, pp. 55–68)

The most important of several differences between the Correlation Hypothesis and such theses of supervenience, which are of course related to it, has to do with the general characterization of the latter, however well that characterization may be deserved. The Correlation Hypothesis *cannot* in any ordinary way be said to represent the mental as *determined by* or *dependent on* the neural. What it in itself asserts is no more than precisely the specified nomic connections. It will be as well to recall them specifically, in a different order from before. What it asserts of a mental event *m* and a neural event *n* is as follows.

> The events *m* and *n* were simultaneous; since *n* occurred, so did *m*, and still would have despite certain changes; if or since no neural events of certain other types occurred, if *n* had not occurred, neither would *m*, despite certain changes; since *m* occurred, and certain neural events other than *n* did not, *n* occurred, and still would have despite certain changes; if *m* had not occurred, neither would *n*, despite certain changes.

Clearly enough, these claims would be misleadingly summarized by saying that the mental is determined by or depends on the neural. For one thing, there is involved *no* causal connection, in the fairly standard sense defined earlier. (1.5) In particular, *m* is a *simultaneous* necessary condition of *n*. Secondly, any proper talk of *n*'s determining *m* will have to do most importantly on something else: causation by prior things of *n*, or *m*, or both. For example, if we were to add the proposition that *m* was to be explained as the effect of a prior circumstance, and that *n* was not an effect, that would be crucial. It would make it proper to say that the *neural* was determined by or depended on the *mental*.

The matter of such causal connections is of course one to which we shall come. (Ch. 3) It is only when we have considered them, in connection with the second hypothesis, that we shall have a full account of mind and brain. More precisely, what will be called the Theory of Psychoneural Union, or the Union Theory (pp. 165 ff.), will provide a full characterization of what we have been concerned with so far, the fundamental psychoneural relation itself. According to the Union Theory, the principal fact about the psychoneural relation is nomic connection, but it is not the only fact—a mental and a neural event also constitute what will be called a causal pair. Our coming reflections will also involve views that may have come to mind already—for example that the relation between mental and neural events consists in no more than the causation by neural events of mental events, the latter necessarily being later in time, and the causation by some mental events of neural events, the latter being later in time. (pp. 151 ff.)

Taking into account only what we now have, the Correlation Hypothesis, it is indeed clear that what is to be said in a word or two of it is that it makes mental and neural events into nomic correlates rather than items in a determination-relation or a dependency-relation. Similarly, the hypothesis cannot with reason be called *reductionist.* The term is used variously and loosely. Reductionist hypotheses, by one definition, deny the existence of mental events realistically conceived. (P. S Churchland, 1986, pp. 278 f.) A reductionist hypothesis, in another sense, is one which explains the mental in terms of the neural, or supplants a mental by a neural theory, or takes the neural to be basic or more fundamental. (Cf. Wilkes, 1978, p. 30.) What the Correlation Hypothesis itself asserts, clearly, cannot unmisleadingly be said to be that the neural is basic or more fundamental, or that one theory is to give way to another. Here, as often enough in philosophy, there is more than fruitlessness in generic labelling.

More importantly, it is unenlightening or worse to describe the Correlation Hypothesis as mind–brain *dualism,* for at least five reasons. It has nothing much to do, first, with what is generally called Cartesian Dualism, which asserts the existence of a mental substance, *res cogitans,* out of space. It is to be distinguished, too, from what is sometimes called bundle-dualism, which, as it seems, takes thoughts, desires, and so on to be independent entities, in fact substances, whether or not in space. On the present view of mentality, which informs the Correlation Hypothesis, talk of mental events is strictly speaking talk of individual properties. They are, if it is useful to say so, dependent rather than independent. (1.1) Secondly, as has not been remarked explicitly until now, they are to be taken as individual properties of such structures as the Central Nervous Systems of our species. More plainly, in our case, they are properties of our brains. They are properties, then, of exactly that which also has neural properties. Perhaps it is such a view, incidentally, that has sometimes been described as a dual-aspect view (T. Nagel, 1986, pp. 30, 32) but that description has traditionally carried metaphysical implications that are no part of the Correlation Hypothesis. (Schaffer, 1967)

Thirdly, as remarked before, mental events are not excluded from the category of physical events. Nothing said earlier of their fundamental nature, having to do with the interdependent existence of subject and content, excludes them from the physical as defined. They are in space, and causally related to space-occupants with further properties. If dualism is understood as by definition taking mental events to be non-physical, as generally it is, then the Correlation Hypothesis on this ground alone is not dualistic.

Fourthly, it is commonly said or implied that dualism makes the mental in a certain sense irreducible: it puts the mental outside of the reach of law, which is to say not in nomic connection, or not in nomic connection with the neural. This, it is sometimes said or implied, is really what makes dualism into dualism. (P. M. Churchland, 1984, p. 12; cf. P. S. Churchland, 1986, pp. 323 f.) It is the fundamental thing about dualism. Evidently this irreducibility-claim is no part of the Correlation Hyothesis. Very definitely, it does not make mental events into somehow novel properties beyond prediction or explanation by science. On the contrary, the Correlation Hypothesis is in the given sense a *reducibility*-claim. 'Dualism, the idea that minds (unlike brains) are composed of stuff that is exempt from the laws of physical nature, is a desperate vision which richly deserves its current disfavour' (Dennett, 1984, p. 28)—that firm sentence, whatever else is to be said of it, has nothing to do with the Correlation Hypothesis.

Fifthly and finally, as became clear earlier, typical Identity Theories or monisms are as dualistic as the Correlation Hypothesis. In fact all arguable Identity Theories, those involving a realist conception of the mental, propound the existence of two categories of properties. They are only misleadingly given the name they have. Given this consideration and the four previous ones, there is little or no point in referring to the Correlation Hypothesis as asserting or presupposing a dualism—as I must own up to having spoken of a predecessor of it in the past. (Honderich, 1981a, b) It is best labelled as asserting just what it does, which is psychoneural nomic correlation.

Something related is to be said of doctrines of what is called mind–brain parallelism. Traditionally conceived, and as the name suggests, these are views of sequences of mental events, and neural and other events, say the two sequences associated with me over the past half-hour, which deny any *connection* between them. That several touches on my shoulder came together with two identical or similar mental events is not to be explained by any connection between the former pair and the latter pair. Causal connections may or may not hold within each sequence, but none holds between sequences. The views are associated with Leibniz's doctrine of pre-established harmony, and with Malebranche's occasionalism: all events are caused not by other events but directly by God.

It is true that parallelism has traditionally been conceived as opposed to psychoneural *causal interraction*. (pp. 151, 161) In that, the Correlation Hypothesis is like it. Parallelism is, in its fundamental nature, as much opposed to all psychoneural nomic connection. All such connection raises the problem which parallelism seeks to avoid by its extravagant speculation of no connection. There is thus little to be said for the idea (Addis, 1984; Wilson, 1981, p. 306) that the Correlation Hypothesis is to be regarded as a doctrine of mind–brain parallelism. It is, rather, anti-parallelist. (Mackie, 1981, p. 347)

To say a word now about the problem which parallelism seeks to avoid, and which is raised by the Correlation Hypothesis and also the other two hypotheses to come, it is the problem of explaining the *nature* of the connection between what are obviously two different sorts of events and the like—at bottom two sorts of individual properties, the neural and the mental. In one familiar form, the problem has been stated in terms of the physical laws having to do with the conservation of energy and momentum. On the assumption that mental events are non-physical, and the assumption that any nomic connections with the neural into which they might be thought to enter would be subject to the laws in question, it appears that they in fact could not enter into any such connections. The issue, which has been considerably disputed (Broad, 1925, pp. 103–9; Spicker and Engelhardt, 1976, p. 144 f.; Schaffer, 1968, pp. 66–7), may be thought not to arise at all for the Correlation Hypothesis, which allows for the physicality as defined of mental events.

Still, whatever is to be said of that, it is very plain we have nothing like an account of the ultimate nature of psychoneural connection, and more particularly of psychoneural nomic connection. What we lack, if the connection is not in a sense primitive, would be supplied by an adequate idea as to a sequence or mechanism joining the mental and the neural. This problem, as fundamental as any, arises for *all* accounts of the mind which are starters at all, by which I again mean all accounts which conceive adequately of the mental. It arises for the principal Identity Theories we have considered, as it does for epiphenomenalism, for causal pictures and other freer suppositions we shall be considering in connection with the second hypothesis, and also theories of mental supervenience. It is difficult to suppose that it can be escaped even by that kind of supervenience noticed earlier in this section (p. 108) which supposes that there is no nomic connection between the mental and the neural. Also, the problem arises in scientific contexts, where indeed, contrary to the views of some philosophers, an adequate conception of the mental is commonly assumed. Neuroscience itself, in its assumption of ordinary psychoneural connection, must recognize the problem. It is faced directly by any scientific theory or speculation as to the emergence of consciousness in the evolution of species and in the development of an individual from a fertilized cell.

As will be anticipated, it is no argument against the Correlation Hypothesis that it offers no solution. The principal philosophical attempt at a solution which respects the reality of the mental consists in panpsychism, the doctrine that all that exists somehow already includes the mental. If it continues to have resourceful proponents (Sprigge, 1983) and also reluctant friends (T. Nagel, 1979, Ch. 13, 1986, pp. 49 ff.), it is difficult not to side with its critics. (McGinn, 1982) No doubt, if this were an ideal book, more would be said of it.

2.5 MIXED OBJECTIONS

The Correlation Hypothesis calls up a good deal of questioning, resistance, and objection. We shall consider some of it, mainly since to do so will clarify the hypothesis. One question that will be left unconsidered is that of the relation of the hypothesis to another matter: the explanation of mental and neural events by earlier facts. In looking at Identity Theories we did take into account their entailments or implications for the explanatory question. The like entailments or implications of the Correlation Hypothesis will not be considered here, but in the next chapter.

It is natural enough, as already remarked, to be inclined to accept some proposition about nomic connection between brain and mind. The Correlation Hypothesis, however, may strike some as extreme. Is there not something *moderate* which satisfies our natural inclination? The idea of such a thing may derive from a selection of small ideas. One is that people generally, because they are physically similar, behave in similar ways—they walk and talk similarly—but, because they are not physically identical, they do not behave identically. A second idea is that a particular individual, given his or her physical constitution, does or would behave in the same ways in identical circumstances. It is not a great step to the *General Correlation Hypothesis* and the *Individual Correlation Hypothesis*, both distinct from our Correlation Hypothesis. The Individual Correlation Hypothesis, to give it first, is as follows.

> For each mental event of a given type *in the life of a given individual*, there exists a simultaneous neural event of one of a set of types. The existence of the neural event necessitates the existence of the mental event, the mental event thus being necessary to the neural event. Another neural event of one of the given types will stand in the same way to another mental event of the given type.

My having had what I describe as a mental image of Charlotte Street. was necessitated by one or another of certain neural events, and necessary to it, but, given the Individual Correlation Hypothesis, an identical neural event in you might not be accompanied by that image, or indeed any mental image. The General Correlation Hypothesis, which is not at all tied to individuals, is as follows.

> For each mental event of *similar* types, there exists some simultaneous neural event of one of certain *similar* types. The existence of the neural event necessitates the existence of the mental event, the mental event being necessary to the existence of the neural. Another neural event of one of the types will stand in the same way to another mental event of the given types.

When the two of us standing under Admiralty Arch have what we report as a view of the Mall, there exist certain neural events. Mine is of one of a certain set *s* of types, which set is similar to a set *s'*, including the type of which your neural event is an instance.

Anyone who takes up the Individual Correlation Hypothesis, and has the natural inclination to accept some non-accidental relation between minds and brains, is likely to want the General Correlation Hypothesis as well. How well the inclination is satisfied by the latter hypothesis is wholly uncertain until more is said, however, since the hypothesis as it stands is entirely indeterminate. *How* similar are the mental events in question, and *how* similar the two sets of types of neural events?

What might be said to make the Individual Correlation Hypothesis persuasive, or more persuasive? An attempt can be made to explain in a general way how it comes to be true, as perhaps it is taken to be. (Thorp, 1980, pp. 53 f.) At a certain time, young Albert is aware of a bun, and, as it happens, there is occurring the neural event or process *n*. What happens, we are told, is that his awareness comes into nomic correlation with *n*-type neural events. Thereafter there is nomic correlation in his case between *n*-type neural events and a type of mental event pertaining to buns. We can say, metaphorically, that neural events of the given type were blank until a certain time, and thereafter were bun-imprinted. Young Bertie, on the other hand, since an *n*-type neural event occurred in his case at a time when he was aware of his thumb, came to have another type of mental event nomically correlated with the given neural events. In him, *n*-type neural events are thumb-imprinted.

The Individual Correlation Hypothesis is surely hopeless, and not helped by the imprinting story. *Can* we suppose that *n*-type neural events in Albert's case do *necessitate* bun-type mental events? No, since if this were so, the same type of neural events would be accompanied by such mental events in the case of Bertie. It is equally true that no mental event of the given type is *necessary to* such a neural event, since such neural events turn up in the case of Bertie without the given mental events. These conclusions follow directly, of course, from the analysis of nomic connections to which we have come. (Ch. 1) In fact, there is no remotely arguable analysis of nomic connection which will save the Individual Correlation Hypothesis, or anything relevantly like it.

What may now come to a hopeful mind is roughly this: that *n*-type events plus something else, *x*, constitute the nomic correlates of mental events concerning buns, and *n*-type events plus something different, *y*, constitute the nomic correlates of mental events concerning thumbs. This is not the Individual Correlation Hypothesis, but something suggested by it and the imprinting speculation. There is an immediate problem. What are *x* and *y*? It would be pointless to take them as types of neural events themselves. The result would be tantamount to accepting what we are considering replacing, which is the Correlation Hypothesis. Can *x* be taken to be just (a) Albert's history, or a bit of his history, or the past fact of imprinting? Can it be (b) the non-neural 'storing' in him of the initial bun-awareness? All of this is vague, and the first alternative is surely incomprehensible. It cannot be that the older Albert's mental events of the given type are in fact owed in part to the *direct* or *unmediated* influence across time of an event in the past. If non-neural 'storing' is opted for, we have already abandoned any recognizable hypothesis of psychoneural correlation. What we have is a mysterious bundle: the later neural events of the type of *n*, the non-neural *x*, and the mental events. To succumb to yet another temptation, to take *x* as another type of mental event, and to provide a neural correlate for it, would merely complicate matters and be of no help.

It is not only the Individual Correlation Hypothesis, certainly, that pays attention to the plainly true idea that learning and an individual's past environment somehow play a fundamental role in the occurrence of mental and neural events. We shall, as already remarked, be much concerned with the matter in due course. The Correlation Thesis in itself neither entails nor excludes any arguable account of this background. As for the imprinting story, it in fact does no more than raise terrible questions. Was Albert's initial awareness of the bun not or not at first correlated with *any* neural process? Was it 'free-floating' before the link with *n*-type neural processes was established? If that is so, to ask but one question, why should all mental events not be free-floating in the same way?

Thus we do not have a more moderate alternative to the Correlation Hypothesis. The failure of the Individual Correlation Hypothesis affects its partner, the General Correlation Thesis. The latter, like the former, needs a partner. Given our natural inclination to require some considerable psychoneural uniformity in the life of a given individual, the general hypothesis needs the companionship of the individual hypothesis. It must do without it. On its own, however, the general hypothesis will not suffice. If we have the common inclination to accept that psychoneural relations are not accidental or coincidental, we cannot regard the General Correlation Hypothesis as sufficient. Or rather, to speak more sensibly, we cannot so regard the various hypotheses of correlation which would fall under it, since it is indeed indeterminate, no more than a schema.

So much for one kind of resistance to the Correlation Hypothesis. Other kinds have to do with its neural side. One possible doubt can be dealt with by the reassurance that the hypothesis is not tied in any way to a discredited or indeed *any* particular theory of brain localization: a theory which assigns mental events to locales in the brain. It is of course independent of the 'narrow localization' of some 19th Century neurophysiologists, notably Gall and Spurzheim. (1810) Evidently, and consistently with what was said earlier of psychoneural intimacy (2.3), the hypothesis is consistent with different neuroanatomical and other accounts of the brain and also with speculations about it, including various contemporary ones. It is consistent, I think, with very nearly all such theories and speculations, which is to say that vast majority of them which are not specifically devised to preserve indeterminism or the freedom of the will.

The hypothesis does not assert that your brain is very like mine. Or, to say something with more content, it is wholly consistent with facts that have sometimes been produced against psychoneural correlation—for example, that Byron's brain weighed 2,200 grams but Anatole France's only 1,100. (Wilkes, 1980, p. 115) That the hypothesis is consistent with such facts is owed in part to its making the brain–mind relation many–one rather than one–one. Also, the hypothesis does not assert that a type of neural event ever recurs exactly, although it properly allows this to be a possibility. (Cf. Wilkes, 1980, p. 114.)

The hypothesis does not consist in a proposal to neuroscientists to investigate the brain or psychoneural correlation in a particular way: guided by the ordinary categories we use in speaking of consciousness—the various emotions and so on. It is sometimes said that there is no reason to think that these ordinary categories have neural counterparts, since the the categories are of a 'practical' rather than a 'theoretical' kind, the categories of 'folk psychology'. Such sceptical opinions are held by very different sorts of philosopher, some committed to what can be called the autonomy of mental life and some committed to neuroscience. (Hampshire, 1972a, pp. 16–7; P. M. Churchland, 1984, pp. 43 f.) As it seems to me, it is an unlikely idea that our ordinary categorization of consciousness is not in some good relation with neural categories. All that is to be said here, however, is that the Correlation Hypothesis does no more than assert that what is in fact discriminable in consciousness, with the aid of folk-psychological categories or any others, has a neural counterpart. This is no proposal to neuroscience, or a constraint being imposed on it, but, rather, close to being an assumption of it.

To begin now on a larger matter, effectively the subject of the remainder of this chapter, it consists in doubts and objections which have to do with consciousness and with mental events.

2.5.1 Atomizing Consciousness

Does the conception of a mental event carry the idea that consciousness is somehow 'atomistic', that it consists only in 'a sequence of elementary ideas', that one's stream of consciousness consists just in an ordered series of events? It has been supposed that psychophysiological isomorphism and the one–one theory of psychoneural correlation mentioned earlier (p. 107) involves such an atomism, and is therefore mistaken. (Popper and Eccles, 1977, pp. 88 f.) The same objection, whatever its worth, might be made to the Correlation Hypothesis. No doubt any view of the mind which, so to speak, denied relations between mental events would indeed be absurd. Any view would be absurd, that is, which denied the unity or continuity or connectedness of consciousness, at a time or over time. It is plain enough, however, that distinguishing mental events within consciousness in no way commits one to denying its unity. To discriminate is not to deny continuity or connectedness, in which, it may be argued, its unity consists. (Parfit, 1984, Pt. 3) By way of the obvious analogy, a stream of water is not made any less a stream for us by the fact that we pick out stages, parts, variations in flow, eddies, and ripples. Nor, of course, are these discriminations in any way illusory or not based in the given reality.

2.5.2 Resemblance of Nomic Correlates

Can it be that the mental is *simple*, and that the neural is extremely *complex*, and that this is a barrier to nomic connection, or a barrier to nomic connection on the assumption that such connection is not mediated by a mechanism or sequence? (Cf. K. Campbell, 1970, p. 51.) It is hard to suppose that there is any serviceable generalization on which the objection can depend, to the effect that all nomic correlates are of similar complexity, or of similar complexity when they are unmediated. In any case, the initial assumption that the mental is simple and the neural complex is itself open to objection. Clearly, despite the great complexity of the brain, there are neural events which are in relative terms extremely simple. All of them, further, are of a common character or of common characters. Also, despite the unitary nature of many mental events, and the common character of all consciousness, there is the richness and great information-content of, say, ordinary visual experience.

2.5.3 Counting Mental Events

If one thinks of any stretch of one's experience, the question can be asked of how many mental events were in it, or of the number of mental events in which it consisted. Consider just two seconds of my visual experience. A great many answers seem possible. One is that my perception of my hand was one mental event, and my perception of the pen in it a second, and my lesser visual awareness of such things as the Post Office Tower out the window a third. However, it is possible to say, differently, that my whole visual experience during the two seconds was a single two-second event, or that it was two one-second events. It is also possible to say a good deal else—for example that the number of mental events was very large indeed—the number of items that logically could have been discriminated by me, the number of items of visual information. An objection that may arise from this is that the idea of a mental event is somehow indeterminate. Do we not in fact have no decent idea of a set of things if we have no settled rule as to counting them, whether or not we are able to act effectively on the rule? We would have no decent idea of a human being, surely, if we had no rule for counting them. Do we, despite all that has been said, have no decent idea of a mental event?

The simple reply to this, albeit one that skirts several problems, is that we *do* have a rule for counting mental events. It is that anything discriminated in consciousness is to count as one mental event. The rule does not tell us *how* to discriminate—there are indeed many ways—but that is nothing to the point. There is a large category of adequate counting rules which are in this way indeterminate. There is thus no objection here either to the idea of a mental event or to the Correlation Hypothesis. Consider an analogy of the idea of a mental event and an analogy of the Correlation Hypothesis—the idea of a solid thing and the generalization that all the solid things on my table are inflammable. What of the book, say Wittgenstein's *Zettel*? Does it with its paper jacket count as one solid thing or do they count as two? It would be unusual but certainly not something that counts as a mistake to take each page as one solid thing. Evidently there are more possibilities. That this is a matter of decision, of a kind, does not make indeterminate either the idea of a solid thing or the generalization that all the solid things on my table are inflammable. The generalization's truth, if it is true, is not affected by how we count the things in question, and neither is its falsehood if it is false. So with mental events and the Correlation Hypothesis. What it comes to is that each thing logically discriminable in consciousness, no matter how the discrimination goes, has a neural correlate.

2.5.4 Linguistic Individuation

Is the claim that we can discriminate mental events put in doubt by something said of them originally, that a single piece of language may describe occurrents of two or more different types? It was said that an ordinary sentence, such as 'I was thinking that my daughter is a quick thinker', can be used truly of different mental events. Does it follow that an ordinary piece of language cannot be used by me to discriminate or single out a mental event? The matter touches on many others, but the short answer appears to be no. The sentence in and by itself does not discriminate a mental event. It can be used by me, however, in my act of singling out the mental event. (Cf. Mackie, 1981, p. 350.) It is a marker or pointer, a marker of that of which we have fuller knowledge than the marker itself expresses. It is of use to me despite the fact that it is not so specific as to prohibit its use of a related but different mental event. Consciousness is in no way unusual in this respect. I can today use the description 'the velvet tie he is wearing' in my own entirely successful enterprise of picking out a certain material object. I can use the same description tomorrow, of another tie, in picking out that somewhat different thing.

2.5.5 Conceptualization

If I can use the same description for different things that I pick out as having occurred in my consciousness, is it none the less necessary that in order to do so I must in some sense *conceptualize* them? Must I in some sense put a discriminated thing under an idea, or, more likely, ideas? The question is not clear, but perhaps clear enough. We may feel tempted to reply that we cannot know what we cannot in some sense put under a concept. We may feel tempted to reply that if thereafter an ordinary piece of language may serve as a marker for something we experience, we require in the first instance that it be put under fuller private description, in a superior 'language of thought'. Having been tempted to this, we may then succumb to a certain vague uncertainty. Are our capabilities of private description not very meagre? To take only simple examples, I may be in doubt as to whether my feeling of a moment ago was nostalgia, in the ordinary sense, and have no confidence as to any closer description. I am hard-pressed, to say the least, to provide for myself some conceptualization, even an image, of the sensation in my ankle as I sit in this chair. Is there the general upshot that mental events escape us, that we cannot single them out?

Just one of several replies has to do with the first part of this sequence of reflection, the temptation. It is *not* always necessary that in order to discriminate something within my experience I must conceptualize it at all. It is not necessary, in order to have a hold on it, that I put it under a description of any kind, however my own. Instances of two shades of colour, for which I use no names whatever, are clearly different to me. I can indeed tell the one from the other. I can say of a third colour-patch that it is the same shade as one of the other two. There is a longish history of psychological research concerned with such discrimination—notably research into Just Noticeable Differences in sensation and perception, and also allied phenomena. Very rightly, Fechner's psychophysical law about sensation-magnitudes and stimuli has not been doubted on the grounds that sensations are not conceptualized. (Marks, 1974)

2.5.6 Reidentification

Will it be supposed that there is a particular difficulty about the reidentification of mental events, the assigning of two tokens to one type? It may be thought that this is necessary in connection with the

Correlation Hypothesis. Consider a thought of mine a moment ago, that reflection on the nature of the mind is frustrating. Can I be confident, in a few minutes, that a second mental event of the same type, which of course is to say an identical mental event, has occurred? The answer is that it is most unlikely. Memory is not up to the task. On the other hand, it is plain that we can contrive to feel or think precisely the same thing twice. If I touch the top of the table twice in the same way in a very short time, or if I repeat a sentence to myself in a very short time, it may well be that I can confidently judge that the two mental events were indiscriminable.

None the less, it needs to be admitted that our capabilities in this respect are small: they are effective with very little of consciousness. For the most part, indeed almost always save for contrived situations, we cannot confidently suppose that the same type of mental event has obtained again. Nothing much follows from this. In particular, it does not follow that the Correlation Hypothesis cannot be confirmed, as we shall see.

2.5.7 Richness and Transience

Nor is there a threat in what has already been in view, what can be called the richness and transience of consciousness. If I ask myself, when walking in Hampstead High Street, what I have been aware of during even the past few moments, I shall be wholly defeated if I try to produce anything that could be called a complete answer. There is no possibility whatever of my doing so. Such facts, however, do nothing to make the Correlation Hypothesis unclear. That a generalization covers a multitude of items, such that it is impossible to nab each one, in no way makes the generalization unclear. The theory of evolution is not made unclear by the fact that we cannot enumerate the multitude of properties of species which fall under it. So with the law of gravity and indeed innumerable other examples.

2.6 WITTGENSTEIN AND OTHERS

Wittgenstein declares in *Zettel*, entirely in opposition to what has been suggested here, that 'no supposition seems . . . more natural than that there is no process in the brain correlated with associating or thinking . . .' (1967, 608) None of us, of course, is without the convictions that make one view or another seem natural. The question is whether they issue in effective argument. Let us consider four objections offered by or derived from Wittgenstein, and after that two of a somewhat similar kind.

2.6.1 Mechanism

In one place in *Zettel* he is concerned with the matter of how it comes about that we think and act as we do as a result of a certain visual experience. He mentions the idea of 'a physiological explanation' and writes:

But must there be a physiological explanation here? Why don't we just leave explaining alone?—But you would never talk like that, if you were examining the behaviour of a machine!—Well, who says that a living creature, an animal body, is a machine in this sense? (1967, s. 614)

Wittgenstein is to be taken as speaking the first and second sentences in this dialogue, and the last. It would be unreasonable to suppose that there is not an objection in what he says. One is intimated, at any rate, and it applies as readily to the Correlation Hypothesis as to the idea of physiological explanation. So applied, it is that the hypothesis gives us a falsifying picture of men, a picture of men as being or being like machines. It gives us no more than the doctrine of mechanism.

We need to make certain distinctions in reply, since, as has often been noted, there is more than one doctrine of mechanism, or mechanism so-called. (Gregory, 1981, Ch. 3; Glover, 1970, pp. 28 f.) We need to distinguish at least three kinds of pictures of our existence. If the Correlation Hypothesis were a picture of the first kind, or necessarily a part of one, it would be hopeless. We are all certain, to speak very briefly, that men are not relevantly like solar systems, the reproductive mechanisms of plants, clocks, Turing machines, neurons, or perhaps computers. All of these are at least arguably mechanistic in a certain primary sense: they have *characteristic functions or activities, to which they are limited*. We are all certain that men by contrast have a different kind or different kinds of *control* over their existence, that they *direct* their own activities in a different way or ways, and hence that they do not in the same way have a characteristic function or activity, or even a discrete set of these. This conviction, despite its imprecision and the several kinds of controversy in which it is involved, rightly defeats any picture of men as mechanistic in the given primary sense: being limited in a certain way to a characteristic function. (Cf. Dennett, 1984.)

We must judge that most talk of men as *being machines*, talk of them as being in some exact and complete analogy with clocks, solar systems, and so on, which are limited to characteristic functions, is fundamentally mistaken. The conviction thus defeats a certain amount of declaration by some neuroscientists. The conviction, or so it seems to me, is unlike others in being not only an attitude but a guide to truth. However, to come to the principal proposition here, the

Correlation Hypothesis is not at all reasonably regarded as such a mechanist picture. It says only what it says, that there is psychoneural nomic connection, and in itself it does not fight with the given conviction. Nor, as it will turn out, is it necessarily part of such a picture.

A second mechanist picture is of men as machine-like in that all that they do can be described and explained simply by the principles of mechanics, that part of physical science which deals with the behaviour of matter under the action of force. A different but related proposition is that every biological event, including every human event, is in fact also a non-biological event. That is, each biological event is subject to nomic connections which are also exemplified by non-biological systems. There is perhaps slightly more room for uncertainty about the rejection of some mechanist pictures of this kind. What is to be said of them here is just that their falsity, if they are false, is no danger to the Correlation Hypothesis. It does not assert them. Indeed, it may be thought to conflict with them. (Cf. Trusted, 1984, p. 129.)

It must be allowed that the Correlation Hypothesis does give us a part of a third and different picture. Again briefly identified, it is a picture of human consciousness and indeed human life as having about it an *order* or *regularity* related to that of machines and the natural world. It is an order and regularity of overwhelming variation and complexity, and in this quite different from that of all machines, but an order or regularity none the less. It is owed to no more than nomic connection. What is to be said of *this* picture, as distinct from its predecessors, is that the fact that the Correlation Hypothesis is part of it is no problem—since *this* picture is not in conflict with the mentioned truth-guiding conviction about our control and direction of our lives, our not having a limited function or activity.

That is, we are rightly certain we are self-directing and not subject to certain kinds of limits, in some sense or senses yet to be made clear, but we are not at all convinced in the same proper way that our lives do not have the order and regularity asserted by the Correlation Hypothesis. No doubt there are persons, including philosophers, who have different convictions, convictions which *are* affronted by no more than the order of the Correlation Hypothesis. These are convictions not in the same way fundamental. In the case of conflict between *these* convictions and the hypothesis, there can be no prediction of the outcome. It is far from certain that the hypothesis must bow to the attitudes. If it is defeated, its defeat is unlikely to be owed to them.

2.6.2 No Conceptual Necessity

The second objection from Wittgenstein, which is open to briefer reply, consists in certain rhetorical questions. They follow on his proposition quoted above, to the effect that it is a pre-eminently natural supposition that no process in the brain is correlated with associating or thinking. *Why*, he asks, should there be a process in the brain correlated with my thinking something or other? Why, so to speak, should there not be 'chaos' instead? Is it not perfectly possible that *nothing* physiological corresponds to certain psychological phenomena? When I remember something, why must there be something or other in the nervous system which corresponds to my remembering? (1967, ss. 608–10)

The intent of the rhetorical questions is uncertain. There are two possibilities as to their interpretation. They may be taken, first, as conveying the idea that the Correlation Hypothesis is not a logical or conceptual truth, that our ordinary notions of thinking, remembering, and the like have in them nothing about neural processes. That the hypothesis is not such a truth is undeniable, but not to the point. That our ordinary ideas of thinking and the like are as claimed is probably also undeniable, but not to the point. It has not been supposed that it would be self-contradictory or in any related way incoherent to deny psychoneural nomic connection. Nomic connection, as we know from our inquiry into it, (Ch. 1), is certainly not logical or conceptual connection. What we are considering is the question of whether psychoneural nomic connection is an empirical truth, whether as a matter of contingent fact there is such connection.

Do the questions rather convey the absolutely relevant proposition that there is not in fact nomic connection of the given kind? Or, perhaps better, that really there is nothing in life to suggest that there is? These are propositions whose truth or falsehood needs establishing. Certainly falsehood or improbability is not to be discerned on their faces. There can be no possibility whatever of settling by fiat or declaration that the Correlation Hypothesis is true, or well supported. There can be no more possibility of settling by fiat that it is false, or ill supported.

It may be that the first interpretation of the rhetorical questions, that there is not a conceptual connection between the mental and the neural, is the gist of a remarkable remark in another collection.

Thinking in terms of physiological processes is extremely dangerous in connection with the clarification of conceptual problems in psychology. Thinking in physiological hypotheses deludes us sometimes with false difficulties, sometimes with false solutions. The best prophylactic against this is the thought that I don't know at all whether the humans I am acquainted with actually have a nervous system. (1980, Vol. 1, s. 1063)

One might try to make sense of this, that is, by understanding the mentioned 'clarification of conceptual problems in psychology' as being the conceptual analysis of ordinary notions of thinking, remembering and the like. It is charitable to take the last sentence as referring to the truism that my ordinary layman's experience of others does not logically entail that they have nervous systems.

2.6.3 Explaining a Prejudice

The third objection begins by offering an explanation of what is called the 'prejudice' in favour of such views as the Correlation Hypothesis. (Wittgenstein, 1967, s. 611, cf. 1959, ss. 304–8, s. 412) We are said to be inclined to suppose that thinking and feeling is 'a process in the head', some 'occult' stream in which earlier parts somehow give rise to later. If we try to think of this as a plain matter of cause and effect, we become puzzled, since we are taking ourselves to be dealing with 'gaseous' stuff. At this point we are inclined to try another path in order to account for the seeming process or stream. We accept the existence of correlations between mental and neural events, and we add what has not been mentioned so far, the idea that there are ordinary causal relations between neural events. These ordinary connections underlie and help explain the seeming connection between mental events.

What, according to Wittgenstein, is to be said against this course of reflection issuing in a 'prejudice', is that it begins with a 'primitive' idea. It is 'primitive' to take thinking and feeling as 'a process in the head'. They have to do, rather, with behaviour. We thus need not ever start on the given course of reflection which issues in the Correlation Hypothesis and like things.

If we abstract from the rhetoric, what we have in these remarks first of all is a behaviourist account of consciousness. It is an account which remains in ways obscure in Wittgenstein's work, but certainly goes against any ordinary account of consciousness and also the account in terms of interdependent subject and content. As will be anticipated, given what has been said of behaviourism (p. 73), there seems to me no compelling reason to reopen that subject. Secondly, there is in the given remarks the idea that if we do suppose that consciousness *is* what can be called ongoing experience, we do not find it easy to accept that there are causal connections holding it together. This is more persuasive than the behaviourism. It has to do, in part, with the admitted difficulty we have in getting consciousness into focus, and also something hitherto unmentioned, the evident *conceptual* connections within it. However, to come to a third crux, there is no reason to think that this is what leads us, in the way outlined or any other way, to the Correlation Hypothesis. We do not come to it by way of a frustrated search within consciousness for what holds consciousness together. That story cannot count as serious. Furthermore, it would not much matter if the story were true, or if it had somehow influenced us without our explicit awareness. What does matter is whether the hypothesis is conceptually adequate and true. No light is shed on either question, the first of which is our question at the moment, by the speculation as to the origin of the hypothesis. Well-formed truths can have disreputable pasts.

2.6.4 The African Lad

The fourth and last Wittgensteinian objection, given shape by another philosopher, will appear to be more telling. (Anscombe, 1972; cf. Wittgenstein, 1967, s. 608 ff.) It needs more attention, as do several related contentions. It depends on the fact that if the Correlation Hypothesis is true, then, in virtue of one part of it, if a certain neural process occurs, someone has a certain belief—let us say he believes at a certain moment that the art galleries in Paris close at 7 p.m. He has a belief which he might so express. It is not near to being true, of course, that it is within the capabilities of neurophysiologists or others to bring about *artificially* a neural process of the given type, perhaps by electro-stimulation of the cortex. It is perhaps as good as certain that no such thing will ever be accomplished in the future. It will remain a factual impossibility. none the less, there is no logical or conceptual barrier to our engaging in speculation about such a thing: the speculation is about something logically possible, something not self-contradictory or incoherent. We can conceive, as the objector invites us to do, of neurophysiologists producing such a neural process, and also draw a conclusion which follows from the Correlation Hypothesis. It is that *if* the process were brought about artificially, someone would then be believing that the art galleries in Paris close at 7 p.m. It follows from the hypothesis, and this is the essential premiss of the objection in hand, that *if* the process were brought about in the given way, or indeed in any way whatever, someone would be having that belief.

Let us now add some more supposing. We add that the person is, say, an African lad who has never heard of art galleries, Paris, or clock-time. His life-experience has been such that he has had no conception of these things. This, we are told, brings us to *absurdity*. It is quite absurd to suppose that if the neural process in question were made to occur in his brain, he would then have the belief in question. That this *would* happen, however, as already allowed, *is* a consequence of the Correlation Hypothesis. The hypothesis is not partly to the effect that if a certain type of neural process obtains, *and* a person has a certain history, a certain mental event will occur. It is partly to the effect that if a type of neural process obtains, a certain type of mental event will then obtain. What follows from the hypothesis is absurd, and so the hypothesis must be mistaken.

To state the objection in a general way, it is not enough, for a person to believe certain things—or to think of them, want them, or decide or intend to get them—that a certain neural description is true of him. He must be a person with a certain history, a certain learning-history. The Correlation Hypothesis may seem arguable or even true, and certainly not lacking in sense, until one actually fixes one's gaze upon one of its consequences. It is then seen for what it is, naīve or worse, since it leaves out of account something essential to belief, which is the experience that precedes it.

Things are not so simple. By way of a preliminary, it is important to keep in mind that the hypothesis does not commit us to anything whatever about the origin or development of neural or of any physical processes. It says nothing about what neuroscience will or may ever do. We are not forced or even inclined by the hypothesis to think that it will ever be a practical possibility to do what has been imagined. We need not give up the thought noticed already (p. 91) that in the world as it is, and as it is likely to be, a man believes something only if he has a certain history. There *is* the fact of mental indispensability. That is quite consistent with the hypothesis. We *are* committed to thinking, as rightly said, that *if* a certain neural process were made to occur in the given way, a certain belief would be had—*if* an African lad, with no previous experience of art galleries, Paris, or clock-time, were to be the subject of a certain neural process, he would then be believing that the art galleries in Paris close at 7 p.m. The neural process would necessitate the believing.

To come to the point, are we then committed to thinking a thing which in fact is absurd? Given what has been said, there is only one kind of absurdity that is a possibility. Although the objection does not make it clear, and perhaps trades on the fact, it is the absurdity of thinking something that is self-contradictory or incoherent. But *is* it self-contradictory or the like, whatever the practical possibilities, to suppose a person might have a certain belief without having had the experience that always, in the world as it is, goes with it? Is there a connection of a logically necessary or conceptual kind between beliefs and past experience? Until the objection is made specific, by the spelling-out of the kind or kinds of supposed logical connection between beliefs and past experience, and hence the kind or kinds of supposed self-contradiction or the like, it is entirely incomplete. Still, it may be thought to be persuasive none the less.

What is it that moves us, as I allow that something does, in the direction of supposing that the speculation is somehow self-contradictory or incoherent? Most importantly, I think, it is the truth that for the African lad to believe the given thing when he does, *he would also have to have many other beliefs*. To have the belief, to make use of a favoured phrase, he would have to have an awareness of certain *forms of life*. It *is* somehow incoherent to suppose that he could have the belief about 7 p.m., as we understand it, without also having beliefs about the number of hours in the day and so on. It is incoherent to suppose that he could have a belief about art galleries without also having a grasp of the existence of painters and dealers.

What importantly moves us towards supposing that our speculation is incoherent, then, is a particular implicit aspect of it: the suggestion that a belief might be had, so to speak, *in isolation*. What is objectionable is this, and not anything to do with the African lad's lack of ordinary experience. This can be brought out by doing some different imagining. Let us, in the unrestrained way suggested by the original objection, imagine that neurophysiologists, in a long campaign of activity, produce in an African lad very many neural structures which go with an appreciation of all the relevant forms of life. That is, they manufacture a great number of dispositional beliefs on his part. They make possible a great number of occurrent beliefs, mental events in our standard sense. We may imagine they begin with fundamentals and go forward from there. In the end, they produce a neural process which is the correlate of a certain mental event, the lad's believing that the art galleries in Paris close at 7 p.m.

This speculation, although there is more to be said of it, does not raise in us the idea that it is somehow self-contradictory or incoherent. The absence of experience on the part of the lad, the absence of anything that could be called a learning-history, does not make it so. Of course, to be tedious, it is possible to confuse this truth with a certain falsehood: that the whole business is in another sense imaginable, which is to say something like a foreseeable practical possibility. The Correlation Hypothesis, to repeat again, has nothing to do with the falsehood. It requires only that something is a *logical* possibility, and indeed it is. If the hypothesis did have the consequence that the belief could be produced 'in isolation', it would be untenable. It has no such consequence.

It may occur to readers that our latter piece of imagining is open to a certain question. We have imagined the artificial production of a corpus of dispositional beliefs. Is it possible to think realistically of this sequence of production without also bringing in mental events along the way on the part of the African lad? Let us suppose the answer is no. (Cf. pp. 146 f.) To be brought to have a dispositional belief about artists and dealers, he must as a matter of fact, let us say, consciously ask certain questions, contemplate this and that. If this is so, it is no harm to the Correlation Hypothesis. What is true, rather, is that the speculation as to the artificial production of the neural correlate of a belief must be complicated, in a way that undercuts the objection. If it is true that any speculation about the artificial production of neural correlates will be a speculation about something somehow approximate to a learning-history, that in no way affects the hypothesis. It is about such neural events, not how they might or might not come about.

Will someone still object, in an emphatic tone of voice, that the simple fact is that the African lad 'would not have a belief'? What this insistence may reveal is a determination to conceive of beliefs and other things in a certain way—as what we can call *learned mental episodes*. Beliefs, in this conception, necessarily are learned, learned in the course of taking part in forms or ways of life. Anything that is a belief does by definition have this kind of origin. The reply to this must be that if one so chooses to define beliefs, it does indeed follow with iron certainty that the African lad does not have the belief that the art galleries in Paris close at 7 p.m. But it hardly needs saying that there is no necessity whatever of defining beliefs in this way. Certainly it cannot be said to be their ordinary definition, to the extent that such a thing exists.

But further, this is not greatly important. We have in our inquiry arrived at a conception of mental events, including beliefs. It does not have the feature in question. Mental events were not defined as the products of a certain kind of experience. What matters is that mental events, understood as we have understood them, are indubitably the stuff of our fundamental subject-matter, which is human experience, our conscious lives. It would not seriously matter if there were some divergence between our conception of mental events and an ordinary conception of beliefs.

2.6.5 Standardly Based Mental Episodes

There is a related argument by a philosopher who is in fact sympathetic to something close to the Correlation Hypothesis, but who would, in a sense, restrict its application. He takes the view that such a person as the African lad, or, to be yet freer in our science fiction, a newly manufactured physical replica of me, could not count as *thinking of Vienna*, as I can, since *ex hypothesi* he lacks 'a certain historical and cognitive relationship' with it. (Kim, 1982, p. 57) That he cannot count as thinking of Vienna, we are told, can be seen more clearly by way of another speculation. When I think of Vienna at some moment, I have a visual image of a church and thoughts of hot summer weather. Someone else we can imagine, or try to imagine, the Iowan, has never been to Vienna, but at some moment has just the same image. *His* image can in fact be traced to a church in Iowa which he saw in his boyhood. We would hardly say of this person, it is maintained, that he was thinking of Vienna.

Indeed we would not, if he had only the image, which is all that is mentioned in the speculation. What, however, if his mental event is in *all* its parts and aspects identical with that one of mine which I describe as thinking of Vienna? He speaks of 'Vienna' in connection with his thought and does not speak of 'Iowa'; the right word occurs *in* his thought; his momentary experience has a recollective aspect; it involves precisely as much internal confusion, if any, as my own; and so on. Differently from the case of the African lad, however, his mental event is owed in part to his past ordinary experience, rather than neurophysiological manufacturing, although not to any experience of Vienna. Does the fact that the Iowan's mental event has this bizarre history—it is owed in part to a church in Iowa—make it other than a case of *thinking of Vienna*?

It is possible to maintain that he *is* properly said to be thinking of Vienna. What else could he be said to be doing? But, as must be admitted, we also have an inclination to say that he is not doing so. That is, we have the inclination to require, of anyone who counts as thinking of Vienna, that his thought has some kind of *standard* history. It would be difficult indeed to specify such a history—certainly the thinker does not have to have been to Vienna—but we are inclined to such a requirement.

This does not affect the Correlation Hypothesis. Let us give the name *standardly based mental episodes* to such things as my thinking of Vienna conceived partly in terms of the history of the thinking. Such an episode by definition requires a standard history. Evidently no neural process by itself can guarantee such an episode. It would have to reach into the past to do so. But the Correlation Hypothesis is in no way weakened, or made insufficient to any end of the present inquiry, by being limited to mental events. The fact that it does not cover standardly based mental episodes is perfectly consistent with the fact that it covers *all* of consciousness or experience—choices, decisions, formings of intention, and all of what is relevant to our inquiry.

2.6.6 Twin Worlds

A sixth objection was made by a critic of an earlier version of the Correlation Hypothesis but applies as well to it. (Stich, 1981; cf. Putnam, 1985a; Stich, 1983) It includes two contentions, of which the first, with which we will be concerned for a while, is that there is a categorization of consciousness which has to do with one or another *context*, and this categorization gives us the common-sense or ordinary conceptions of thoughts, beliefs, and the like. We can see this, it is claimed, by considering several examples.

The first one involves a thought experiment about twin worlds. Harry on Earth, and another Harry on the very similar planet Yon, which has on it some *doppelgängers* of Earth folk, have wholly indistinguishable experiences. These consist in mental events of one type, which each Harry might report by saying that he was thinking that President Reagan had been shot. If either Harry had also had the other's experience, he could not distinguish the two in any way whatever. However, since there are also two Reagans in question, it follows in the contextual conception of consciousness that Earth Harry and Yon Harry were having different thoughts.

Is this true of the ordinary conception of thoughts? On the assumption that there is something clear which deserves the name, I doubt it. Certainly it does not follow from the falsehood of what Yon Harry thought—on the assumption that Yon Reagan was *not* shot—that his thought in an ordinary way of speaking was not the same as Earth Harry's. We shall not change our mind about what thought I had a moment ago, when I reported that my mother's maiden name was Armstrong, if it turns out that I was mistaken. As for the different fact of there being two referents in question, Earth Reagan and Yon Reagan, that will make common sense hesitate about whether there are two thoughts or one, since, so to speak, common sense has not thought much about twin worlds and the questions they raise.

Suppose common sense concentrates on the supposition that the two items of consciousness really are wholly indistinguishable. The two mental events, in their representative contents, are identical. Evidently these representative contents, as they are, are not fixed by anything whatever outside of them, or outside of Earth Harry and Yon Harry. Their *aboutness* is of their own nature—as is the aboutness of my present thought of the wine in the refrigerator, which is independent of whether the bottle contains vinegar and whether there is any bottle in the refrigerator at all. Common sense, if it fixes on the indistinguishability of the mental events of Earth Harry and Yon Harry, will surely be inclined to conclude that there is the same thought in question, and allow that on different occasions it is made true or false by things that are numerically different. Or, perhaps, common sense will not merely hesitate but be stymied by the question in hand. The contrary assumption, that common sense will take wholly indistinguishable mental events to be different thoughts, strikes me as remarkable.

A second example is different, although it has the same aim: to show that identical mental events may be different thoughts. A young girl has an experience which she reports by saying that she has been thinking that President McKinley was assassinated, and she gives evidence of understanding many relevant things, including the fact that assassination involves death. Grown old, and suffering from senility, she has an experience which issues in her saying again that she has been thinking that President McKinley was assassinated. But she gives evidence that she does not connect assassination and death, does not know the difference between a president and a senator, and so on.

Here, evidently, we might well speak of two thoughts in an ordinary way. But it is surely far, far too brave to suppose that her two experiences would be indistinguishable, that they could be described as identical mental events. It *might* be that we would accept that there was *an* aspect of both experiences that was indistinguishable, an aspect having to do with the use of certain words. The experiences at the two times would then involve one type of mental event which was instantiated both times. But this is far from granting that the two experiences were alike in terms of all the involved mental events.

I conclude, despite various ancillary problems, that the contextual conception of consciousness, in terms of what can be called *contextual mental episodes*, is no closer to the ordinary one, if such a thing can properly be spoken of, than the conception in terms of mental events as we have them. The relative closeness of talk of mental events to ordinary talk of consciousness is of some importance, but my principal rejoinder to our critic has to do with the second contention in his objection, which we have not yet considered.

Certainly the mental *can* be characterized in the contextual way, and we shall then say that Earth Harry and Yon Harry have had different contextual thoughts athough their experiences are indistinguishable. On the assumption that Earth Harry and Yon Harry are physically identical, we shall nevertheless not have any evidence that the contextual thoughts and the neural processes are in a nomic connection of the kind specified by the Correlation Hypothesis. Certainly—this is the second contention of the objection—it will not follow from nomic correlation between mental events and the neural that there is such a correlation between contextual thoughts and the neural.

It is true, since a contextual mental event is partly a function of other things than the thinker, but it is hard to see why on earth (or Yon) it matters. The claim that there are nomic correlations between contextual thoughts and the neural is not part of the Correlation Hypothesis, or of course a premiss of it. Nor is the falsehood of the given claim about contextual thoughts and nomic connection inconsistent with the hypothesis. It is not the case, either, that any relevant subject-matter is left out when we characterize consciousness in terms of mental events. There are not two subject-matters, but two classifications of one subject-matter, which is all of consciousness. That the subject-matter is open to *a* classification such that all of it falls under the Correlation Hypothesis is entirely sufficient to raise the questions to which we are coming.

So much for six objections. Let us finish here by glancing back and bringing together what are best described as clarifications of the Correlation Hypothesis, not qualifications of it or concessions with respect to it, and by making a principal point more explicit. The hypothesis is concerned with mental events, the discriminable constituents of consciousness, and hence, as was remarked in the beginning, not with all of what constitutes a *personal epistemic fact*. (p. 85) It is not concerned, that is, with all of what is the case when I know, believe truly, remember, or see something. Secondly, it is not concerned with all of what constitutes beliefs, thoughts, wants, decisions, or intentions where these are taken as *learned mental episodes*, explicitly defined as somehow owed to participation in certain forms of life and not to neurophysiological construction. The hypothesis, further, is not concerned with all of what is needed for *standardly based mental episodes* as defined or with all of *contextual mental episodes* as defined.

What the Correlation Hypothesis *is* concerned with is all of consciousness or experience. *None* is left out. Further, to conceive of consciousness in terms of mental events is to conceive of it in an ordinary way, arguably the most ordinary way. It is not essential to its role in a theory of determinism that consciousness is so conceived, but certainly it simplifies matters. A final and related consideration—the principal point to be made more explicit—begins from the fact that the inquiry of this book can properly be described as an inquiry into the *explanation* of our choices, decisions, and actions. Speaking *very* generally, we wish to know which of two sorts of explanation is the true one, a determinist explanation or another, and what follows from that. It is clear that whichever of the two sorts of explanation is true, it will have to do only with mental events as defined, and not with personal epistemic facts or any of the mentioned sorts of episode.

A man may be said to have shot his mistress either because (i) he *believed* her to be unfaithful to him, or because (ii) he *knew* her to be unfaithful to him. The explanation given in each case is precisely the same. If I produce (i) as an explanation today, and tonight become convinced that it is a fact that she was unfaithful and he had good reason to believe it, and so produce (ii) tomorrow, I will none the less then be giving no *further* explanation. The personal epistemic fact about him gives me no further explanation of his action. I *may* have a further answer as to why he believed what he did, but no further answer as to why he did what he did. (Kim, 1982; Stich, 1978) So with what we labelled as learned mental episodes—beliefs and so on which by definition were acquired by participation in certain forms of life. To cite something beyond the contained mental events is not to give a further explanation of ensuing mental events or actions. The same is to be said of the other defined episodes.

2.7 DEFINITIONAL PRECEDENTS, HOLISM, INDETERMINACY OF TRANSLATION, DOMAINS

2.7.1 Definitional Precedents

Several bodies of work which are technical, partly as a result of their connection with linguistics, but also schematic and in a rather free way speculative, contain or suggest four objections against the Correlation Hypothesis. (Quine, 1960, 1970; Davidson, 1980).

The first begins from the falsehood of behaviourism. When we speak of mental events we do not *mean* only something about bodily movements. There is no such synonymy. Behaviourism, sometimes rightly called 'definitional behaviourism', may be thought to be one of a number of futile definitional enterprises in philosophy. It is not possible either, it may be said, to define 'right action' and other terms of the language of morality in a naturalistic way—that is, by way of such non-moral terms as 'distress-minimizing action'. Again, it is not possible, as is widely agreed, to carry out the phenomenalist programme of translating our physical-object language into a language of sensations or sense-data.

This set of definitional failures, it is said, sharpens our understanding of the issue of psychoneural nomic connection. The set of failures are precedents. They give us hints, *some* reason to expect that in fact there is no psychoneural nomic connection. (Davidson, 1980e, pp. 215–18) The objection, applied to the Correlation Hypothesis, is this: (i) the various definitional enterprises are hopeless; (ii) the Correlation Hypothesis is such a definitional enterprise; (iii) the hypothesis can be expected to be as hopeless, or at any rate must be regarded with scepticism.

It is of course the second premiss that calls out for support by the objector. Many remarks made about nomic connection generally are germane to it. First, a distinction is made between statements of nomic connection, which is to say laws and lawlike statements, and statements of causal connection. The former, it is suggested, are in a particular way language-dependent. Certainly this is not the distinction we have, which makes causal statements a species of nomic or lawlike statements, but a distinction having to do with facts noticed earlier in connection with Anomalous Monism. (2.3.2) It is said, secondly, that the lawlikeness or nomologicality of a statement is much like the analyticity of a statement, since both are linked to meaning. Lawlike statements, thirdly, are such that the predicates in them, for the things stated to be in nomic connection, were 'made for each other', as we can know a priori. Fourthly, and much the same, judging a statement to be a law or to be lawlike is in some sense an a priori matter. More plainly, nomic connection is not established by empirical inquiry but by a priori means. Finally, it is to be expected that a given statement will be lawlike within one theory and non-lawlike in another.

The audacious remarks are not intended as, and, more important, are not near to being, *arguments* for the startling proposition that nomic connections, say between a hammer blow and the shattering of a goblet, or between the temperature and pressure of a gas, are connections which depend on language, on the synonymy of terms. It is a proposition not much heard of since Hume dispatched it. We have in the remarks no reason to think that the fact of two things being in nomic connection is a fact in the same category with the fact that all bachelors are unmarried, or any sufficiently like category. The remarks may owe something to the known fact that laws of nature may give rise to related analytic truths. (Ayer, 1963a, p. 215 f.; E. Nagel, 1979, pp. 52 f.) and also to a general theory about language and conceptual schemes. (Quine, 1953, 1960; cf. Grice and Strawson, 1956) We shall not pursue these matters. In fact it is difficult to believe that the startling proposition, which cannot seem other than a plain falsehood, is what is really intended. What is really intended remains uncertain.

We earlier arrived at an account of nomic connection—an account having affinities with other widely accepted accounts—which is fundamentally unlike what might be christened the linguistic or conceptual speculation. Further, our own account and those like it are within the broad tradition of philosophical and other belief to the effect that causal and like connections are not, despite certain caveats, a matter of definitional connection, synonymy, analyticity, language-dependence, the a priori, or like things. It is difficult to entertain the idea that the broad tradition, together with the common assumption of science and ordinary belief and practice, is seriously misguided. The linguistic or conceptual speculation is the more distinctive the closer it approaches to the proposition that there would be no nomic connection in a world without language. As it becomes more distinctive, then, it comes closer to what no one can believe.

2.7.2 Holism

We have it that to ascribe a belief to someone, as in the case of the African lad, requires that one also ascribe many dispositional beliefs to him. There is the same requirement with a great many mental events. It is true of thinking, memory, emotion, desire, choice, decision and intention, or at any rate standard cases of these. It is true, it seems, of all mental events whose contents are representative of other things, or those where the representation brings in language. (p. 81) It is not true, then, or at any rate not at all widely true, of sensation and perhaps perception. This general if not entirely clear fact, although use was made of it in dealing with the African-lad objection, may well be taken as itself more enemy than friend to the Correlation Hypothesis.

Consider a man's feeling at some moment that more than a half-bottle of wine at dinner is excessive. It follows, we can say, that he cannot be ignorant about certain things—he has dispositional beliefs about grapes and about whole bottles of wine. We can add that he has attitudes other than the one explicitly given in his thought. We could not ascribe the particular attitude to him if we believed, for example, that he was in the grips of some general Dionysian conviction that all standards and self-restraints are nonsense, or that he would in no conceivable circumstances feel self-critical or embarrassed. That is a small beginning. Moreover, each of the dispositional beliefs and attitudes we mention is itself like the initial mental event. It also somehow entails that he has other beliefs and attitudes, and so on. In short, to ascribe the mentioned feeling to the diner is to be committed to attributing to him a large and indefinite corpus of beliefs and attitudes. This is so because there are logical or conceptual connections or relations between concepts, and nomic relations between most things conceived. It is sometimes claimed, more strongly, that a concept has its very existence in these connections or relations. This is unclear. If what is meant is the idea that a concept cannot in some good sense stand on its own, be itself individuated, it is certainly mistaken. If the latter were true, logic as we conceive it would presumably be impossible, since it would be impossible to specify relations between single thoughts. The diner's feeling, it is as true to say, was not other or more than the feeling that more than a half-bottle of wine at dinner is excessive. That is consistent with the fact that concepts do stand in relation to others, and hence that it would be mistaken or even senseless to speak of someone having but a single belief. To have one belief a person must in some way have more.

All of which is one thing that can be meant by the dictum that mentality is *holistic*, that it is to be thought of in terms of wholes rather than either items or collections of independent parts. The same facts can be put into other lights. They can be regarded, as has been popular, in several epistemological ways, as fundamental to the answering of certain questions or the development of certain theories. A first question posed is that of how we understand another person, how we come to judge or know that he or she is thinking, meaning, feeling, or intending something. Or, what comes to much the same thing, there may be considered the philosophical problem of devising or constructing a theory of how we do go about ascribing mental events to others. This will in part be a theory of the translation or interpretation of utterances. Secondly, there may be posed the different question of how we explain the non-verbal actions or behaviour of another. Or, there may be considered the problem of constructing a theory of how we do it.

In each case it is rightly said, for example, that we cannot hope to proceed by taking an utterance or action singly and reading off its mental significance: that it expresses or derives from a single desire or whatever. Rather, we make sense of a particular utterance or action by making further judgements or assumptions about the person, to the effect that he has more beliefs, desires, and so on than the one in question. We assign a complex system of beliefs and the like. As we must, we consider a whole rather than any single part in isolation. Our theory-construction cannot be piecemeal.

Mentality is *holistic*, then, which is essentially to say that if we are to ascribe a mental event of common sorts to a person, we must also ascribe a good deal else to him—a whole, a related body of things, of indefinite size and shape. More follows from this, it may now be supposed, than that a neurophysiologist could not conceivably manufacture an 'isolated' mental event in someone, without doing a lot else first. There is a more fundamental consequence. The fact of mental holism undercuts our second speculation, having to do with the supposedly successful neurophysiological procedure, and it undercuts more than that. This is so, it may be supposed, because the objection from holism *ruins the very ideas of a mental event and of the neural correlate of a mental event*, whether the latter is artificially produced in whatever way, or entirely natural. The fact that consciousness is holistic thus ruins the Correlation Hypothesis. The hypothesis asserts specific connections between determinate items, but there are no determinate mental items to be had.

Furthermore, it may be said, any neural correlate of the mental event when the latter is taken seriously, which is to say the event and what necessarily goes with it, must be a correlate for that whole indefinite body or network. The neural correlate in question must be something obscure, a large and indeterminate neural something-or-other. There would be no point in merely labelling it, for example, as *the general type of neural correlate related to thoughts about wine and excess*, or anything of the sort. We would not thereby get a grip on anything.

That is the objection from holism, or rather, one explicit objection that can claim the name. It cannot in fact be assigned to any philosopher associated with the doctrine of mental holism, but is derived from their reflections. (Quine, 1960; Davidson, 1980e, f, g; cf. Peacocke, 1979b) It or something like it has been an unofficial helping hand with nearby objections to the Correlation Hypothesis to which we are coming.

Should the objection from holism in fact persuade us to abandon hope of such views as the Correlation Hypothesis? It seems, on reflection, that anyone persuaded will have been persuaded by a kind of illusion, which owes something to the distraction of large theories of language and meaning. The illusion is dependent at bottom on the failure to respect an essential distinction, one we already have on hand, but which it is possible to forget. Let us reflect on the crux of the objection from holism. *Why* is it that if the diner cannot be assigned his particular feeling, m, without also being assigned a collection of other items, then we cannot suppose that m is determinate, and further, that it had a determinate neural correlate n? To consider the question, I think, is to be at least uncertain of the answer.

To consider the question, what certainly needs to be remembered is that the further items are not themselves mental events but *dispositions*. As we have already assumed, such dispositions are neural structures. (pp. 86, 128) They are neural structures such that persons who have them are disposed to certain mental events. In certain circumstances, they will have certain thoughts, feelings, or whatever. These dispositions, incidentally, are therefore exactly like dispositions generally. That is, if we put aside a certain amount of mystery about 'powers' and the like, as we shall, they are sets of standing or persisting conditions. Any disposition is a set of standing conditions such that when it is conjoined with a further condition or conditions, the resulting whole is in fact a causal circumstance for something. To say that a cube of sugar will dissolve in water, that it has that disposition, is to say that the sugar has certain properties such that when they are conjoined with certain properties of water, the resulting causal circumstance has the effect of the sugar's dissolving.

Thus, if we begin with the diner's mental event, and then ascribe to him a great deal else, we do not thereby ascribe a corpus of mental events to him. It is not as if his mind were flooded with ideas, or that his feeling occurs in a setting or surround of other mental events. It must be such an image that takes the place of an actual argument in the reflections we are considering. Certainly nothing else, and in particular no argument, suggests itself. As for the neural correlate of the mental event, there is no cause to try to think of anything indeterminate, a correlate for some indefinite corpus of conscious items. The additional items we ascribe to the diner are in fact neural things which are in some way necessary to the mental event. The question of their exact relation is a matter of the second hypothesis of the theory being set out, the subject of the next chapter. What is to be proposed now is that the objection from holism collapses or disappears as soon as dispositions are rightly located—at the neural end of the Correlation Hypothesis. They are no complication of the mental end. That end has to do with a single mental event, which mental event is perfectly determinate and can have a perfectly determinate neural correlate.

2.7.3 Indeterminacy of Translation

The third possible objection, or sort of objection, derives from something left out of the sketch of mental holism just given. This is best known as it is presented in the consideration of the problem of translation of a wholly unknown language—hence called the problem of *radical* translation. There is the alleged fact of *the indeterminacy of translation*. Consider two linguists, native speakers of language *l*, who set out to translate language *m*. They succeed, and each comes to be able to speak *m* as it is spoken by native speakers. However, they do so by devising *different* translation manuals, where such a manual includes a dictionary and a good deal else. As a result, consistently with the fact that each linguist has mastered *m*, there is the upshot that their translations of certain theoretical rather than observational sentences of *m*—the sentences they give as translations in *l*—diverge and indeed conflict. They may differ in truth value or be logically incompatible.

It is maintained, secondly, that no sense can be attached to the idea of one of these translations being correct and the other incorrect. It is said, famously to some and notoriously to others, that there is no 'fact of the matter' as to whether one sentence or the other of *l* is *the* translation of the given sentence in *m*. There is not an objective matter to be right or wrong about. (Quine, 1960, Ch. 2, 1970) The description 'the indeterminacy of translation' is sometimes used for the alleged fact of the two translation manuals, sometimes for the alleged fact that neither can be said to be correct.

Despite what has been said so far, the aim of these reflections is not an understanding of the nature of translation but of something larger, language and meaning. Essentially the same two views are taken of the *interpretation* by one speaker of a given language of certain sentences uttered by another speaker of that language. Extremely little flesh is put on the bones of the core doctrine of indeterminacy when it is expounded in terms of translation, and yet less in connection with interpretation. The same is true when the core doctrine is stated in the nearby terms of *meaning*. We are to understand, however, that the situation is as with translation. With respect to certain sentences of our language, as it seems, you and I may understand them differently, and there is no fact of the matter as to which of us is right. There does not exist *the true understanding* or *the meaning*.

As in the case of the doctrine of holism, the doctrine of the indeterminacy of translation, interpretation, and meaning is not made to issue directly in objections to the Correlation Hypothesis and like views. Again, however, it is likely that the doctrine in itself has predisposed some philosophers against such views.

(i) One of two possible objections is that indeterminacy puts insuperable difficulties in the way of confirming the Correlation Hypothesis, since there must be indeterminacy in the ascription of beliefs and the like. There are several brief replies. (a) The doctrine, as it must be in order to get consideration, is limited to the translations or interpretations of certain sentences, the ascribing of certain mental events. It is limited to theoretical as against observational sentences and the like. Thus, if we were to accept all of the doctrine, it remains true that we would be in no doubt about many, indeed countless, ascriptions of belief and the like. To take an entirely rudimentary example, we can have a fully warranted belief, about a man reporting his experience at two moments as he sits facing a red screen, that his perceptual experience was indeed identical at the two moments. It is also possible to have an odd view of what is needed to confirm the Correlation Hypothesis. It is a matter of inference from several kinds of evidence, of which more in due course. In short, although the matter is a large one, the doctrine of indeterminacy is not inconsistent with the confirmation of the Correlation Hypothesis. (b) If we need a certain grip on a man's experience in order to answer a question about certain mental and neural events, and we have not got that grip, then we are unable to answer the question. To be in trouble is to be unable to see which of *two* answers is correct, one for and one against nomic correlation. It is not to be able to see that one answer is mistaken. If exact and final judgements about some mental events are impossible, then this contributes precisely as much to uncertainty about the non-existence of psychoneural nomic correlation.

(ii) The other possible objection, in brief, is that a part of a man's experience simply is not determinate, but is itself in some way indeterminate, and therefore cannot be conceived in terms of the Correlation Hypothesis. The notorious conclusion of objective indeterminacy, to say the least, has not been shown to follow from its premiss. That is, to speak again in terms of translation, it has not been anything like shown that the presumed fact of two translations entails the impossibility of a correct translation. The conclusion, of course, is in analogy to what is perhaps better known: for example, the idea that something in certain ways not open to measurement lacks certain dimensions. The notorious conclusion, if taken seriously and as having significantly wide application to mental states, would undercut virtually all reflection and theory about the mind. It would dissolve a part of the subject-matter, to the benefit of no view in particular, including the Identity Theories to which the proponents of the view are inclined. (Quine, 1960, p. 264; Davidson, 1980e)

2.7.4 Domains

We come finally to the objection from domains, the different domains of the mental and the neural. It is specifically advanced against psychoneural nomic connection. (Davidson, 1980e) It has the doctrine of indeterminacy in its background, but more so the doctrine of holism. That, we have supposed, is that it cannot be that any of many mental events can be ascribed to a person without also ascribing to him, most importantly, a body, system, or network of dispositional beliefs. To say this is in fact to assume something touched upon above, that the given dispositional beliefs are in accordance with at least a certain rationality. It consists in facts of consistency, coherence, truth, evidence, cogency, plausibility, and so on. The diner, as we know, has the disposition to have the thought, in certain situations, of a half-bottle of wine. He must then also have a disposition pertaining to a whole bottle. Moreover, to come to a rudimentary instance of the main idea which we are now considering, this pair of dispositions is such that he will not suppose that two wholes make a half, or that he can take away half of something and have the whole left, or anything else along such inconsistent lines. He rightly takes certain things to be inconsistent, and hence others to be logically necessary.

It is essential, if we are to assign beliefs and the like to him, that we credit him with at least a certain rationality. If a man's utterances showed no consistency from moment to moment, to speak only of that, and we took his utterances to reflect his actual ongoing consciousness, it would follow that he had nothing worth the name of a belief. Certainly we could not assign an occurrent or a dispositional belief to man who said, without hint of a secondary meaning, that he was in favour of drinking only a half-bottle, since that left him with the whole one to start on next time.

Consciousness and the dispositions to it, then, are a matter of at least some rationality. The domain of the mental is a matter of standards or criteria of rationality. In order to ascribe many or most mental events to others, we are under the constraint of taking them to be rational. Consider now our descriptions and theories of the non-mental, including the neural. This subject-matter will include, incidentally, mental dispositions taken as neural, or mental dispositions in their neural properties. We do not ascribe or impute rationality to what is in this domain. There are instead very different ruling ideas or principles in our description of the non-mental. To describe it we depend, for example, in the measurement of length, temperature, time, and a great deal else, on the postulate of transitivity. (Davidson, 1980e) If x is greater than y, and y greater than z, then x is greater than z. Thus we have two domains or systems, the mental and the non-mental, depending upon disparate assumptions or principles. They do not make up a single closed domain which is necessary for nomic connection.

That is all of the premiss. In brief, beliefs and the like are a matter of rationality, and the non-mental is not. Criteria of rationality do not have counterparts in the non-mental, the description of which is governed by quite different things. Therefore—to come to the conclusion—there is reason to think that there are no nomic connections between the mental and the neural. The argument depends in the end on a quick transition, and is open to a direct reply.

Certainly there is no gainsaying that there are differences between the mental and the non-mental, and the account given of the mental does indeed rest on a truth. It is apparent, too, that the requirement of ascribing rationality to someone in order to understand him would place a certain wholly acceptable constraint on certain conceivable attempts to confirm the Correlation Hypothesis in a direct way, by ascribing certain mental events to him. Still, there remains a certain fact: we have not actually been given a reason for thinking that the given conclusion follows from the disparateness of the two domains. (Cf. Mackie, 1981, p. 350; Lycan, 1982; Kim, 1979)

Certainly there is no clear general truth to the effect that there cannot be nomic connection between items in different domains, items whose investigation or description depends on different assumptions. There appears to be no relevant sense in which the two domains of the mental and the non-mental can be said to be other than a single 'closed' domain without begging the question at issue. What we need is a true premiss, something independent of the conclusion that there cannot be the given nomic connection. No reason is actually supplied for precluding nomic connection between the two domains.

Two further comments need making. The argument from domains, as we know, is restricted to mental events that have what can be called representative or perhaps propositional content. It does then pertain to a great deal of consciousness, but not to all. It is not advanced as relevant to sensations, for example. Thus it is *not* denied that there is nomic correlation between the neural and some of consciousness. This discontinuity is unreassuring, to say the least. If anything can count as a single domain, it is *consciousness or experience as a whole*. It alone, all of it, has the character of interdependent subject and content. It is most curious to suppose that it falls into two truly fundamentally different parts, and, it might be added, two parts distinguished in the given way. If nomic connection is allowed in one part, that must necessarily cast doubt on any argument against its possibility in the other part.

The other comment, which can be made in connection with more objections than the present one, can be brief. We are presented with what are properly called, somewhat sceptically, *theoretical considerations*. They are to persuade us of the falsehood of the Correlation Hypothesis. There is evidence relevant to this empirical hypothesis, to the empirical question of whether the connection between the mental and the neural is nomic, which does not consist in theoretical considerations in anything like the same sense. It is the evidence provided by direct research into the subject-matter, which is to say neuroscience. It is no blinkered empiricism to regard such evidence as superior to the given theoretical considerations.

The objection from domains, then, appears to be like its predecessors in failing to prevent further consideration of the Correlation Hypothesis. The hypothesis may not be true, but it does not fall victim to the philosophical and theoretical objections we have considered. It is not the whole story of the psychoneural relation, as we shall see, but we have found no reason to doubt that it is the main part.

3

The Causation of Psychoneural Pairs

3.1 PERSONAL INDISPENSABILITY, CAUSAL SEQUENCES

The question we have been considering was expressed in this way: How is mind related to brain, or, what is the relation between mental and neural events? More particularly, we have been considering the question of *exactly* how a mental event is related to the particular neural event to which it is intimately related (p. 90) and thus with which it is simultaneous. The Identity Theory in its various forms attempts an answer to this question of the psychoneural relation. The Correlation Hypothesis gives a better answer, but one which will be supplemented. It will be supplemented, thereby giving us what will be called the Union Theory, in the course of our considering the question to which we now turn. It is this: What exactly is the explanation of mental events? In this *diachronic* rather than synchronic question our concern is with facts prior to a given mental event, those which provide an answer to the question of why it occurred.

Answers to the first question partly depend for their worth on, and can be tested by, what they include or entail by way of answers to the second question. Hence, in considering Identity Theories of the psychoneural relation, we took into account their entailments with respect to the explanation of mental events. We thus did in effect consider certain answers to the diachronic question of the explanation of mental events. The Correlation Hypothesis, however, was not considered in terms of its implications for the question of the explanation of the mental. The answer to be given to that question in this chapter, as will be anticipated, consorts with the Correlation Hypothesis—indeed supplements it—and raises no problems for it.

One criterion which must be satisfied by any arguable answer to the question of the explanation of the mental, as we have safely assumed, is that the answer must satisfy the conviction or axiom of mental indispensability: earlier mental events are essential parts of the explanation of typical mental events and of actions. (pp. 91 f.) A second criterion, as we also know, is psychoneural intimacy: mental events co-occur with neural events in particular neural locales as a matter of some sort of necessity. (p. 90) A third criterion, since not relevant to the question of the psychoneural relation, has so far not been considered.

Take any decision or choice, above all any decision or choice reached or made in a reflective or deliberative rather than a spontaneous way. If it would be intolerable to leave earlier mental events out of the explanation of a person's choice, it would be as intolerable to leave out something else—the person. Certainly I am convinced that *I* am part of the explanation of my resolution to resist certain temptations, perhaps to put certain things out of mind. This conviction of *personal indispensability*, as we may call it, evidently is in connection with the conviction of mental indispensability, but is distinct. Consider a person's decision, and the earlier mental events taken as indispensable to it, say a certain perception and desire. We can feel no assurance that earlier mental events of roughly the same types, or of the very same types, if they occurred in the life of another person, would have been followed by the given decision. Evidently, and to say the least, the person makes a difference.

What is a person? The question most often addressed in this neighbourhood is that of what makes a person at any earlier time identical with a person at a later time. What are the logically necessary and sufficient conditions of the fact that the person who wrote earlier pages of this book is the same person writing later pages? Presumably the correct answer to this question, although it sometimes seems to be implied otherwise (Parfit, p. 202), will contain an answer to the question of what a person is. To know what is logically necessary and sufficient for something's being the same person as another is to know what is logically necessary and sufficient for being a person at all. I could perhaps be sure something is a person without being able to identify it with any previous person—having the concept of a person may not give one a criterion of personal identity. But having the correct criterion of personal identity, which is to say an account of what is necessary and sufficient to make *x* now the same person as *y* before, is surely to have the concept of a person.

The controversy about personal identity (Parfit, 1984; B. Williams, 1973; Perry, 1975) has perhaps not been finally resolved, and will not be resolved here. What is to be said here will be consistent with what can be taken as the arguable views. These do not include the view that a person is to be identified with an ongoing simple mental substance. The best-known version of this idea is of course that of Descartes, which makes the substance into the non-spatial pure ego. Any such idea has its origin, no doubt, in our perception of what was earlier called a subject, one part of the linked duality which is the nature of any mental event. (2.2) There is no adequate reason to elevate or reify a subject into an ongoing mental substance. The objections to any such idea are several (pp. 80 f.), as are the objections to any such idea's being adequate to our conception of a person. The principal objection of the latter kind is that the idea of a simple substance taken by itself entirely leaves out *continuities of experience*. It leaves out, that is, what is sometimes called psychological continuity and connectedness or mental relations. Such a simple substance, if one existed, might exist as or in an entity of no memories, character traits, or ongoing feelings. If such things are somehow taken as added in, we do of course have something quite different from the official view.

Whatever the correct account is of what makes the writer of the earlier pages identical with the writer of this page, the account must somehow include continuities of experience. It is essential to the fact of identity, as it is to the nature of a person at a time, that he is the individual distinguished by certain memories, dispositions of character, tendencies, persistent inclinations and desires, hopes, commitments, intentions, plans, and the like. It is no doubt true that my identity is not dependent on my settled political convictions, or my uncertain optimism, or my memory-image of a farm in a dell on the road to Kitchener. It is inconceivable that my identity could be independent of all or perhaps even most of my memories, tendencies, hopes and so on.

The arguable views of personal identity respect this fact to one degree or another. The better known views, very broadly speaking, fall into two categories. The first actually centres on continuities of experience. By far the most impressive account of personal identity available (Parfit, 1984), which may in fact be true, is at bottom an account in terms of continuities of experience.

> We are not separately existing entities, apart from our brains and bodies, and various interrelated physical and mental events. Our existence just involves the existence of our brains and bodies, and the doing of our deeds, and the thinking of our thoughts, and the occurrence of certain other physical and mental events. Our identity over time just involves (a) Relation R—psychological connectedness and/or psychological continuity—either with the normal cause or with any cause, provided (b) that there is no different person who is R-related to us as we once were. (Parfit, 1984, p. 216)

The other category of views gives a fundamental place to bodily or brain continuity. I today am identical with the writer of a previous page of this book in virtue of a spatio-temporal line between an earlier living body or brain or brain-part and a present living body or brain or brain-part. It seems true, although perhaps in need of more argument than will be provided here, that this category of views in fact depends for its considerable recommendation on its at least implicit inclusion of the idea of continuities of experience, taken as rooted in the body or brain. If the stuff of my brain persisted despite the total absence of such continuities, I would surely not persist. If it is conceivable to suppose that the organization or structures of my brain might also persist in the total absence of such continuities, surely I would still not persist.

It may be that both of these categories of views, and also views which somehow amalgamate the categories (Nozick, 1981, Ch. 1), require an addition. Perhaps the addition will make for a further category of views. It has to do with an elusive continuity of consciousness in a sense so far unmentioned. By way of minimal and less than helpful description, a person is a continuing *locus* or *tunnel* or *frame* of consciousness or mental events. This seeming fact, which stands in *some* relation to the subject within a mental event, and which is not to be dramatized into a mental substance or ego, would account for what has often enough been considered by philosophers, a particular apprehension one would feel at the prospect of, say, future pain—after and despite an ending of one's continuities of experience. The apprehension has to do with the pain, not the loss of the given continuities. The seeming fact carries the conclusion that there is more to an ongoing person than continuities of experience with one or another cause.

The possible addition to views of personal identity, or third category of view, is not a matter to be pursued here. What is to be concluded, rather, is that answers to our question of the explanation of mental events must indeed satisfy a criterion in addition to the criteria of mental indispensability and of psychoneural intimacy. It is the criterion of the indispensability of the person, or of personal indispensability. It is that a person, conceived as an entity involving at least continuities of experience, is essential to the explanation of at least typical mental events and actions. My thoughts and desires, and my actions, are in general owed to such things as my memories, character, hopes, and intentions. A decision or act of mine is owed to more than the obvious mental events which precede it, typically a certain desire and belief.

So much for one preliminary to consideration of the question of the explanation of mental events. The second, quite different, has to do with causation. The answer to be given to the question of explanation, and other answers to be considered, and also the answers looked at above (2.3)—although we did not pause over the matter—have to do with *causal sequences*. These are related to what are ordinarily called

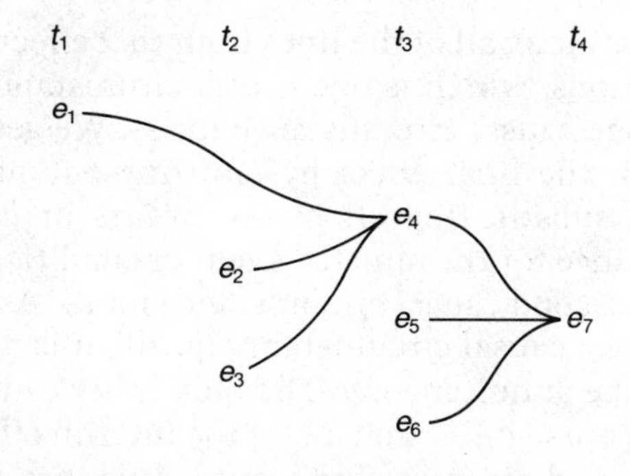

FIG. 9. Causal Sequence

causal chains. The car stopped, we may carelessly suppose, because of a causal chain or chain or events which began with the driver's stepping on the brake pedal. The vague picture we may have is of some single line of mechanical events involving the pedal, mechanical connections, brake drums, and the like, linking the driver's action with the stopping of the car. However, as we shall of course grant, more was needed than such a single line of events. Had the car been coming down Highgate Hill, and it was icy, the car would not have stopped as it did, despite such a line of events.

A causal sequence is a sequence of events or conditions—individual properties or collections of them, strictly speaking—each of which, save those at the beginning or beginnings and at the end, is both an effect and also an element in a causal circumstance. Thus there are no gaps. Figure 9 is of a causal sequence for event e_7, perhaps the stopping of the car. e_1 might be the driver's braking. The items in it are of course particular events or conditions rather than types, and are to be taken as having occurred at the given times. Wherever a line joins an earlier and a later event or condition, say e_1 and e_4, or e_1 and e_7, the earlier was cause of the later. That is, in part, if the earlier had not happened, the later would not have—the earlier was required for the later. The situation in question was then an ordinary one, not involving items which were only alternatively required for others. (p. 19)

To recall our earlier brief description of a causal circumstance (p. 47), we took it that one consists in no more than a set of events or conditions such that the set (1) necessitated the effect, (2) preceded it, and (3) was dependently-necessitated by it. In virtue of these three conditions it can be said that the set made the effect happen and explained it. Figure 9 is to be taken in its converging lines as specifying two different causal circumstances for e_7, and one for the interim effect e_4. The rule for interpreting such diagrams is partly and roughly as follows. We get a causal circumstance for the culminating effect by moving leftwards along all of the lines from that effect to the first set of events or conditions, which is one causal cirumstance for it. Thus e_4, e_5, and e_6 was one causal circumstance for e_7. We get a further causal circumstance for the final effect by omitting one of these events or conditions, and substituting all of the events or conditions in the causal circumstance for the omitted event or condition. Thus e_1, e_2, e_3, e_5, and e_6 was a second causal circumstance for e_7. As remarked in our earlier account of a causal circumstance (p. 20), it is not necessary that all its elements be simultaneous. The only other causal circumstance in Figure 9 is of course e_1, e_2, and e_3, for the interim effect e_4.

Each item in the diagrammed sequence, save the last, is given as an element in one or more causal circumstances. The only two effects are e_4 and e_7. The diagram does not convey, about the other items, either that they are or that they are not effects. The question is left open. All events and conditions save those just mentioned, e_4 and e_7, can be labelled *initial elements*, which is to say that their causal backgrounds, if any, are not within the specified causal sequence. It may be, of course, that they are effects in terms of another causal sequence, perhaps one which includes the one we have and also has e_7 as culmination. It is at least generally a matter of choice, governed by our own interests, what the initial items of a causal sequence are taken to be. There is also choice, evidently, in the matter of the final effect. We pay attention to a course of events up to a certain natural point and not beyond, leaving open the question of whether the final effect also has a causal role.

Whether or not we wish to pay attention to the fact, is every event or condition in the universe an effect? That is not our question here, and it will not be a fundamental question at any point in this inquiry into determinism. We shall not be concerned with universal or LaPlacean determinism, which asserts that every event or condition without exception—as we may say, every individual property—is an effect. Our fundamental concern is not with all events or conditions, but only with those, to speak quickly, which are of direct relevance to our own lives. That is, we are concerned with those which give rise to feeling, intention, decision, choice, action, and the like.

A causal sequence, then, putting aside its initial and final elements, consists in events or conditions each of which is both an effect and a part of a causal circumstance or of several such circumstances. We might go considerably further in elaborating the account of a causal sequence, but there is no need to do so.

Our ultimate concern will have to do with one principal fact which certain causal sequences share with all others. It is that the final element, and all elements save initial ones, are necessitated events or conditions. All elements but the initial ones are necessitated by the initial elements, and typically also by circumstances internal to the sequence. Hence, and above all, the final effect of a causal sequence is necessitated by its initial elements.

A second important concern will have to do with the fact, about causal sequences however defined, that each earlier circumstance in an ordinary sequence is dependently-necessary to all that follows. (p. 29) This is of course entailed—to recall the summary mentioned above—just by specifying the effect of a causal circumstance as dependently necessitating it. Such facts about necessity, of course, are to be understood as they finally were in our inquiry into nomic connections. For a circumstance to necessitate an effect is for roughly this to be true: given the circumstance, the effect would still have occurred as it did, even given most conceivable changes in the universe.

3.2 CAUSAL INTERACTIONALISM, NEURAL CAUSATION WITH PSYCHONEURAL CORRELATION, THE PSYCHONEURAL RELATION AS MACROMICRO

It is a natural idea that both external or environmental events and also bodily events somehow cause mental events, and that mental events somehow cause bodily and external events. The salt dish I see causes retinal events and these cause the mental events in which some of my visual experience consists. My wants cause my actions and thus their effects, from the state of my shoe-laces to the state of my garden. The natural idea pertains to several subject-matters, including a large one to which we have not yet come, which is that of the relation between mental events and action. The natural idea evidently contributes to a number of answers or types of answers given to the question of the explanation of mental events.

The simplest of them can have the name of *Causal Interactionism*. It is, as usually expressed, that neural events somehow cause mental events and mental events somehow cause neural events—which latter causal fact gives us mental events, by way of intervening neural events, causing later mental events. It would be unkind to suppose that it makes *all* neural events causal with respect to mental events, since it is safe to say that many are not. Nor would it be kind to take it as making all neural events into effects of mental events. Many are not. Are all mental events to be regarded as causal with respect to neural events? And all mental events to be regarded as effects of neural events? We might answer yes in both cases, as is contemplated with another doctrine of mind and brain noticed earlier, Anomalous

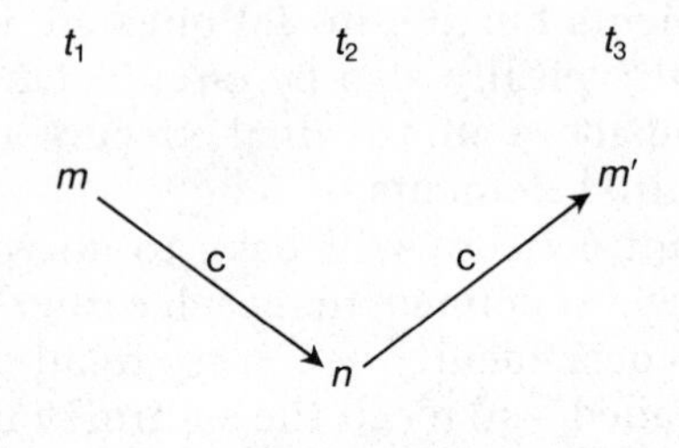

FIG. 10. Causal Interactionism

Monism. (2.3; Davidson, 1980e, p. 208) However, we need not settle the matter. Let us simply consider the view that some mental somehow cause some neural events, and some neural somehow cause some mental events.

It is to be noticed that this allows that the given events, of both sorts, may both be causal—causes or causal circumstances—and also be effects. Causal Interactionism has occasionally seemed to be regarded as the view, to speak only of mental events, that some of them are causal and only that, and some are effects and only that. None are both effects and also causal. Mental events that are causal may be characterized as the ones we regard as *active*. Those that are effects are the one we regard as *passive*. (Cf. pp. 174 f.) Decisions go in the first category, and sensory experience, or part of it, in the second. The general idea is decidedly odd, to say the least, although in part like things to which we shall come—certain indeterminist doctrines of the mind. Causal Interactionism is better taken as allowing mental events to have the ordinary dual role: both effects and also causal. The view as we have it, then, as already anticipated, also allows a further thing, that a given mental event may be the effect of a neural event and, by way of that neural event, also the effect of an earlier mental event.

Are we to suppose that a mental event is a cause as distinct from a causal circumstance for a neural event? And so with our causal neural events? Let us start with that assumption, which gives us a first instance of Causal Interactionism. What we have is the view modelled in Figure 10, where the arrow with a 'c' above it is to be read as 'is a cause of'. The view as it stands is either unacceptable or incomplete. As it stands, taken as complete, it leaves *m* (and *m'*) free-floating. That is, *m* is not given in *any* relation with a simultaneous neural event. As it stands, to speak differently, the view makes *m ghostly*: it has the nature of a ghost, in the popular sense of something mental or spiritual not in the ordinary simultaneous connection with a body. More precisely, the view as it stands, taken as complete, offends against the axiom of psychoneural intimacy. It needs remarking, as well, that in *m* and n (and n and m') we supposedly have a standard case of cause of effect. We thus do also need a causal circumstance for n (and m'. There are no causes without causal circumstances, as we know. (Ch. 1)

We can deal with this last matter by making the other assumption—by taking the view that a mental event may itself be a causal circumstance for a neural event, and a neural event may itself be a causal circumstance for a mental event. This gives us a second instance of Causal Interactionism. (There is no conceptual barrier, of course, to what we call a single event functioning as a causal circumstance.) Or, indeed, we could try something else that would be in the spirit of the present enterprise, and thus get a third instance of Causal Interactionism. We could take it that a mental event may be a part of a wholly mental causal circumstance for a neural event, and a neural event may be part of a wholly neural causal circumstance for a mental event. However, if either of the second or the third variant is taken to be a complete view, it is as unacceptable as its predecessor, for the same reason. It offends against psychoneural intimacy. Mental events are again left free-floating or ghostly.

There is a second objection, not much less serious, to all these three versions of Causal Interactionism. To speak in terms of the first, and Figure 10, it is incredible that the causation of the *neural* event n includes *no* neural fact at t_1. The objection is an empirical one, certainly, as its predecessor also is, but surely irresistible even in the absence of detailed evidence. It cannot be that n comes from *nothing*, neurally speaking, at t_1. It cannot be sufficient that n can be said to have a neural causal antecedent which is cause of m, at some time prior to t_1. While that gives us some neural causation for the neural event n, it leaves us with the neural gap at t_1.

It will be recalled, of course, that we have not excluded mental events from the class of *physical* events. They are within the physical realm as defined essentially in terms of spatiality. (pp. 87 ff.) Thus, n *does* have a *physical* causal antecedent at t_1. This does not much affect the objection. Our characterization of mental events, as it needs to, distinguishes them sharply from other physical events, including neural events. With respect to almost all neural events, or at any rate typical neural events, it is impossible to deny them immediately prior neural causes. Just this is done by the views in question.

Causal Interactionism as we have it also faces a third objection, that it does not allow for personal indispensability, but it may be that what we have could be interpreted so as to allow for the role of memory, character, and the like. Let us not pursue this. We can instead take it that the intended views have been incompletely stated. There is more to them. They are to have added to them a proposition about mental events being in relation with simultaneous neural events. This acceptance of psychoneural intimacy, which will also deal with the neural gaps, will then require something more: elaboration of the causal story. The completion or enlargement of the views, and the causal elaboration, will turn them into quite different things. These will be noticed briefly at the end of this section.

If we set out to secure psychoneural intimacy in dealing with the question of the explanation of mental events, and we are also persuaded of a proposition noticed a moment ago and of which there is more to be said (pp. 164 f.), the neural events are somehow or other owed to neural or other bodily causation involving no mysterious gaps, we are likely to come to a particular answer. It is one of several answers taken as suggested by the practice of neuroscience, and like Identity Theories. Certainly, as we shall see in more detail in Chapter 5, it is in general true that research into the brain and Central Nervous System proceeds on the assumption that neural events have explanations that are *somehow* neural or other otherwise bodily. The particular answer to the explanation question is also suggested by fairly common usages in neuroscience, and by occasional philosophical reflections on the part of those engaged in it—although the reflections are often enough accompanied by others than do not consort well with them. (Rosenblueth, 1970)

The answer, which is also an answer to the question of the psychoneural relation, consists in two things: the Correlation Hypothesis and the plain proposition that all neural correlates are elements in causal circumstances within causal sequences which are *wholly* neural or bodily—none of the elements of the contained causal circumstances is mental. What we have is ongoing non-mental activity, to be explained wholly in terms of non-mental causation, with mental events as nomic correlates of some of this activity. They are correlates in just the sense specified by the Correlation Hypothesis. That is, a neural event necessitates a simultaneous mental event, the mental being necessary to the neural. Also, the mental event dependently necessitates the neural event, the neural thus being dependently necessary to the mental. (pp. 47–8, 107–8)

This view, *Neural Causation with Psychoneural Correlation*, is as modelled in Figure 11. The vertical bar joining mental event m and neural event n (and m' and n') represents them as in nomic correlation of the specified kind. Neural event n, and another neural event n^*, form a causal circumstance for n', the neural correlate of m'. The model thus represents something of the natural assumption that the occurrence of such a neural event as n' is owed to more neural facts at t_1 than just the occurrence of the neural correlate of m. The model can

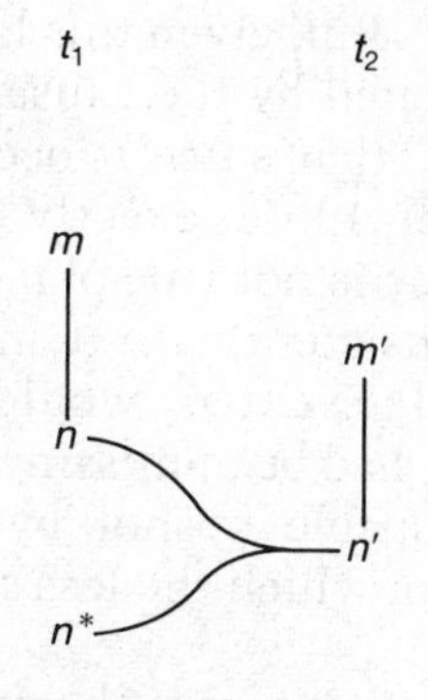

FIG. 11. Neural Causation with Psychoneural Correlation

be taken as modelling someone's noticing olives on the table (*m*), and his intending to have one (*m'*). The view, or something close to it, seemed to me for a considerable time to be the best answer to the question of the explanation of mental events. (Honderich, 1981a, b, 1984d) The fact of the matter is that the view, despite its having neuroscientific and other support, is for a particular reason untenable.

It proposes to make mental events indispensable to the explanation of later mental events and actions by making them nomic correlates. With respect to the indispensability of *m* to *m'*, we have *m* necessary to and simultaneous with *n*, and, by way of the other two nomic connections, we have *n* dependently necessary to *m'*. Is the fact that event *m* is at least dependently necessary to *m'* enough to secure what we want, that *m* is an essential part of the explanation of *m'*? In ordinary cases of causal and other nomic connection, there is not overdetermination. (p. 21) We can take it, safely, that at least most neural and psychoneural connections are ordinary. Thus, with respect to Figure 11, we take it that there existed no causal circumstance for *n'* additional to the circumstance consisting in *n* and *n**. Similarly, we take it that there existed no neural correlate for *m'* additional to the correlate *n'*. In this situation it is true that any full explanation of *m'*, perhaps the mental event of intending to have an olive, includes *n'*. And any full explanation of *n'* includes *n*. And, finally, there is no full explanation of *n* that does not include *m*, perhaps the mental event of noticing the olives.

An attempt can be made to maintain, then, that the answer of Neural Causation with Psychoneural Correlation does respect the axiom of mental indispensability. Alas, some of the utilized facts enter into what makes the view no less than self-contradictory.

Event *m*, as remarked, is necessary to its simultaneous correlate *n*.

So too is m' necessary to n'. But, given this latter proposition, n' in fact simply cannot be necessitated by the causal circumstance n, n^*. It is precisely *not* true of n, n^* that since *it* occurred, n' was guaranteed. More precisely, to recall more exactly the nature of a causal circumstance, the following is not true of n, n^*: that since it occurred, whatever else logically consistent with it and with n' had occurred, n' would still have occurred. Event n' would not have occurred if its simultaneous correlate m' had been missing. An attempt was made by me to escape this inescapable upshot by way of an unacceptable doctrine about causation, of which the less said the better. (Honderich, 1981b, cf. 1986)

It is worth remarking, perhaps, that the self-contradictoriness of the view in question is made clear by the particular account of causation and of nomic connection to which we have come, but that it is not dependent on the distinctive features of that account. Almost all accounts of causation include what are often called causally sufficient conditions, akin to our causal circumstances. And any at all arguable account of nomic connection, applied to the psychoneural relation, is likely to make m' in a substantial way necessary to n'. The result is inconsistency.

The short story as we have it about the Theory of Neural Causation with Psychoneural Correlation is that if we have the mental correlates explanatorily indispensable, we cannot have the non-mental causation of the neural correlates. To add the non-mental causation of the neural correlates is to fall into contradiction. The same short story, put the other way on, is that if we have the non-mental causation of the neural correlates, we cannot have the mental correlates explanatorily indispensable. Of the general problems raised by the mind, perhaps none is more fundamental than the one of which this is a clear formulation. It is a formulation, of course, which derives from and presupposes a proper rejection of Identity Theories as tenable accounts of the psychoneural relation, and puts the Correlation Hypothesis in their place.

Several related theories have the great recommendation of proceeding from such a clear view of the problem. One, which owes much to the course of modern science, above all physics, is the theory of *the Psychoneural Relation as a Macromicro Relation*. (Kim, 1984) It can be put into terms with which we are familiar. Its basic proposition is that neural events are effects of wholly neural or other bodily causal sequences. The neural realm, however, consists in micro-events of which the macro-events are mental. That is, mental events stand to neural events in a macromicro relation—mental events are in a certain sense *reducible* to neural events. Mental events thus stand to neural events as, say, events of temperature stand to molecular events, or molecular events stand to atomic events. As a consequence of this nature of the psychoneural relation—a nature of which more needs to be said—and also the previous proposition about neural causation, a certain conclusion is drawn about mental events causing mental events. There is, in some sense, mental causation. The previous view we considered, *Neural Causation with Psychoneural Correlation*, sought to secure the indispensability of the mental by linking an earlier mental event to a later or to an action indirectly, by way of neural events. The present view, despite what may be a moment of uncertainty on the issue (Kim, 1984, p. 268), proposes a kind of direct connection.

To proceed again through the sequence of argument, we must accept as basic that neural events have only non-mental explanations, that they are effects of prior neural and other bodily events. Nothing, we are told, will pre-empt such an account of why a limb movement takes place, and nor will anything pre-empt such an account of why any neural event takes place. As for the second step, taking the psychoneural relation as a macromicro relation, this is of course not to *identify* mental or macro-events with the related neural or micro-events. The macromicro relation is one such that the micro-events are constitutive, simpler, finer, or at a lower level. To say that, however, is not actually to specify the relation.

The relation is so understood as to reflect the general belief that a type of macro-event may go with more than one type of micro-event, perhaps many. To use a common term, a type of macro-event may be *multiply realized*—realized in or by more than one type of micro-event. For example, fractures or breaks in materials go with many types of events in underlying molecular structures. To come to the nub, any family *b* of events is macrocausal with respect to a family *a* of events under the following two conditions: (i) if there occurs an event of one type within family *b*, there necessarily occur events of one of some set of types within family *a*, and (ii) if there occur events of any one of the set of types in *a*, there necessarily occurs an event of the single type in *b*. The macromicro relation, so specified, is also referred to as a relation of supervenience or dependency. Events of family *b* supervene on or are dependent on events of family *a*.

So—to speak more particularly—mental events are macrocausal with respect to neural events, which is to say: (i) if there occurs an instance of one type of mental event, there necessarily occur neural events of some or other type, and (ii) if there occur such neural events, there necessarily occurs such a mental event. Suppose now, as in Figure 12, that neural events n together with neural event n^* cause

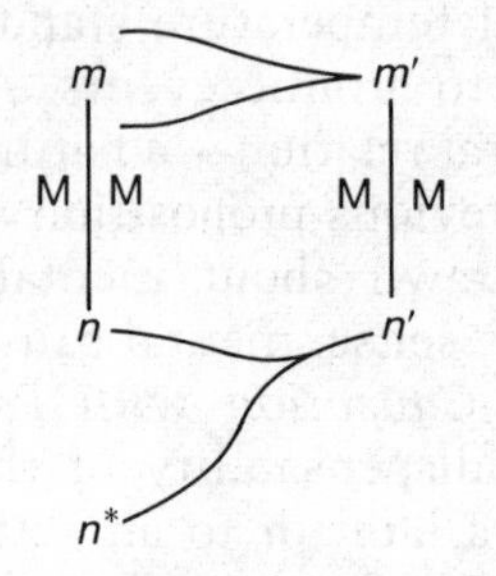

FIG. 12. The Psychoneural Relation as Macromicro

neural events n'. Events n are micro-events of which m is the macro-event, as represented by the vertical line with an 'M' on each side. So with n' and m'. In virtue of the microcosmic causal connection between n and n' we may conclude something about causal connection between the macrocosmic events m and m'. It is not certain exactly what conclusion is proposed to us.

It is said that the causal connection between m and m' is real, that m is not inert. Something along the same lines but more definite—very definite indeed—is also said. It is said that m is as much a cause of m', in virtue of the relations of these to n and n', as wholly familiar events are causes of their effects. Event m is as much cause of m' as fire indubitably is of smoke, which things are also macro-events whose micro-events are indubitably in causal connection. *However*, it is also said in various ways that the microcausal relations are fundamental. Microcausal relations underlie, ground, and explain the macrocausal ones. Macrocausal relations are in a certain sense epiphenomenal and *less* real. Ultimately the world is as it is because the microworld is as it is. Above all, to speak in terms of Figure 12, or rather, *despite* what appears to be conveyed by the figure, it is said there is only the neural causal path. There are not two causal paths to m'.

The first thing to be remarked about this view is that it is difficult to take it as determinate. This is a matter of what has just been mentioned, its conclusion. There are two possibilities. *If* the view is that there is in fact only *one* causal path or connection, the one holding between n and n', then the view is fundamentally identical to the one we have just considered and rejected, Neural Causation with Psychoneural Correlation. They evidently share the proposition that neural events are the effects of wholly neural or at any rate non-mental causal sequences. It is not so immediately clear, but clear enough, that they they share the proposition of psychoneural nomic connection. The macromicro account of the psychoneural relation does certainly imply that mental and neural events are of certain different kinds, such that the neural are somehow constitutive of the mental, which is not to imply that they are identical. What *can* it mean, to say the more important thing, in definition of the macromicro relation, that if an *a*-family event occurs, a *b*-family event *necessarily* occurs, and if an *b*-family event occurs, some *a*-family event *necessarily* occurs? The matter is not attended to, but there is no room for choice. The necessity is evidently not logical, which rightly is not suggested, and therefore must be nomic. The macromicro relation is fundamentally nomic connection, and, as will have been noticed, its one–many character brings it wholly into line with the Correlation Hypothesis.

It is worth remarking in passing that the theory of the Psychoneural Relation as a Macromicro Relation, while it has the great distinction of starting in the right place, does exemplify a certain weakness of many theories of mind and brain. In the aim of making use of scientific advance and conceptions of it, various mind–brain theories pay insufficient attention to what must be their own more fundamental constituents, such as an idea of necessary connection. Giving these constituents attention is essential. One profit is seeing the weaknesses and strengths of elaborated theories more clearly as a result of seeing their likeness to starker theories and hypotheses.

To repeat, *if* the theory we are considering allows but one causal connection—between n and n'—it is at bottom identical with what we have discarded. However, as noted, the theory does speak of a causal connection between m and m' which is as indubitable and substantial as perfectly ordinary causal connections not involving the mind. Let us therefore make the other assumption, and remove any question mark about the causal connection between m and m'. Doing so does give us a distinctive view. It is also a view which has great problems.

(i) The least of these is that little reason is given to assert the mental causal connection. We must not suppose, of course, that it is guaranteed by the axiom of mental indispensability. That axiom is not the proposition that mental events are the effects of wholly mental causal circumstances, or of wholly mental causal sequences. It is rather the proposition that prior mental events enter essentially into any full explanations of later mental events and actions. The two are not identical, as will already be apparent. What then *is* the basis for the proposition that m is causal circumstance for m'? (To take m as merely *cause* of m' would of course require us to provide a containing causal circumstance. To make it other than wholly mental would lead us off in the direction of a quite different theory.)

Obviously the intended basis for the proposition that m necessitates m' is the microcausal relation between n and n'. However, it is not clear that it *is* a basis. We are invited to see the psychoneural relation, and the relation between an earlier and a later mental event, as respectively like the relation between, say, temperature and molecular movement, and the relation between temperature and melting. But there are evident discontinuities. I do not conclude that the heat is melting the candle *because* of a proposition of science about a causal relation between micro-events. The latter proposition is not premiss for the former. Further, any argument from microfacts to macrofacts with respect, say, to heat and molecular motion, is unlikely to carry over to the neural and the mental. Or, to speak more carefully, the relevant *likeness* of the two domains needs first to be established. Clearly it has not been. To be brief, mental events are not much like temperature or other macrofacts of scientific theory.

(ii) The view as we now have it, although quite distinct from Neural Causation with Psychoneural Correlation, has precisely the great disability of that view. It is contradictory in virtue of claiming that a neural event has a wholly neural causal explanation, and also claiming that something simultaneous and non-neural is necessary to that event. In Figure 12, we have a neural causal circumstance for n'. Also, the macromicro relation involving n' is such that n' necessitates m', and hence that m' is necessary to n'. If we hold on to the macromicro relations, we must give up the idea that there is a neural causal circumstance for n'. If we hold on to the causal proposition, we must give up the macromicro relations. By giving up the m–n macromicro relation we lose mental indispensability. (It may be supposed, carelessly, that the view preserves mental indispensability by way of the direct causal connection between m and m'. But the axiom of mental indispensability, as just remarked, is that earlier mental events are indispensable parts of *any* explanations of typical mental events and of actions. And we are supposing that there is a full explanation of a later mental event, m', which leaves out any earlier mental event. We have it that m' is necessitated by n', and n' is necessitated by a causal circumstance which does not include m.)

(iii) Lastly, the view is involved in overdetermination, as unpersuasive as overdetermination generally or very often is. We have a full explanation of m' by way of n' and the causal circumstance for it. No doubt there are difficulties in the way of formulating satisfactorily a general principle to exclude suppositions of overdetermination, as there are difficulties in satisfactorily formulating the wider principle of simplicity or parsimony. 'Plurality is not to be assumed without necessity', 'What can be done with fewer assumptions is done in vain with more'—neither of these has sufficient edge. (Moody, 1967) Still, there can be no doubt of the strength of simplicity. It is rightly taken, not merely as an end in itself, but as a means to truth. In part, this is owed to the pragmatic consideration that complexity has the possibility of concealing error. More importantly, it is owed to well-founded inductive beliefs about reality generally.

The theory of the Psychoneural Relation as Macromicro is therefore not a satisfactory answer to the question of the explanation of mental events. Nor are any other views of the same basic structure. To mention but one of these, it centrally involves the ideas that mental events stand in a basic kind of nomic correlation with neural events, and hence that mental events somehow share any causal roles of the correlated neural events. (Mackie, 1979) Hence an earlier mental event *m*, the correlate of neural event *n*, is said like *n* to be causal with respect to *n*′, the neural correlate of *m*′. Mental indispensability, it is supposed, is thus achieved.

This view, *Psychoneural Causal Transfer*, although its distinction between basic and less basic nomic correlation is consistent with our own view of nomic correlation, is open to objection with respect to the idea that one of two nomic correlates, of any kind, somehow transfers its causal power to the other. (Honderich, 1981b, p. 425) It is also open to the objection of which we know. There is contradiction in asserting, with whatever elaboration having to do with basic as against less basic sorts of nomic connection, that a given neural event has a mental correlate which is necessary to it, and that the neural event is the effect of a neural causal sequence.

Finally, and very briefly, the objection of self-contradiction is also to be made to certain views not of the same structure as those we have latterly been considering. Earlier in this section, in connection with *Causal Interactionism*, it was mentioned that the doctrine might be completed or supplemented in certain ways. Three possible completions are modelled in Figure 13. What we have in *Supplemented Causal Interactionism*, despite the addition of the correlates missing from Causal Interactionism as previously defined (Figure 10), is recognizably of the same sort. All of the modelled views, however, whatever else is to be said against them, are involved in the given self-contradiction. All of them, further, are open to the objection that although they avoid neural gaps and ghostly mental events (p. 152), they leave some neural events without a proper neural cause. This is true of the event *n*″ in each of the three versions of Supplemented Causal Interactionism.

The third version (13c) introduces what we have not had until now, a mixed causal circumstance—made up of a mental and a neural element. There are other conceivable views involving such circumstances but which do not involve the causal alternation of interactionism—sometimes a mental and sometimes a neural event being the

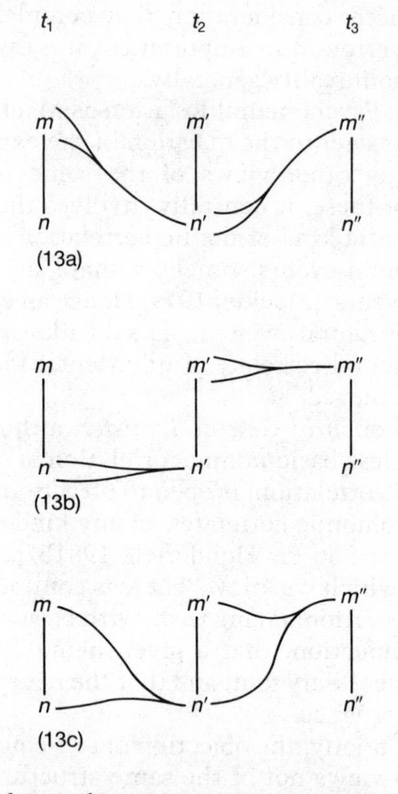

FIG. 13. Supplemented Causal Interactionism

effect. Two are modelled in Figure 14. Both suffer from the same self-contradiction. In 14a, event n' has a necessary condition m' outside the supposed causal sequence for it. In 14b there is not the same but a related truth about m'. The difference has to do with the fact that m' is necessary to n', but n' is dependently necessary to m'.

To return for a moment to interactionism, it is to be added that all of

FIG. 14. Mixed Causal Circumstances

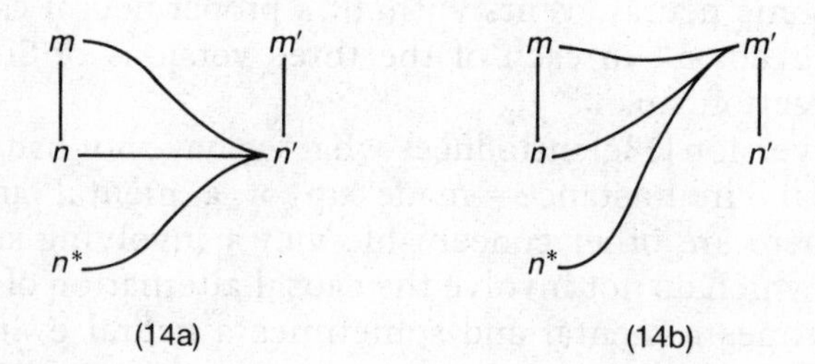

the versions of Supplemented Causal Interactionism, like all of the versions of plain Causal Interactionism (p. 151) are very unreassuring for a particular reason just touched upon. It is that in one way or another they involve a truly fundamental switching back and forth between brain and mind: sometimes it is neural events that are causally efficacious and mental events that are effects, and sometimes it is mental events that are causally efficacious and neural events that are effects. Proponents of interactionism, as it seems, are untroubled by this. One favours a view which in part is approximate to 13b, and writes of it as follows.

It would ... perspicuously represent the difference between spontaneous or stream-of-consciousness thinking and deliberate attentive thinking. It is plausible to suggest that when, for example, we walk in the country and allow our thoughts free rein, it is the neurology beneath them that, according to its own laws, directs their course. On the other hand when we force our thought on to some track, to perform a deduction, or to solve a problem, the stream of mental events is directed by some mental (logical) laws, and the mental descriptive level is hegemonic: here the mental descriptions drag the neural descriptions about according to the laws of sequence which belong to the mental.... (Thorp, 1980, pp. 91–2)

Whatever the support given by slight reflections about walks in the country and solving problems, and passivity and activity generally, of which more will be said (pp. 174 f.), few things seem to me less plausible than the idea that mind and brain are involved in this sort of fundamental alternation or toing and froing, with the brain running the mind at one moment, the mind running the brain at the next. (Cf. P. F. Strawson, 1985, p. 67.) There is room for disagreement here, but it may be difficult, as it is for most philosophers of mind, to regard the idea as other than startling, an offence against the principle of simplicity or parsimony. It is not peculiar to the various interactionist pictures at which we have looked, all of which are of a determinist character. It is also a part of indeterminist interactionism. (3.4)

3.3 THE HYPOTHESIS ON THE CAUSATION OF PSYCHONEURAL PAIRS

A satisfactory answer to the question of the explanation of mental events can only be arrived at by constructing it under the guidance of constraints with which we are now familiar. Indeed it is not far off the mark to say that a satisfactory answer—as satisfactory an answer as we can have—will not be greatly more than a sum of these. Our situation is like that of someone inquiring into the nature of a cell, organism, or machine to which he has no direct access, or only imperfect access.

Still, he has a good deal of constraining knowledge of it. Something of this sort was once true of research into the gene, as is often noted. Our inquiry so far has made clear or suggested a number of constraints, which is not to say that we can have a completed list of such things. Some are the constraints on rational inquiry generally, of which it would be unwise to attempt an enumeration, despite the fact that we can make an easy beginning, very relevant to our recent reflections, with non-contradiction—the avoidance of one contradiction in particular.

To bring together what we have, however, any answer to the question of the explanation of mental events must of course be in accord with mental realism. To conceive of the mental in some merely relational, formal, or abstract way is not to be in touch with the subject-matter. Secondly, any satisfactory answer must satisfy the principle of psychoneural intimacy, which will get further detailed support in our coming consideration of neuroscience. Thirdly, it must satisfy the conviction of mental indispensability, and fourthly, that of personal indispensability. Fifthly, a satisfactory answer to the question of the explanation of mental events should not involve any general proposition of overdetermination, or in other ways offend against simplicity or parsimony.

A sixth constraint has so far been no more than implied in passing. The Dualistic Identity Theory with Neural Causation (2.3) includes the proposition that neural events have *wholly* neural or bodily causes—which proposition was left undiscussed. Causal Interactionism was subsequently judged to be incredible in its suggestion that a neural event at a given time might be owed to *no* neural or other bodily event a moment before. (3.2) The proposition that typical neural events are owed to *wholly* neural causes was also part of the theory of Neural Causation with Psychoneural Correlation, and perhaps the theory of the Psychoneural Relation as Macromicro. (3.2) We thus tacitly accepted with respect to these views, or anyway did not dispute, that neural events are *somehow or other* owed to neural or other bodily causation—notably bodily causation in the case of neural events at the very periphery of the nervous system, close to the environment. Particularly given the neuroscientific reasons to which we shall come, it is impossible to deny the proposition that typical neural events are somehow owed to neural causation. It is an imprecise version of our sixth constraint, that of *the neural or other bodily causation of neural events*, which will get more attention shortly.

A seventh and final constraint on answers to the question of the explanation of mental events is as important as it is obvious. The existence of the constraint is a large part of what made necessary our

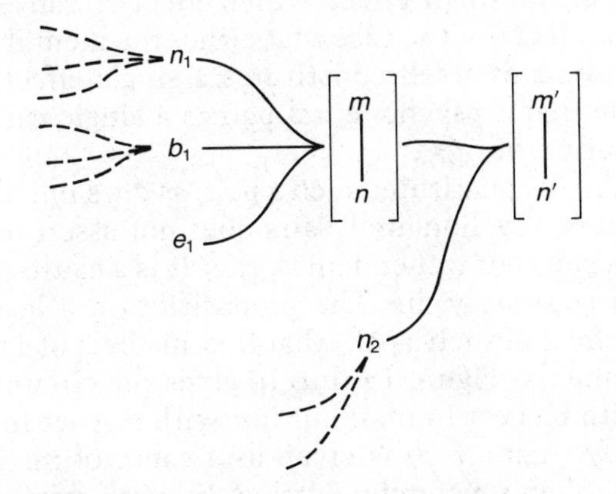

FIG. 15. Hypothesis of Psychoneural Nomic Correlation and Hypothesis on the Causation of Psychoneural Pairs

initial inquiry into causation. (Ch. 1) There can be no hope for any answer to our question of explanation which does not rest on and indeed incorporate an adequate conception of causal connection. Any answer which does not make use of, and, it might be said, is not formed by such a conception, cannot properly have attention.

The answer which issues from these seven constraints is the *Hypothesis on the Causation of Psychoneural Pairs*. The hypothesis, which involves a subordinate proposition of importance, is roughly that mental events are explained by certain causal sequences. The very end of such a sequence, involving two mental events, is modelled in Figure 15. The figure also models the Correlation Hypothesis. It is represented, as before, by no more than the vertical bars joining *m* and *n*, and *m'* and *n'*.

The side brackets [ ] in the figure represent the subordinate proposition, which is in fact essential to the present view of the mind. The Proposition of Psychoneural Pairs is that a mental and a neural event constitute what is to be called a *a psychoneural pair*. For two things of any sort to be a pair in one sense of the word is for them to function together or as a unit. One dictionary definition of a pair in this sense is this: an article consisting of two corresponding parts not used separately, such as a pair of scissors, tongs, or trousers. (*Concise Oxford Dictionary*) A psychoneural pair—if the comment is useful—stands in *some* relation to a pair in this ordinary sense rather than any other.

A psychoneural pair, more precisely, consists in a mental event and neural event, nomically correlated in the way of which we know, and

(i) constituting only a single effect, which effect (ii) can also be regarded as a single cause. It is not the case that either the mental event by itself or the neural event by itself constitutes a single effect. If it is, so to speak, less true that a psychoneural pair is a single cause, it is, so to speak, sufficiently true.

In Figure 15, *m, n* constitutes such a pair, as does *m′, n′*. To repeat, (i) the Proposition of Psychoneural Pairs does not assert or entail of any cause of such a pair, but rather denies, that it is a cause of a *member* of the pair taken separately. (ii) The proposition in a lesser sense also denies, of any effect of such a pair, that it is an effect of a member of the pair taken separately. Figure 15 thus (i) gives the circumstance n_1, b_1, e_1 as causal with respect to *m, n* but not with respect to either *m* or *n* taken separately. Also, *m, n* is given as a cause of *m′, n′*, but not of either *m′* or *n′* taken separately. Further, (ii) with respect to the latter connection, involving a pair as causal, it is only the pair that is shown as a cause, not either of its elements, *m* or *n*. The Proposition of Psychoneural Pairs, to make it clear by a contrast, thus is *not* modelled by Figure 16. There, the mental event *m* in itself is given as both effect and cause. So is the neural event *n*.

Figure 15, by way of another description, gives us *m*, say noticing the olives, as correlated with a neural event *n*, the two being members of a psychoneural pair, which is to say a single effect within a certain causal sequence, and also what can be taken as a single cause of a later pair, the pair of which one member is *m′*, say intending to have an olive. All of this, as remarked, is the very end of a causal sequence. The items n_1 and n_2 are neural events or persisting neural structures, item b_1 is a non-neural bodily event, and e_1 is an environmental event. More will be said of each category.

The Proposition of Psychoneural Pairs may strike one as surprising, particularly in its first part. How can it be that something is an effect

FIG. 16. *Not* the Proposition of Psychoneural Pairs or the Hypothesis on the Causation of Psychoneural Pairs

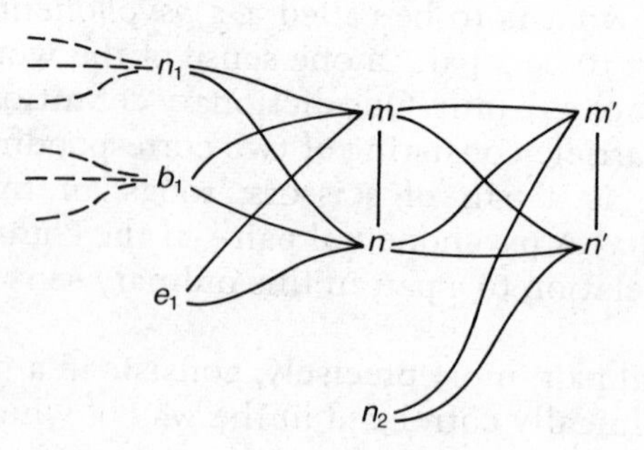

without its elements taken individually also being effects? How can something cause x without also causing each part of x? How can *m*, *n* be an effect of n_1, b_1, e_1 without *m* being such an effect and *n* being such an effect? In fact this is not only possible, but a logical consequence of what we have already accepted. On reflection it is also intuitively correct.

Recall (1) that for a circumstance causally to necessitate something else is for it to be true, roughly, that given the circumstance, even if there were to occur anything other than logical excluders of circumstance and effect, the effect would still occur. (1.3) Recall as well (2) that in virtue of the Correlation Hypothesis, a mental event is necessary to the occurrence of the simultaneous neural event. In Figure 15, *m* is necessary to *n*—*n* would not happen in the absence of *m*. (3) However, the non-occurrence of *m* would not *logically exclude* the occurrence of *n*. The absence of *m*, that is, would not logically entail the absence of *n*. It follows from these three propositions that there can be nothing that necessitates *n* by itself. There can be no thing of which *n* alone can be an effect.

There is roughly the same story about *m*—that nothing can necessitate it by itself. In virtue of the Correlation Hypothesis, *n* is dependently necessary to the occurrence of *m*, but *n*'s being missing would not logically entail *m*'s being missing. Hence there can be nothing which necessitates *m* by itself. There can be no thing of which *m* alone can be an effect.

Will it be thought that these conclusions about *n* and *m* are overstated? Will it be thought that all that the Correlation Hypothesis requires is, roughly, that if there occurs something which necessitates *n*, there must also be something which necessitates *m*? (And that there is a related truth beginning from the hypothesis and there being something that necessitates *m*?) The thought ignores what it is for *a* to necessitate *b*. It is for it to be true that if *a* occurs, so will *b*, despite any changes short of logical excluders of both. But with respect to the supposedly possible causal circumstance for *n* by itself, there logically *can* occur what will prevent the occurrence of the supposed effect *n*—to wit, the absence of *m*. That *m*, by a certain supposition, must be caused to happen, does not make its absence logically inconsistent with *n*.

Not that it is needed, there is a further reason why there can be no causal circumstance for either *n* or *m* taken separately. If causal circumstances necessitate their effects, they are also dependently necessary to them. Can there be something that is dependently necessary to *m* by itself? There is a related argument against the existence of such a thing. Given the Correlation Hypothesis, *n* necessitates m. Can it then be true that something is dependently necessary to m—that in the absence of the thing in question, and like things, m would not occur? Can it be true, more explicitly, that there is something such that if it and like things did not occur, m also would not occur, no matter what else were to occur other than logical excluders of both non-occurrences? The answer must be no, since the presence of n, which is not a logical excluder in the relevant sense, would guarantee the occurrence of m. There is a related argument against anything's being dependently necessary to n.

The idea that there can be effects whose elements are not effects, on further reflection, has application to things more ordinary than psychoneural pairs. Consider a car door—or anything else—shaped from a a sheet of steel by a press. It will be true that parts or properties of the door, say p_1 and p_2, are in nomic connection. If, say, each is necessary to the other, and neither logically excludes the other, then there can be nothing that necessitates either on its own. There can be no press, and no function of a press, which is such that given it, whatever were also to happen short of certain logical excluders, p_1 would be the effect. The same is true of p_2.

So much for the matter of a mental and a neural event as a single effect. The other part of the Proposition of Psychoneural Pairs, which is of less importance, is that a pair is best regarded as a single cause. To return to Figure 15, it needs to be allowed that it is not false to assert of each of m and n that it is a cause of the later pair, m', n'. However, these are unusual causes, what might be called *bound causes*. Given the Correlation Hypothesis, one of them (n) necessitates the other, and the other dependently necessitates it. Equally, one of them (m) is necessary to the other, and the other is dependently necessary to it. It is peculiarly fitting to speak of two such things, if they have a common effect, as a single cause. If no new proposition is added to what we have by this way of speaking, it does at least give emphasis to what we have.

Before more is said of the recommendations of the Hypothesis on the Causation of Psychoneural Pairs, and its subordinate proposition, let us attend to some unfinished business. As has already been clear, and remarked upon, it is impossible wholly to separate the synchronic question of the psychoneural relation and the diachronic question of the explanation of mental events. We thus found ourselves, in connection with Identity Theories taken as answers to the first question, also considering them in connection with the second question. Now, in taking the Proposition of Psychoneural Pairs as part of the answer to the second question, we also have a further answer to the first question. We add the Proposition of Psychoneural Pairs to the

Correlation Hypothesis and so come to a *complete answer* to the question of the nature of the psychoneural relation.

That is, what is true of a mental event and a neural event when they are intimately related is that they are nomic correlates, in the sense of the Correlation Hypothesis, and also that they constitute a single effect, and, as can be said, a single cause. This complete answer to the question of the psychoneural relation can do with a name. The answer resembles but is different from Identity Theories, so-called, and it is distant from traditional dualisms. (pp. 110 f.) It does not make a mental and a neural event into a unity but it does bring them into a certain union. The answer, therefore, is tolerably named the *Theory of Psychoneural Union* or the *Union Theory*.

To be a bit more explicit about the comparisons, this picture of the psychoneural relation, the Union Theory of mind and brain, is of course akin to but none the less distinct from a number of particular views we have considered or passed by. It is akin to the particular Identity Theory, so-called, which we passed by, and which mistakenly takes the relation of identity to consist in shared causes and effects. (p. 93) So too is it akin to the Two-Level Identity Theory (2.3), and the theory of Psychoneural Causal Transfer (p. 161). It would be tedious to lay out all the specific differences between these several views and what we have. They are differences, as may easily be confirmed, which enable the present view to escape all the difficulties faced by its predecessors.

Let us rather return to our main business and consider the recommendations of the Hypothesis on the Causation of Psychoneural Pairs—whose full statement can be delayed for a while—taken in conjunction with the Correlation Hypothesis. What must come to mind as the first virtue of this picture is that it does not come near to falling into the contradiction we have been noticing. (p. 155 ff.) By way of the Correlation Thesis, which is essential for psychoneural intimacy, we still have it that a mental event is necessary to the simultaneous neural event. However, that does not give us a necessary condition for the neural event which, so to speak, conflicts with a causal sequence for the neural event. The mental event, rather, is within the sequence. Event *m* is as much within the given sequence as is *n*. This is the result of our conclusion that *m* and *n* constitute a single effect. In escaping the given contradiction, further, we escape any difficulty about mental indispensability. As is clear, this picture of the mind secures mental indispensability in the most satisfactory way: by making mental events causes of later mental events and presumably actions.

As a consequence of having *m* as much within the causal sequence as *n*, to come to the second distinguishing feature of this picture of the mind, we do not embrace the proposition that a typical neural event is the effect of a *wholly* neural or other bodily causal sequence. It is the effect, rather, of a causal sequence which includes psychoneural pairs, very likely very many. It is of fundamental importance, of course, that this gives us no neural gaps in the history of any neural event. No neural event in the sequence comes out of nothing, neurally speaking. *Every* neural event is in a way owed to an immediately prior neural or at any rate a bodily event.

Is there any weakness or embarrassment in our exchanging (i) the proposition that a typical neural event, putting aside any environmental or pre-bodily history, is the effect of a wholly neural causal sequence for (ii) the proposition that a neural event is the effect of a causal sequence which is at every moment neural and also at some moments mental? There is none, but rather, as we shall come to see, the satisfaction of embracing what is surely a truth. A part of the argument—its empirical content—must wait until we come to our consideration of neuroscience. What can be said now on the point is partly that there is no tolerating the contradiction we have been noticing, that the proposition to which we have come enables us to escape it.

Also, the proposition is one of a general kind accepted in *all* the pictures of mind at which we have looked—save the least arguable, which are Eliminative Materialism and Local Idealism. That is, it is a proposition which asserts nomic connection—diachronic and in particular causal nomic connection—between the mental and the neural. Proponents of the Theory of Neural Causation with Psychophysical Correlation (3.2), to speak only of them, are disinclined to have partly mental causation of the neural, diachronic psychoneural nomic correlation. In this they perhaps are like a number of neuroscientists. It is difficult to see what general ground they can have, since they *already* have nomic connection between simultaneous mental and neural events, synchronic psychoneural nomic correlation. There can be no more reason of a general kind against diachronic nomic connection. Both connections, and *all* arguable accounts of the mind, encounter the still intractable problem of the ultimate nature or mechanism of connection between the mental and the neural. (p. 112)

Further, it was remarked above that the sixth constraint on answers to the question of the explanation of mental events is that neural events are *somehow or other* owed to neural causation. A part of this constraint, to be somewhat more precise about it, and if we put aside first events at the periphery of the nervous system, mut be that there can be no neural event which has a full explanation which leaves out prior neural or other bodily events. This is a counterpart, for the neural, of the axiom of mental indispensability. Another stronger and more important part of the constraint is that no neural event, putting aside any environmental or pre-bodily history, has a causal history which contains neural or other bodily gaps—times at which there occur no neural or other bodily events. The view to which we have come absolutely satisfies the constraint of neural causation so specified, in both parts. It is not in accord with the excessive requirement that each neural event, post-environmentally, is the effect of a merely or wholly neural causal sequence.

To pass on to another constraint, the view we now have evidently does not conflict with psychoneural intimacy. As remarked several times, psychoneural intimacy is achieved or explained by the Correlation Hypothesis. It was argued in the last chapter that the unacceptability of Identity Theories, and the absence of logical or conceptual connection between the mental and the neural, made the Correlation Hypothesis the only possible means of satisfying the intimacy requirement. It is essential. We now have a proposition that was not on hand before, that a mental and a simultaneous neural event form a psychoneural pair. Might it be that the intimacy requirement is satisfied by the proposition of psychoneural pairs *taken by itself*?

The argument is impossible. The core of the idea of psychoneural intimacy is that there holds a *direct* connection of some necessary kind between a mental and a neural event. The Proposition of Psychoneural Pairs is that they derive from a single causal sequence and are in a sense one cause. But this falls short of being necessary connection directly between them. Indeed, the Proposition of Psychoneural Pairs is consistent with the doctrine of parallelism (p. 111) correctly named, or pre-established harmony, which *denies* direct connection between the mental and the neural. (The distance between the Proposition and the Correlation Hypothesis, incidentally, is also made clear by the fact that the hypothesis is logically consistent with mental and neural events not being effects at all.) Thus we shall certainly persist with the view that the Correlation Hypothesis is essential for psychoneural intimacy. It would be mistaken, even, to take it that the Proposition of Psychoneural Pairs even contributes to the achievement of psychoneural intimacy.

The Hypothesis on the Causation of Psychoneural Pairs—we *shall* arrive at its full statement shortly—is of course consistent with the constraint of mental realism. How does it stand to the prohibition on general assumptions of overdetermination? We do not have overdetermination—that is, an event's having two existing causal circumstances or a single circumstance containing an element plus a substitute for it. We do have something distantly related to it. In terms of Figure 15, it was allowed that it is not false to say n is a cause of the later causal pair m', n'. Also, by the Correlation Hypothesis, n necessitates m, which can be said to be a further cause of m', n'. Thus it can be said there are two routes from n to m', n'. Different but related remarks can be made about m, which dependently necessitates n. Greater detail might be provided, but is unnecessary. That the view we have has this feature distantly related to overdetermination is no significant objection to it.

The constraint of mental indispensability, as we know, is simply and effectively satisfied, without contradiction, by the placing of earlier mental events within the causal sequences for later mental events. Our hypothesis *has* the recommendation at which Identity Theories *aim*. The satisfaction of the constraint in so far as it has to do with actions will be considered in the next chapter. What of personal indispensability? The constraint having to do with persons is in essence that the explanations of typical mental events must include references to such things as memories, beliefs, intentions, resolutions, tendencies, commitments, plans, hopes, and a good deal more. A choice is certainly not to be explained just by the desire and belief that immediately precede it, or by these with their neural correlates. It is also to be explained by the continuities of experience which are at least partly constitutive of a person.

What all these continuities involve, as was in fact remarked earlier in connection with a wide conception of the mental (pp. 86 f.), are *dispositions*. Dispositions in general are persisting conditions or sets of conditions which may be parts of causal circumstances. (2.2) The particular dispositions which are or are involved in a man's memories, tendencies, and so on cannot be regarded as other than neural facts about him. Such dispositions, in brief, are persisting neural structures. The Hypothesis on the Causation of Neural Pairs evidently accommodates them. It does not explain mental events only in terms of preceding mental events and their neural correlates. It allows a role to personal dispositions. The fact is represented by n_2 in Figure 15.

The hypothesis, to move now to a full statement of it, has to do with causal sequences of a certain length, and whose initial elements are of certain kinds. If we were to contemplate certain short causal sequences, or perhaps even long ones, with certain sorts of beginnings, we would fail to have a determinism. An *indeterminist* picture of the mind can of course include certain causal sequences for mental events and actions—sequences which begin with, or leave open the possibility of beginning with, acts of Free Will or the like, the sorts of thing often taken as needed to guarantee freedom and responsibility.

Still, the Hypothesis on the Causation of Psychoneural Pairs will not be a claim about the whole causal background or history of each mental event. If it were such, it would be a yet larger proposition than it is. It would concern not merely a person and his relevant background, so to speak, but also a history consisting in parents, ancestors, and each thing that touched on his life and in so doing contributed to the mental event in question. If Brown's thought was a result of Green's words, then such a large claim as to Brown's mental events would include some of Green's life, and of course antecedents of that. The Hypothesis on the Causation of Psychoneural Pairs, rather, specifies that mental events are owed to more limited causal sequences, whose initial elements fall into two kinds.

The first kind consists in neural and other bodily elements *just prior to the first mental event in the existence of the person in question*. We need not exercise ourselves about the question of exactly when, after fertilization, that event occurs. No initial elements of this kind are shown in Figure 15, of course—as remarked, what we have there is no more than the final upshot of a causal sequence.

The second kind of initial elements consists in environmental events then and thereafter (such as e_1 in Figure 15) which directly affect the person. For an environmental event to affect a person *directly*, let us say, it must be the last event in an environmental causal sequence whose effect is a bodily or neural event. That is, the causal sequence as specified contains no environmental event later than the given event. All of these *direct environmental elements* will of course be non-mental. The thoughts and feelings of others certainly affect me, but by way of their bodies and actions and the non-mental effects of these. This is part of the large fact of sense-perception and sensation, the only channels between our environments and ourselves.

The Hypothesis on the Causation of Psychoneural Pairs, to come to a full statement, is thus as follows:

> Each psychoneural pair, which is to say a mental event and a neural event which are a single effect and in a lesser sense a single cause—each such pair is in fact the effect of the initial elements of a certain causal sequence. The initial elements are (i) neural and other bodily elements just prior to the first mental event in the existence of the person in question, and (ii) direct environmental elements then and thereafter.

The hypothesis is evidently related to familiar beliefs, rightly entrenched or anyway sturdy, as to a person or an aspect of a person being a product of heredity and environment. It is no grander than these beliefs, but more specific. The hypothesis, given our understanding of causal sequences generally, of course excludes the possibility that the mentioned sequences contain any gaps (3.1)—that is, each event within them, whether neural or mental, is both an effect and an element of a causal circumstance. Thus the mentioned sequences cannot contain acts of Free Will as traditionally conceived—these, as we shall see more fully, are or closely involve items which by definition are *not* effects. The hypothesis does not exclude the possibility, of course, that the causal sequence for a psychoneural pair includes other psychoneural pairs. It allows the possibility that typically just this will be the case, as in Figure 15. The hypothesis, as already remarked, and to speak of another kind of possible gap, is of course consistent with the proposition that there is no *non-neural or non-bodily* gap in the causal background of a psychoneural pair.

One final characterization of the hypothesis is of some importance. It is that it is consistent with what has several times been noticed, the distinction we make between active and passive mental events. This might, perhaps, have been added to our list of constraints. There is no doubt that such a distinction is in order. There may occur a sequence of active mental events when I engage in a little habit: I look from my college window first at the top of the Post Office Tower, then at the lowest part that is visible, then at a cornice of the college in the foreground, and then at a pillar. So too are there active mental events when I wonder if something read in a book is consistent with something read earlier, look back unsuccessfully at a previous chapter, try to recall the earlier item, succeed in thinking of it, turn to the right chapter to check. As against such sequences of active events, there are episodes when, as I say, feelings and inclinations come upon me, thoughts intrude, or will not go away. The passive parts of consciousness also include sensation and sense-perception, but not, as it seems, attending to a sensation or perception.

It is common in indeterminist accounts of the mind to regard at least some active mental events, in the above sense, as *unnecessitated* events of a certain kind. Whatever can be said in characterization of these latter events—we shall be considering the question fully—there is no possibility that the ordinary distinction between activity and passivity has to do fundamentally with such an idea. Nor need we despair of analysing activity, as some do, or say something so limited as that passivity is to be distinguished from activity by the fact that in the former a man is a 'bystander' or 'victim' with respect to his mental events. (Thalberg, 1978)

What is fundamental is that active experience, as it can be called, is purposive or goal-directed. It involves a reason. When I was looking out the window and ran my eye over the four items, I did so in the desire to complete a little ritual. The episode of my thought about the book and inconsistency was similarly goal-directed. In each case, to be clearer but brief, my experience was subject to a certain ongoing desire. There was, in a certain uncontentious sense, quite different from one to which we shall come (3.6), a *teleological* explanation of what occurred.

There is nothing in this familiar account of activity, which can be much elaborated by way of such further ideas as those of interim goals and a plan, that conflicts with the Hypothesis on the Causation of Psychoneural Pairs, or any other account of all mental events as effects. The hypothesis allows, evidently, that a sequence of mental events might be owed, importantly, to a dispositional desire. It also allows what may be necessary, that a sequence of mental events may involve an ongoing conscious desire. The latter fact may perhaps be doubted, as a consequence of having a certain image of the import of the given hypothesis together with its predecessor, the Correlation Hypothesis. The image is roughly, that the two hypotheses together make consciousness into a succession of somehow unrelated events. This may be taken to conflict with the idea of active experience as a matter in part of an ongoing conscious desire. However, the import of the two hypotheses is *not* caught by the image.

As remarked in connection with the Correlation Hypothesis (p. 117) it in fact does not 'atomize' consciousness or deny its continuity, despite the fact that the hypothesis has to do with discriminable mental events. Further, we are in no way committed to the absurdity that successive occurrents are necessarily different in character—that, by way of an example, aches do not persist. What we have is wholly consistent with regarding a sequence of mental events as incorporating a sequence which can be described as an ongoing conscious desire, indeed a governing desire.

3.4 THE DEFINING GOAL OF INDETERMINIST THEORIES OF THE MIND

We have already considered quite a number of objections that can be or have been attempted against the Hypothesis on the Causation of Psychoneural Pairs. This is so since various propositions raised against the Correlation Hypothesis can also be raised, sometimes with little or no alteration, against its companion. Some are those having to do with its contained ideas of consciousness and of a mental event. (2.5) Others are the Wittgensteinian objection to mechanism (2.6), and the objection having to do with the different domains of the mental and the neural. (2.7) We shall not return to these disputes.

The remainder of this chapter, rather, will be given over to a consideration of a mixed set of radical alternatives to the picture of the mind given by our two hypotheses. We shall consider the plethora of indeterminist theories and ideas of the mind. They attempt to give an account of choosing, deciding and like things which is wholly different from the account we have in our two hypotheses. It is the burden of our hypotheses, or rather one burden, that choosing and deciding are fundamentally on a level with other mental events—if choosing and deciding are of a unique character (p. 222) and if they are bound up with what has just been noticed, active as against passive experience (p. 174), they remain with other mental events a matter of nomic correlation and a certain causal background. Acceptance of one of the opposed plethora of indeterminist theories and ideas of the mind, which *elevate* choosing and deciding, is implied by very many of the objections mentioned above, which fact is quietly passed over, and by a good deal of inexplicit philosophical scepticism about determinism.

If the indeterminist theories and ideas have negative sources—sources in objection—they are also to be characterized in terms of positive sources. One is the conviction of mental indispensability, as well described as the conviction that epiphenomenalism is false. (Popper, 1982b, pp. 28, 41; Popper & Eccles, 1977, pp. 72–5, 86–8) But to suppose that the conviction of mental indispensability cannot be respected by a determinist view is of course mistaken. Certainly the determinism that has been expounded here, essentially because of its contained idea of psychoneural pairs, excludes the possibility, say, that Mozart's devising of his music, his creative imagination, could in principle be explained in a wholly neural way, without reference to Mozart's mental events.

Indeterminisms have a source too in religious commitments or aspirations. (Popper & Eccles, 1977, pp. 557–8) At least some of these, however, can coexist with determinism, as is made clear by episodes in the history of religion and theology. Indeterminist theories are also owed to a more vague spiritual inclination to have the mental somehow above or beyond the ordinary physical universe, or really different in kind. There is the inclination to have the mental *emerge* from all else, to have yet more than a unique character.

Indeterminist theories and ideas have their main sources in certain commitments and attitudes having to do with the nature of our ordinary unspeculative and unspiritual lives. The most discussed of these, whether or not it should be, is a certain commitment to moral responsibility, a commitment to moral responsibility conceived in a certain way. It is enough to remark, for the present, that indeterminist pictures derive from the attitude, among others, that we are responsible for at least some of our decisions and actions in the sense that *we could have decided or acted otherwise than we did, given all things as they then were, and given all of the past as it was*. To speak in a currently favoured way, but one which adds no further idea, it is true of my decision or action in this actual world that there is a possible world in which I do not so decide or act, and everything is the same in this possible world, save for my decision or action and its consequences. (Van Inwagen, 1974, 1983; Boyle, Grisez, Tollefsen, 1976) I can in this sense *credit* myself and others with responsibility, and *hold* myself and others responsible.

These true sources of indeterminist theories and thoughts are noted not idly but for a particular reason. The main sources determine the natures of the theories and the like, fix their definition—or what definition they can have. If the theories and the like are often enough implied by objections to determinism, and by inexplicit scepticism about it, attempts to set them out explicitly and completely are rare.

We need to try to clarify libertarianism in order to have a better view of the picture of mind given by our own two hypotheses. That picture, like any conception which competes with others, is partly to be judged by way of the strength of its competitors. This is so with respect to both conceptual adequacy, which is our present business, and of course also truth. There is—to look forward—also another equally large reason for trying to clarify and for considering indeterminist theories. Indeterminist assumptions and beliefs are not the property of philosophers, but of all of us. They are entrenched in our culture—by which I mean at least European or Western culture—and inform many of our attitudes, practices, and institutions. Indeterminist assumptions and beliefs thus enter into attitudes, practices, and institutions—including the practice of regarding people as responsible in the way noted—which will be our concern in considering the consequences of determinism. (Chs. 7–10) These attitudes and the like, and the assumptions and beliefs within them, are further understood by way of what can be called their best underlying theories, or so it must be hoped.

Before proceeding, we need to attend to one further matter having to do with those attitudes, assumptions, and so on, in particular the aspiration to regard ourselves as in a way responsible. It is possible to understate or to have too small a conception of the goal of indeterminist theories, and hence of their necessary nature, and hence of the difficulties they involve. It is also possible, however, to overstate our attitudes and hence the goal of theory. (G. Strawson, 1986, Ch. 2; cf. T. Nagel, 1986, Ch. 7) The result of this is the conclusion that indeterminist theories and thoughts are yet more conceptually doubtful than they have often been taken to be. A second and connected conclusion would be alarming if true, in that it would make a good deal of this book, like a multitude of others, in one way otiose. It is, however, a conclusion which is unlikely—sufficiently so as to put into doubt the distinctive and original argument for it.

The conclusion is that there is actually no need, in connection with the dispute about which of two kinds of freedom we have, one of them a matter of indeterminist theories, to consider the conceptual viability and the truth of determinism. All of the traditional debate about determinism, in connection with the two kinds of freedom, can be ignored, since the utter futility or even fatuity of indeterminist thinking can be fully established without considering the worth of determinism. That cannot be to say, incidentally, as is not made clear, that even on the given argument, determinist pictures of the mind can *in general* be ignored. Indeed, on the given argument, they must call for *more* concern since, so to speak, they have the subject-matter of our lives entirely to themselves. They must, on the given argument, be the only theories in which to locate the freedom we actually have. Still, it would be at least unsettling if time spent on determinist pictures just as means to the refutation of indeterminist pictures was time wasted.

The argument for that conclusion proceeds, as remarked, from a particular conception of the goal of indeterminist theories. It is said to be the goal of true responsibility, or responsibility in the strongest possible sense. (G. Strawson, 1986, p. 1) This responsibility is taken to consist in a certain self-determination, wholly different from self-determination or responsibility in a more familiar sense, where it has to do with one's actions flowing from one's own desires and the like. Self-determination of the required kind consists in *determining one's own self*, or *determining how one is*. (G. Strawson, p. 26) It is a creative power akin to that assigned to us by Existentialism in the dictum that existence precedes essence: that in my life I make my essential nature, that my life is at no point a product of my nature. (Sartre, 1957; M. Warnock, 1965, Ch. 5, 1973) No theory, it is said, can possibly give us this, as can be established by the following argument.

(1) One's rational action is necessarily a function of, or somehow determined by, *how one is, mentally speaking*. (2) If one is *truly responsible* for the action, one must therefore be truly responsible for how one is, mentally speaking. (3) To have the latter responsibility, one must have *chosen* the way one is, mentally speaking, at least in certain respects. (4) To have made that choice, one must have had *principles of choice*, reasons in the light of which one made the choice. (5) One must also have been truly responsible for these principles, and hence (6) have chosen them. (7) But for this choice, one must have had

prior principles of choice. (8) One must also have been truly responsible for these principles, and hence .... We must thus conclude that true responsibility for one's actions, a determining of one's self, would require what is logically impossible, one's actually completing an *infinite regress* of choices of principles of choice. Indeterminist theories of the mind are therefore attempts to secure or elaborate the logically impossible. They are exercises in fatuity. (G. Strawson, 1986, p. 52; cf. T. Nagel, 1986, Ch. 7)

The argument, even when further filled in by its proponent, remains in several ways too rapid. Its fundamental idea, however, is not complex, and more or less independent of the matter of rationality. It is that since a certain conception of responsibility requires that I stand in a certain creative relation to an action, and since that action has a certain source or ground, I must also stand in the given creative relation to the source or ground, and so on back through an infinite regress. The essential weakness of the argument can be approached by way of an *ad hominem* route. A parallel argument can be attempted against conceptions of responsibility opposed to the indeterminist conception.

These, in fact noticed a moment ago and to be more fully considered in due course (Ch. 8), may briefly be said to take a man's responsibility for an action to consist in the fact that it flows from his own decisions and the like. It would not be enough, evidently, for the action just to be *not contrary to* desires of his. If *only* that were true, indeed, it could not be an action, or at any rate his action. But now, it can be argued, it must also be the case that the given desires must be desired by him. (Cf. Frankfurt, 1969, 1971, 1975) To speak differently, they must not be disconnected from his personality, nature, or identity. It may be remarked in this connection that there are drug addicts, say, who truly desire not to be subject to their cravings, and whose responsibility for their drug-taking is therefore at least in doubt. To allow that the given conception of responsibility has such a need to incorporate *desires for desires* evidently makes conceivable an argument about regress.

However, as can be anticipated, proponents of the conception are unlikely to be disturbed by the argument. They will reply, in short, that they do not require of *all* desires antecedent to an action that they themselves are desired. Enough is enough.

Similarly, proponents of indeterminisms can about as easily assert that the responsibility which they pursue faces no regress. There are several possibilities here, of which we need note only one. It can be asserted that the responsibility in question requires that I stand in a certain creative relation to an action, and hence in a certain creative relation to what it certainly has, a certain ground or source, but that either this ground or source or an earlier one is *primitive*. That is, *it* has no ground or source. An indeterminist theory of the mind may centre on this primitive episode. There will certainly be great difficulties about its description—as we shall see in what follows—but there need be no capitulation to a fiat, the fiat that any such episode must have a similar antecedent. To return to rationality for a moment, it can of course be said that there is no more difficulty about the primitive episode in this connection than with primitives in any logical system, morality or whatever, each of which must leave something unproved.

3.5 INDETERMINIST THEORIES—FIRST ELEMENT

Over past centuries of their existence, indeterminist accounts of the mind have pretty well ignored the brain. This, of necessity, is no longer true. Philosophers who now espouse or are inclined to such accounts give some place to propositions about the psychoneural relation and claim to include a somehow neural explanation of the occurrence of neural events. It is of the essence of such theories, however, that they do not accept both of the Correlation Hypothesis and also the Hypothesis on the Causation of Psychoneural Pairs, or any related pair of propositions of nomic connection. The denial of one or both kinds of nomic connection—which denial is the first of three elements of an indeterminist account of the mind—is now based on physics, or rather, a certain customary interpretation of Quantum Theory.

That interpretation is of course to the effect that there are micro-events which are random or a matter of chance, which is to say not necessitated. Such an individual or token event is not to be explained as an effect of a causal circumstance or in terms of other kinds of nomic connection as we have defined it. It is not that it is merely inexplicable in the sense that it is impossible to know a causal circumstance or explanatory nomic correlate for it, which thing may exist. Rather, it is a truly random event, inexplicable in the sense that there actually exists no causal circumstance or explanatory nomic correlate for it. The 'uncertainty' about it is not epistemological but ontological. In a traditional usage, if not one that is congenial to its advocates, it is a *miraculous* event. It is true that it is allowed that there is an antecedent probability of its occurring, perhaps a high probability, and this as a matter of necessity or law, but that it actually occurs is something which can have no standard explanation whatever. Our present concern, however, is to consider indeterminist pictures of the mind rather than to consider their supposed basis in Quantum Theory. (5.6)

The *first* element of any indeterminist picture, to repeat, is a denial or one or both of our two hypotheses of nomic connection, or any like hypotheses. Macro-events which are the subject of these hypotheses are inferred to be random on the premiss that certain micro-events are random—which inference distinguishes the picture from near-determinisms. (pp. 2, 9) This is rightly taken as a logically necessary condition of the given sort of responsibility. A *second* element of any satisfactory indeterminist picture is what we shall call an *originator*, some persistent entity that originates certain mental events. We might, more traditionally, but less descriptively, or indeed misleadingly, name it a self, ego, mind, self-conscious mind, faculty of mind, active power, the will, the faculty of the will, rational capacity, agent, or person. As we shall see, some indeterminist pictures attempt to give relatively little place or role to such a thing. A *third* element, not entirely separable from the second, is an account of the relation of origination holding between the originator and the mental events.

All indeterminist accounts worth considering, to speak very summarily indeed, are thus to the effect that mental events stand in some relation to neural events, which neural events may have a certain non-nomic neural history and are in a certain connection with actions—and that certain of the mental events also stand in a certain relation to an originator. Only the terms of these relations are represented in Figure 17—the figure thus does not give all of an indeterminist theory. The originator *o*, whose nature is one of our problems, is some kind of persistent thing, represented by the arrow as moving through time. It stands in some relation to the mental event *m′* (as perhaps it previously did to *m*), which relation is another of our problems and is not represented. The mental event *m′* also stands in a certain relation to *n′* and hence to the non-nomically related neural event *n* and by way of it to the prior mental event, and also in some relation to *a*, which is an action. These unrepresented relations are also problems for us. As

FIG. 17. Elements of Indeterminist Theories of the Mind

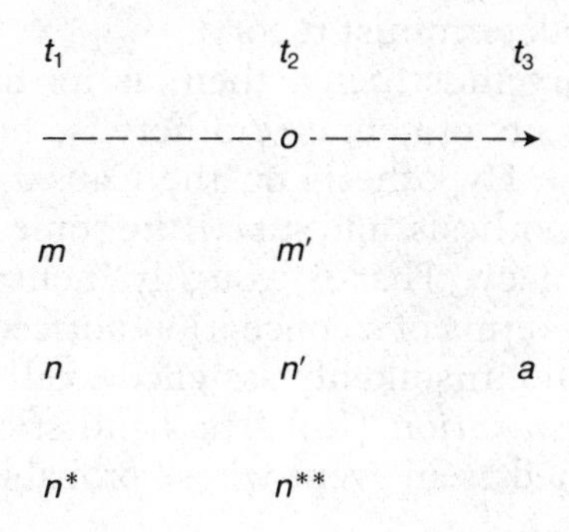

before, m can be taken as someone's noticing olives and m' as his intending to have one. The action a can be taken as his reaching for one. The items n^* and n^{**} are included only to represent the fact that such a neural event as n' and any action will somehow be owed to more than a single preceding neural event—such as n or n'.

Typical indeterminists, if they now give some attention to the brain, do not go into relevant detail about what is necessarily the first element of their view, either a non-nomic neural history for neural events or a denial of the Correlation Hypothesis. One idea is a denial of the Correlation Hypothesis and any like proposition, and the retention of some proposition of neural and bodily causation. That is, roughly, the brain and Central Nervous System and body are left to run on nomically, as there is good reason to suppose they do, but nomic connection with mental events is denied. This would certainly free such mental events as decisions from constraint by the brain, and leave them as products of some kind of the originator. The idea is natural and common, but fatal to the enterprise in question.

One reason is that it would stand in the way of securing the desired kind of responsibility for a person's *actions*. In terms of Figure 17, if action a is left as the effect of a standard causal sequence of a wholly neural and bodily kind, we shall close off the possibility of the agent's having the desired kind of responsibility for it. *Whatever* is said of the level of mental events, and that of the originator, we shall not have it that the action could have been otherwise in the strong sense that is desired. It is neurally fixed. Let us take it, then, that an indeterminist theory of the mind includes the Correlation Hypothesis. Certainly any tolerable theory of this kind must include something like the hypothesis. It has sometimes been supposed, incidentally, that an indeterminist picture can make use of an Identity Theory or of Interactionism for the given purposes. (Popper & Eccles, 1977; Thorp, 1980) What was said earlier against Identity Theories (2.3), and Interactionism (3.2), when we were not considering indeterminism, is as fatal here. It is the Correlation Hypothesis or something like it that must enter into an indeterminist theory.

What such a theory must deny, then, as its first element, is that neural events are effects of neural or otherwise bodily causal sequences. It must deny the Hypothesis on the Causation of Psychoneural Pairs and any like hypothesis, and substitute some or other Proposition of Neural Indeterminacy. That is, roughly, neural events enter into sequences defined in terms of a conception noticed in our inquiry into nomic connection, and mistakenly assigned to all of us—what can be called *probabilistic causation*. (1.6) A probabilistic effect, roughly, can be stipulatively defined as an event whose probability of occurrence is raised by antecedent events and conditions, including the 'probabilistic cause'. As our earlier inquiry showed, improving such a rough definition will involve difficulties in connection with probability itself and the theories of it. Improving on the definition of 'probabilistic effect', however, will also involve a particular act of stipulation of an unnerving kind—a decision as to *how much* the probability of a neural event must be increased by antecedents if it is to count as their probabilistic effect. Let us consider this second matter.

This decision will have to be taken not simply on general grounds, having to do with the nature of reality generally, but with reference to the formulation of an indeterminist account of mind and brain. The decision will have to be taken, more particularly, by attending to a number of constraints. One of these will be truth to neuroscience. However much science in general may be inclined to defer to physics, it is a plain fact, of which we shall see more, that neuroscience does not suggest, let alone propound, that the brain and Central Nervous System are prone to indeterminacy, let alone a great indeterminacy: that the most that can be said of neural events is that they have a low probability. On the contrary, it seems that the denial that neural events are necessitated can only be replaced by some proposition to the effect that their antecedents give them a very high probability.

A second constraint cuts in the opposite direction. If, so to speak, the originator is to be left with room for effective action, it seems that neural events *cannot* be near to being necessitated by their neural antecedents. If they are near to being necessitated in this way, and mental events are nomic correlates of them, then how are we to achieve a sufficient explanatory role for the originator and hence achieve the goal of a certain responsibility for mental events and actions? The two constraints therefore produce a dilemma for indeterminists. They are pulled towards too much and too little neural indeterminacy.

There is also a third constraint, which also fights with the second. If the originator is to act effectively, a secure link is needed between its mental event m' and the neural event n'. But there is the same requirement with the link between n' and the action a. The originator's plan for the olive runs the risk of failing at that stage. If we now have in mind the second constraint, that of leaving n' open to the originator's influence by reducing its probability relative to its neural antecedents, and also the third constraint, we will face a certain temptation. Or rather, the indeterminist will.

To speak in terms of Figure 17, an indeterminist may or indeed must be tempted to conclude (i) that the probability of the occurrence of n', given its neural antecedents, is p, (ii) that the probability of the action *a*, given its neural antecedents, is p', and (iii) that p' is greater than p, or indeed that *a* has a probability of *1* or even is an event necessitated by its neural antecedents. The first proposition would leave room for activity by the originator, and the second would secure its efficacy with respect to *a*. It is very clear, however, that the familiar and well-founded scepticism about *ad hoc* stipulations must at least work against the conclusion.

There are other large problems about what no one has ever specified—the Proposition of Neural Indeterminacy—but let us press on.

3.6 SECOND ELEMENT, WITH ORIGINATOR TO THE REAR

It was supposed by a few philosophers, and more scientists, when Quantum Theory was first seen as an ally to indeterminist views of the mind, that a denial of nomic connection and in particular of causation is all that is needed in order to secure the desired responsibility for decisions and actions. The Proposition of Neural Indeterminacy by itself may indeed seem to promise the conclusion that a vicious or virtuous decision at *t* could have been otherwise than it was, given everything as it was at *t* and before. However, the proposition provides at best only a logically necessary rather than a logically sufficient condition of what is desired. How could an unnecessitated or chance event be something for which the person in question could be responsible in the given way? How could anyone be disapproved of or censured in the given fundamental way for a mere chance event? Indeed, how could anyone be responsible in *any* way for a mere chance event? (Ayer, 1954c)

The matter is made a bit more complicated, but the question is not answered, by the postulation of high-probability chance events—as will be anticipated from what has been said already. In the end, however probable it is in advance, the fact that one thing happens rather than another—that a man on a given occasion decides to torture or kill rather than not—is something which has no cause whatever. Thus there is *nothing* that can provide a standard or fundamental explanation of the event, that particular outcome. (p. 59) Moreover, there is surely *nothing* that in *any* sense explains that particular outcome. It is, as it seems, easy to lose sight of this proposition. In fact, it is nothing other than the essence of the main contention on hand: that the event is no more than probable, however highly probable. It bears repeating that what we are to understand, at bottom, is that things might have been exactly the same up to *t* in two situations: one in which the event happens at *t*, and one in which it does not. Hence it seems there can be nothing that in any sense explains the event if it does occur.

Indeterminism by itself, then, cannot secure an indeterminist theory's goal of a certain responsibility. If indeterminist views of the mind are also aimed, as indeed they are, at securing or safeguarding other things, the postulation of undetermined neural events is again logically necessary but not sufficient.

All serious attempts to formulate an indeterminist theory of the mind thus have more in them than a denial that neural events are neurally necessitated. What is also included is some non-neural account of how it is that a mental event occurs. Inevitably this is again an account such that the mental event is not necessitated. It remains a chance event in the sense of being unnecessitated itself or having a suitable unnecessitated antecedent. However, the non-neural account does not leave a decision, say, as *merely* or *only* a chance event—as it would do, of course, if the decision were made a matter of only probabilistic causation, which is no part of the present story. The non-neural account of the decision attempts to explain it as arising in such a way that it is *grounded or non-arbitrary*. We shall take that to mean, just, that the decision or an antecedent of it is such that the agent *can* be responsible for it in the given way. In sum, then, an indeterminist theory will be to the effect that a decision's neural correlate is not necessitated, and the decision is not necessitated by non-neural facts, but somehow it is rendered grounded and not arbitary by them. As result of these non-neural facts, further, partly by way of the Correlation Hypothesis, a resulting action is somehow explained by the grounded decision. There are great problems with the last part, about the action, as we know, but let us press on.

The non-neural facts include some persisting entity, what we earlier labelled an originator, what is also named a self, mind, the will, active power, a person, and so on. However, such an entity is often and rightly taken as problematic, and, as a result, there are indeterminist theories which seek to have little to do with one. The originator is at least kept well to the rear. We shall first consider three of these, having to do with self-causing decisions, self-justifying decisions, and teleology.

3.6.1 Self-Causing Decisions

The first idea as to how an unnecessitated decision as so far understood may be non-arbitrary has to do, as we can say, with a way in which such a decision can in another understanding *be* necessitated. We have taken it, certainly, that the standard concept of a necessitated event is the idea of an event necessitated by other things than itself, and that the idea of an unnecessitated decision is the idea of a decision unnecessitated by things other than itself. Necessitation like all other causal relations, as we well know, involves two items. There can be no doubt whatever that this is fundamental to our standard conception of causation. A decision may be said to be unnecessitated in this sense but the idea added—an idea with a long history—that it *is* necessitated by a set of conditions and the like *which includes itself*. The other conditions are presumably desires, beliefs and so on. (Cf. Boyle, Grisez, Tollefsen, 1976, p. 11, p. 83, p. 90.)

A decision of this kind is therefore to be understood as having something like the nature that has sometimes been attributed to God. It is to a significant extent cause of itself, *causa sui*. Any such proposal, about decisions, God, or anything else, is deceptive. It is deceptive in that it is likely to be taken as making use of a clear idea with which we are familiar—the idea simply of a cause, or of necessitation, a causally sufficient condition, or the like. We are at least tempted to think that the proposal involves no more, so to speak, than a novel application of a satisfactory and settled idea. The temptation is to be resisted absolutely.

To repeat, whatever else is true of causation as we understand it, it involves two items, numerically different items. Causation as we understand it involves relations whose terms are non-identical. To speak of a self-caused thing is not to make use of a satisfactory and settled idea of a certain relation in a new way, but to introduce a wholly unfamiliar and different relation. To say of a thing that it is cause of itself, and to insist that the given idea of a cause is our standard idea, would be to fall not into mere obscurity but into self-contradiction. What must necessarily be intended by any consistent talk of a thing as cause of itself is a quite different relation which is consistent with its own terms being strictly identical.

Any of the standard causal relations noted in our own inquiry can be used to illustrate this. Suppose it is said that *c* caused *e*, and it is added that *c* was identical with *e*. Can we partly understand this by way of the idea that a cause is *required for* its effect—that if *c* had not happened, *e* would not have happened? Very evidently we cannot. That conditional, given the addition that *c* was identical with *e*, becomes the logical truth, to express it one way, that if *c* had not happened, *c* would not have happened. What is therefore needed is some quite different account of '*c* caused *e*'. To my knowledge, no explicit account has ever been supplied, whether of decisions, God, or anything else. It is difficult to avoid the suspicion that no such account is possible.

That is, it is difficult to avoid the suspicion that it is impossible to give sense to talk of explanation of *b* by *a*, where the explanation is significantly similar to an explanation in terms of causation, and it is none the less true that $b = a$. It is easy enough to conceive, say, of a thing's being self-changing or self-governing, where that is the fact of one part of a thing being in causal relation to another part of the thing, or of one event in the history of a thing causing another such event, but this of course is nothing to the point. Self-guided missiles and the like are not near to being instances of the relevant sort of self-causation.

We must therefore resist the inclination to think that what is proposed, in order to make certain chance events other than arbitrary in the given sense, is the novel application of a settled idea. What we rather have is a no longer novel application of an unsettled idea. We are thus nowhere near having a clear idea of a grounded decision. In fact the conclusion that we have nothing clear is indubitably implied by what else must be said in or of the proposal. One thing already said is that a decision is to be understood as *not* standardly necessitated by either neural or non-neural facts. It follows that talk of its *being* necessitated, partly by itself, is to be understood quite differently. It is typically also said that the causation in question is to be distinguished from 'the causality which obtains between events' (Boyle, Grisez, and Tollefsen, 1976, p. 12) or 'event-causation' or 'Humean causation', where what is intended by these misleading descriptions is standard causation. Again, the explanation of something which involves self-reference is allowed to be a unique kind of explanation, an appeal to a unique kind of innate intelligibility. (Boyle, Grisez, and Tollefsen, 1976, pp. 83–5)

3.6.2 *Self-Justifying Decisions*

Let us look at a second attempt to clarify the idea or perhaps what can only be called the words 'chance or random but nonetheless grounded or non-arbitrary decision'. The attempt is one of a family, all the members of which try to explain something by way of the idea of the thing's 'turning back upon itself' or reflexivity. (Nozick, 1981)

A decision or choice, we may agree, is a matter of arriving at a view as to which of a number of reasons, for or against an action, have most weight. Or rather, we may agree, since it seems that reasons do not come with weights attached, the process is importantly one of assigning weights to reasons, of *weighting* them. This assigning or bestowing of weights, we are now told, is not necessitated by anything. However, to come to the principal point, it is grounded or non-arbitrary in that it refers to itself.

> The decision may be self-subsuming; the weights it bestows may fix general principles that mandate not only the relevant act but also the bestowing of those (or similar) weights. The bestowal of weights yields both the action and (as a subsumption, not a repetition) that very bestowal. For example, consider the policy of choosing to track bestness: if the act weren't best you wouldn't do it, while if it were best you would. The decision to follow this policy may itself be an instance of it, subsumed under it. (Nozick, 1981, p. 300)

Is there the difficulty that some *different* bestowal of weights might have occurred, and hence that the given bestowal must count as arbitrary? To repeat,

> . . . can anything be said about why that one self-subsuming decision is made rather than another? No, the weights are bestowed in virtue of weights that come into effect in the very act of bestowal. This is the translation into this context of the notion of reflexivity: the phenomenon . . . has an 'inside' character when it holds in virtue of a feature bestowed by its holding or occurring. (Nozick, 1981, p. 304)

Further, such decisions bring into existence a self.

> . . . the self can synthesize itself around this bestowing: 'I value things in this way'. If in that reflexive self-reference the I synthesized itself (in part) around the act of bestowing weight on reasons, then it will not be arbitrary. . .that *that* self bestowed those weights. (Nozick, 1981, p. 306)

The fundamental objection to this is that what we are given is a certain account of how decisions may be *justified*, which we do not need, and we are given no account of *how they come about*, which is what is needed and was promised. In any relevant sense of explanation, we are given *no* explanation of the occurrence of decisions. The greatly elaborated idea in what is actually said of decisions is a simple one: that a decision, like some other mental events, may apply in a justificatory way to itself. To take a clear example, a man in the grips of some idea of naturalness may make the decision *d* that only those decisions are to be trusted which are spontaneous ones, and he may have some tolerably clear conception of the spontaneity in question. If decision *d* was itself spontaneous in the given sense, then decision *d* is to be trusted. We can enter absolutely into his idea and outlook but have no answer at all to this question: Why did *d* occur? But precisely that *is* the question with which we are concerned.

Our question is indeed what was earlier called the diachronic question of the explanation of mental events, and to which we have been considering various answers, latterly the Hypothesis on the Causation of Psychoneural Pairs. The very heart of any Indeterminist Theory must also be an answer to this question. We surely get *no* answer to this question by having any account of how a decision is one to be trusted, a good or a rational one, one which 'tracks bestness' or fits a proper conception of one's own life, or any like thing. What we do get in what is offered is in analogy to explanations of the self-verifying nature or the incorrigibility or the like of, say, Descartes's proposition 'Cogito ergo sum', as distinct from an explanation of how it comes about that there occur mental events which can be individuated as being thoughts of that proposition.

Is there any reason to doubt the distinction between explanation in the relevant sense and justification, or, more generally, the rightly entrenched and pretty well universally accepted distinction between explanation and evaluation? Is there any reason to doubt the distinction between answering the question of why or of why and how something occurred, and of what value it is? It is perhaps no surprise that none is supplied in the given account of decisions.

Obviously we typically explain the occurrence of decisions by citing evaluations by the agent, values of his. That is, we explain decisions by taking them as somehow owed to mental events and dispositions of an evaluative kind. It is part of the Hypothesis on the Causation of Psychoneural Pairs that such mental events and dispositions are precisely *causes* of decisions. The Indeterminist theory denies this account and seeks to give some alternative explanation of the occurrence of decisions and the like. The idea now under consideration, like the previous one in terms of what is called self-causation, sets out to make decisions somehow self-explanatory. However, it in fact offers an account of their self-justification, which is not on a level with and does not compete with either our typical account of the occurrence of decisions as owed to evaluative mental events and dispositions, or the causal hypothesis.

It may be that something called the self can be regarded as somehow brought into being, synthesized, delineated, developed, or whatever, by one's decisions. Certainly it is true that one's personal identity, being in part a matter of what were earlier called continuities of experience, is affected by at least certain of one's decisions. That a self can synthesize itself around a bestowing of weights, however, is nothing to the point. There is then at least guaranteed consistency, so to speak, between self and decision. What we are pursuing, however, is not a reason for thinking that self and decision do not conflict, or that the latter is not 'arbitrary' with respect to the former, but an explanation of how the decision comes about. If the justification or evaluation of a decision is nothing to the point, so too some sort of logical consistency or the like between decision and decision-created self is nothing to the point. Finally, with respect to what is said of decisions themselves, can the relevant question be thought to come into view in the second passage quoted (Nozick, 1981, p. 304) despite the persistent ambiguity? In fact the truth seems otherwise. In any case, no answer is given to the relevant question.

Both the previous account of decisions as self-causing and the account now on hand of decisions as self-justifying attempt to explain the supposed groundedness or non-arbitrariness of chance decisions by attending mainly or entirely to the decisions themselves—their own very nature. The central ideas about groundedness, it may appear, do not in fact much involve an explanatory self, ego, mind, self-conscious mind, faculty of mind, active power, rational capacity, agent, person, or the like—although much is said of the self, somehow conceived, in the present account. Both accounts, if they were to succeed in their aims, *would* have to give a real place to a somehow explanatory persisting entity and *would* have to clarify its nature effectively. The existence of such a thing could be no mere corollary but a central proposition. If the accounts are to achieve their goal of establishing a certain responsibility, they need not only mental events but such an entity. In the relevant sense, it is not decisions that we blame.

3.6.3 Teleology

Let us turn to a third attempt that may be made to give an account of the groundedness of decisions and choices, without facing the difficulties of a self. In answer to the question of why I decided to stop smoking, I say that I did so for health. Clearly this claim of mine is perfectly meaningful. It makes my decision intelligible, and evidently it may be true.

It is ordinary to think that such claims somehow are or reduce to ordinary causal explanations. They are so taken when they occur in biological science, where they are common and indeed fundamental. (Woodfield, 1976; Braithwaite, 1953; Sommerhoff, 1950) However, it may be denied, as in the present account of decisions, that they are or reduce to ordinary causal explanations. It is said, specifically, that such propositions as those about my stopping smoking are *not* to the effect that my desires and beliefs or the like caused my decision. Rather, it is said, such claims explain my decision by reference to a goal, end, or telos. These *teleological claims*, as we shall call them, are in fact as others describe them: *teleological explanations*. This view of them, dating back to Aristotle (Sorabji, 1980, Ch. 10), can be and often has been pressed into use to give an account of the groundedness of unnecessitated decisions.

Do teleological claims not reduce to ordinary causal explanations, and are they nevertheless explanatory? If we can persuade ourselves of this, we shall be enabled to see how an unnecessitated decision can none the less be grounded or non-arbitrary.

It will be as well first to look at the matter of teleological claims generally, independently of our present concern. Here are more examples. *Birds have hollow bones because that enables them to fly better. We perspire in order to reduce bodily temperature. The gas is turned on because that will make the kettle boil. Shoe factories grow in size because such growth produces economies in production.* In trying to think of them in the required way it is essential, to repeat, to exclude the ordinary idea that they are or reduce to ordinary causal explanations.

In each case we are given what supporters of the doctrine of teleological explanation regard as an *explanandum* and an *explanans*, an *explanans* that *by itself* is no less than complete. In the first case, the hollow bones of the birds are the *explanandum*, what is to be explained, and their better flight is a full *explanans*, that which explains. In each case, further, as of course must be allowed and sometimes is, whatever else is said, the *explanandum* does in fact have a causal role with respect to the *explanans*. Hollow bones are parts of causal circumstances for better flight. What we have, then, involves precisely a reversal of ordinary explanations, where causes and causal circumstances explain effects. In what is called teleological explanation we appear to have effects explaining their causes or causal circumstances.

The appearance cannot be the reality. Effects do not in any ordinary sense explain causal circumstances or causes. They do not explain them in the precise sense set out in our inquiry into nomic connection, having to do with temporal order and necessitation as against dependent-necessitation. (1.5) Nor are effects taken to explain causal circumstances, or any parts of them, by those who rely on an intuitive and unanalysed conception of explanation. In answer to the question of why something occurred, it is not possible to give as the sum total of explanation that it had a certain effect, whatever else may be said. This much is now explicitly allowed by defenders of teleological explanation, whatever the inclinations of their forebears.

Contemporary defenders of teleological explanation do maintain, as they must if they are to claim anything distinctive, that there is some related way of looking at the supposed *explanandum* and *explanans*, and at nothing in addition, and thereby getting an explanation of the supposed *explanandum*. If causes cannot be explained by effects, it is supposed, there do exist related possibilities. Let us look briefly at two ideas, suggested in what is the most distinguished recent attempt to defend true teleological explanation, or, as it is called, functional explanation. (G. A. Cohen, 1978, Ch. 9)

It seems sometimes to be suggested in this defence that a *fact* in the sense of a *true proposition*, as distinct from what makes the proposition true, does the explaining. The true proposition that the growth of shoe factories gives rise to economies in production, as distinct from anything else, explains the growth of the shoe factories. The true proposition that perspiration has the effect of reducing bodily temperature, as distinct from anything else, explains perspiration. An effect does not explain its cause, but the proposition that the cause has that effect explains the cause. This will not do. Certainly true propositions may be said to explain events, things, and the like—in general, spatio-temporal items. But what this must come to is that the events and the like are explained by *what makes the true propositions true*, which is to say events and the like.

Might something of that relevant kind really be the intention of saying that a fact does the explaining? Whether or not that is so, something of that kind is suggested in the second idea about the supposed *explanans* and *explanandum*. What explains the hollow bones of the birds? The answer given here is that it is their *causal disposition* to give rise to better flight. Why is the gas turned on? Because that state of affairs has a causal disposition to make the kettle boil. This will not do either, again for a clear reason.

To speak of a thing's causal disposition, as we know, is to speak of a property of that thing—a property such that when it is conjoined with certain other things, the resulting whole is a causal circumstance for the mentioned effect. But then to attempt to explain something by citing one of its causal dispositions is to attempt to explain part of a thing by just that very part. That claim, as we also know, is wholly obscure. Perhaps it does not require audacity to say that it is impossible. If it is impossible to explain the occurrence of a thing by citing its effect, or by citing merely a proposition, it is no less impossible to explain it by citing itself.

These two attempts to make sense of the idea that teleological claims are rightly called teleological explanations deserve more attention, and have been given some elsewhere. (Honderich, 1982a) However, without lingering, let us now briefly consider another view of teleological claims. There is in fact no mystery about them and they can be understood in such a way as to account for their appearance of explaining a cause by its effect.

To make a teleological claim may often be to intend, and to be understood as intending, an abbreviation of a certain undoubted explanation. Teleological claims, when they are not abbreviations, are in fact not explanations, let alone full or complete explanations, but *explanation-claims*: claims that explanations of a certain kind exist. (Cf. Woodfield, 1976, p. 206; Mackie, 1974, p. 283, E. Nagel, 1979, p. 401) To say birds have hollow bones because that enables them to fly better is to say, here, that there does exist a standard causal explanation of the hollow bones, an explanation which *includes* the fact that hollow bones have a causal role with respect to better flight. To say that shoe factories expand because that issues in economies is to say, here, that there exists a standard causal explanation of their expansion, which explanation includes the fact that expansion is part of a causal sequence for the economies.

What are the explanations? In the case of the hollow bones the explanation is of course an evolutionary one. This explanation, having to do with the persistence of some species and the disappearance of others, evidently includes the fact that hollow bones have the effect they do. In the case of the shoe factories, although there is the possibility of an explanation which is Darwinian in character, the best explanation is in terms of men's desires and beliefs. Men, wanting the benefit of economies in production, increased the size of shoe factories. The fact that greater size does have the given effect is again part of the explanation. It is such Darwinian and purposive explanations, of course, which are typically abbreviated by teleological claims.

The principal point in this view of explanation-claims, which has been given without elaboration, is not a terminological one, a restriction on the use of the term 'explanation'. It can be allowed, however misleadingly, and despite what was said a moment ago, that we give an 'explanation' in saying that birds have hollow bones because that enables them to fly better. Certainly it can be said that teleological claims make events somehow intelligible. What *cannot* be allowed is that we have anything remotely like a full explanation of the hollow bones if we mention only the hollow bones, the better flight, and the causal role of the bones with respect to the flight. Partly as a result, what cannot be allowed is that what is in question is anything other than standard causal explanation. Both the Darwinian explanation and the given purposive explanation are of course of this form. In consonance with this, so-called teleological explanations are indeed better called explanation-claims.

The relevance of all this to indeterminist pictures of the mind will be clear. There is no possibility of explaining a decision *d*—no possibility of giving anything like a full explanation of it—simply by referring to its effect or by any related method. The ideas pertaining to causal *facts* and *dispositions* cannot help. However, given the proper understanding of teleological claims, there is more to be said. The teleological claim that I decided to stop smoking for my health, far from providing an explanation of my decision which does not involve necessitation, is to be taken as claiming the existence of precisely what indeterminist pictures of the mind deny, which is to say a standard causal explanation of the decision. Or, of course, the claim merely abbreviates the explanation. Should these analyses be denied, our situation is that we have *no* clarification whatever of the supposed groundedness of decisions.

It hardly needs remarking that there is a further disability of any view which seeks to explain groundedness in the given teleological way. It has exactly a disability shared with its predecessors, having to do with self-causing and self-justifying decisions. Were the teleological view to succeed in explaining a decision in the intended way, it would none the less not achieve the goal of safeguarding responsibility of the desired kind. It cannot be that we could hold a *decision* responsible. Some ongoing entity is required.

3.7 SECOND ELEMENT, WITH ORIGINATOR TO THE FORE

We turn now to theories which with good reason are less reluctant about an originator, or which in fact fully accept the necessity of one and are largely about it. To have that recommendation, of course, they must involve themselves in difficulty. Some involve themselves in more than difficulty.

3.7.1 Homunculus

What some seem to offer is an homunculus, a little person located within a person. This is suggested by the use of ordinary mental verbs. The originator perceives, chooses, and so on. It may even have 'a personality, something like an ethos or moral character', partly formed by its own actions. (Popper & Eccles, 1977, pp. 472–3) Thus an originator is first located somehow at the centre of a person conceived in an ordinary way, and then, it may seem, the originator is itself regarded as such a person. The given picture of the mind, when made explicit, is absurd, and almost certainly involved in an intolerable regress. The originator will in turn require its own originator, and so on.

3.7.2 'The Mind'

A second idea is that what stands as the non-necessitating ground of a particular decision is no less than 'the mind' or 'the self-conscious mind' or some such, where this is one series of mental events, the one experienced by a particular person, or, better, such a series together with mental dispositions. (p. 86) The idea of the mind as mental events and dispositions, makes use of what is evidently the clearest and by far the most defensible conception of 'the mind'. It is in fact a conception at which we arrived without comment in considering consciousness and mental events, and which has informed our subsequent inquiry. Can the grounding originator which the Indeterminist Theory needs be identified with the mind in this sense? It cannot. A part of one of several reasons is that not anything like all of the sequence of mental events which is my conscious history, or anything like all of my mental dispositions, can be thought to enter into the explanation of a given particular decision. Something more precise is needed, analagous to a causally relevant property or a set of such. Other objections to the view, as serious, will be noted in connection with the next one, where they also arise.

3.7.3 Selected Elements of 'the Mind'

Suppose we quickly and radically amend the view that what we need is all of a mind, and select certain items to constitute our originator. We thus come to a different idea about groundedness, which can also have brief treatment. There is the first difficulty that we already have these items within indeterminist theories as so far outlined. In terms of Figure 17 (p. 181), and thinking of mental event *m'*, they are represented by the earlier mental event *m* and what can be called the neural structure *n**. Are these now to be *identified* with the originator *o*, supposing that to be possible, and to be taken as entering into some *new* explanatory connection with *m'*, over and above the connection we already have? The connection we already have involves the Proposition of Neural Indeterminacy, probabilistic causation, and the Correlation Hypothesis. We cannot *discard* that connection, if we are stay in touch at all with what is known empirically of the brain and mental events.

There is also a second great difficulty about the course contemplated. If we take it, there will again be no chance of achieving the goal of an indeterminist theory, holding a man responsible in the given way. Here a certain distinction needs to be made clear. There *is* a conception of responsibility, to which we shall certainly be attending at a later stage, which has much to do with an agent's ongoing dispositions. The conception which gives definition to the Indeterminist Theory, however, is different. The theory sets out to secure that we could have decided otherwise than we did at *t*, given all things as they were at *t* and before. If one has this in mind, it becomes starkly clear that what grounds a decision *cannot* be the given dispositions or past mental events. Precisely these are central to what the responsible agent, so to speak, *rises over* or *leaves behind*. The whole enterprise of the theory is to achieve this escape or transcendence, importantly by way of the Proposition of Neural Indeterminacy.

3.7.4 The Person

These last two ideas as to the grounding of mental events have been noticed not only for themselves, but in introduction to something rather more popular which in a sense includes them. As indicated earlier, some indeterminist philosophers have the credit of being at least sceptical of an originator as traditionally conceived—a self, ego or whatever. One is thus scandalized by the supposition that an indeterminist view necessarily involves the 'dreadful and bizarre' idea of an Ego or Self. (Taylor, 1966, p. 134) A second hastens to say he will have no dealings with the 'extravagant' idea of a Self or Mind. (Thorp, 1980, p. 120) Both of these philosophers, and some others, take or contemplate what they regard as the far happier course of grounding mental events in, simply, a *person*, a living man or woman. When Tom decides to give up the unequal struggle, it is Tom himself who brings this decision about in some unnecessitated way. (van Inwagen, 1983, p. 4) In our way of speaking, the originator is no part or element or the like of Tom, but simply Tom.

It is very notable that none of these latter speculations includes an account of the nature of a person. That subject is fatally passed by. It was remarked earlier (p. 148) that a finally satisfactory account of the nature of a person will include continuities of experience and also, very likely, some idea of a *locus*, tunnel or frame for these. It can be speculated, further, that a person is also to be understood in terms of bodily or brain or brain-part continuity. Without attempting to press on to a specific answer to the question of personal identity and of the nature of a person, several things are clear.

Any tolerable account of a person, first, will be an account of greatly more than can with any reason at all be included in an entity which is to be explanatory of a given particular decision. To propose a selection of items faces the objection that we already have these relevant items in any tolerable indeterminist theory. The objection is a repetition of one made to the previous ideas as to groundedness, the attempted identification of the originator with 'the mind' or with selected elements of it. Another objection also transfers itself readily to the present fourth idea. To attempt to identify an originator and a person, since a person consists at least in part in a psychological past, a past of experience, is to tie an originated decision to precisely that of which it is to be independent, that which it is to transcend.

3.7.5 Originator

The outcome of our reflections as to the second element of an indeterminist theory is that it requires something which is indeed best suggested by the technical term or name of *originator*. How are we to attempt to regard this entity? It has traditionally been described as a substance or independent entity as distinct from an attribute, which is to say it enjoys a kind of independent existence. Also, it is a substance not in space. Despite Descartes, it seems unkind to proponents of indeterminist pictures of the mind to insist on either of these. An originator, we shall take it, may be an attribute of a person, rather than a substance, and may be in space. Certainly it must persist in time, be ongoing, for reasons we know. It would be yet more unkind to assign immortality to it, let alone an early choice of which side of the brain to concentrate on. (Popper & Eccles, 1977, p. 507)

It is natural and certainly traditional to take an originator as having a particularly fundamental unity and being simple rather than composite. This is presumably the same proposition as that the idea of an originator is primitive, not to be reduced to anything else—certainly it is no bundle or collection of things. Perhaps this is in some accord with what we are also to understand, that an originator is that to which, at least sometimes, we refer in using the first person pronoun, 'I'. It is said that it is something of which each of us necessarily has a pre-theoretical grasp. It is that which is misconceived by those who seek to give a unity to consciousness by speaking of merely a *locus*, tunnel or frame. It is that which is misconceived, also, by anyone inclined to the conception of consciousness and mental events set out earlier in this inquiry, and in particular to the minimal idea of a subject as distinct from content. (pp. 80 ff.)

To what extent can an originator be understood by way of that minimal idea of a subject as distinct from content? An originator, mainly because of its fundamental explanatory role with respect to mental events, cannot be identified with such a subject. An originator is not internal but external to mental events. The postulation of an originator requires, or at least makes it natural to take up, a quite different conception of mental events. It will be one which subtracts the subject from a mental event and makes it entirely a matter of content. Further, an originator as against a subject, even if it is attribute rather than substance, is not necessarily understood by way of an idea of interdependent existence—the mutual dependence of subject and content.

There is one other question of characterization which is uniquely difficult and very likely intractable, another question that proponents of indeterminist theories fail to get on to their agendas. What is an originator's ontological kind? Is an originator mental, or neural, or something else? Is it or is it not physical in the sense defined earlier, where physical events include both mental and neural? (2.2) To remember again our own conception of a mental event, if the subject is subtracted from it and elevated into an originator, and the event left as content, what is the nature of the originator?

It is of course customary, and perhaps essential to the goal of indeterminism, to locate it on the mental side of things, but in fact, it is also customarily *distinguished* from mental events—such as the decisions and choices it originates, or, on many views, other mental events which occur *to it*, or *of which* it is somehow the proprietor. Further, if it were mental, it presumably would have to be subject to something like the Correlation Hypothesis. It would then lack exactly the autonomy and creativity with which it is credited, which we shall be considering in a moment. The originator thus appears to be a different order of thing from neural things and also from mental things as usually conceived. What order of thing is it? For want of help from its proponents, and want of fortitude on my part, let us abandon this question. No answer of any useful content has ever been proposed.

In short, what has been and can be said of the sort of originator which the indeterminist requires is not satisfactory. If we put aside the suppositions that make it into a soul—immortality and so on—we have a conception of hardly any positive content. What we have is the thin idea of a unity of a wholly unspecified ontological kind. Needless to say, a principle of simplicity or parsimony must count against the supposition of this third category of being. Nor is there any effective argument to the effect that we have need of the complexity. It is nonsense to say we have a pre-theoretical grasp of something of this nature—neither mental nor otherwise physical. The supposition of an originator also raises further questions, of whether an indeterminist view satisfies the constraints of mental indispensability and personal indispensability. This has to do with the autonomy of an originator—its independence of desires, beliefs, and so on. Let us leave these matters unconsidered.

All of this is critical, but, from the point of view of precisely an indeterminist theory, as distinct from the philosophy of mind generally, it is not central. The central characterization of an originator must be in terms of its relationship to decisions and choices, its unique explanatory and responsibility-preserving role, its autonomous creativity. We have not been paying much attention to that. Let us turn to it, the third element of an indeterminist theory.

3.8 THE ORIGINATOR'S RELATION TO MENTAL EVENTS

3.8.1 Mental Verbs

In one indeterminist theory, the self-conscious mind is said to 'read out' from the brain, select from it at will according to its attention and interests, scan and unify it, probe and control it. (Popper & Eccles, 1977, pp. 282–3, 355, 363, 427) These are all relations between self-conscious mind and neural structures or activity. In other indeterminist speculations, a self or ego is presented as contemplating, evaluating, and judging mental events and dispositions, and of course somehow giving rise to them. One traditional theme is that such an entity considers and weighs up a conflicting desire and a feeling of moral obligation, and sides with one or the other.

One clear and fundamental objection to any such account is that the relation between originator and either neural or mental events is described in terms of *ordinary mental verbs*, principally verbs having to do with perceiving, controlling, choosing, and acting. Whatever else is to be said of these usages, there is the objection that they are *neutral* or *uninformative*. What we want to know is the nature of this contemplating, attending, choosing and so on. A determinist analysis of, say, my contemplating my desires, is given in terms of nomic connections, those specified in our first two hypotheses. It is no objection to this analysis merely to assert the *analysandum*. More importantly, we get *no alternative analysis or account whatever* of such an *analysandum* when what is offered to us is precisely that *analysandum*. What we are promised by the indeterminist is an account of contemplating, choosing and so on which presents them as unnecessitated but grounded pheneomena. We get no such thing by a recourse to the ordinary verbs themselves.

3.8.2 Self-Causing Cause

Talk of self-causation is used by indeterminists other than those noticed earlier, who speak of self-causing decisions. It may be said that an originator causes an event in or of itself—a change *c* in or of itself. This event enters into a wholly standard causal connection with, say, a decision *d*. Thus, it may be supposed, we have an answer to the question of the originator's relation to its decision.

We need to ask, however, for more precision about the self-causing. Is it the case that what causes *c* is some other event in or of the originator? Suppose the answer is yes. If so, since we presumably here have another standard causal connection, how can it be that decision *d* is unnecessitated? To gain that end, the indeterminist may instead say of *c* that it is self-causing. What we have, then, is a self-causing event *c* of an originator *o*, which event enters into a standard causal connection with the decision *d*. That is, either *c* is itself a causal circumsance for *d*, or *c* is an element of a causal circumstance for *d*. On the latter assumption, the other elements of the causal circumstance for *d* will be mental items relevant to it, desires and the like. The latter assumption seems at least preferable, in that it includes essential elements within the story of the emergence of *d*.

The view might be claimed to lack one disability of taking decisions themselves as self-causing. That is, it does have an originator firmly on the scene, a persistent entity to the subject of judgements or feelings as to responsibility. Precisely the claimed superiority, however, rests on unexplained propositions. In what sense is change *c* 'in' or 'of' the originator *o*? An answer is not easy to come by, or even to move towards, partly because an originator seems best conceived, as we have, as a simple unity.

Whatever is to be said of this, the view does have other disabilities. One is that to speak of self-causation of any kind, whether of decisions themselves or events which cause decisions, is, as we know, to speak of nothing clear. A second disability attaches to taking the self-causing originator-event as an element of a standard causal circumstance which includes other mental items, such as desires. These items already turn up in the attempted explanation of the decision by way of the Correlation Hypothesis and the Proposition of Neural Indeterminacy.

3.8.3 The Originator Itself as Causal

Another view goes further in the direction of making the originator a suitable subject of judgements and feelings as to responsibility. This it does by making the originator itself somehow causal, as distinct from any event of it or in it. Further, it is the originator on its own that is somehow causal, as distinct from a set of items including the originator. Of course some fitting account must be attempted of this causing, an account which makes the resulting decision other than necessitated but none the less grounded.

The substitution of the originator for a causal event, to begin with that, produces what is at least a certain surprise, given either of two assumptions. We may assume, first, an originator that is in fact unchanging throughout its existence. It is permanently eventless. The idea is in some accord with our description of it as a simple unity, and its being named a pure ego and the like. Or, we may assume an originator that is unchanging during a certain period of time. It is for this time eventless. In either case the surprising supposition is this: an originator *o* which is unchanging between t_1 and t_3 somehow causes a decision at t_2.

What then is the so-called causal relation? Clearly it cannot include what is fundamental to standard causation, which is necessitation. It cannot be that originator *o* necessitates decision *d* at t_2. Given that *o* at the earlier time was *identical* to *o* at t_2, then, on the assumption of necessitation, *d* ought to have occurred at t_1, or indeed *d*-type decisions ought to be being produced through the period from t_1 to t_2. Whatever is to be said of the mysterious relation, incidentally, it will carry a certain corollary satisfactory to the indeterminist. It will not make sense to suppose that something else in any way caused *o* to produce *d* at t_2, since there occurred no change in *o*.

To repeat, what is this so-called causal relation between *o* and *d*? We may be told, perforce, that what is in question when *o* causes *d* is not what we have called standard causation. That is no news. What *is* in question, it may be said, is a somehow more basic or older or less controversial conception of causation, or in fact *the* conception of causation. It is a conception of causation, so-called, encountered earlier in this inquiry. (pp. 57 ff.) A reminder is in order, and can be had by way of a typical exposition and defence of the conception, offered by a strong defender of an indeterminist theory.

Suppose someone throws a stone at a window and that the stone strikes the glass and the glass shatters in just the way we should expect. . . . Suppose further that God reveals to us that the glass did not *have* to shatter under these conditions, that there are possible worlds having exactly the same laws or nature as the actual world and having histories identical up to the instant at which the stone comes into contact with the glass, but in which the stone rebounded from the intact glass. . . . Could this revelation really lead us to say that, despite appearances, the stone didn't cause the glass to break? That this is a *logical consequence* of the revelation? Wouldn't it be more reasonable to say this: that, while the stone did cause the window to break, it was not *determined* that it should; that it *in fact* caused the window to break, though, even if all conditions had been precisely the same, it might not have?

This case convinces me that whatever the *facts* of the matter may be, it is at any rate not part of the concept of causation that a cause—or even a cause plus the totality of its accompanying conditions—determines its effect. (van Inwagen, 1983, pp. 139–40)

Philosophers have a history of discovering scandals in the thought of their opponents. It seems to *me* a small scandal that this sort of persuasion, without even any colour given to it by talk of probability, should convince anyone to contemplate the idea that we have any such ordinary idea of causation. If we entirely put aside all of our firmly based causal preconceptions about stone and glass—the argument more or less depends on not doing so—what we have is the supposition that today there occurred an event, and it might not have occurred despite the fact that everything was as it was up to the moment of its occurrence. We have the related supposition that taking into account all relevant factors, we might get an exact and total repetition of them tomorrow, or daily for the rest of time, without the glass breaking. If so, it is to my mind absurd to speak of today's breaking of the glass as an effect, as something caused. For a clear reason, the event was no effect in any ordinary sense and we have no case of causation in any ordinary sense. The reason, expressed one way, is that there exists *no explanation whatever* of today's event. Anything and everything we can contemplate as a possible explanation, as we know, might have been followed by *no* event. Should we still be persuaded that the stone caused the given event since, although it did not necessitate it, it did 'produce' it? (van Inwagen, 1983, p. 4) Most certainly we should not. The stone 'produced' the breaking only in some wonderful sense consistent with the proposition that stone and all else might have been exactly as was, without a breaking. It is notable that the philosopher attracted to the given line of thought, earlier in his often admirable book, has the following to say: 'Causation is a morass into which I for one refuse to set foot. Or not until I am pushed.' (van Inwagen, 1983, p. 65) That reluctance, one might say, came home to roost.

To return to the originator *o*, and the occurrence of the decision *d*, we thus have no account of how the decision is more than a chance event. To say that the originator caused the decision, in the given sense, is in fact to add nothing whatever to the proposition we have, that the decision was unnecessitated. Needless to say, things do not become better, but still worse, when we contemplate what else is proposed, that what somehow causes something may in a further sense do so out of nothing. In the story of the stone we at least have a so-called causal *event*. In the story of the unchanging originator, no event at all occurs. If there is nothing that explains *why* the so-called effect occurs at all, there is also nothing that explains why the so-called effect occurs *when* it does.

3.8.4 *The Will*

It can be said of the originator, greatly more persuasively, that it is the will. (Kenny, 1975, 1978) It is one of the two great faculties of the mind, the other being the intellect, both of them being bound up with the creation and use of symbols. More specifically, the will is a capacity or power at least to intend, choose, or decide to do things. Further, it is a rational capacity or power, and thus to be distinguished from a natural or irrational capacity or power.

> Aristotle . . . drew a sharp distinction between rational powers . . . and natural powers like the power of fire to burn. If all the necessary conditions for the exercise of a natural power were present, then, he maintained, the power was necessarily exercised: put the wood, appropriately dry, on the fire, and the fire will burn it; there are no two ways about it. Rational powers, however, are essentially, he argued, two-way powers, powers which can be exercised at will: a rational agent, presented with all the necessary external conditions for exercising a power, may choose not to do so. (Kenny, 1975, pp. 52–3)

It must be, I take it, that we are to derive the idea of a rational power from a natural power, but this is not easy. It is not easy, in part, since we are not to understand that a natural power is a property of a thing such that the combination of it with other items forms a causal circumstance for some event. A natural power is nothing so clear or palpable. Some natural capacities are *of* our bodies, but they are not to be reduced to the structural parts and features in virtue of which we possess those capacities. They themselves are something other or something more than properties, or at any rate ordinary properties, of our bodies. This, it may seem, is other than fully clear.

We may, for want of something fully explicit, start from what we know, the clear idea of a natural power. We can then suppose that the rational power which is the will is in part this: a property of a person such that when it is conjoined with other things, the combination is such that it may or may not issue in a decision or whatever. There may be *no difference whatever* between a situation where the power is exercised, so to speak, and a situation where it is not. In fact, we are to understand, there is no situation whatever such that in it a rational power will have a standard effect, a necessitated event.

What must be said of this, in line with earlier remarks, is that we are offered no explanation of why the will gives rise to a decision when it does. In fact, it is at least arguable that no adequate content is given to speaking of the will as *giving rise to* a decision. We have no adequate idea of what relation is in question. It is not as if we have an articulated account of how a decision may be more than chance or random, which account we may dispute. Rather, it seems, we have no such articulated account.

What is implied by the description of the will as *rational* power? It may be conjectured that what enters into the exercise of the given power is a set of *reasons* pertaining to the decision or whatever. A reason may be specified by its propositional content—'The train leaves at 5 p.m.'—and a decision may also be specified by its content—'I'll try to catch it'. There is thus what can be called, in a loose sense difficult to characterize, a logical connection between reasons and decision, a connection with which the will has something to do. Do we here have an account of the relation between the will and its decision?

It has been clear since Hume that causes do not give rise to their effects by entailing them. One thing does not cause another in virtue of the fact that a statement of the occurrence of the cause entails a statement of the occurrence of the effect. No such entailment is any part of the explanation by the cause of the effect. If, in what is said of reasons, there were intended a description of the occurrence of a decision in terms only of *logical* relations, the intention could not succeed. Certainly just this could not be carried forward into anything of the right sort. No amount of reflection on logical relations in themselves could bring us closer to having an explanation of why a certain event occurred. That is what we are after.

To look quickly at a further doctrine, it may well be true that in what is called practical reasoning, premisses which stand in some *logically* generative relation to a conclusion may cease to do so if a consistent addition is made to them. My conclusion that I will try to catch the train no longer stands when I learn that the London meeting to which I was going has been cancelled. Thus there is a certain want of analogy between a set of reasons and a causal circumstance. The reasons are defeasible. (Kenny, 1975, pp. 22–3) To see just that, however, is not to see how the logical relation between the unsupplemented premisses and the conclusion gives anything like an explanation of an event, however defeasible. The point is made clearer by the fact that a determinist has no concern whatever to question or deny the given logical relations. What he will be sceptical of is quite different: that sense has been given to talk of a kind of explanation of events which is somehow akin to causal explanation and yet lacks its fundamental proposition, about necessitation.

The doctrine of the faculty of the will is impressively elaborated, more so than has been indicated, and it should have more attention. Some, not all well-judged, has been given elsewhere. (Honderich, 1980b) The doctrine is indeed ancient, but it is difficult to suppose that it owes its longevity to its own recommendations, as distinct from what it promises to us in terms of our aspirations.

3.8.5 Primitive Relation

It will not be surprising, given our survey, that some philosophers inclined to indeterminism take a certain step. They abandon the unequal struggle, and conclude that no analysis is possible of the relation between an originator and a decision. Their view is that there does exist a relation which satisfies the goal of indeterminist theories, but that no more can be said of it than that. What can be said, that is, is no more than this: a decision is not necessitated by an originator, but is related to it in such a way that it is a grounded or non-arbitrary event. The decision is such that the agent is responsible for it in the required sense—he could have decided oterwise given things just as they were.

It is typically the case that this unnerving recourse to an unanalysable relation is somewhat concealed by certain talk. It may be said, for example, that the originator 'acts' with respect to the decision, or possesses 'active power', or of course is 'causal'. Taken together with the absence of any analysis of the relation, and the proposition that it is in fact unanalysable, it follows that we must divest the terms 'act', 'active power', and 'causal' of any ordinary sense. To say the originator acts, or has an active power, or causes a decision, can mean nothing whatever other than that it does not necessitate a decision but is so related to it as to make the agent in the given way responsible for it. There is exactly the same situation with various other terms and locutions that may be pressed into service.

There are also overt attempts to make more acceptable the claim of an unanalysable relation. Some are to the effect that there is nothing extraordinary about the given recourse. Certain of the things said here are themselves extraordinary. It is claimed that the assertion of an unanalysed and unanalysable 'causal' relation is in no way peculiar, since *all* talk of causation must involve such a primitive idea, *all* talk of causation is in the given way 'unclear'. The indeterminist, we are told, is in just the position of any philosopher concerned with standard or 'event' causation. 'He is not so much introducing mystery into the world as introducing more mystery into the world; the event causality with which we seem so comfortable is itself unfathomably mysterious, as any glance at a freshman metaphysics text will show.' (Thorp, 1980, p. 106, cf. Chisolm, 1966, pp. 21–2; Taylor, 1966, Chs. 2, 3)

The reply is not difficult. Despite the existence of problems with respect to our standard conception of causation, and dispute about it, the comparison with the given unanalysable 'causation' must surely be regarded as absurd. Our opening inquiry into our standard conception of causation, if it establishes nothing else, establishes that that conception is not unanalysable or in any remotely comparable sense 'mysterious' or 'unclear'.

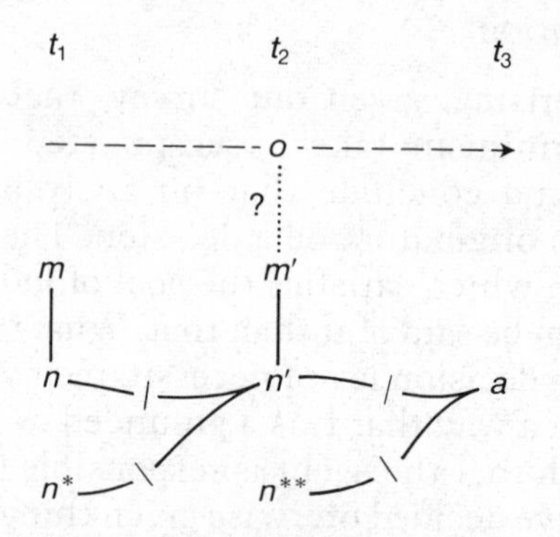

FIG. 18. Further Elements of Indeterminist Theories of the Mind

Putting aside other like attempts to coat the pill of a primitive and unanalysable relation (Thorp, 1981, pp. 108–9), it needs to be said that those who wish to swallow the uncoated pill cannot be deterred. More plainly, there is no possibility of rejecting by philosophical or conceptual means the proposition that there exists a certain unanalysable relation. Since, so to speak, nothing is said of *it* in itself, as distinct from what it secures, a certain responsibility, there is no account of it on offer to be examined for shortcomings.

At the same time, it is difficult to give attention to the supposition of a primitive relation between originator and decision. It is all very well to say, in a different sort of attempt to make the supposition tolerable, that all explanation must begin somewhere, with something taken as somehow primitive. (Boyle, Grisez, Tollefsen, 1976) However, there is all the difference in the world between beginning with a *clear* or *contentful* proposition taken as primitive, and beginning with something quite otherwise.

Figure 17 on p. 181 gave elements of indeterminist theories of the mind. Figure 18 can hardly be said to complete the picture, since, as we know, there are obstacles in the way of that enterprise. (Cf. G. Strawson, 1986, p. 32) It does add the Correlation Hypothesis, modelled by the vertical bars joining *m* and *n*, and *m'* and *n'*, and indicates the obstacles. One is the relations of probabilistic causation, holding between *n* and *n** and their 'upshot' *n'*, and between *n'* and *n*** and their 'upshot', the action *a*. The further obstacle, over and above the mysteriousness of the originator *o* itself, is the relation between it and *m'*, where *m'* is taken as a decision or like mental event.

3.9 CONCLUSION

The second hypothesis of the determinism being expounded, the Hypothesis on the Causation of Psychoneural Pairs, if opposed, must have something put in its place. The worth of the hypothesis, and the subordinate Proposition of Psychoneural Pairs, and the Union Theory, are partly fixed by what is available instead. If the earlier alternatives we considered—Neural Causation with Psychoneural Correlation and so on (3.2)—were open to objection, they were in a general way superior to the alternatives we have lately been considering. The indeterminist theories and ideas of the mind can hardly be regarded as conceptual challenges to the Hypothesis on the Causation of Psychoneural Pairs. More precisely, these theories and ideas cannot be regarded as alternatives at the same level of contentfulness, clarity, coherence, and development.

This is a matter, first, of their first element, the denial of nomic connection and the substituted probabilistic account of the neural background of the correlates of mental events and actions. That account, or rather the promise of that account, is replete with difficulties. It is in internal tension, if indeed it can escape contradiction; it includes *ad hoc* stipulation, and so on. To remark on its *ad hoc* nature in particular, perhaps, is to move in the direction of what is not yet our official subject, the question of the empirical basis of our determinist theory and of indeterminist theories. However, there is not quite so sharp a line between empirical basis and conceptual adequacy as might be desired.

The second element of any indeterminist theory, an originator, is yet more unsatisfactory. There is no avoiding the need for an idea of an ongoing entity of a certain kind, and no avoiding the fact that the idea is rebarbative. It offends dramatically against the principle of simplicity. It has never escaped widespread philosophical and other suspicion.

Finally, in so far as the distinction between the second and the third elements can be maintained, the account that can be given of the originator's unique activity, the activity itself, is thin nearly to the point of non-existence. At bottom we are told no more of this activity, so-called, than that it is such as to secure that the person is in a certain sense responsible for decisions and actions.

Indeterminist theories and thoughts of the mind are thus conceptually unsatisfactory. They are rightly described, in all their three elements, as consisting in 'obscure and panicky metaphysics'. (P. F. Strawson, 1962, p. 211) The conclusion would stand if we were to carry our inquiry further—if we were to consider the possibility of combining indeterminist ideas we have treated separately, and so on. There is very good reason to think the conclusion will be true of their successors in the future.

That is not all that is to be said of indeterminist theories and thoughts, however. It must be said for them, however reluctantly, that they are not without content. It comes to this:

> In each of us there exists an ongoing entity or attribute which originates decisions, and hence actions; this origination is such as to give rise to a certain responsibility; our decisions are nomically correlated with neural events, but these neural events, like the subsequent actions, are unnecessitated.

We shall in due course look at what is said to be empirical evidence for the proposition. (5.5, 5.6) One other fact about this indeterminism, a larger fact to which we shall also come later, explains why its has a future. It is a response to what is itself a large and indubitable fact, a fact of our deep attitudes and beliefs, of which much will be said.

4

The Causation of Actions

4.1 ACTIONS AS BODILY, AND AS COMPOUNDS

A man remarks to his wife, 'That will do the trick', signs his name to the letter he has written, and stands up. The episode, we say without hesitation, includes his *action* of standing up, meaning of course his intentional action. We cannot naturally say his utterance or his signing his name was an action. The reason is that the ordinary idea of an action, as distinct from the ideas of activity and acting, is the idea of something which is in a way simple. It will be convenient, however, to use the term 'action' to include not only standing up, but also uttering a sentence and signing one's name.

We are unlikely to be misled by our wider usage. Certainly we need a term for the wide category of items which includes not only standing up, winking, pulling the trigger, and the like, which are relatively simple, but also—not that the distinction is sharp—uttering a sentence, signing one's name, writing a letter, keeping someone company for a day, going abroad to avoid income tax, getting divorced, having a child and the like. Primarily, as usual, the term 'action' will be used for a token or particular, something with a single although perhaps diverse and extended spatio-temporal location or extent, rather than a type.

What, in general, do we take such intentional actions to be? What is our general conception of them? The question, as it turns out, is a disputed one. A defensible account is essential to a fully articulated human determinism, which necessarily concerns actions. If the foundation of any human determinism is an account of the mind, no determinism could be complete without a consideration of actions. One of its goals is an explanation of them, and them as distinct from any lesser or more general thing, perhaps behaviour or behaviours. It gives an answer to the question of the explanation of actions. If there were no clear and effective conception of an action, there could be no clear and effective conception of determinism.

Also, the controversy about what freedom we have and what we lack, the subject of a later part of this inquiry, arises as naturally in connection with actions as with decisions and choices. We need to attend to actions, as well, since it is possible to suppose—some have confidently done so—that there are truths in 'the philosophy of action' which stand in the way of the formulation or truth of determinism. Finally, it is possible to suppose, with a very different intent, that our general conception of action provides a kind of evidence for determinism.

4.1.1 Neo-Behaviourism

Actions, including the action of standing still and the like, somehow involve sequences of bodily events, typically movements, those of which we are aware when we are aware of actions. But we do not suppose that our actions *are* such bodily events, or rather, are *merely* such sequences of bodily events. More precisely, we do not suppose that actions can be fully described as such sequences. If it were possible for physiologists, by direct stimulation of the muscles involved in speech, to bring it about that a person to his surprise emitted just those sounds involved in the man's saying 'That will do the trick', the sequence of bodily events and their sounds would be no intentional action of the person. The same kind of consideration applies to all actions.

It is natural to think at this point that actions involve sequences of bodily events and somehow also involve mental events. However, another sort of thought has been canvassed. Is it possible that actions involve or indeed *are* those particular bodily sequences about which there is the added fact that they stand in some relation to something else, which other thing is also not mental? Better, is it possible that actions are bodily sequences which stand in a family of relations to other non-mental things? There has been a good deal of writing and assumption, inspired by Wittgenstein, which is of uncertain tendency but which at least half-invites an understanding along these lines. (Melden, 1961; Anscombe, 1957) These views are with reason called neo-behaviourist, since they are motivated by a kind of general scepticism about the existence of mental events realistically conceived (2.2), or at any rate a scepticism about the need to refer to them in the analysis of action.

One such view of the mind and action, certainly stronger than others, draws not only on Wittgenstein but also Aristotle and Aquinas. It is the doctrine lately noticed (pp. 203 ff.) which has the Faculty of the Will at its centre. (Kenny, 1975, 1978) The mind is conceived, fundamentally, as consisting in capacities or powers, with these firmly distinguished from anything neural. If these capacities, including the

Will, are obscure, it is clear that something different, consciousness, is but a peripheral if not a dispensable part of this conception of the mind.

It is supposed that the idea of consciousness is given a kind of place by allowing that we have a disposition to use language. It is also explicitly granted that there are *some* mental events, as ordinarily conceived. For the most part, however, the mental appears to be reduced to behaviour. If there do occur some mental events, it is also true that when I am coming downstairs, '[my] belief that there is another step may simply be my lurching forward when arriving at the landing . . .'. (Kenny, 1975, p. 119)

The following paragraphs are perhaps fundamental to the given doctrine of intentional action. They have to do with what traditionally has been described as willing an action.

> On the introvert view the will is a phenomenon, an episode in one's mental history, an item of introspective consciousness. Volition is a mental event whose occurrence makes the difference between voluntary and involuntary actions. For an overt action to be voluntary is for it to be preceded and caused by a characteristic internal impression or conscious thought.
>
> The contrasted, extrovert view starts with the observable behaviour of agents and asks for the external criteria by which to distinguish between voluntary and involuntary actions. It may see the mark of voluntariness as being a certain style of activity, or, more likely, as simply the absence of certain untoward features which would render the action coerced or the agent misinformed. On this view to say that human beings have wills can only mean that they can perform, because they sometimes do perform, voluntary actions. (Kenny, 1975, p. 13)

It is a view of the second type that is advanced and defended, rather than the view which involves what are put aside as 'mythical acts of pure willing.' (Kenny, 1975, p. 24)

This, while of an original cast, cannot in the end be satisfactory. Can we suppose, along the lines of the second paragraph just quoted, that actions do not necessarily somehow involve the mental or consciousness as ordinarily conceived and as characterized earlier in this inquiry? (2.2) Does action, which presumably involves belief, sometimes or always involve belief only in the remarkable sense that my belief that there is another step *is* precisely no more than my lurching? What all of this disregard of the axiom of mental realism in fact relies upon is surely insufficient to support it: the many facts of the elusiveness of parts of consciousness, its varying nature and content, different degrees of specificity of content, different levels of awareness in habitual activity, the owing of learned sequences of movements to a single intention, and so on. (p. 300)

All of that may indeed be the case with coming downstairs, playing the piano or typing, and a good deal else, but cannot make persuasive a denial of the mental, in this case such a denial in connection with action. In particular, the absence of a determinate mental event or a mental act of attending with respect to each item describable as an action does not go far towards establishing the non-existence of the mental as properly conceived in so far as actions are concerned. What is said relies as well on a traditional and abstract characterization of willing or volition, certainly not the only possible characterization, as we shall see.

4.1.2 Unambiguous Exclusion of the Mental

Without further reflection on Wittgensteinian or partly Wittgensteinian views of action and the mind, let us simply contemplate an unambiguous exclusion of the mental properly conceived. One possible characterization of an action which remains, in brief, is that of *a movement which somehow coheres with other movements into a pattern of movements, a large or small pattern.* (Cf. Melden, 1961, p. 102.) After the man writes the letter, he does indeed sign it, and then produce the sounds involved in saying 'That will do the trick'. Consider the episode only in terms of relations between movements or bodily events. Such relations, we try to suppose, are sufficient to make the signing and speaking what they are, which is actions.

If sense can be made of the proposal, perhaps it needs to be altered into the true idea that an action is a movement or sequence of bodily events which is open to a kind of explanation partly by way of prior movements, and which enables us to explain subsequent movements. The man's producing of the sounds in 'That will do the trick' is an inference-basis for us with respect to his past and future movements. Movements which stand in no such connections—this is true of tics, spasms, fallings-over—are partly for this reason not taken by us to be actions.

This truth about explanation, even if made clearer, gives us nothing like an adequate account of the nature of action as we conceive it. Certainly it needs to be granted that we generally take a movement to be this or that action on the basis of the evidence of other movements. If the man's movements did not hang together, it would be difficult or perhaps impossible to ascribe actions to him. This fact of evidence is intimately related to the fact that particular actions are identified by way of ascribing reasons and the like to agents. (Cf. Danto, 1973, pp. 46–50; L. Davis, 1979, pp. 12–14.) That is *not* to say that in the truth about explanation we have a full account of the nature of an action. We do not conceive of actions merely as movements or bodily events which stand in certain explanatory relations to other such things, which relations do not bring in mental events as ordinarily conceived. By way of very distant but yet proper analogy, my evidence that it is a cherry tree in front of me in the botanical garden is indeed the sign on it, but it is not this evidence that makes it a cherry tree.

There is no escaping a proposition on which we in fact depended initially in connection with the speculation about movements artificially produced by physiologists. It is that we conceive of our actions as being or involving sequences of bodily events of which a certain thing is true: they are somehow related to the mental realistically conceived. That is as much an axiom as those noticed earlier pertaining only to the mind. (pp. 163 ff.) The axiom can be used to exclude *any* conception of an action which deals only in non-mental events. There is at least one other, and it has more support, at any rate tacit support, than those mentioned so far. It is a consequence of certain Identity Theories of mind and brain.

4.1.3 Identity Theories

Suppose we conceive of a bodily sequence which was the counterpart of the one which occurred when the man signed his name, and also explain it as the effect, by way of certain connections, of a certain *neural* process, that process being related to others. We wholly exclude from our explanation of the bodily sequence any relation of the neural process to the mental realistically conceived. This we may do in connection, say, with the Dualistic Identity Theory with Neural Causation or of course Eliminative Materialism. (2.3) It is at least arguable that we would not have *an intentional action*. No doubt this claim is not so immediately persuasive as its predecessors, for the simple reason that by ineluctable habit we associate the neural with the mental. If, however, we do absolutely exclude the mental realistically conceived from the explanation of the bodily sequence, so too do we exclude an intentional action.

To think, by the way, as some have, that there is an embarrassment for this claim in the fact that we *could* speak here of *action*, as we also speak of the action of a kite, a lever, or the mechanism of a gun, is surely to have lost sight of rudimentary facts about the distinctness of related concepts. Action in this wide sense is not intentional action. Further, to depend on Identity Theories of the mind with respect to intentional action would be to depend on defeated theories. It is a further weakness of some of these theories, although not noticed by their proponents, that they have the consequence that it is arguable that there are no intentional actions as ordinarily conceived.

4.1.4 Actions as Compound, and as Dual-Aspect

It is possible to try to move from something like the axiom we have, that actions as we ordinarily understand them involve the mental, to something absolutely opposed to any idea of the kind we have been considering. It is the conclusion that actions are in fact to be identified with *mental* events somehow conceived, that they are *no more than* such events. (Hornsby, 1980; cf. Prichard, 1949; Foley, 1977) That proposition, if taken as having to do with mental events realistically conceived, is such than any argument for it, however intriguing, must be mistaken. Actions involve both the mental and the bodily. They involve mental events and they involve bodily movements and states as ordinarily conceived.

Shall we then say that an action is a psycho-bodily compound, something with a mental and a bodily part? Most of what we ordinarily call events are in fact compound events. That is, they consist in a number of sub-events, often readily distinguishable. The opening of a flower, a match's lighting, the running of a motor—examples are easily come by. The sub-events may or may not be simultaneous. They are often related as cause and effect, although they need not be. An action, on this view, may be taken as consisting in a volition or trying, or in an intention, and in that in which volition or intention issues, such a sequence of bodily events as occurs when someone stands up, speaks, or whatever. (Mill, 1961 (1843), p. 35; cf. McGinn, 1982, pp. 89–90; Thalberg, 1977) Evidently much might be said in elaborating this clear conception of an action as a compound event, in particular of the nature of volition or intention, and the nature of the connection with the series of bodily events. These matters will get attention below in other contexts.

We can take this view of actions as compounds together with another original and richly elaborated one. (O'Shaughnessy, 1980; cf. Danto, 1973) Here, an action is described as an act of will or a trying or a volition, which thing, however, may be a whole sequence of events which begins with an event in the interior of the mind or brain, and subsequently includes neural and muscular events, and finally includes the overt bodily movements which we see. One distinction between this and the previous view, thus, is that what is called an act of will or whatever is not an initiating mental event, but rather the entire sequence of events or process. The process is dual-aspect. Its nature is what is called psychological and also bodily. That willing is 'spirit in motion' is a kind of literal truth.

The Compound and Dual-Aspect views of action are evidently different. The intricate and impressive development of the second raises many problems. The views are alike in a fundamental respect, which fact opens them to two objections. It is only these that we shall consider.

The first objection, expressed in one way, is that we do not suppose that an action of the ordinary kind does itself have a part or parts which are absolutely or ordinarily hidden from sense-experience. Nothing is more familiar than the fact that what we can call *the springs of action*—motives, intentions, desires, beliefs, decisions and so on—are not open to our sensory awareness, and may take a great deal of divining, but that fact is not to the point.

I may be entirely puzzled as to why a woman nodded when she did, or walked in the way she did, or took her own life. So a judge may be in a quandary as to why the defendant killed someone, and much may hang on his decision as to the springs of the defendant's action. But I do not suppose that there is *a part of the nod* that is in every way unperceivable. Nor does the judge suppose this of the defendant's action. There was not *a part or stage of the killing* that was such that there is no possibility that anyone might have witnessed it. With respect to the defendant's action of *shooting someone* or the like, as distinct from anything else, the judge does not engage in two kinds of inquiry, the first into a part which was or might have been seen, heard and the like, and the second into an earlier hidden part. The judge does not take ordinary eye witness testimony or the like as to a part of the killing, and *necessarily* hear inferences, or make his own, as to another part of the killing. Nor *must* I infer the existence of part of the nod, walk, or suicide. If an action was itself a compound including a mental and a bodily element, or a whole process such that bodily movements were only its end-stage, this would necessarily be exactly our situation. We would take a part of the action to be unperceivable and known only by inference.

There is a second and related objection to the view of actions as compounds and the view of actions as certain dual-aspect processes. It is that they have the consequence that actions begin before we standardly take them to begin. To speak of the Compound view, presumably the intention or volition, or a part of the intention or volition, precedes the movement which it causes. On the Dual-Aspect view, similarly, parts of an action precede the movement. If this is not always true on these views, it is sometimes true. But then, if the action of a nod includes the intending or volition, *the nod may begin before the head moves at all.* The action of firing the gun may begin before the finger begins to tighten on the trigger. The same point is to be made against the Dual-Aspect view. It can hardly be escaped by saying we are not absolutely sure when actions begin. A nod begins when a head begins to move, which moment may be difficult to determine. But on the present views, the nod may begin before that moment. The objection is no footling one. A conception of something which mislocates it in time is a seriously flawed conception.

4.2 THE MENTAL TERMS

What we require in answer to the question of the general nature of actions is a view which brings in bodily events of which we are ordinarily aware, and also the mental, and which, to put it one way, does not partly or wholly make actions into other than the ordinarily public occurrences they are. Let us reassure ourselves about the availability of such a view, by way of certain analogies. A full description of a certain bicycle, in a certain ordinary sense of 'full description', would not include a description of it as my bicycle—that is, Ted Honderich's. The full description in question would bring in all of its parts, but that description would leave open the question of its ownership. There is no part of it which is its being owned by me. It is a truth, none the less, and this is the essential point of general relevance, that the bicycle can be distinguished from others as being within—and, as it happens, the sole member of—a subclass of bicycles. That subclass is the subclass of bicycles owned by me. Its being in that subclass is evidently a distinguishing matter of fact about it but not one of its parts.

Again, we can evidently distinguish a subclass of events, say fallings of wickets, by knowing that they were effects of bowlings by Botham. The bowlings were not parts of the fallings. Such facts may well bring to mind a number of traditional philosophical inquiries, having to do with substances, essences, internal and external properties, and parts. The several facts, and like ones, do not depend on those inquiries, but must rather be guides to them.

There is the possibility, then, as was tacitly assumed earlier, of characterizing actions as a subclass of sequences of bodily events, which subclass stands in a certain relation to the mental. This characterization will have the recommendation of having bodily events constitutive of actions, as evidently they are, and of having the mental properly included as that which distinguishes actions from other bodily events, but it will not have the disability of making actions inaccessible to us, wholly or partly. To clarify such a conception we need to attend to its three parts.

(i) What exactly is the mental element or fact? Or, more likely, what are the several mental elements or facts? Volitions and intentions have been much mentioned, but their nature is less than clear. Are they of very different characters? In any case, it is not evident that they are the only candidates for the mental elements or facts. However related they may be to volitions and intentions, a good many other things have been mentioned: reasons, desires, beliefs, choices, and decisions, not to mention items evidently closely related to or identical with volitions, such as tryings and acts of will.

(ii) What exactly is the relation that holds between the mental element or fact, or several such, and the other term? Some causal connection, as already suggested, must come to mind, but can that be the whole story of the relation?

(iii) What, exactly, is the other term? We have assumed it to be a bodily event or bodily events, typically a movement. As it turns out, there is room for further disagreement here, about a number of questions, of which this is one: How many actions are performed by a single movement? There is also room for an objection. If my actions are taken to be sequences of bodily events, although ones which necessarily have mental elements in a certain relation to them, are my actions not mistakenly made *external* to me? Are they not mistakenly deprived of what they clearly have, an *inner* side or character?

The first question, which will occupy us for a time, has been given the answer that an action is some bodily event or event-sequence which is caused, in a certain sense, by a person. (Chisolm, 1966, 1971, 1976; cf. Taylor, 1966) The answer is bound up with something already considered, a certain indeterminist picture of the mind, that one which assigns some sort of non-standard causal efficacy, or creativity or whatever, to an ongoing and unchanging entity named a person. One thing to be said against the answer is that we shall not have to wait until the issue of whether there exists anything like agent-causation before being certain as to whether or not there exist actions. Or, if that issue about the mind has now been resolved, we did not have to wait until that past moment in order to conclude that we do or do not act. It can be added that it is very far from being the case that we do not take actions to derive from mental *events*, but only from a somehow ongoing and perhaps somehow mental thing. We *do* take actions to derive from mental events. At the very least the story of agent-causation would have to incorporate wants, beliefs, or whatever.

It was remarked earlier that it has sometimes been supposed that 'the philosophy of action' contains truths which stand in the way of the formulation or truth of determinism. The account of actions in terms of agent-causation provides an example. If actions *were* ordinarily taken as involving agent-causation, then a determinism could not be stated in terms of the ordinary conception of actions. There would be an argument, further, from the existence of actions to the falsehood of determinism. In fact we face neither of these troubles.

Let us start elsewhere. Let us contemplate the man's action of *writing the letter*, and, more particularly, the question of what mental events *may* have been involved. By doing so, we may come to an answer to the general question of what mental events are logically necessary to an action. Let us begin with the period before his first movement. It is at least natural to think of a *want or desire*. What was true of him, we may suppose, is that he wanted *to write the letter*. He wanted what can be so described reasonably precisely—that a certain state of affairs, the letter's being written, be brought about by him. If it would be unreasonably conjectural to say more, it is reasonable to assign to him a want not greatly more specific than this. This point will be of some importance to us. It is, to repeat, that actions are typically preceded by desires for end-states *generally* described, to be brought about by the agent.

We can indeed think of this want as a mental event, or as a series of events of more or less the same type. It is natural, as well, to think of a related dispositional want, a neural rather than a conscious matter, also prior in time to his first movement. It is the particular dispositional want which may issue or reissue in the conscious want to write the letter. Many have objected, however, to the general proposition that all actions involve prior wants or desires, or indeed wants or desires at all. The given reason is that some actions are done, as may be said, very much against the desires of the agent. These are actions traditionally described as done out of a sense of duty. Here, it is said, agents may in fact want to do otherwise than they do.

We here touch on large issues, traditionally located in moral philosophy. Still, it seems possible to reply, quickly but without discomfort, that there exists an important sense in which all actions, including the ones in question, *are* wanted. Evidently a distinction between kinds of wants or desires is called for, and one is surely possible. We may distinguish between what can be called appetitive and volitive desires. (W. A. Davis, 1983b) An appetitive desire is typically an appetite, craving, inclination or the like. We have appetitive desires for what is pleasurable to us. A volitive desire is a want, perhaps a kind of commitment, which is not so rooted in pleasure or the like. If these characterizations are not wholly adequate, it seems clear that it must be possible to improve them, since there evidently is a distinction to be made. It is possible that the man's writing of his letter is a disagreeable prospect, even one he finds abhorrent. But that cannot be the whole story. He also proposes to write it. Thus it is somehow true both that he desires not to write, or lacks a desire to write, and also that he does desire to write.

Let us suppose, whatever importance we will later pay to the fact in a completed characterization of action, that our correspondent's movements involved a prior volitive desire to write the letter. What more is to be said of it has to do with its content. As with any mental event, we can try to specify it in terms of its content, in this case its representational and indeed propositional content. We might suppose, following one line of thought (Hare, 1952, 1971; Kenny, 1963), that the content of any want can be expressed as an imperative—perhaps as 'Let it be the case that *p*', where *p* is a proposition. Our correspondent's prior want would then be expressed, perhaps, as 'Let it be the case that I write the letter'. But *do* wants express imperatives as ordinarily understood? There are reasons for doubt. (Pears, 1971a, Gauthier, 1971)

Surely wants generally are such as to be prone to *issue in* imperatives, rather than be expressed by them. Better, wants are prone to issue in something expressible in imperatives. To want something is somehow to be attracted to it, which attraction may issue in or enter into something different, something in what might be called the executive mode. Secondly, is it not possible to want what there is no chance of having, and, more relevantly, to want what is such that there is no point in any imperative, addressed either to oneself or others? I can want what is lost, my brother's RAF medal, when further search is pointless and there is nothing to be done. There seems no adequate reason to attempt to set aside such cases as not wants, but something else, wishes or mere wishes. It is often more natural to speak of wishes rather than wants when the object in question is unlikely to be attained, but there seems no barrier—certainly none is provided by the dictionary—to wanting what we cannot have, and can see no means of getting. Wants typically subside or decline in such situations, but that is not to say that they are never in existence when ordinary imperatives, however internal, seem out of the question.

There are several traditions of thought which connect wanting and evaluation, for evident reasons. Not to look into them, but rather to come to the point, it seems that the best expression of a want is indeed some evaluative judgement to the effect that something is in some way *desirable* or *good*. (Davidson, 1980f) The wide range of evaluative predicates, both positive and negative ones, allows for wide variation in the expression of wants. What needs to be added is that what is wanted is not ever evaluated only in some way which does not touch one's own interests, but also as somehow desirable or good *for oneself*. We shall not try harder than to say that the content of my want, when I want *x*, is that *x* or the doing of *x* would be good. What is not to be added, of course, given what has been said, is that the evaluation is itself to be further understood in terms of ordinary imperatives.

If it is reasonable, with respect to what precedes our correspondent's movements, to think of a desire to write the letter—which desire has a content no more specific than that—it is also reasonable to think of a belief. If having a want does not require that I have a belief as to how to satisfy it, there is also the truth that a want by itself is typically accompanied by such a belief, an *instrumental belief*, and that it too contributes to action. The content of our correspondent's instrumental belief—to what extent it is about such matters as the availability of pen and paper, to what extent about his own determination or whatever, and so on, can of course be no more than guess and speculation. What the belief comes to is that certain means are available, that something is for certain general reasons within the power of the agent. The belief will not be greatly specific—pertaining, say, to wrist or finger movements. We can leave it unsettled to what extent the belief is dispositional rather than within consciousness.

To move quickly towards further possible antecedents of the writing of the letter, it is clear that what we have so far does not allow us to say that man *intended* or had an or the *intention* to write the letter. This is so for a number of reasons. The most obvious is that his volitive desire to write it might be accompanied by a stronger appetitive desire not to do so. His desire to write it might also be accompanied by a contrary volitive desire. Suppose, however, that we make the additional assumption, whose analysis we shall not attempt, that his desire to write was his only relevant desire, or stronger than any relevant desire. Some have at least seemed to suppose—if their brevity does not mislead— that we can now regard him, with no more said, as intending to write the letter—and thus that the antecedent of action is the sum of want and instrumental belief, or somehow produced by them. (Davidson, 1980f)

This is mistaken. Having an intention to do something requires having a belief other than an instrumental or how-to-do-it belief. It requires a *predictive belief* to the effect that what is intended is not merely possible but probable. I cannot intend to do what I believe that I will not do, even if I think that for certain reasons I can do it. It is true I can intend to do what I am not certain and do not know that I will do, but it is necessary, to speak loosely, that I believe that it is at least likely that I will do the thing.

Do the want, the instrumental belief, and the predictive belief guarantee that our man intends to write the letter? It may be tempting to suppose so, taking an intention to be identical with the sum of such items, a kind of unity composed of them. An enlightening inquiry (W. A. Davis, 1984), although not one to agree with wholly, shows the incompleteness of this. What is also required, to speak generally, is at least that the predictive belief somehow be *based on* or *derive from* the desire and the instrumental belief. To see the necessity of this, suppose that our man has an instrumental belief as to how to write the letter, the untroubled single desire to do so, and the predictive belief that he will do so—but that something else is also true of him. He has the fourth belief, wholly extraordinary, that it is in a way not the case that he will write the letter *because of* his desire and his instrumental knowledge. Perhaps he has the belief instead that his wife will in fact guide his wholly untrying hand through the requisite movements, or that he will fall into a trance and his hand will make what he will merely observe: the requisite movements. These extraordinary speculations are not essential. All that is essential is the supposition that our man lacks what we can call a *dependency-belief*: that his movements will occur in a way because of his desire and his instrumental belief. If he lacks it, he cannot be said to intend to write the letter.

More needs to be said of the dependency-belief, which must be only one of the things that might have that name. Suppose John wants to kiss someone, knows how, and believes that he will. Further, he has a certain general belief, got from his past experience, to the effect that whenever he has a like desire, and like instrumental knowledge, he *willy-nilly* kisses the person. The explanation, perhaps, is that he always falls into a kind of momentary terror, out of which the requisite movements emerge. Not to struggle for a better positive description of his general belief about terror, it is *not* what is needed for an intention. The proper dependency-belief he must have is that his want and his instrumental belief will so *motivate* him that the requisite movements occur. This idea of motivation is essential to intention. (Cf. W. A. Davis, 1984.) What is it for someone to be *motivated* in the given way? Let us put aside reflection on this important matter for a time.

What we now have is the supposition that our correspondent, before he gets into action, may have a volitive desire to write the letter, an instrumental belief of a certain kind as to how, the predictive belief that he will move so as to write the letter, and the dependency-belief that the movements will come about at a particular time because he will then be motivated by his desire and his instrumental belief. Having accumulated these four items, can we draw a further conclusion: that in virtue of them, the man has the intention to write the letter? We can so conclude either if the four items *constitute* having the intention, or if the four items are in some way *logically sufficient* for what is distinct from them, having the intention. It seems one or the other must be true.

Can the issue be resolved by reflecting that we *form* intentions, that there does occur the active as distinct from passive mental event of forming an intention? Does that suggest that an intention *follows on* rather than is *identical with* the process of someone's coming to have a predictive belief taken as dependent on want and instrumental belief, or, as we can perhaps say, someone's coming to have a desire in accordance with which he believes he will act, on account of the desire and an instrumental belief? There is a related consideration, involving considerably more pervasive concepts than the concept, understood narrowly, of forming an intention. They are also concepts fundamental to any inquiry into determinism. They are those of *choosing* and *deciding*, the former having to do with preference for one thing over others and the latter with a process of reasoning. Are choosing and deciding, or at any rate central cases of choosing and deciding, to be identified with forms of the given process? Or are choosing or deciding a subsequent mental event?

The latter is perhaps arguable. (McGinn, 1982, pp. 94–5; cf. Harman, 1976, p. 441) It is possible to suppose that our conception of choosing or deciding, like that of forming an intention, is something subsequent to and different from the given process, certainly, without taking choosing or deciding to be the mysterious acts of origination talked of by philosophers in terms of agent-causation. However, it seems at least *as* arguable or satisfactory, indeed rather more so, to take choosing, deciding, and the forming of an intention to *be* episodes of coming to have a predictive belief of the given kind. Or, relatedly but differently, and perhaps better, choosing and the like can be taken to be the culminating events *within* such episodes. Such an episode can be said to be replete with what is true of active as against passive mental events, which is goal-directedness. (pp. 174 ff.) This is a matter not only of the want, but also of the instrumental belief.

The question of whether a predictive belief of the given kind is what constitutes an intention, or what gives rise to an intention, this being something distinct, is not easy to settle definitely, and we need not pursue it further. It clearly is possible and useful to take a predictive belief of the given kind as itself *being* an intention, and we shall do so in what follows. (For reasons to which we shall come, however, such a thing will be renamed an 'inactive intention') It is difficult to suppose that in concentrating on an intention in this sense, and leaving out of consideration intentions in any other sense, we are overlooking anything of great importance with respect to the mental antecedents of action—antecedents, that is, that may exist in advance of more or less the very moment of action.

Something like intentions as we have them, incidentally, are sometimes spoken of as *reasons* for action. Other things can also be so named. Each of the want and the instrumental belief can`be so named. Further, reasons can be quite other than mental events—as when someone says there was a reason for my doing something, although a reason unknown to me. A reason in this sense is a proposition taken as somehow supporting another. It will be better to avoid the usage. So too with a curious hybrid usage where something like intentions as we have them are spoken of as *rational causes*. Exactly what claims the usage abbreviates is far from clear.

An intention, then, is a predictive belief taken as based in a certain way on an instrumental belief and a want, all of these being in a sense general rather than specific—or perhaps a want which with the help of an instrumental belief issues by way of a certain assumption in a predictive belief. One of these, we can conclude, typically if perhaps not necessarily precedes an action. Shall we attempt to produce a kind of summary characterization of such an intention as the man's intention to write the letter? Is it perhaps accurate to think of an intention as a kind of undertaking or a commitment?

So much for being in the state of intending, or having an intention. As for the matters of choosing, deciding, and forming an intention, we shall of course take related lines. We shall take them to consist in, rather than to be products of, episodes of the given kind. To choose, decide, or form an intention, is to come to have a want, an instrumental belief, and a belief that these will motivate one to do something in the future.

An intention, as we have now characterized one, is of course something that can persist through time. There is nothing in our account of what it is to have an intention which conflicts with this. We may take it that the man's intention to write the letter, if he had one, consisted for much of its duration in a dispositional want and dispositional beliefs, and persisted until about the moment when he actually starts writing. Let us concentrate on that time. Might there then have occurred a different mental event?

It has been the view of a majority of philosophers, as it seems an overwhelming majority, including Descartes, Hume, and Bentham, that it is reasonable to say yes. The majority view, that there occur volitions, tryings, willings, conations, or whatever, is perhaps best recommended by an appeal to everyone's experience. Do we not sometimes have a direct awareness of an executive episode, of something distinctive which finally issues in our own actions? Can it not be agreed that it is at least possible to ascribe to our correspondent, just before or at the moment when he begins to write, what can be called a volition to write the letter?

Is it not likely that he then *sets himself to do it*? May there not then occur something different in kind from, so to speak, one last undifferentiated moment in the more or less uniform course or history of his intention? May he not then *try to do it*—which is just to say attempt to do it, no doubt with absolute confidence, rather than exert himself or struggle to do it? If we would not ordinarily say of him that he tried to write the letter unless he had failed to do so, it is surely meaningful and may not be false to say that he tried. Indeed we very ordinarily say of someone that he tried and succeeded. If the verb 'to try' is capable of several uses, it is not obvious that 'He tried and he succeeded' is somehow senseless or false unless it means something like 'He struggled and he succeeded'—and hence refers to something, struggle or exertion, which certainly is not a fact about every action or indeed many actions.

It must seem rash to maintain that volitions or tryings never occur. Before coming to the fundamental question, about whether they and intentions are logically necessary conditions of a movement's being an action, let us consider what can be said of their nature. They share some features, it must seem, with intentions.

Consider the man's starting to write his letter, his forming of the letter 'D' in 'Dear Timothy'. It seems impossible to suppose that his volition to do so does not involve a volitive desire. When, certainly seeming to refer to a volition, he reports that he made his hand and fingers move in a certain way, so as to form a 'D', we rightly take the episode to have included a particular desire. Further, the episode included a particular predictive belief, that his hand and fingers would move in a certain way, so as to form a 'D'. Did he also have an instrumental belief: that some activity on his part would have the effect that his hand and fingers so moved as to form a 'D'? It is plausible to say yes. It would be a mistake, evidently, to suppose any such instrumental belief could be neatly or effectively caught in language. One cannot supply a specific description, independent of the effect in question—a letter's having been formed—of how one moves so as to secure that effect. What is true, rather, is that one's knowledge of how to do it has a content of imagery and of recollected kinaesthetic and proprioceptive sensations. (Cf. A. I. Goldman, 1976.) Shall we say, on the assumption that the man's volition included an instrumental belief, that it also included a belief that he would form the letter *because of* the instrumental belief and the desire? There is perhaps as much reason to do so as with intentions.

Volitions, then, have in them elements like some or all of those in intentions. A few philosophers who have spoken interchangeably of volitions and intentions (Mill, 1961 (1843), p. 35) have had reason to do so. Volitions, however, appear to have at least one further feature which distinguishes them from intentions. It is a feature which must appear fundamental to them: what is weakly and obscurely described as their *active nature*. They have a nature different from the nature of all other active as against passive mental events. They are, in a way in which intentions are not, *executive*. They are *executions* or *carryings-out*—indeed executions or carryings-out of intentions. It is in virtue of this feature that it is at the very least tempting to characterize volitions, as against intentions, in an imperatival way. They are perhaps better regarded in this way than as undertakings, commitments or resolutions, as seems sometimes to be suggested. (Sellars, 1976) They are, as it seems enlightening to say, although certainly annoying, not imperatives addressed to oneself, but *imperatives addressed to one's body*. If there seems little point in the quite different conception of volitions as actually including bodily events (p. 214), it is not difficult to understand its motivation.

The likeness of intentions and volitions, and the difference between them, can be well marked by renaming volitions—as *active intentions*. The name serves the purpose of reminding us that these events are not to be conceived as somehow wholly *sui generis* and mysterious, as has often been supposed, but as open to some characterization and indeed analysis which makes them akin to what have so far been called intentions. There seems little doubt that resistance to them on the part of hard-headed philosophers has been much owed to what they are mistakenly assigned: a kind of paralysing ineffability. As for what have so far been called simply intentions, we may refer to them more carefully as *inactive intentions*, or *forward-looking* or *pending* intentions.

We have so far been concerned with what mental events *may* have preceded the man's action of writing his letter, and what mental event *may* have occurred at about the moment he began. There is a third time-period to be considered, the period of his movements. It comes to mind partly because of an account of actions which contains a proposition that is bound to give one pause. (Frankfurt, 1978; cf. MacIntyre, 1972) It is that if movements can be known to be actions only by way of their antecedents—which is what is suggested by what we have so far—then no agent can know he is acting just on the basis of facts simultaneous with his movements. If movements are actions somehow necessarily related to prior intentions and volitions, I cannot know I am tying my shoe on the basis just of what is then happening. But, it is said, surely I *can* know, from what is going on *then*, that I am tying my shoe—I am not restricted to the knowledge that my fingers are going through certain movements.

What is said in this account of actions, very roughly, is that they are movements which are under the *guidance* of the agents. Or, they are movements which throughout their course are in a certain sense *purposive*. Further, behaviour 'is purposive when its course is subject to adjustments which compensate for the effects of forces which would otherwise interfere with the course of the behaviour, and when the occurrence of these adjustments is not explainable by what explains the state of affairs that elicits them'. (Frankfurt, 1978, p. 160) It is not difficult to accept that actions may or do involve guidance or purposiveness, while they are occurring, but it is difficult to accept that they involve it only in the given sense, which appears to exclude desire and belief.

There is a more persuasive conclusion to be drawn from both (i) the idea that when we are acting we have knowledge of this that is derived from something other than events prior to the action, and (ii) the idea that simultaneous guidance or purposiveness is involved. That conclusion is that it may often or always be mistaken to conceive of an active intention as something prior in time, or more or less prior, to the action. Rather, an active intention is to be taken as something which may persist *during* the course of an action. This is in accord, incidentally, with accounts of volitions as being precisely other than just items antecedent to actions but rather as being 'volitional processes'. (L. Davis, 1979, p. 22) It is indeed wholly unacceptable to think of a conscious element with respect to actions as being entirely prior to the moments, as somehow setting in train a sequence of events which is wholly bodily. This could be true, it seems, of only simple ballistic movements. Let us have it, then, that active intentions may not only precede but also be simultaneous with movements.

Here and elsewhere, brevity excludes detail. It needs to be allowed that it may be misleading to speak, simply, of *an ongoing active intention*. There is no single and unchanging mental event, or any sequence of identical mental events, that can reasonably be thought to go with any ordinarily considerable sequence of bodily movements that is an action. Even a simple bodily sequence may involve what it is possible to call a changing and developing active intention, or a sequence of different if related active intentions. To speak of an active intention is not to be taken as denying, for example, the following of a plan with respect to an action, or the existence of a superior and also subordinate strategies. Somewhat similar remarks might be made of inactive intentions. It is near enough true that what has been provided with both active and inactive intentions is more a *model* than anything closer to the facts.

We have so far been officially concerned with the question of what mental events *may* be involved in action, prefatory to considering the fundamental question of what mental events are logically required for a movement's being an action. If some or many such movements are preceded by inactive intentions, and more or less preceded and accompanied by active intentions, are either of these *necessary* to make a movement an action? Are both necessary?

A certain amount of recent philosophical writing assumes, and sometimes includes the explicit claim, that in so far as mental events are concerned, nothing needs to be added to an intention, somehow conceived, in order to have a movement count as an action. The intention which is taken to be all that is needed is near to being—it is less than rather than more than—what we have named an inactive intention. (Davidson, 1980f, pp. 87–8) A more traditional view of action, although in some cases the difference may be more terminological than real, is the opposite: that for an action we need to have a volition or active intention, and no more than that. It will be agreeable if either of these views is true, as agreeable as it always is to arrive at an economical account. To consider the matter, let us begin with the question of whether an *active* intention is logically necessary to a movement's being an action.

(i) An inactive intention consists, we took it, to describe it in one way, in a predictive belief as to future action, dependent on the belief that desire and instrumental belief will *motivate one to move* in a certain way. It is necessary, surely, to take the necessary reference to being motivated as in fact a reference to the occurrence of a suitable *active intention*. What made it necessary to include talk of motivation-to-action in the account of an inactive intention to write a letter was the thought of cases of trance-writing and the like. Exactly what it missing from these is a suitable active intention. To *be* motivated in the given way *is* to be subject to an active intention. We thus have this argument: there do commonly occur inactive intentions; their proper analysis calls for the existence of active intentions; therefore there commonly exist active intentions.

(ii) Are there not cases where someone intends to do a thing within a certain period of time, can be said to persist in that intention during the period, and fails to act? It is not that he does not really have a proper intention—rather it is the case that he *does* have a single volitive desire and the requisite beliefs—but that he somehow fails to get into action. Is it not too doctrinaire to insist that in fact he must lack the inactive intention? Is it not preferable to say that what explains his failure to act is the absence of his actually willing to do the thing? He lacks a kind of power as distinct from desire and beliefs. If so, is it to be concluded that when agents *do* act, this necessarily involves an active intention? Is there some impropriety in arguing towards a conceptual or definitional proposition—about what is logically necessary to an action—by way of a consideration of this kind or the previous one? Much might be said in resisting the suggestion. One thing is that conceptual or definitional propositions in this area are not wholly a matter of discovery, but in part a matter of decision. In so far as decision is in question, conformity to such considerations as the present and the previous one is evidently of importance.

(iii) It is well known that in extraordinary cases of paralysis and the like, where the person is initially unaware of his condition, he may be told to do something, perhaps clench the fist that is behind his back, and have no doubt whatever that he has succeeded when in fact he has not. What leads him to his mistake, it is reasonably said, is that *something* decisive occurred. What occurred was an active intention, in this case an unsuccessful one. It might be argued, further, that what leads him to the error is evidently something distinct from an inactive intention—say his intention in advance to do what he is told, or indeed an intention to clench his fist at a time. He does not fall into error in virtue of the inactive intention, since such intentions very commonly are not acted on, for several reasons. He falls into confident error in virtue of the occurrence of something of distinctly *greater* predictive value, something more tightly bound up with action.

(iv) I intend inactively to have another drink when the approaching wine-waiter, making his way along the line of party guests, comes to me. My inactive intention is fully formed before he comes to me, so why does my arm not go out before he comes to me? Is the answer that the tray is not in range? Further, my hand does go out when it does because the tray then *is* in range. But there surely remains the question of *why* it goes out when the tray is within reach. No doubt it will be said by some that the answer is just that I believe, at the moment the tray is within reach, that it is within reach. More fully, it may be said that my hand goes out at a certain time because of only these facts: I inactively intended to have another drink, which intention involved the instrumental belief that the tray would be within reach at some time, and it is now that time. But is there not room for a further question? Why, given all that, did my hand go out?

(v) There is a fifth and more decisive consideration. To return to our correspondent, his inactive intention was very naturally characterized as his intention to *write the letter*. The elements within it, such as the volitive desire, were characterized in conformity with this *general* description. This is not to say that his intention was in fact greatly more particular, and was for brevity generally characterized by us, but that the intention itself had only such a general content. How is his *active* intention to be characterized? As already implied, any attempt to do so fully and explicitly would fail. This is particularly evident when we keep in mind that the active intention is to be taken as continuing during the course of his movements. It is that which gives content, rich and detailed content, to the idea of *guidance* of his movements.

What can be said quickly and confidently is that the contents of related inactive and active intentions are different, and, crucially, that the *specific* content of the active intention appears necessary to the action. No action, it seems, can be taken as owed only to the mere *general* content of an inactive intention. Someone who intended to write a letter in the ordinary way, but had no predictive or instrumental knowledge as to movements of the hand and fingers, say for forming a 'D'—if the speculation is coherent—could surely not write a letter.

(vi) It *does* appear to be logically necessary for an action that there occurs what was described as an imperative addressed to one's body. It does appear to be necessary that there occurs an event of such peculiar *activity*, an activity so directed. Here, surely, we have something that seems necessary to action and which very clearly is no part of an inactive intention. The point is bound up with the particularity of active intentions but not identical with it.

It is nevertheless to be allowed, before we arrive at the conclusion towards which we are headed, that there are objections to taking actions as necessarily being movements involving active intentions.

One is that it rings true to say that sometimes 'we just act'. That is, more carefully, we sometimes have no sense of actively intending to act, or no recollection of having done so. It is hard to take this as forceful. One reason is that it seems clear of other indubitably existing kinds of mental event that we cannot both experience them and also attend to them. I cannot, as it seems, even have ordinarily attentive visual experience and also have that experience itself as an object of attention. As for recollecting active intentions, which is more important, the recollection of mental events generally is uncertain, and to a great extent a function of the saliency of those events within one's past experience. When I act, surely, what is salient in my awareness is more likely to be the external rather than the internal world. Thus there is every likelihood of my not being able to recall what may nevertheless have occurred. There is a further reply to the objection. One may be taken aback by the thought that sometimes 'we just act', and led thereby to doubt the occurrence of an active intention *prior* to a movement. *Can* one doubt, however, that an action is a movement that is under one's guidance during its duration? It is difficult to do so, and difficult to conceive of this guidance other than in terms of an active intention.

A second and related objection to the logical necessity of active intentions to actions has to do with practised or habitual activity: playing the piano or walking. It is not convincing, we are told, to assert the existence of an active intention for each of what may seem to be many actions—single finger movements or steps—contained in a stretch of playing or walking. A part of a reply is that it is unclear what is to count as one action and what is to count as a second one. It may be that certain items in the playing or the walking are not rightly regarded as actions but as *action-parts*, and hence do not require separate active intentions. Much the same point, put differently, is that single active intention may be sufficient for a somehow automatic sequence or set of events.

Thirdly, there is a long-running objection about regress. It fits what we have less well than it fits some traditional accounts of volitions. It may be objected, still, that if a movement requires a preceding active intention to be an action, so too the active intention itself, which is akin to an action, requires some predecessor in order to be what it is. It requires another active intention, and so on. Here we have a regress in some way vicious. (Ryle, 1949; cf. MacIntyre, 1971) The objection is not formidable, perhaps even when directed against a traditional view. Still, let us think only of the present view. A reply is that actions as we so far have them, which is to say sequences of bodily events related to active intentions, are wholly different in nature from active intentions themselves. Why should it be supposed that what is required for a bodily sequence to be an action has a counterpart in what is required for an active intention to have an active character? The bodily sequence and the intention, after all, are on different sides of a fundamental division, the division into the bodily and the mental.

The objections, I take it, cannot stand in the way of the conclusion now to be drawn, on the basis of the previous considerations, that it is a logically necessary condition of a movement's being an action that it somehow involves—we have not yet looked closely at the relation—an active intention. The alternative, that what is required with respect to mental events is only an inactive intention, is not tolerable.

Our second question is this: do inactive or forward-looking intentions, as distinct from active intentions enter into our concept of an intentional action? More particularly, is it a logically necessary condition of a movement's being an action that it is preceded by an inactive intention as defined? Part of what may move us in the direction of saying yes, of course, is that what we take as actions at least typically—indeed very often indeed—are in fact preceded by inactive intentions. It seems to be true even of many actions that cannot possibly be said to have been contemplated, planned, or premeditated that they do involve inactive intentions. The crucial features of inactive intentions for present purposes are that they typically concern end-states generally described, and that they lack the active character of active intentions—that feature which may lead us to speak of active intentions as imperatives addressed to our bodies. A man who defends himself against a wholly unexpected attack in the street, as it seems, does so partly out of a momentary inactive intention.

Still, if inactive intentions are to be logically necessary to actions, there can be no exceptions, and it does appear arguable that there are. If, while thinking of how to continue this paragraph, I idly touch the flower on the table, it can hardly be doubted that my movement is an action. Does the nervous speaker on television, repeatedly moving his hand or otherwise fidgeting or fiddling, not act? Do we not occasionally 'find ourselves doing things', perhaps pacing up and down in a room? (Cf. Frankfurt, 1978; O'Shaughnessy, 1980) However, it does seem impossible to avoid the proposition that we cannot know, of every one of what we take to be actions, that they involve inactive intentions.

We shall take it without further ado that an inactive intention is not logically necessary to an action. With this conclusion about inactive intentions, as with the previous one, it is less than clear to what extent it is a departure from certain common philosophical views. It is certainly likely that some who espouse the logical necessity of 'intention' to action, and speak of it in ways which suggest antecedent mental events, do in fact wish to insist only on the necessity of what we have defined as active intentions. As noted, intentions generally share most of the features of active ones.

In what follows, although actions will not be taken as logically requiring inactive intentions, we shall have them as much in mind as active intentions. This is so for the reason that inactive intentions are very nearly as fundamental a part of our subject-matter as those which are yet more closely bound up with actions. There is good reason for the attention that has been given to them.

4.3 THE RELATION AND THE BODILY TERM

We need now to look in particular at the relation between the active intention and the bodily sequence, and then at the bodily sequence itself. Both of these, but particularly the relation, have in recent philosophy had persistent attention from philosophers influenced by a certain strong view. (Davidson, 1980) We shall not be so attentive, partly for the reason that the foundation of the determinism being expounded is its first and second hypotheses together with the particular account we now have of the springs of action themselves—inactive and active intentions.

An active intention has been conceived as consisting in a volitive desire, an instrumental belief, and a predictive belief deriving from them, all of which unity has a peculiarly active character. What is mainly of importance to us, in connection with a first part of what is to be said of the relation between intention and bodily sequence, is the volitive desire and the instrumental and predictive beliefs. As assumed in all that has been said so far, all of these are *representational*. (Cf. Danto, 1973.) They are somehow united mental events which have contents of the kind which are representational (p. 82), as of course is implied simply by calling them desire or belief. They are by nature *of* or *about* something. (Cf. Pears, 1968, pp. 113–15)

The relation may involve a propositional content, which is to say a content dependent on language, and to a significant extent expressible in it, but it may also have a lesser explicitness or determinateness. Consider my instrumental belief with respect to my active intention to form the letter 'D'. It is a belief that certain hand and finger movements, conceived independently of their effect, will in fact have the effect of forming the letter 'D'. It would be mistaken to suppose that this belief has propositional content in the sense just mentioned. It would also be mistaken to suppose that it is a mental event of no representative content. (Cf. p. 211). As for the predictive belief, to the effect that a 'D' will be formed, it is more natural to speak of its representative character as being propositional. So too with the desire—to form a 'D'.

Is there any room at all for doubt about the need for an active intention to represent the ensuing movements in the given predictive way, if the movement is to be an action? Suppose, through some bizarre disability, that a man truly intended certain movements there and then, and had no thought of others, but his intention caused quite different movements. His active intention had in its several contents the movements involved in kissing his bride. He then shook hands with her. We would perhaps be baffled as to whether the movements counted as an action. It may well be that our conception of an action does not give us a decision about such a case, and that we should be uncertainly inclined to speak of 'a kind of action' or 'something like an action'. Evidently it is possible for movements to go *somewhat* wrong—to be somewhat other than as represented—and still to be an action. But how far wrong can they go? Whatever is to be said of that, undecideable cases cannot stand in the way of the conclusion that a standard action must involve an active intention which does indeed represent—let us say represent tolerably well—the movements in question. That is not to say, as needs to be kept in mind, that the movements will always be represented only in terms of themselves. We may conceive of our movements in terms of their effects.

To move now towards the second part of the relation between intention and action, would we have an action if we believed of a person only that there occurred the active intention to stand up, and there also occurred the movements which the intention represents, those involved in standing up? Alas it is possible to have written as if with the belief that the answer is yes. (Honderich, 1981a, pp. 282–3; cf. Weil, 1980) Cases of several kinds establish what will already have been assumed, that more is logically necessary to an action, with respect to the relation, than that the relation is representative to a decent degree.

The simplest case is one in which the movements are random events, in no nomic connection with antecedents, and in particular in no nomic connection with the intention, which does nevertheless represent them. The representing intention is not a cause of the movements, which movements have no cause or causal circumstance. We should regard the person in question as lucky, but not as having performed an action. Do we think differently if we concentrate on the fact that the person did in fact have the correct predictive belief, that he did *expect* the movements, which is implied by what has been said? What if it is the case, additionally, that he has a good record of such predictions? This last supposition would certainly give us pause, but cannot demand serious attention. The case now involves an assumption of wholly mysterious precognition which makes it near enough to irrelevant. It is most plausible to say of such a case that it would move us in the direction of a second and different conception of action.

Nor would we suppose a standard action had been done in a case different from one imagined earlier for another purpose, but also involving an intervening physiologist. Here, a physiologist causes certain of a man's muscles to contract, producing movements, and, although there had occurred a representing intention—an intention representative of the movements—there was no causal connection between the intention and the muscles. The physiologist, that is, was not caused to act by his knowledge of the intention, and thus by the intention. At least we would not take the movements to be an action if the man had no expectation of the intervention of the physiologist. No doubt we should be puzzled if something else were true: that the man for good reason expected and counted on the intervention. Again, however, it is most plausible to say of such a bizarre case that it would prompt us towards a second and different conception of action.

There is the case, as well, also involving a properly representative intention, where the movements are owed in a simpler way to external forces. The soldier who actively intended to get his head down out of the line of fire was pulled down before his active intention took effect. There is also the final conceivable case where the right intention occurs, as do the movements, but the movements are in fact owed to something of the order of a tic or spasm. I intend to make a certain movement typical of Huntington's chorea, but the movement occurs not because of my intention but because I am subject to Huntington's chorea.

It thus appears that for an action we require an intention which stands to a movement in a relation which is not only representative but also *causal*. Is the intention a cause of the movement, or a causal circumstance? It is impossible that an intention might be a causal circumstance for a bodily movement—the initial circumstance of a causal sequence for the movement. Much more is needed for a bodily movement. The intention could not itself *necessitate* the occurrence of muscle-events, since such events require the existence of muscles, which cannot be ascribed to the intention. Leaving aside the objection that we might require less than that the intention be a cause—that it be *some* nomic condition of a movement—let us conclude that an intentional action requires a movement *caused* by a representing intention. It is difficult to believe we depart at all significantly from the ordinary conception.

Before turning to one final matter having to do with the causation of movements by intentions, it is worth noting a possible resistance to what has been said. It is a resistance of a kind that has certainly inspired some doubtful 'philosophy of action'. To think of the experience of acting is perhaps to think first of all of the active character of the intention. Having an active intention, it may be said, is not an experience of causal connection. It is an experience, as remarked, such that something of its reality can be caught by talk of imperatives directed to one's body. In thinking of the action, one's thought or image of this activity may fill the mind, and spread itself over the action. In particular, the thought or image may somehow encompass or spread itself over the relation between intention and movement. Thus one does not think of the relation as being one of causation. One may instead resist this characterization.

This inclination, to say the least, needs itself to be resisted. A first point to be made is that whatever else is to be said of the active character of intention, it is not in any way tied to a denial of causal connection. Attempts to represent it rightly do not include any denial of causal connection. If the experience is not one *of* causal connection, nor is it an experience of its absence. A second point has to do with the spreading of one's thought or image of activity to exactly the relation between intention and movement. It is certainly possible to concentrate clearly on this and to ask about its nature. If one does so, it seems quite impossible to think other than that it is in part a causal relation. As has been forcibly asked, what else *could* it be? (Davidson, 1980)

One final matter needs to be considered in connection with the relation between mental antecedent and movement—the psycho-behavioural relation, as we can call it. Is it possible that refinement is needed with respect to our requirement that an action be *caused* by an active intention? A number of philosophers advocate or accept accounts of the psycho-behavioural relation related to the account we have. They take it that what is needed is not merely causal connection between a certain mental antecedent and a movement, but a particular kind of causal connection. A movement, it is said, may fail to be an action although it is caused by a mental antecedent which represents it or matches it or is a reason for it. The movement fails to be an action if it is the effect of a *deviant* or *wayward* causal sequence. (Davidson, 1980g)

The given accounts of the psycho-behavioural relation do not involve an active intention as we have conceived one, but only something very distantly approximate to an inactive or forward-looking intention. Typically we are invited to consider someone of whom it is said only that he has a desire and a belief that do represent an ensuing movement. The belief is sometimes conceived as what we have called an instrumental belief, sometimes as a predictive belief, sometimes as both. Certain cases, it is said, establish that more is needed for an action than that a movement is caused by a representative desire and belief.

A mountain climber in trouble wants to be free of the weight and danger of holding another climber on a rope, and believes that by letting go he can do so. Being a decent fellow, the want and desire *unnerve* him so much that he does let go. They cause him to do so, but by way of his consternation or shock. He does not choose to let go, or do so intentionally. He does not perform an intentional action. A related case involves the resentful servant who wants to drop his master's prize dish, certainly knows how, but does the thing out of fright or nervousness which is itself the effect of his want and belief.

In both cases we are said to have a deviant or wayward causal sequence linking desire and belief with movement. What needs to be done, it is supposed, in order to improve on the given accounts of the psycho-behavioural relation, and thus to exclude the the cases in question from the category of intentional actions, is to specify the *right sort* of causal connection. Some of the philosophers in question despair of doing so. Others offer accounts of great interest and ingenuity (Peacocke, 1979a, b), but which are certainly open to objection (Bishop, 1981).

In fact it must seem wholly unlikely that the given cases of non-action can be explained simply by attention to *causal* connection. To suppose that they can is the result of insufficient attention to the nature of causation and in particular causal sequences. (Ch. 1, pp. 148 ff.) Consider the causal sequence which has in it an initial circumstance including the climber's desire and belief, a subsequent circumstance including his becoming unnerved or frantic, and the upshot of his letting go of the rope. Compare it with another episode and another causal sequence, which sequence has in it an initial causal circumstance including desire and belief, a subsequent circumstance including whatever else is taken as necessary to make into an action the ensuing letting-go of the rope, and the letting-go. Considered as *causal sequences*, as distinct from sequences whose stages are distinguished by their non-causal properties—say by being states of alarm as distinct from active intentions—it must be impossible to distinguish them. Certainly, in any sense having to do with causation, each upshot is as much as the other a 'function' of the initial causal circumstance or elements within it. In each case, further, there are the same theoretical possibilities of reasoning backwards from the upshot to the initial causal circumstance or forward from circumstance to upshot. It is in fact the vain hope of making distinctions of such kinds that is basic to the attempt to separate deviant from non-deviant causal sequences.

What is mainly to be said of the cases, from the perspective of our own view of the mental antecedents of action, is obvious enough. The mountain climber fails to act, as does the servant, for the reason that neither has an active intention to act. What is missing is an active intention in its imperatival character, and also, arguably, despite what is said to the contrary, a correctly representative desire and belief. (Cf. Pears, 1975a.) What gives rise to the movement in each case is what is so described as effectively to exclude an active intention—great consternation or the like. The cases do indeed establish the need for an addition to the given accounts of the psycho-behavioural relation. They do not establish any need for an addition to our own account. They can rather be regarded as indicating precisely the need for an account in terms of active intention.

Do certain different and more bizarre cases establish a need for some addition to our account? They too are taken as posing a problem for the mentioned accounts of the psycho-behavioural relation. These cases also involve what are called deviant causal sequences, but of an external rather than an internal kind. That is, they have to do with facts not prior to a moment, like nervousness, but with facts subsequent to a movement. We suppose a man desires to kill his enemy, has a certain belief, shoots at him and misses, and the shot causes a church steeple to collapse, killing the enemy. We are to suppose that what the gunman does must count as the intentional action of killing his enemy unless the causal part of the given accounts of the psycho-behavioural relation is somehow amended.

It is difficult to see that the case, although it is in several ways puzzling, can pose any fundamental problem for anything like an adequate account of the antecedents of an action and their relation to the action. The reason, of course, is that any such account of the given case must involve representative mental antecedents which in fact do not represent or match what happens. The collapse of the church steeple was no part of the plan. Certainly our own account, in terms of active intention, does not have the embarrassing consequence. We need not struggle to distinguish between kinds of causal sequences externally considered.

That is not to say that our account as it stands is as complete as it might be. The bridegroom noticed earlier (p. 232) who does actively intend to kiss his bride, and hence is not in any state which would exclude that intention, and yet shakes hands with her, can be thought to pose us a problem. With some effort, we might arrive at an account of how far a movement can diverge from a representing active intention and yet count as an action. There is also the problem noticed earlier (p. 233) of someone's movement represented by an active intention but caused by it by way of the intervention of a physiologist, which intervention was anticipated by the person in question. The intention, that is, in its instrumental belief, included a representation of the intervention. More might be said of this case than that it would prompt us to a second and different conception of action. The role of another person evidently raises more of a problem than the use of prosthetic devices by a person—say an implanted mechanism which he uses to contract certain muscles. Thirdly and finally, there are conceivable cases involving an active intention to get into a situation where nervousness or the like will issue in a movement. Here there is the resentful but more foreseeing servant, different from the one before (p. 235), who appears before his master with the prize dish, having anticipated that his state of mind then, which will not consist in an active intention to drop the dish, will none the less result in the dish's being dropped. A fuller account of action in terms of active intention would require elaboration to allow for what seems intuitively right, that he *does* perform the action of dropping the dish.

Let us go no further, but rather conclude summarily that the psycho-behavioural relation, when a movement is an action, is such that an active intention represents and causes the movement, which movement is typically also represented and caused by a prior inactive intention. It is a conclusion, incidentally, in full conformity with what was prominent in our earlier consideration of the causation of mental events, the axiom of mental indispensability: that earlier mental events are essential to any full explanations of certain later mental events and also actions. (p. 91) It is also in conformity with the related conviction of personal indispensability. (pp. 146 f.)

We come now to our third and last question about the nature of actions, having to do with the bodily rather than the mental term of the psycho-behavioural relation. What has been said already implies answers to a number of questions having to do with bodily sequences. What will be offered in what follows, however, will not be proper answers but hardly more than remarks.

We began with an intuitive idea of what bodily sequences are involved in actions, and in particular the actions of saying 'That will do the trick', signing a letter, and standing up. The sequences in question, some of them long and complex, are in principle wholly open to sense-experience. If commonly they are also *in fact* open to sense-experience, they need not be. If a man is capable of tightening certain muscles around his stomach without giving any external sign of doing so, or capable of doing some yet more surprising internal thing, this may none the less be an action.

We have so far concentrated on only certain descriptions of bodily sequences which are actions. The bodily sequences in the original example have been described as those involved in the man's saying 'That will do the trick', in signing his name, and in standing up. That is to say, more perspicuously, that we have spoken only of certain properties of these sequences. It is to say, still more perspicuously, that we have spoken of the sequences by way of only certain properties of them. Each sequence had many other properties. The first had the property of involving sounds which can be described independently of the fact that they constituted an utterance of English. The sequence also had the effect, we may suppose, of making the cat look up from its sleep. The second sequence, involved in the man's signing his name, also had the property of using up ink in the pen, and the effects of identifying him as the writer of the letter and of finally convincing his wife that he had succumbed to a temptation. The general fact that actions have many properties is fundamental to or enters into a number of possible questions about them. Let us glance at five, and then at a final objection.

(i) What is the location of actions? If, as just remarked, the man's signing made his wife think as she did, where did his action take place, in his part of the room, or hers, or both? The puzzlement is better generated, perhaps, by the example of killing, of causing death. Did it happen where the assassin pulled the trigger, or in the distance where his victim stood, or in both places? The general answer to such questions, implied by what has been settled already, is this: any action occurs only within the space occupied by the agent's body, and, more particularly, in the space occupied by a sequence of bodily events represented by the active intention. There is no real reason to think, however tempting it may be to do so, that an action which has effects, intended or unintended, is in part where the effects are, however much they may enter into the agent's contemplations and in particular his forward-looking intention. The sun is not in every place it lights.

(ii) When, or over what duration of time, does an action occur? If an action has ongoing effects, later and later in time, is there not a problem as to when actions end? Some have thought so, but one's puzzlement must surely be brief. An action ends exactly with the ending of the bodily event-sequence represented by the active intention. That sequence, in the case of the signing of the letter, may later become that which identified the writer of a letter to its recipient. That is not to say that the action lasted longer than the bodily sequence. It is indeed true that actions may be said to acquire some of their properties after they end. Whatever else is to be said of this quite general matter, the point is unexceptionable—as unexceptionable as the point that the avalanche which ended at noon yesterday became a fatal avalanche only today when a victim died. As for the beginning of an action, as we know, it begins when there begins a bodily event-sequence represented by the active intention. Earlier it was remarked that a nod begins when a head moves. That is true, but it is a consequence of something else. As it happens, with respect to a nod, or any other ordinary action, the beginning of the ordinarily visible movement is the beginning of the intended movement.

(iii) There is more of a problem about numbers of actions. The fact that each movement which is an action has many properties allows us to say, when someone acts, that he or she has done more than one thing. It was true about the man, in signing his name, that there were many things he did. In addition to signing his name, he used up some ink, did what would prove the identity of the writer, convinced his wife that he had succumbed to a temptation, and put on to paper some lines and marks describable without mention of his name or signature. The five things, to repeat, were *things he did*.

Was each of the five things he did an action? Let us assume that the man's intention was indeed to sign his name. He was not seeking to use up ink in his pen, or consciously making sure that the recipient of the letter would know its sender, or intent on making his wife think something, or on making marks and lines for a geometer to describe. Still, these were not only things he did, but *were* actions. Take his using up some ink in his pen. That was a bodily sequence caused by an active intention, although not an intention which represented that particular thing he did. The sequence, we can say, was not an action in so far as it was *that* thing he did, using up some ink. It was an action anyway. It was an action in virtue of being identical with a sequence which *was* represented by an intention which caused it. It was identical with the sequence represented and caused by the man's intention to sign his name. So with the other things done by the man in signing his name. That answer, that each of the things he did was an action, in virtue of facts of identity, can give rise to mystification.

It may lead one to the wrong answer to another question. How many actions did the man perform? To say that each of the things he did was an action is not to say that he performed more than one action. He performed exactly one action, which action had a number of different properties and so is open to a number of different descriptions. An action is an event-sequence caused by a representative intention. There was one of those, whatever the number of its properties and hence descriptions. To say he performed the action of signing his name and also the action of making his wife think such-and-such—not that we would ordinarily say such a thing, which itself is indicative—is *not* to say that he performed two actions. There is no more reason to think so than to think, since the avalanche both rumbled and crushed daisies, that there were two avalanches. For a closer analogy, consider the girl offered two boiled sweets, a plain one, and one with a red stripe, a blue stripe and a green stripe. It may be true to say that she took a plain sweet, a sweet with a red stripe, a sweet with a blue stripe, and a sweet with a green stripe. That is *not* to say she got five sweets.

There is a very different doctrine about our conception of an action. It has the consequence that our man in signing his name performed five different actions. (A. I. Goldman, 1970) That is not the full consequence of this doctrine. We have mentioned but five properties of the bodily sequence. It is safe to say that it had innumerable properties: it was a signing quickly done, it produced certain shadows, and so on. The doctrine in question then has the consequence that our man performed innumerable actions. He was very busy indeed.

The doctrine is based on the idea that an action is an instantiation of a kind of general property by a person at a time—in effect, the idea that any distinct description of a bodily sequence gives us a numerically separate action. The doctrine is supported by argument, meticulously developed, and faces up to its own counter-intuitive nature. The central upshot, that whenever there occurs a bodily sequence owed to a representative intention, *innumerable* actions are performed, nevertheless establishes the error of the doctrine. It cannot be said to give us our standard conception of an action.

(iv) A further matter has to do with there being two kinds of action. (Danto, 1973) Some of the things we do are *done by doing other things*. Others of the things we do are *not done by our doing other things*. Certainly that is in some sense true. I move the paper-clip on the table by flicking my finger. I do not, it may be said, flick my finger by performing another action. More particularly,,whatever is to be said of an inactive or forward-looking intention to move the paper-clip, my moving it is owed to a movement whose representative intention is not the intention to move it—it is owed to a movement whose representative intention is to flick my finger. My flicking my finger is not owed to a movement whose representative intention is other than the intention to flick my finger. My moving the paper-clip is a *non-basic action*. My flicking my finger is a *basic action*.

Given what was said above, it can be entirely misleading to say there are two actions, or—to stick to the present matter—that actions divide into two classes, the basic and the non-basic. There are not some event-sequences that are only basic actions and other event-sequences that are only non-basic actions. Every action is a basic action, and perhaps almost every action is also a non-basic action. Each event-sequence has a property such that the sequence is owed to an active intention which represents that property. This, it may be said, is the case with that property of the particular event-sequence which is its being a flick of my finger. Almost every event-sequence is owed to an active intention which does not represent properties had by the sequence. This is the case with the property of the particular event-sequence which was its being a moving of the paper-clip. There are certainly further questions and obscurities here, but let us leave them and pass on to a final and somewhat similar matter.

(v) We have been concerned throughout with intentional actions, but do actions not divide into those which are intentional and those which are not? In some sense that must be true. What is not true is that there are some event-sequences which are only intentional and some event-sequences which are only or wholly unintentional. Anything that is an action is an event-sequence caused by some active intention, necessarily a representative intention. Therefore *anything* that is an action is in some way intentional. It is true, in just that sense, that we *always* act intentionally. Equally, every event-sequence has properties not represented by the intention which causes it. It is true, in just that sense, that we *always* act unintentionally.

To ask if an action was intentional, at bottom, is to ask if a certain property of an event-sequence was represented in the intention which caused the sequence. The question can be very difficult. To ask if an action was unintentional is, again, to ask of one of its properties whether it was represented by the intention which caused the action. The question can be very difficult. To say that every action is intentional, certainly, is not to say that every action is such that the agent is to be held responsible for it, that he can rightly be praised or blamed for it. When I tie up my girl-friend, because I believe that the armed burglars will shoot us if I do not, my action is evidently intentional. Given the genesis of my intention, I am not to be blamed for the action. More will be said of such matters in due course.

(vi) We come now to the objection, mentioned at the start of this chapter, having to do with identifying actions with certain movements and the like. If my actions are sequences of bodily events, although such sequences in a relation with mental facts, are my actions not made *external* to me, deprived of what they possess, an *inner* side or nature? The objection, if impressionistic, needs to have attention paid to it. It is strongest against a view of actions which takes them to be movements or the like related only to *prior* mental facts. Such a view, certainly, does seem to leave out what might be called *agent-participation*. Our own view, of course, has it that the duration of an active intention includes more or less the duration of the movement.

None the less, the objection can be raised against our stronger view. It derives, certainly, from *a fact of experience* that can be further described. We do in a sense *enter into* our actions. My experience of writing a line is certainly not properly described, even in part, as the experience of observing effects. I am not related to my writing the line as I am, say, to an event-sequence which is the effect of one of my actions—say the printing out of a line, caused by my touching a key on a keyboard. What seems to be true, to have resort to an over-used manner of speaking, is that I *identify with* my actions. (Cf. McGinn, 1982, pp. 87–8.)

The somewhat indistinct fact of experience can be accommodated within the view we have. One of the means of doing so is the plain fact of kinaesthesis. The event of my writing a line is certainly something with an inner nature in a clear and indubitable way: there are the sensations of the movement, in the main sensations of muscular effort or tension. The sensations, of course, are located within my hand and fingers. It is possible to say that they are within the movement. There are also ordinary tactile sensations, of which the same can be said. I am not 'external' to these sensations and certainly I do not observe them. Through them I can indeed be said to participate or enter into the movement.

A second consideration which gives content to such usages is of course the proposition of an ongoing active intention. If we insist on an inner nature, or on our internality or participation, we can be taken to be insisting on what is certainly true of any *episode* of acting, a stretch of one's experience which is not identical with but includes the movement. It is that the episode has within *it*, saliently indeed, a mental episode. The episode of acting has within it, therefore, what is as internal and the like as dreaming or thinking about a problem. It has within it, some might say, what is yet more internal, which is the execution of one's own past intention. Would it actually make sense to object that the innerness and our internality and particiption are not facts about the episode of acting, but about *an action itself*, and thus that precisely the active intention is irrelevant to the matter? It is doubtful.

Thirdly, with respect to our identification with our actions, to the extent that that can be prised apart from other things mentioned, it appears that to identify with them in the relevant sense is in part to have one's attention upon them in a particular way, to be bound up with them. Clearly enough, as in the case of what has just been said of sensation and of active intention, such a claim of identification is wholly consistent with our fundamental proposition, that actions are effects of representative intentions. It may be supposed, further, that our identification with our actions is in some intimate relation with some persistent fact of our identification with our whole bodies, having to do with what is called a body-image. Here again there is no inconsistency.

There is also the likelihood of what can be called projection. Hume's account of causation in its psychological part, as distinct from the analytical part considered earlier (1.3), depends on the proposition that we project our feeling of necessity, a fact about us, onto things in causal connection. There is a related proposition about the projection of moral feelings on to the external world. It is likely that there is such a fact with action. As was remarked earlier (p. 234), we may in acting somehow extend or feed into the psycho-behavioural relation, and the sequence of bodily events itself, that experience of activity which is a fundamental feature of an active intention.

With this consideration of sequences of bodily events—the bodily terms of certain relations, of which the other terms are mental—we come to the end of the question of our conception of an action. The answer is that we take an action to be a sequence of bodily events, necessarily a sequence of many properties, one or some of which are represented by an active intention which causes the sequence, which sequence is typically also caused and represented by a prior inactive intention.

4.4 THE HYPOTHESIS ON THE CAUSATION OF ACTIONS

Neither the hypothesis that mental and neural events are in nomic correlation, nor the hypothesis on the causation of psychoneural pairs, can be said to be generally accepted, to be part of common sense or of 'folk psychology'. Of the first hypothesis, although judgements in favour of it can certainly be elicited, it can be said there is no settled ordinary belief in it or against it. As for the second hypothesis, it does in part actually conflict with ordinary belief, or, to speak more accurately, with ordinary attitudes, fundamental attitudes to be examined in a later stage of this inquiry. Things are different with what we now come to, the answer to the question of the explanation of actions, the Hypothesis on the Causation of Actions.

It is surely impossible to resist the proposition that our standard conception of an action is, in short, a conception of something as an effect. To take an action as something having a certain cause, as undoubtedly we do, is necessarily to take it as being an effect. (p. 61 f.) Whatever fundamental attitudes we may have to the *springs of action* and their genesis, which attitudes are in part related to indeterminist pictures of the mind, we do suppose that an action itself is a bodily sequence which is the upshot of a certain causal sequence, partly bodily, with an initial term that is at least bound up with the mental. We have a great deal of common-sense knowledge of these latter causal sequences, those which issue in action, and, so to speak, of their care, maintenance, and improvement, and their breaking-down because of injury, sickness, disease, or old age. Needless to say, we also have scientific knowledge of several kinds.

It is indeed the burden of our inquiry so far that in our ordinary conception, more fully, *an action is a sequence of bodily events caused by an active intention which represents it*. This may be said to be an analytic or conceptual truth: in speaking of actions, with whatever degree of awareness we may have of our own usage, and to whatever degree that usage is unsettled (p. 232, 233, 242), we do speak of such bodily sequences. Thus there is contradiction in speaking in the ordinary way of an action and saying of it that it was in a certain way uncaused or that it was not an effect.

The Hypothesis on the Causation of Actions is in a certain close relation to the analytic proposition. If it were actually identical with it, we should have the upshot that the third hypothesis of the determinism being expounded was radically different from its predecessors. Those, certainly, are not analytic or conceptual truths. The idea that a theory of determinism is in part analytic, or else moves in that direction, is not new. The idea is implicit in much discussion of mental concepts which has nothing explicitly to do with determinism. It is often implicit in discussions of desire and belief which assume or assert certain logical connections between the two things and action. It may be implicit, for example, in proposals to the effect that if a person does have certain beliefs and an unopposed pro-attitude to something, it follows logically that in a certain circumstance he acts—that if he does not act, it follows that he lacks a belief or such an attitude. This in effect includes a conception of action as by definition the effect of certain mental antecedents.

It would be unsatisfactory, however, simply to put forward an analytic proposition with respect to the causation of actions. This is so since the existence of the analytic truth cannot be a bar to the raising of a large question of fact, one to which any complete determinism must give an answer. It is a question to which an indeterminist answer is logically possible. There is a weak analogy in the fact that a sunrise might be taken as being by definition a rising of the sun over the earth, and that it is none the less possible to ask, of a certain daily phenomenon, whether it does in fact involve the sun's rising over the earth. The most natural answer, as we know, is no. Although an action is by our ordinary conception a bodily sequence of which a representative active intention is a cause, it must be possible to raise something like the question of whether what we can call our behaviour *is* such that our conception of an action is true of it. As anticipated earlier (p. 210) the existence of that conception is surely itself a significant argument for an affirmative and hence a certain determinist answer. The situation here, it can be argued, is like the situation in connection with our conception of causation (p. 50), and unlike the situation with sunrises. Whether or not the existence of the conception is an argument for its truth, so to speak, the existence of the conception certainly does not close off the question of fact. It is a question, as will have been anticipated, to which an answer is given by the Hypothesis on the Causation of Actions.

What is the question of fact? It cannot be (i) *Are actions bodily sequences such as are caused by the active intentions which represent them?*—or it cannot be that, as standardly or properly understood. Given the result of our conceptual inquiry, that is not a question of fact. It is not an open question, but one which, on a common view of analyticity, contains its own affirmative answer. The analytic truth which is its answer needs more careful statement than it was given above. It is that *if* anything is an action, it is a bodily sequence caused by a representative active intention. That, however, is a truth logically consistent both with determinism and with an indeterminist proposition that there are no actions as defined, that no bodily sequence is an effect. It also also consistent with a good deal else of an affirmtive kind, for example that our behaviour in its genesis is a matter of agent-causation or of some sort of non-standard causation by wants and beliefs.

Nor should we have something satisfactory in the plain existential question (ii) *Are there actions?* The answer yes, that there are some things properly so called, is consistent with the wholly indeterminist proposition that very nearly all bodily sequences which are represented by active intentions are uncaused. An affirmative answer is also consistent with the proposition that some or even very nearly all somehow represented bodily sequences are caused in the mysterious way by persons.

(iii) *Are what we believe to be actions in fact actions?* That is nearer to what we want, but it will not do. Its answer, which cannot be a matter of argument, is certainly no. This is so simply because we do occasionally make mistakes in believing something to be an action.

(iv) *Are all bodily sequences represented by active intentions caused by them?* That question, again, is easily answered no. As we have seen, in arriving at a view of our conception of action, there are or can be represented sequences which are not caused in the given way. One example is the soldier who actively intended to get his head down, but was pulled down in the right way before his intention took effect. Another is the man who actively intends a movement and is afflicted by Huntington's chorea.

The question we want can best be posed by the introduction of a stipulated and other than standard use of the term 'action'. It is a use perhaps related to a careless or unreflective ordinary use or misuse of the term. The stipulated use gives a different sense to the interrogative sentence first mentioned above: (i) *Are actions bodily sequences such as are caused by the active intentions which represent them?* It is the same use of 'action' as must be involved in other factual questions that might be asked. 'Are actions bodily sequences only non-standardly caused by persons?' 'Are actions bodily sequences caused only by wants and beliefs?' Neither of the two questions *must* be taken in such a way as logically to require the answer no, in such a way as to encapsulate a contradiction. That is, the first is not necessarily, and is not intended as, this question: 'Are the bodily sequences standardly caused by representative intentions something other than that—bodily sequences only mysteriously caused by persons?' It is not a question of the form 'Is *A not-A*?' So with the question about wants and beliefs.

Evidently the three questions just mentioned, and others, tacitly involve a use of the term 'action' which makes them pointful. In them an action is conceived imprecisely, as roughly this: *a bodily sequence somehow owed to the mental*. Thus the factual question we want, expressible as 'Are actions bodily sequences such as are caused by the active intentions which represent them?', has this sense: *Are the bodily sequences which are somehow owed to the mental in fact bodily sequences caused by the active intentions which represent them?* Or, in part more punctiliously, the question has this sense: *Are the bodily sequences somehow owed to the mental in fact bodily sequences which are effects of causal sequences whose initial elements are or include the active intentions which represent the bodily sequences?*

The Hypothesis on the Causation of Actions gives an answer to the question, one which also uses the stipulated sense of 'action'. As just implied, the hypothesis involves the idea of a causal sequence as defined with the second hypothesis. (3.1) Such a sequence consists in causal circumstances such that each element of the sequence—save the initial elements and the last—is both an effect and also part of a causal circumstance. It will be best to give the hypothesis together with its companions. The Correlation Hypothesis, to recall, is this:

> For each mental event of a given type there exists some simultaneous neural event of one of a certain set of types. The existence of the neural event necessitates the existence of the mental event, the mental event thus being necessary to the neural event. Any other neural event of any of the mentioned set of types will stand in the same relation to another mental event of the given type.

This account of the psychoneural relation was supplemented by the proposition that a simultaneous mental and neural event form a pair—which proposition enters into the Hypothesis on the Causation of Psychoneural Pairs, which is this:

> Each psychoneural pair, which is to say a mental event and a neural event which are a single effect and in a lesser sense a single cause—each such pair is in fact the effect of the initial elements of a certain causal sequence. The initial elements are (i) neural and other bodily elements just prior to the first mental event in the existence of the

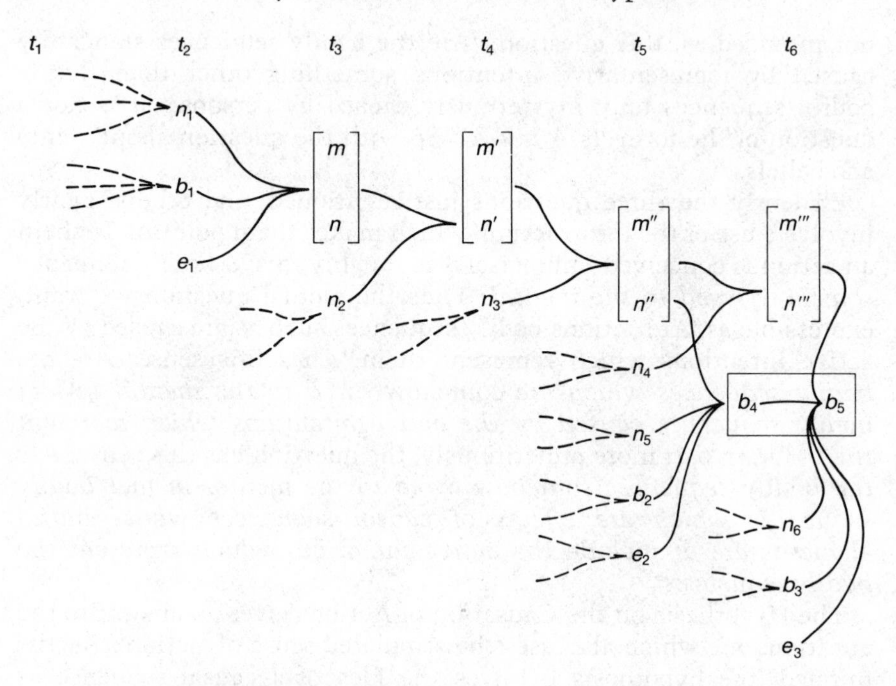

FIG. 19. Hypothesis of Psychoneural Nomic Correlation, Hypothesis on the Causation of Psychoneural Pairs, Hypothesis on the Causation of Actions

> person in question, and (ii) direct environmental elements then and thereafter.

The Hypothesis on the Causation of Actions is this:

> Each action is a sequence of bodily events which is the effect of a causal sequence one of whose initial elements and some of whose subsequent elements are psychoneural pairs which incorporate the active intention which represents the sequence of bodily events. The other initial elements of the causal sequence, at or after the beginning of the active intention, are neural events, non-neural bodily events, and direct environmental events.

As remarked, 'action' in the hypothesis is to be taken in the stipulated imprecise sense, as it is in 'The Hypothesis on the Causation of Actions'.

Figure 19 gives a model of all three hypotheses. The model is of the whole of the theory of determinism being advanced. What is modelled is part of the very end of a causal sequence, involving four psychoneural pairs and an action. The sequence contains neural events (n_1, n, n', etc.), non-neural bodily events (b_1, etc.), last or direct environmental events (e_1, etc.), mental events (m, m', etc.), the psychoneural pairs (m, n, m', n', etc.), and the action consisting in the boxed bodily events b_4 and b_5.

The given psychoneural pairs model the Correlation Hypothesis—and also, of course, the proposition that mental and neural events form pairs in the sense that they constitute a single effect and in a sense a single cause. (3.3) Hence they also model the Union Theory. (3.3) The first pair m, n includes the mental event m, say noticing the olives on the table. The second pair m', n' can be taken in its mental part as being the forward-looking intention to take one.

The sequence issues first in the the psychoneural pair m, n, and subsequently in three other pairs, thus modelling the second hypothesis, on the causation of such pairs. That hypothesis, then, is modelled by events at all the times t_1 to t_6.

The third and fourth psychoneural pairs, in their mental elements m'' and m''', can be taken as stages of an active intention to move so as to have an olive. The boxed bodily events b_4 and b_5 constitute a movement, say taking an olive. The first event, b_4, is given as the effect of a causal circumstance which includes the pair m'', n'', the first stage of the active intention. The causal circumstance also includes a neural event n_5, a bodily event b_2, and an environmental event e_2. The second event b_5 within the movement is given as the effect of a causal circumstance which includes the pair m''', n''', the second stage of the active intention. The causal circumstance also includes three other events. (These two causal propositions, about the two events which constitute an action both being effects, indicate what is to be understood by speaking in an abbreviated way of an action as an effect.) In the various items mentioned, we have modelled the Hypothesis on the Causation of Actions. That hypothesis, then, is modelled by events at t_5 and t_6.

As will be clear, the three hypotheses overlap to a significant extent in subject-matter. The psychoneural pair m'', n'' is an instance of what is claimed by the Correlation Hypothesis and the proposition about a mental and a neural event as a single cause and effect. The pair is an effect within a causal sequence which models the second hypothesis, about the causation of such pairs. The pair, finally, if regarded as involving an active intention, falls under the hypothesis about the causation of actions.

The burden of the model—some would say the gravamen—is in one part that *mental events* are necessitated events in the sense we know, necessitated by earlier and later initial elements of a causal sequence going back to at least the birth of the person. The burden of the model in another part is that *actions* are necessitated events, necessitated by short sequences whose initial elements are at the time of the active intentions. (The latter contingent rather than analytic proposition, like many others that will follow it in this inquiry, uses 'action' in the stipulated imprecise sense.) Since the two claims overlap, it is of course possible to conjoin them. What we then have is roughly that an action is the effect of a causal sequence whose very earliest initiating elements occur at roughly the time of birth of the person.

One of two things which remain to be done here is to note that our third hypothesis satisfies various criteria we have accumulated and which apply as much to it as to its predecessors. The argument for the hypothesis, so far as it is needed, consists in that fact. It satisfies the criterion of using a clear and effective conception of causation, and the criteria of mental realism, psychoneural intimacy, mental indispensability, personal indispensability, simplicity, and the somehow neural causation of neural events. (pp. 89 f., 163 ff.) The last criterion might be enlarged so as to require the gapless neural and bodily causation of actions. The hypothesis is also consistent with the axiom not satisfied by certain conceptions of an action—that it is a logical or conceptual truth that an action standardly conceived stands in relation to the mental realistically conceived. On a certain view of logical necessity the hypothesis might be said to be in the background of the truth.

What else is to be said is that we have already in effect considered various alternatives to the third hypothesis, partly for the reason that it is impossible wholly to insulate the subject-matters of the three hypotheses. We have in effect considered all the alternatives to the third hypothesis worth considering, some of them of a determinist character, some not. Various Identity Theories of mind and brain, such as the Variant of Anomalous Monism and the Two-Level Identity Theory, were found to give an unacceptable account of the antecedents of actions. (2.3) Various accounts primarily concerned with the diachronic explanation of mental events also fail to be acceptable in what they state or imply of the explanation of actions. The most prominent of these are the view involving wholly neural causation with psychoneural correlation and the view involving a macromicro relation between the mental and the neural (3.2), and various indeterminist views (3.4–3.9). With the indeterminist views, centring on an originator, there is the particular difficulty of securing a mental explanation of action by way of probabilistic sequences. (3.5) Finally, we have latterly given consideration to alternative explanations of action which involve different and unacceptable conceptions of action. These are the several Wittgensteinian conceptions, the compound view, and so on. (4.1)

4.5 OBJECTIONS

It is not only alternatives to the Hypothesis on the Causation of Actions that we have in effect already considered. We have also considered a number of objections to it—they apply not only to the previous two hypotheses, but also to it. The Correlation Hypothesis, to recall examples, may be opposed by associating it with famous definitional failures, by means of the doctrines of holism, indeterminacy of translation, and the radically different domains of the mental and the neural. (2.7)

All of these four items were considered by us as objections to the existence of nomic connection between mental events and simultaneous neural events. Being quite general objections, to *any* nomic connection involving the mental, they necessarily also pertain to the hypothesis on actions—as they pertain to the second hypothesis, on the explanation of mental events. The objections can be no more effective in these locales than in their original one. Let us instead consider one somewhat related idea, about our knowledge of psycho-behavioural laws, and then several objections of a somewhat Wittgensteinian kind.

It is commonly claimed that we know no laws connecting mental events and actions. It is claimed, that is, that we have confirmed no clear and adequate statements of psycho-behavioural nomic connection, or none at a certain level of generality. Consider the 'law of action' which runs as follows: 'If *A* wants *G* and believes *X*-ing is a way to bring about *G*, and *A* has no overriding want, and knows how to *X*, and is able to *X*, then *A X*'s.' (P. M. Churchland, 1970) Various questions can be raised about this, one being whether the agent might in fact have both the want and the belief and yet somehow fail to 'bring them together', and so not act. It can be maintained, further, that if the 'law of action' is rectified to meet this difficulty, as perhaps it can be, there are more difficulties to come.

There is a recurrent theme in such objections. It is that our supposed failure up until now to state nomic connections—which supposed failure we can assume to exist for present purposes—gives us a premiss for the conclusion that we shall not and cannot ever be successful. The rough generalizations such as the 'law of action' which we have—to assume that all our psycho-behavioural generalizations are no more than rough—will not and cannot ever be improved into anything better, which is to say laws of the kind found within the wholly non-mental sciences. There is a further and logically separate theme which understandably is not made quite so explicit. It is that if laws are not in fact discovered in a domain, not discovered until some time or other, there are in fact no nomic connections there to be stated. (Davidson, 1980g)

We shall shortly be considering the state of our knowledge as to psycho-behavioural nomic connection, but a general comment on the objection is possible now. The question of whether there are such connections is and always has been routinely taken to be an empirical question, whose answer is a contingent proposition, neither logically true nor logically false. It is the question of whether mental events and bodily sequences are related in the way specified by certain conditional statements of which we know, contingent statements, those investigated in our inquiry into nomic connection. (Ch. 1) It is a question precisely on a level with other such questions, not involving the mind, which can properly be described as the very stuff of science. It is a question about the nature of reality, more particularly about our nature.

What *could conceivably establish* that ongoing empirical inquiry into mind and behaviour—if it has not yet done so—*cannot* ever succeed in specifying laws? The only thing that could do so would be a truth to the effect that the given psycho-behavioural investigation is aimed at establishing what can in advance be seen to be logically impossible or to face some related conceptual barrier. What has always been taken to be a possibly true contingent proposition open to empirical inquiry, that there is psycho-behavioural nomic connection, must in fact be a logical impossibility or the like. That large unlikelihood, as it seems to me to be, needs to be supported. It needs to be supported by something greatly more substantial than large and in part elusive philosophical theory.

Certainly it is not made seriously arguable by considerations of which we know, having to do with the supposed analogy with such definitional failures as behaviourism, with holism, indeterminacy in translation, and the disparateness of the mental and bodily domains. The evidence of past failures is relevant *only* if it can be shown to give us the conclusion that some non-empirical barrier stands in the way of progress. Simple failure, not shown to have the requisite non-empirical explanation, is in fact of no great significance, and not in the required way predictive. The history of mankind, and in particular of science, would be absolutely and totally different if simple failure in empirical inquiry, over some span of time, did indeed guarantee logically necessary eternal failure.

Let us turn to certain objections that have often been relied on by Wittgensteinian philosophers and are summed up in the declaration that reasons cannot be causes. As indicated by examinations of these objections (MacIntyre, 1971a; Gustafson, 1973), they are diverse, but they centre on the ideas that there cannot be causal connection between items that stand in a certain logical or conceptual connection, and that reasons and causes do stand in this connection. (Melden, 1961; cf. Pears, 1967, 1971b) It is usually supposed that the general proposition about causal connection is the one which was established by Hume in his inquiry into causation. The objections, whatever their worth, can be applied as well to active intentions and actions. Active intentions have the relevant features of the mentioned reasons and may indeed be taken as identical with them.

Our diner mentioned in connection with the model in Figure 19 had the active intention to have an olive. It was remarked earlier that it is tempting to characterize active intentions in an imperatival way. (p. 225) Let us say that the intention in question had the content 'Let me so move here and now as to bring it about that I *have an olive*'. The Hypothesis on the Causation of Actions puts the intention in causal connection with the ensuing action, which in so far as it was intention was precisely his *having an olive*. There is thus a certain 'logical connection' between intention and action: the same concepts, those involved in speaking of having an olive, turn up in the descriptions of both intention and action.

If we now turn to Hume and causation, he asserts that if two things are cause and effect they are 'distinct events'. (1902 (1748), pp. 25 ff., 60 ff.; 1888 (1739), pp. 86 ff., 155 ff.) A part of what he assumes in saying so is that cause and effect are not identical: they are indeed two things and not one. What he mainly has in mind, however, is the consideration that the relation between the two things is in a certain sense not 'logical': the existence of the first does not logically entail the existence of the second, or, more precisely, it is logically possible or non-contradictory, despite any fact of causal connection, to suppose that the first exists and the second does not. To say the first caused the second is not to ascribe to the first a more or less incomprehensible logical power to bring the second into existence.

To come to the crux, is the 'logical' connection we noted as holding between the diner's active intention and his action a relation of the kind that precludes causal connection? Evidently not. That the intention has the content it does, and thus is to be described in terms of certain concepts, and that the action is also to be described in terms of those concepts—this fact comes nowhere near the propositions that intention and action are not two things, or that it is inconceivable that the first should exist without the second. As has been remarked, there is no more than confusion here. (Davidson, 1980h; Mackie, 1974) There is no similar confusion, but not great credibility, in a different form of the argument. It moves from the premiss that some mental events *are* actions—that a belief of mine may *be* my lurching while coming downstairs—to the conclusion which certainly follows, that mental events so conceived cannot cause actions. (Kenny, 1975, pp. 117 f.; 1978, pp. 27–8) The premiss consists in the neo-behaviourism of which we know. (4.1)

There is an objection closely related to the main one we have been considering, having to do with confirming the occurrence of actions and intentions. It is true that it is conceptually necessary, in order to identify an action, that I arrive at a view of the intention in question. (pp. 212, 232) To take a particular kind of instance, we may suppose that whether a witch-doctor's action counts as murdering someone does necessarily turn on the content of his intention, the content having to do with beliefs he may have had as to whether a potion kills only witches, whether a certain person is a witch, and so on. (Kenny, 1978, p. 12 f.; von Wright, pp. 94 f., 103 f.; cf. Mackie, 1974, pp. 288–93) It does not follow, however, that intention and action are not two things, or that we cannot conceive of one existing without the other, and hence that there cannot be causal connection.

Finally, let us look quickly at a number of possible objections from a philosopher encountered above and earlier, objections which may be taken as establishing the incoherence of a determinism such as ours, and have to do with certain supposed 'loosenesses' between the mental and action, and have certainly been influential.

(i) The diner, we can say, acted as he did *because* he wanted an olive. We can say as naturally, and more fully, that he acted as he did because he intended to, or because of his active intention. But, it may be objected, these ordinary uses of the word 'because' are only mistakenly taken as conveying the belief that a mental event caused an action. 'Because' does not indicate a causal connection, but something quite different, something very much looser, having to do with a *reason* and an action. (Kenny, 1975, pp. 118, 120; 1978, pp. 25–6) One reply is that it is less than clear that 'because' does in general mean what is supposed, but leave that. Suppose it *is* true that by a certain sort of ordinary utterance we do not *mean* that a mental event caused an action. Does that throw into doubt the Hypothesis on the Causation of Actions? It can hardly do so, since the hypothesis is not about usage, or dependent on it.

(ii) The diner's intention, in brief, was to have an olive, and his action was having one. What we have in the third hypothesis is that the action was the effect of a causal circumstance which included the intention. However, it may be said, what we ordinarily think of intentions and actions suggests in another way a far looser connection between them. The diner's intention, it may be said, was indeed to have an olive. But we ordinarily suppose that such an intention can issue in a variety of actions: reaching for an olive with the left hand, reaching for it with the right, doing it slowly or quickly, and so on. The third hypothesis claims that the given intention was involved in the necessitation of a single wholly determinate action, perhaps reaching for an olive slowly with the right hand. (Kenny, 1978, pp. 101–2) A reply to this objection is that we may well suppose, of a *type* of intention, or a *class* of intentions, in loose senses of the words, that they may issue in a variety of actions. Do we ordinarily think, however, that the particular active intention with which a man acts, an intention different from other members of a class, may well issue in *either* a movement of his left hand *or* a movement of his right, and so on? It seems to me plain that we do not.

(iii) There is the fact of the weakness of the will. We come to the conclusion that we ought to do a thing, and we do not do it. Here too, it can be said, there is a looseness between the mental and actions. (Kenny, 1978, pp. 102–8, 114–15) That can be said, but it does not affect the third hypothesis. I may well pass through a sequence of reflection, on a particular occasion, to the belief that I must not have another glass of wine. Out of a contrary active intention, I then have one. That there is the given looseness or rather disconnection between my belief and what I do does not establish that there is any such looseness between things connected by the Hypothesis on the Causation of Actions, which is to say a causal circumstance involving an active intention, and an action.

(iv) We can take it that the diner's action had in its background what can be represented in terms of *practical reasoning*, which is not to say that he carried out explicit steps of reasoning. Very roughly indeed, the premiss of this reasoning might be to the effect that the olives look succulent, and that they are in reach. The upshot of the desire and belief, we are also to take it, was the action. But, as noted in another context (pp. 204 f.) it is the nature of such reasoning that it is defeasible: that the addition of a further premiss might have got in the way of the upshot. Such a premiss, in terms of the example, might have been the diner's reflection that he had already made free with the olives and they were out of the reach of others. But these truths, to come to a first point of the objection in hand, count against an explanation of the action in terms of a causal circumstance, something which necessitated the action. What we must do, we are advised, in order to *attempt* to safeguard the third hypothesis, is to take up a certain general view of causation. We shall then get closer to the picture of the episode in terms of practical reasoning. The general view is, very roughly, that there is a causal connection between two things if there is a true generalization that things like the first are followed by things like the second *unless there is interference*. The view is related to one considered in our inquiry into causation—the 'usual background' or 'usual environment' view. (pp. 57 ff.) The attempt to save the hypothesis, we are to suppose, in fact fails, essentially for reasons connected with the general view. (Kenny, 1975, pp. 22–3, 115–17; 1978, p. 29)

We need not take up the general view of causation, in fact rejected earlier, in order to escape the objection. Nor do we need to look into various difficulties in connection with practical reasoning. We need only keep in mind the first part of the objection and the Hypothesis on the Causation of Actions. Let us grant that the diner's action, which did follow on his desire and belief, would not have done so if he had also had an additional belief. Nothing follows about any looseness in the connection between (a) a causal circumstance including a certain active intention and (b) an action. That there would have been no action if a *different* causal circumstance had existed, one involving a further belief, can go no way towards undercutting the claimed relation of necessity between (a) and (b).

4.6 THE COMPLETED THEORY OF DETERMINISM

It would be fine to be able to believe that the determinist theory of the mind and of action, which is now complete, is free from obscurity and from other conceptual shortcomings and failings. To be able to believe that agreeable thing, also, would be to be all too innocent. Causation and nomicity, the relation between consciousness and the neural events intimately connected with it, the explanation of mental events, and the matter of mental events and actions—each of these four hard subjects becomes complex and mystifying, and pulls one in several directions, so much so as to make it unlikely than any account of it will be crystal-clear and in every way in good order. Optimism about the conceptual adequacy of the theory of determinism which we now possess must also be made uncertain by discomfiting general truths about philosophy and of course other types and styles of inquiry. One of these truths is that one man's clarity may be another man's obscurity, sometimes without detriment to either observer. However, if such chastening reflections about the four subject-matters and about inquiry generally are true, so too is something else.

J. L. Austin, who was in a circumscribed way perhaps as acute a philosopher as any in the middle years of this century, placed himself among those who 'are inclined to think that determinism . . . is still a name for nothing clear, that has been argued for only incoherently'. (1956, p. 131) P. F. Strawson, a philosopher of exemplary and wider judgement, was of like mind. 'Some philosophers say they do not know what the thesis of determinism is. Others say, or imply, that they do know what it is. . . . If I am asked which of these parties I belong to, I must say it is the first. . . .' (1962, p. 187) Strawson allowed that 'of course, though darkling, one has some inkling—some notion of what sort of thing is being talked about', and in fact advanced a substantial argument for the consistency of determinism with certain things which it may be thought to threaten. Austin, for his part, provided an argument against such a consistency. As remarked in the Introduction, others have followed Austin and Strawson in their scepticism. One, David Pears, has impressively removed obstacles to the determinist enterprise but yet writes that 'it is one thing to dream of a deterministic system, and quite another thing to find one that actually works'. (1973, p. 110) A. J. Ayer remarks that we do not yet possess a satisfactory determinist theory. (1984c, p. 3) Others have different kinds and degrees of uncertainty about the conceptual viability of determinism. (Hampshire, 1965, pp. 108–12; Williams, 1963, pp. 106–10)

The scepticism was understandable, but it would be remarkable if it were not possible to bring into at least a general clarity and order a determinist picture of the mind and of action. Or rather, to come to a more careful proposition, which does not scandalously presuppose that the job has never been done before, it would be remarkable if it could not be done again and anew, in a way consonant with the advance of philosophy. It was done, certainly, in the time of Hobbes, by Hobbes, and in the time of Hume, by Hume. So too with Mill in his time. If it is true that the construction and elaboration of any determinism, like that of any large philosophical doctrine, is certain to be at least imperfect, even when the work of a great philosopher, it is equally true that a determinism can be brought wholly into view and seen to be free of such conceptual defects as to be noticed at first sight, even second sight. Thus it is something whose truth and whose consequences need to be investigated. I make no great claim of originality with respect to the determinism now before us—indeed, it seems to me to move towards being the stuff of ordinary enlightened belief—and I make no claim whatever as to its being of the strength of the thought of Hobbes, Hume, or Mill. It does seem to me to satisfy a decently scrupulous demand for clarity and order.

It is worth remarking that what follows, if it does not depend on the unlikely idea that the expounded view of the mind and action is conceptually perfect, does not depend either on just that expounded view. In the course of our inquiry so far, we have looked at a number of determinisms, or at any rate views which can be understood deterministically. I have in mind mainly the Dualistic Identity Theory with Neural Causation and the Two-Level Identity Theory (2.3), and Neural Causation with Psychoneural Correlation and the Psychoneural Relation as Macromicro (3.2). All of these have been found wanting—to my mind, they have been refuted. But, I also think, it has also been clear enough that they are in a way arguable. They come close to satisfying some reasonable requirement as to conceptual adequacy—which requirement is no neat test but rather a matter of judgement. Furthermore, there is the possibility of other determinist theories. If I depend mainly on my own view for the conclusion that determinism is conceptually respectable, it is none the less the case that these other views make a contribution to that conclusion.

Let us then turn to a question which has been near to hand so far, but never directly examined. It is the question of what empirical basis there is for supposing the theory of determinism to be true. The best place to look for such a basis is neuroscience. It is the subject of the next chapter. The subject of the chapter after, which rightly does not have pride of place with respect to the question of truth, considers philosophical rather than scientific claims.

PART 2

The Truth of the Theory

5

Neuroscience and Quantum Theory

5.1 THE NEURON AND THE BRAIN

Among the living cells of the human body, and of the Central and the Peripheral Nervous Systems, it is neurons or nerve cells which are most immediately bound up with consciousness—with the antecedents of mental events, mental events themselves, and their consequents, the latter being other mental events and also actions. Neurons vary from very short indeed, a few millionths of a metre, to a considerable length, perhaps two metres, and are constituted of structures and substructures which are open to study by light and electron miscroscopy and by many other converging means.

The principal structures of the neuron are shown in Figures 20 and 21. In addition to (i) the *cell body* or soma, containing the nucleus, a neuron almost always comprises three other structures. Each is itself elaborate, containing substructures. (ii) A *dendrite* is a root-like formation whose typically short trunk enters the cell body. It may have a few initial branches, or many thousands, a dense arborization.

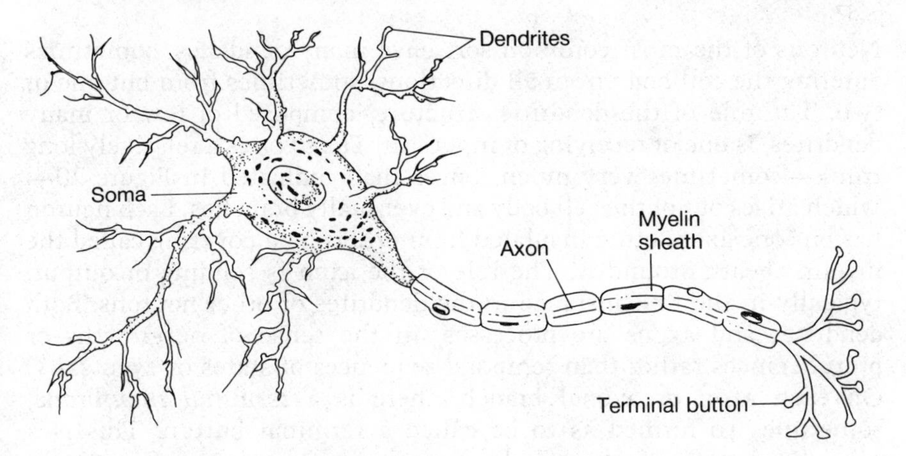

FIG. 20. Principal Structures of the Most Common Type of Neuron, the Multipolar. From Carlson, 1985, fig. 2.3, p. 18.

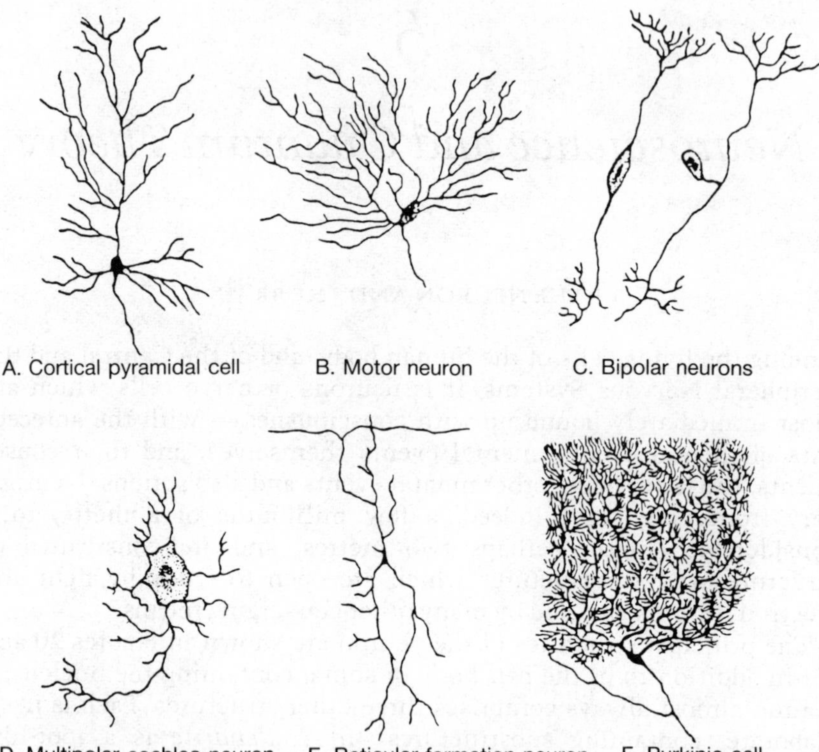

FIG. 21. Various Types of Neurons. From Cotman and McGaugh, 1980, fig. 3.2, p. 67

Neurons of the most common sort have many dendrites, sometimes entering the cell body from all directions, sometimes from but one or two. The role of the dendritic structure, composed of few or many dendrites, is one of receiving or input. (iii) The *axon* is a relatively long trunk—sometimes very much longer than indicated in Figure 20—which arises out of the cell body and eventually branches. Each neuron has but one axon, often insulated from others by a covering called the myelin sheath around it. The role of the axon is sending or output, typically in the first instance to the dendrites of other neurons. Both dendrites and axons are processes, in the sense of outgrowths or protuberances rather than temporal sequences of states or events. (iv) On each axon or axonal branch, there is a *terminal membrane,* sometimes so formed as to be called a terminal button. This is a specialized area of the whole containing membrane of the cell. Terminal membranes are parts of the junctions or *synapses* between a particular neuron and other items, most often the axons and dendrites of other neurons. A synapse, at bottom, typically consists of a terminal button on the axon of a given neuron closely applied to a part of the membrane of a dendrite or the soma of another neuron. A typical neuron enters into very many synaptic connections, from 1,000 to 10,000 or more.

The mentioned receiving or input with respect to a neuron, and its sending or output, are electrochemical in nature. The electrical activity of a neuron depends in part on its resting potential, the electrical property it shares with almost all other cells and indeed with much else, including ordinary batteries. It can correctly if not enlighteningly be described as the property of being an energy reservoir. If a micro-electrode is inserted into a neuron, and another electrode is located in the fluid surrounding the cell, and the two are connected through a voltmeter, it will register a charge across the membrane between the inside and the outside of the cell. While the neuron is inactive or not firing, the inside of the cell is negative and the outside positive. The resting potential—or simply the internal negativity of the cell—is owed to the uneven distribution inside and outside the cell of ions, positively and negatively charged particles of several kinds. Typically they are sodium, potassium, or chloride ions. The distribution is importantly a function of the nature of the membrane.

The special property of a neuron is its electrical excitability—its capacity, in part dependent on its resting potential, to give rise to and conduct an *action potential* or electrical impulse, although not in the way of a metal conductor, such as a telephone cable. Action potentials, despite their name, are themselves *events*, transmitted messages. They are actualities. Given certain other events—events external to the cell, of which more will be said below—an action potential passes down the axon at a speed of something between 1 and 100 metres per second. There is no degeneration of the signal. If recording electrodes are placed, say, in the axon near the cell body, and half-way down the axon, and near one of the terminal membranes, a like episode will be recorded in turn by the three electrodes. Each episode is a rapid changing (over about 1,000th of a second) of a segment of the axon, so to speak, from negative to positive and back to negative. (Figure 22) Each episode, more particularly, is a matter of the changing distribution of negatively and positively charged particles inside and outside the axon. This changing distribution of ions is governed in part by the changing and differently selective permeability of the membrane of the axon, which contains channels or gates. In a different way in each of several phases, the membrane allows through itself some particles but not others. These changes in the permeability of a segment of the axon

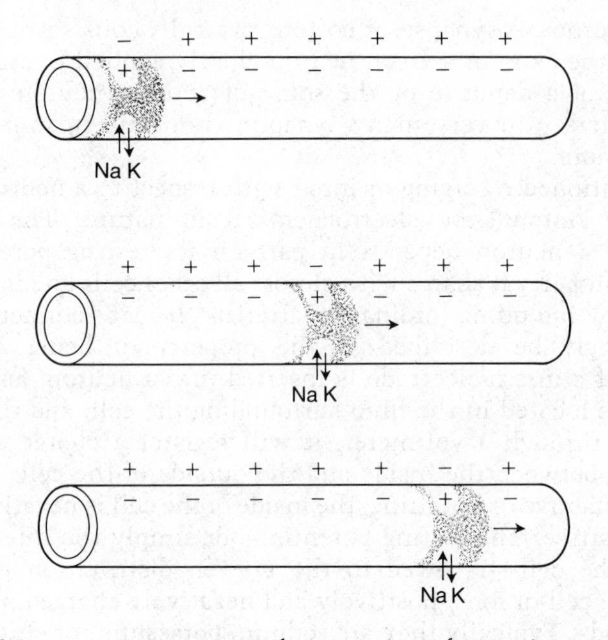

FIG. 22. Schematic representation of propagation of an action potential down part of an unmyelinated axon. Electrical charges across membrane indicated, and movement of sodium and potassium ions through membrane. After Carlson, 1985, fig. 2.34, p. 47

are owed to changes in the segment before it. To speak of *segments* is in a way misleading, however, at any rate with many neurons, since the sequence in question is a smooth or continuous one.

The action potential, no matter the magnitude of the events external to the cell which give rise to it, is in a sense standard. The shape and size of the wave in the oscilloscope-recording of any impulse, with respect to a particular neuron, is identical. It is identical on each occasion the neuron fires, and it is identical at each point in the pulse's progress along the axon. Action potentials are thus said to be subject to the All or None Law. The frequency-patterns of action potentials, however, differ greatly. As recorded at a particular point on the axon, pulses may come in a regular order—that is, may be separated by regular gaps of the same duration—or come in separated bunches, or come in one of many complex frequency-patterns. The stronger the pressure of a touch on the skin, for example, the higher the frequency of firing of sensory nerves innvervating that portion of skin. This temporal distribution of pulses, spoken of as information, or a message or code, is essential to what follows.

When action potentials reach the terminal of a given neuron's axon, they there give rise to what are, in terms of other cells, external events—those events not so far described but already mentioned several times in connection with the input or the inception of the activity of a neuron. That is, action potentials on reaching the terminal cause the release of chemical substances, called transmitter-substances or neurotransmitters, from these terminal or pre-synaptic membranes. Such secreted substances as acetylcholine, dopamine, and norepinephrine cross what is called the synaptic cleft to the membrane of another cell, the post-synaptic membrane, and have an effect on the electrical properties of that cell. More particularly, they affect channels or gates in the post-synaptic membrane and so give rise to the changing distribution of charged particles across its membrane briefly described above. Some transmitter-substances contribute to exciting the receiving cell, to increasing the likelihood that it will fire. Others have an inhibiting effect. The occurrence of an action potential in the receiving cell is a result of the summation or integration of a sum of frequency-patterned inputs through what is typically a large number of synapses of a large number of cells. The action potential in the receiving cell is thus almost certain to have a frequency-pattern different from all of the action potentials which gave rise to it—for which reason, among others, neurons are not fruitfully compared to the logic gates of a digital computer. (P. M. Churchland, 1984, p. 131) It is said to be true of any part of the nervous system, indeed, that what goes in is never what comes out.

So much for the lightest of sketches of the neuron, a handful of generalizations which hardly begins to catch even that part of what is known which is conveyed by standard textbooks of a general kind and more popular introductions. (Blakemore, 1977; Brodal, 1981; Carlson, 1985, Chs. 1–3; Carpenter, 1984; P. S. Churchland, 1986, Ch. 2; Cotman & McGaugh, 1980, Chs. 3–5; Dimond, 1980; Kandel and Schwartz, 1985; Kuffler, Nicholls and Martin, 1984; Rose, 1976; Shepherd, 1983; Stein, 1982, Chs. 1–4; Uttal, 1978; Young, 1978, 1987) A look at them will correct any layman's idea that knowledge of the neuron is impressionistic, speculative, or general.

What the sketch may be said to convey, as the accounts do in detail, and are commonly said to do, is the structure and function of the neuron. With respect to function, in an ordinary standard sense, it is conveyed partly by a great variety of verbs. Some used above are: *produce, generate, conduct, send, receive, secrete, contribute to,* and *govern*. Function is also conveyed by a variety of nouns, including *neuron* itself as defined earlier (p. 52), and by other grammatical means. What the sketch conveys, more precisely, and its point, is what may seem indubitable, that all of what happens within and between neurons is a matter of nothing more or less than standard causal connection.

That the verb 'to cause' is rare in neuroscience, as is sometimes remarked, is irrelevant to this—as irrelevant as like points would be about the verb's incidence in meteorology, engineering, and science generally, and also in common-sense accounts of machines, plants, and so on. Rightly, no one concludes on such a terminological basis that rainstorms, bridges, and electric toasters are not subject to causation. The fact that alternative verbs are used in neuroscience, all of them properly described as *being* causal verbs, is readily explained by their greater informativeness or precision—they convey types and modes of causation. This is so, for example, with 'secrete', 'conduct', and, in a different way, 'produce' and 'contribute to'.

What we have, then, is strong and clear support for the proposition that neural sequences are somehow or in some way causal sequences. More particularly, we have strong support for the proposition of the neural or other bodily causation of the neural—neural events are somehow standardly caused by prior neural or other bodily events. In saying that neural events are *somehow* caused by prior neural or other bodily events, as will be understood, we leave open the possibility that the causal sequences in question, while they are neural or bodily at every moment, post-environmentally, are nevertheless not wholly such. (p. 164)

Might we claim that we have more than *strong and clear support* for the causal proposition? There are reasons for not claiming more at this stage, whatever our final view of the probative value of the given evidence. One is that it may be speculated, certainly obscurely, but as sometimes it has been, that causation within the neuron, or even such internal causation together with causation between any pair of neurons, is somehow consistent with an absence of causation when larger neural sequences are considered, perhaps sequences involving neural systems or levels. We shall in due course look at larger neural sequences. (5.3) A second reason for saying no more at this stage is Quantum Theory. (5.6)

Let us quickly look over the nervous system, preparatory to a survey of some known functions of it, a survey which will be in terms of our three hypotheses. It includes other cells, but, in its relation to consciousness and, more importantly, to our three hypotheses, it is a multiply linked collection of neurons, unimaginably vast. The number of neurons in question, of which estimates vary somewhat, is great. One estimate is that the nervous system contains 10^{12} neurons. There are immensely more synapses, perhaps 10^{15}. A single thought has been said to engage many millions of neurons. As for linkages between neurons, those at the bottom level are of course individual synapses. These bottom-level links are far from being the only ones, but rather enter into many kinds of larger ones. The simplest kind of larger connection consists in tracts of neurons bringing together large parts of the system.

5.1.1 Forebrain

(i) The folded and involuted cerebral cortex covers the left and right cerebral hemispheres or cerebrum. The two hemispheres, which make up the largest part of the human brain, are not completely alike, even from a gross anatomical point of view, and they differ greatly in function. Each has abilities not matched or not had at all by the other. The two cerebral hemispheres, joined by a large bridge of axons, the corpus callosum, are further divided into regions, lobes, and areas, according to anatomy and function. There are areas of the cerebral cortex related to sensation and perception, and areas that participate in the planning, execution, and control of movement. Various other cortical areas are associated with higher mental functions. One of these is Broca's speech area, another Wernicke's auditory association area, both named after their discoverers, and both usually located only in the left hemisphere. With respect to the cerebral cortex, and also other parts of the brain, as implied above, function is importantly a matter not just of the given part but of interaction between it and others. *How* importantly is a large and inevitably disputed question.

(ii) Basal ganglia, primarily located deep in both cerebral hemispheres, are concerned with action or motor control. Just how is uncertain, although it is known that disorders of movement, such as Parkinson's Disease, follow on damage in this area. Basal ganglia and also play a role in emotion, and perhaps learning.

(iii) The limbic system, a collection of various areas for the most part also deep in the cerebrum, is involved in at least emotion, motivation, goal-directedness, and memory. It is the part of the brain most concerned with feelings. One of the areas in it, the hippocampus, as we shall see, is on one theory connected with certain features of consciousness generally.

(iv) The thalamus or inner chamber provides most of the neural input from the rest of the CNS to the cerebral hemispheres, notably sensory information from the PNS. It is also important in affective judgements, pain and pleasure, hunger and thirst, and to consciousness and alertness.

(v) The hypothalamus, at the base of the brain, is of considerable importance to, among other things, the control of the autonomic nervous system (which is concerned with the viscera, smooth muscles, glands, heart and skin), and the organizing of such fundamental survival activity as feeding, fighting, fleeing and mating.

5.1.2 Midbrain

(i) The midbrain has as one of its two principal structures the tectum or roof. It has roles in both the auditory and the visual systems.

(ii) The tegentum, beneath the tectum, mainly consists in a part of the reticular system, which receives and transmits sensory information, and plays a role in sleep and arousal, selective attention, and control of certain vital reflexes. Other parts of the tegentum enter into movement and in particular species-specific behaviour.

5.1.3 Hindbrain

(i) The first of the three principal structures of this most rearward part of the brain is the cerebellum or little brain. It receives muscular and much other information, and to it is owed the co-ordination of posture and movements, including the movements made in speech. The rate, range, and force of movements are regulated by it.

(ii) The medulla, which tapers into the spinal cord, also contains part of the reticular formation. Alertness depends on certain groups of neurons in it. The medulla also contributes to various vital activities, including breath and swallowing, waking and sleeping.

(iii) The pons, in the brain stem, shares a part of the reticular formation. It is important to waking and sleeping, and allows intentional movements to be carried out effectively. It contains neuron tracts that connect the cerebellum and medulla with parts of the upper brain.

This sketch of the nervous system or rather the brain is like the previous one of the neuron. Despite its wonderful brevity, it preserves the character of fuller anatomical accounts of a standard kind, as a look at some of the textbooks and introductions mentioned above will show.. It thus provides a first strong indication that neuroscience provides confirming evidence—despite a tendency to speak of the neural and behaviour—that *the neural and the mental are in intimate connection*. That is, to recall the axiom of psychoneural intimacy, we have an indication that types of mental events do somehow necessarily occur with types of neural events in a locale or locales of the brain. (p. 90) What that comes to is that either mental events are identical with neural events, or there is nomic connection between them. This, like the somehow neural causation of the neural, is a fundamental premiss of our theory of determinism.

As in the case of the neuron and causation, the evidence is not affected by what is also true, that full and explicit assertions within neuroscience of psychoneural intimacy are rare. What we have, rather, is a plethora of easy usages pertaining to *parts* of what we can call the *neural intimates* of mental phenomena. Neural facts are said to be *related to* or *associated with* mental facts; neural facts are said to *play a role in* or be *concerned with* mental ones. We shall return to the matter. (5.2) It needs to be granted that other usages also indicate something else. Neural facts are said to *control* or to *allow for* mental ones, or to be their *substrates*. What is indicated, certainly, is *a* tendency to take mental facts as not merely in intimate relation with neural facts, but as somehow dependent upon them—in a way that neural facts are not dependent on mental. More is to be said of this tendency to epiphenomenalism. It is not the only neuroscientific tendency with respect to the psychoneural relation.

5.2 NEUROSCIENTIFIC EVIDENCE RELATING TO THE FIRST HYPOTHESIS

Let us now see something of what further evidence, over and above the gross anatomical evidence, is provided for the Correlation Hypothesis by the neurosciences—which, as might have been remarked before now, are those sciences which fall under a variety of overlapping labels: *neurobiology*, *neuroanatomy*, *neurophysiology*, *neurochemistry*, *neuropharmacology*, *physiological psychology*, *biological psychology*, *psychobiology*, and others.

The argument advanced earlier and at length for the Correlation Hypothesis, if much reduced, is as follows.

(i) Mental and neural events are intimately connected—each specific type of mental event somehow necessarily occurs with a simultaneous specific type or types of neural event in a given locale or locales of the brain. (p. 90)

(ii) That proposition restricts us either to Identity Theories of mind and brain and like theories or the Correlation Hypothesis, perhaps with the latter so supplemented by the psychoneural pairs proposition as to give us the Union Theory. (p. 169)

(iii) Arguable Identity Theories, partly because they offend against mental indispensability, are unacceptable. (pp. 93 ff.) So with like theories—Neural Causation with Psychoneural Correlation, and the Psychoneural Relation as Macromicro. (pp. 154 ff.)

(iv) Therefore the Correlation Hypothesis is the basic fact of the psychoneural relation. Each mental event of a given type is in a certain nomic connection with a simultaneous neural event of one of certain types. (pp. 173 f.)

Our enterprise now is mainly to survey and then to draw a conclusion about the further evidence to be found within neuroscience for the first and fundamental premiss.

5.2.1 Sensation and Perception

There has been a long history of assigning the various forms of sensation and perception, and indeed all other forms of consciousness and of mental capability, to parts of the brain. The brain charts of the early nineteenth century anatomist and phrenologist, the notorious Gall, have long since been discarded, as has been the work of the 'narrow localizationists' of the late nineteenth century. No part of the brain, it is now said, works alone. However, that is not to come near to allowing that the experiences of sensation and perception can in no way be localized in the brain. On the contrary, localization of a different kind is confirmed. (Luria, 1973, pp. 33–4)

Bodily sensations of many kinds, including sensations of touch, warmth, movement, shape of objects, and internal bodily states, are owed to the somesthetic system. To leave aside the receptors of the system, and two types of neural pathways to the thalamus and the cerebral cortex, of which something will be said later, and to come to what is most germane to psychoneural intimacy, this is the mapping of bodily surfaces onto parts of the cortex. The left hemisphere of the cerebrum, as is well known, deals with the right side of the body, and the right hemisphere with the left side. Figure 23 thus shows the mapping of half of the body's surfaces on to parts of the cortex. As this 'sensory homunculus' indicates, and as might be expected, the brain gives over greatly more of its resources to awareness of hands, fingers, face, lips and mouth than to awareness of the larger trunk and limbs. There has been extensive inquiry into the neural organization of the architecture of the given parts of the cortex—the sensory cortex. The inquiry, into columns of cells, is carried forward by greatly more precise means than the the sensory homunculus. (Dykes, 1983) Figure 24, which is now itself in the distant past of neuroscience, and whose explanation would none the less delay us greatly, gives a small indication of this. The inquiry has involved, among other things, the eliciting of various sensations in human subjects by the direct electrostimulation of the sensory cortex, the insertion of electrodes in

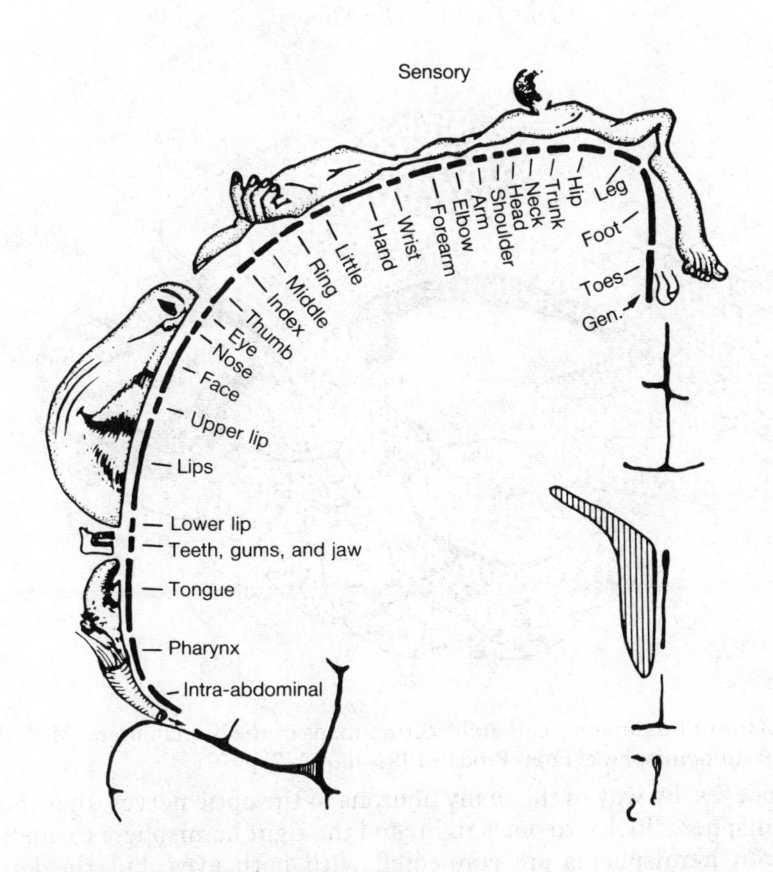

FIG. 23. Sensory Homunculus: diagram of cross-section of cortex showing sensory areas related to various parts of the body. From Penfield and Rasmussen, 1957, fig. 17, p. 44

the brain during surgical operations necessary for clinical or non-experimental purposes, and, recently computer-scanning. (Phelps and Mazziotta, 1985)

The neural relation to sensations having to do with hunger and thirst, and their satisfactions, to pass on to those sensations, has also been the subject of intensive inquiry. Electrical stimulation of the relevant parts of the hypothalamus of the rat results in eating, which under certain conditions stops when the stimulation stops. Taste preferences and aversions of very specific kinds are affected by particular lesions.

The sensory crossing mentioned in connection with touch and the like is also found in a different form in the visual system. That is, it approaches being true that the cerebral hemispheres are so connected

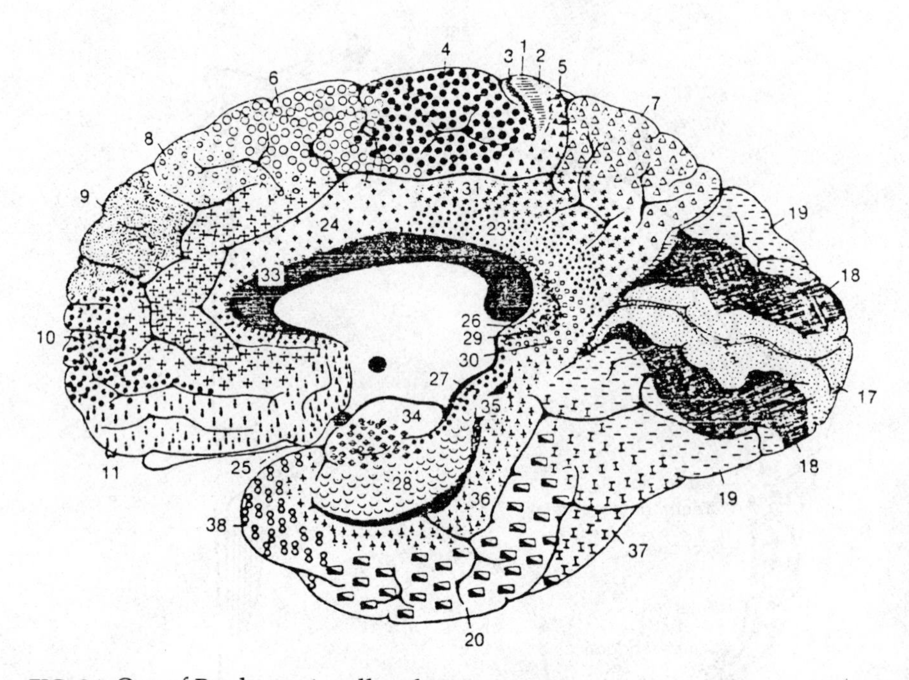

FIG. 24. One of Brodmann's cell-architecture maps of the human brain. Medial view of right hemisphere. From Brodal, 1981, fig. 12–2, p. 791

with the eyes, by way of the many neurons of the optic nerves, that the left hemisphere 'looks' to one's right, and the right hemisphere to one's left. (Both hemispheres are connected with both eyes, but the left hemisphere is connected with the halves of the two retinas that deal with the right half of the visual field. So with the right hemisphere.) Figure 25 shows the connection between parts of the right half of the visual field and areas of the left cerebral cortex.

The organization of the visual cortex, in six layers and in columns of cells, and the relation of particular cells to colours, spots, edges, brightness, contrast, and so on in parts of the visual field has been closely examined. (Hubel and Wiesel, 1977, 1979; Zeki, 1980) Most cells in the visual cortex of a cat or monkey respond best to line-segments at a particular angle and in a particular place in the visual field. Moving the line-segment 10 or 20 degrees clockwise or anticlockwise from the optimal orientation reduces the firing of the cell or stops it. Some cells of the visual cortex are monocular, in the sense that they respond only to a pattern shown to one eye, while others are binocular, responding to stimulation in either eye. With respect to localization and visual recognition, there is the surprising fact that a small brain area in the right hemisphere is involved in the

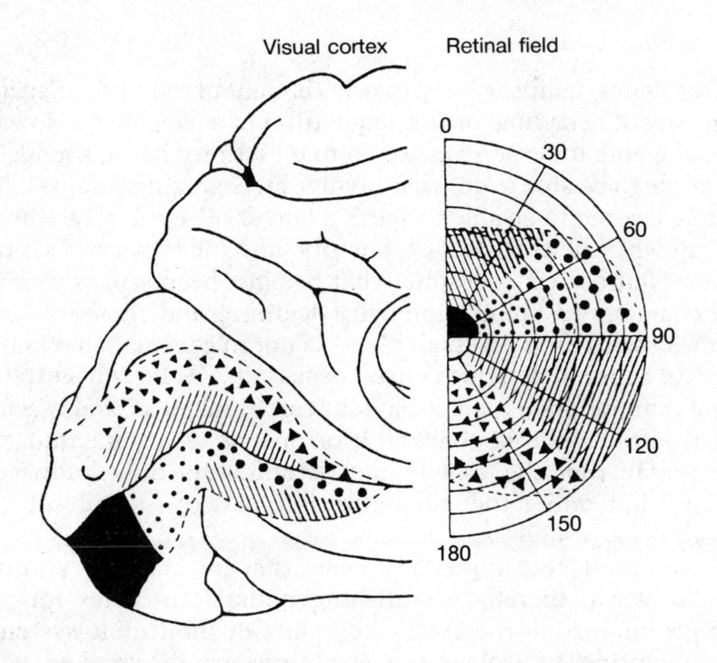

FIG. 25. Areas of the right half of the visual field, and their representation in the visual cortex. From Cotman and McGaugh, 1980, fig. 10.8, p. 424

specific function of recognizing faces. (Alexander and Albert, 1983) Damage to this area results in the disorder of prosopagnosia, a failure to recognize faces.

Here and elsewhere, the reporting of such bits of knowledge in a very general way and the use of rudimentary figures conveys very little indeed of the specificity and extent of what is known about connections between a part of experience, in this case vision, and parts and mechanisms of the CNS. The reporting does a little disservice to the research. On the other hand, if there is much specific knowledge of the connections between seeing and the CNS, there can be no pretending that what the proposition of psychoneural intimacy requires, a specific neural locale, however complex, for a particular whole visual experience as ordinarily understood, has been found. Those who investigate the brain mechanisms of vision are not about to isolate a single cell that responds specifically to just a sight of one's grandmother—even a sight taken as distinct from any accompanying beliefs and so on. Given the organization of visual mechanisms, incidentally, it seems not absurd to wonder if there is such a single cell, rather than something more complex. (Hubel and Wiesel, 1979, p. 112)

5.2.2 Learning and Memory

In neuroscience, learning is ordinarily thought of as the acquisition of a certain way of behaving, or the acquisition of a capability. How do we come to be able to respond as we do to a vast number of sounds? How do we come to be able to do what involves recognizing colours or faces? How do we come to be able to speak a language? Evidently, learning is bound up with the great fact of memory, and the two are often treated together. Neuroscience, despite what has just been said of its concern with behaviour in connection with learning and memory—and in connection with pretty well all else—is not in general behaviourist, in the radical sense of denying consciousness. (p. 73) Its concentration to a great extent on observable behaviour can be taken as, and *is* generally or at least often taken as, concentration on *evidence of mentality*. That is not to say that the goal of the enterprise is only knowledge of mentality, but rather that mentality is a part of the subject-matter. (Carlson, 1977, p. 439)

One common technique has been that of training animals in particular ways, thereby establishing neural structures for altered behaviour, mainly in the cortex and parts of the limbic system, and then attempting to isolate the structures by discovering whether surgical alteration or destruction of particular areas of the brain results in a loss of the acquired capability. Monkeys are trained to perform fairly precise motor tasks, parts of the cortex then ablated, and the behavioural trials run again. There has also been a good deal of investigation of global or gross electrical activity which is involved in the acquisition of the states upon which new responses depend. Electroencephalographic (EEG) activity is a summation of very many electrical potentials in a given area of the CNS, the area with the range of a recording electrode. Animals are trained to make a certain response, such as moving a lever to get food when a light flashes. As the flash grows 'more meaningful' (Cotman and McGaugh, 1980, p. 300) to them because of the connection with food, the EEG waves alter.

In contrast to global investigations of this kind, and the surgical isolation of structures mentioned before, there is the isolation of single neurons, and study of their electrical activity before and after training. There has also been a close examination of the chemical constituents of the brains of trained as against untrained animals, and attempts to establish 'the neurochemical correlates of training' by attempting to transfer memories from one individual animal to another. Extracts taken from the brain of a trained rat or goldfish are injected into the brain of an untrained individual. (Deutsch, 1973) Reseach on human memory has had to do with chronic alcoholics, who may retain old memories but be unable to form new ones, and individuals who have suffered brain injuries or been subjected to brain surgery thought to be advisable in connection with such disorders as epilepsy and psychosis. The loss of very specific parts of the brain results in very specific forms of amnesia.

There is also some rather dramatic if now well-known research bearing on memory and relevant to psychoneural intimacy. Electrical stimulation of the cortex is accompanied by curious 'relivings' of past experience, 'flashbacks', reported by the subject. 'I think I heard a mother calling her little boy somewhere. It seemed to be something that happened years ago... in the neighbourhood where I live.' (Penfield, 1975, p. 21, cf. 1967) Above all, with respect to the question of psychoneural intimacy, there is the claimed fact that when the same electrostimulation is repeated, the recall-experience occurs again, in the same way. Same brain-events, same recall-experience.

5.2.3 Pain and Pleasure

Pain, putting aside receptors and pathways to the brain, has to do with a number of areas. The neural facts of pain involve a special complication in that the sense of pain is affected by emotion and attention. Athletes, soldiers, and ordinary people when excited or enraged may fail to notice wounds or have lesser kinds of pain-experience than ordinarily. However, such extraordinary phenomena are themselves the subject of influential theories as to neural organization and function. (Melzack and Wall, 1982; Wall, 1985) The clinical control of pain has of course been the object of a great deal of research, which also throws light on the general question of psychoneural intimacy in this connection. In biochemistry the discovery of certain peptides—endogenous opiates of the brain—and the known fact of brain receptors for opium, the latter and its chemical derivatives having been used for centuries not only to induce euphoria but to control pain, evidently is relevant to the matter of psychoneural intimacy with respect to both pain and its absence.

It was mentioned above, in connection with thirst and hunger, that electrical stimulation given by way of an electrode implanted in a rat's hypothalamus results in its eating, which under certain conditions stops when the stimulation is stopped. As has long been known (Olds and Milner, 1954), electrostimulation elsewhere in the brain results in what it is not rash to call pleasure or satisfaction. In some experimentation, the stimulation is in the control of the rats. In one fairly early experiment, now succeeded by many others, rats learned to press a lever in order to get or prolong rewarding electrostimulation, and also learned to press a different lever in order to get food. Subsequently the animals were allowed access to a lever—in fact to both levers simultaneously—for only one hour each day. (Routtenberg and Lindy, 1965) Some spent so much time getting rewarding electrical brain stimulation that they starved to death. Not, I think, an experiment that will endear animal experimentation to readers not hardened to its methods, and not an experiment by any means unique in that respect. I uneasily put aside what do certainly exist, serious moral problems about some uses of animals in research, some of them problems that are not easily resolved. They are increasingly recognized by researchers.

5.2.4 Emotions

To turn now to the emotions and psychoneural intimacy, there is less that can be said quickly, and indeed less to be said at any length. Their complex connection with pain and pleasure themselves, as well as perception, belief, and so on, inhibits research, as does their relative lack of distinguishing behavioural manifestations. A considerable part of the frontal lobes of the cerebral cortex, however, appears to be involved in regulating emotion. One of the well-known personal disasters which prompted neurophysiological reflection was that of Phineas Gage, a railway construction foreman of a century ago who survived when a tamping iron was blown by dynamite through the relevant part of his frontal lobes. The great change in him was an emotional and social one. In this century the surgical operation of frontal lobotomy has often enough been performed on patients to relieve psychosis or unbearable pain. One of a number of results is diminished care and anxiety. Another is wholly inappropriate merriment or sadness.

Other emotional phenomena are associated with damage not to the frontal but to the temporal lobes of the cortex, sometimes as a result of epilepsy. They include great placidity, fearful hallucinations, and so on. Stimulation of the temporal lobes during brain surgery gives rise to strong illusions of an emotional kind. Any idea that emotion or particular emotions are based only here, or only in a few circumscribed locales, is of course mistaken.

There has been much study of the physiological mechanisms taken as controlling aggressive and frightened behaviour in a number of animals, with distinctions drawn between types of aggression and fright. It has been shown to be possible to stimulate parts of the brains of cats and rats, notably parts of the hypothalamus, and to elicit irritable aggression and predatory attack. Fear or fear-induced aggression has also been connected, by the same means, with other parts of the hypothalamus in various animals. The septum and the amygdala, parts of the limbic system, are involved in the suppression and the facilitation of emotional behaviour. Rats whose septums have been removed are notoriously aggressive, typically described as vicious.

5.2.5 Consciousness

In effect, a good deal has been said already of this subject. This is so since consciousness consists in perceptions and sensations, memory experiences, pains and pleasures, emotions, and the like. Needless to say, however, that sort of enumerative characterization of consciousness, together with the fundamental characterization of its nature in terms of an interdependent subject and content, do not exhaust the ways of characterizing it. For example, it can be said to contain, as a principal part, a representation of the owner's world, a representation of more than the immediate environment of which he or she is aware. Such a cognitive map, as it is sometimes called, is more importantly a matter of neural dispositions than of consciousness itself, but it does also enter into consciousness. The further characterization of this principal part of consciousness is a matter of the greatest difficulty of several kinds, a matter of competing theories, conceptions, and so on.

To come to the principal point, however, it is possible to engage in empirically based speculation as to the relation of such a conscious representation or map to neural structures and systems. The same is to be said of what can be called, if uncomfortably, other principal parts of consciousness. One is the division of consciousness into foreground and background: those things in sharp focus—whether in perception or thought—and those things which make up a fringe. This division, and the cognitive map itself, are in one experimentally based speculation related to the hippocampus, part of the limbic system. The hippocampus is established as having features that do or would suit it in these connections. (O'Keefe, 1985) If neither this research and speculation nor any other can be thought with confidence to have established 'the site of consciousness', it is as true that neuroscience is very far indeed from being in a state of ignorance as to the subject of consciousness and psychoneural intimacy. (Oakley, 1985)

To consider more fully something that allows for relatively quick exposition, there is the unity of consciousness. (p. 117) The fact enters into 'split-brain' research, as dramatic and now as well known as the research mentioned earlier in connection with memory, having to do with 'flashbacks' or 'relivings'.

Of the two hemispheres of the cerebrum, one is dominant in most situations in each human individual—in most of us, it is the left hemisphere. The dominant hemisphere is so-called, essentially, because it has to do with the comprehension of speech and language. It also has to do with arithmetical operations and with what can be called calculation generally. The other or minor hemisphere has what are sometimes taken as lesser abilities, some of them having to do with shapes and space, and with 'non-verbal ideation', and music. Its abilities having to do with shape and space allow it to do better than the left hemisphere in certain geometrical tasks. The right hemisphere, too, is connected with somewhat more intense emotional reactions.

The research has to do with hemispheric dominance and with another large fact noted earlier, that the left hemisphere deals with receptors for the right side of the body, and largely controls the right

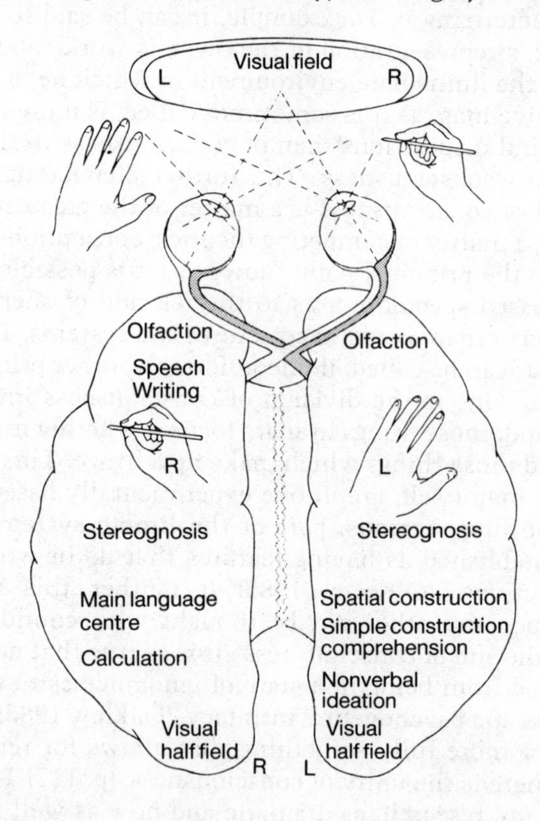

FIG. 26. Diagram showing functions of left of right hemispheres in a split-brain subject. From Stein, 1982, fig. 24.2, p. 339

hand and leg. The left hemisphere is concerned with the right side. (See Figure 26.) Finally ,there is the fact of the corpus callosum, noted earlier, the large bundle of nerve fibres connecting the left and right hemispheres. It is the great communication system between the left and right halves of the brain. In the treatment of severe epilepsy in a small number of patients, the corpus callosum has been surgically severed. There has been extensive study of persons on whom the operation, commissurotomy, has been performed. (Sperry, 1968; Gazzaniga, 1970)

In one experiment, the subject is blindfolded and a familiar object such as a coin is put into one hand for a time—either hand. He can subsequently use that hand to retrieve the object by touch alone when it is put into a collection of somewhat similar objects. He is entirely unable to do so with the other hand. It is claimed that in some sense the mind of the subject has been divided by the operation, that there is 'co-consciousness'. In another experiment, a picture of an object is flashed on a screen so that it falls within the right visual field, connected with the left or speech hemisphere, and the subject is able to name it. When the picture appears in the left visual field, connected with the right hemisphere, he is unable to name it. In a third experiment a word, say 'comb', is projected on a screen so that it falls into the right half of the subject's visual field, connected with the left hemisphere. He is able to say what it is. Also, he can with his right hand select the comb by touch from a collection of objects which he cannot see. If a word is projected on the screen so that it falls into the subject's left visual space, connected with the right hemisphere, he is unable to give the word. He *can* select the correct object, say the comb, by left-hand touch, from a collection of objects. However, it is maintained, he has no awareness that he has made the right choice, is entirely surprised at his success, and truthfully denies that he knew what he was doing. That he does select the right object, unknowingly, is explained by residual comprehension of language in the right hemisphere, but without the power to express conscious appreciation of the spoken word, which is the domain of the left hemisphere. The report of such experiments took them to indicate that the surgery left individuals with two separate minds, two separate spheres of consciousness, with what was experienced in the right hemisphere outside the realm of awareness of the left.

Problems of various kinds, some of them philosophical, are raised by such claims, and have been given acute consideration. (Parfit, 1984) For our purposes, the only conclusion that needs to be drawn is that the experiments can be taken to demonstrate a fact of psychoneural intimacy in persons whose brains have not been split. The unity of consciousness is in such connection with some neural structure or fact. A further and different fact of psychoneural intimacy, having to do with disunity of consciousness, is produced by commissurotomy.

5.2.6 Higher Functions

So much for most of a kind of survey of neuroscience—or rather a collection of pieces of it—in so far as it can be taken to bear on psychoneural intimacy. It is traditional to end surveys of neural matters, undertaken with whatever purpose, with a proposition to the effect that little or nothing has been established about the 'higher functions': problem-solving, reasoning, abstract thought, literary creation, imagination, and so on. All that is possible, it is commonly said, is conjecture, poorly based hypothesis, speculative analogies and models, and so on, however fertile these may turn out to be. Very little is *known*. Such a declaration may be taken to overshoot the mark, or some marks. Certainly to say that nothing or very little is known is to engage only in a kind of rhetoric, as is easily demonstrated. (P. S. Churchland, 1986, Chs. 4, 5)

The higher functions are bound up to a great extent with speech and language, and hence learning and memory. As we have seen, and despite the necessary qualifications, it is true to say that we have significant knowledge of learning and memory. The higher functions also involve goal-directedness, perception, alertness, in some cases feeling, and so on. These are things about whose neural relations a good deal is known, as already conveyed. The main subject-matter in connection with the higher functions, however, must be speech and language, numerical language included.

Broca's area of the cerebral cortex, and Wernicke's area, were mentioned earlier. Damage to them gives rise to dysphasias or speech disorders. Broca's area is close to that part of the motor cortex—of which more will be said below—which controls muscles of the face, tongue, jaw, and throat. However, damage to Broca's area is not to be regarded as merely producing a muscular failure—that is, an inability to control muscles. The muscles involved in speech operate normally in other tasks. Also, the speech of someone afflicted with Broca's aphasia—aphasias being a subclass of dysphasias—has characteristics, including faulty grammar, which could not be explained by muscle failure. Broca's aphasia is in a sense an intellectual failing. Damage to Wernicke's area, to turn to it, gives rise to a dysphasia significantly different from Broca's aphasia. Speech is phonetically and grammatically normal, but there are semantic failings. Wholly inappropriate terms are used, and sentences sometimes include nonsensical words.

Wernicke's area also has a large role in the comprehension of written and spoken language. Damage to Broca's area has a much smaller effect on comprehension. The two areas were discovered in the 1860's, and the knowledge now had of their relations to language and indeed thought is very much greater than can be indicated here. (Mateer, 1983)

To mention now what might have been mentioned in connection with several previous subject-matters, research bearing on the higher functions has often confirmed what is sometimes called *neural redundancy*. The fact lies behind the formulation of the proposition of psychoneural intimacy: each specific type of mental event somehow necessarily occurs with a type or *types* of neural event in a given locale or locales of the brain. A form of neural redundancy can be illustrated by way of the language areas. An area is not organized by a single functional chain, such that when a link is broken, there is general failure. Rather, an area is organized by a number of parallel chains. When a link of a chain is broken, the chain fails, but not the whole system. The system can substantially survive the failure of a part. There is also redundancy in the plain sense that a function previously that of one group of cells can be taken over, at least in part, by others.

Finally, and very briefly, various disorders provide evidence relevant to the higher functions and psychoneural intimacy. Schizophrenia in its various types, and also the affective disorders, including types of depression and mania, evidently involve specific failures of higher functions. Schizophrenia, in recent decades, has come to be effectively treated by the neuroleptic drugs. As a result, to come to a principal point, there is good reason to associate schizophrenia with specific chemical malfunctions in the brain. In contrast, the maintenance of normal mood and rationality is evidently associated with more ordinary facts of brain chemistry. These facts are not a matter of uncertain speculation but of close study and experimentation. It would show an entire ignorance of the sciences in question to suppose that these bases of mood and rationality are the subject of the kind of speculative reflection that occurs within, say, the Freudian theory of the mind. The two are worlds apart. (Henn and Nasrallah, 1982)

Let us now reflect on our look at parts of neuroscience in connection with the Correlation Hypothesis, more particularly in connection with the first premiss of the argument for the hypothesis. The argument again is (i) that mental and neural events are intimately connected, (ii) that we are thus restricted in our characterization of the psychoneural relation to either Identity Theories and like theories or the Correlation Hypothesis, perhaps with the latter so supplemented as to produce the Union Theory, (iii) that Identity Theories and the like ones are out, partly because they do not secure mental indispensability, and therefore (iv) that the Correlation Hypothesis is the basic fact of the psychoneural relation.

The first premiss, as we have been understanding it, of course does not have to do with mental events *identified as* or *defined as* bodily movements or behaviours, but with mental events realistically conceived. As perhaps needs repeating, because of procedures and usages in it, neither does neuroscience identify mental events with bodily movements or so define them. It thus does not in the original behaviourist way (pp. 73, 125) deny the existence of consciousness as ordinarily conceived. Certainly there is an encompassing assumption that the only acceptable evidence of the mental is behavioural, but that, clearly, is not the traditional behaviourism which conflicts with mental realism.

Also, by way of qualification of that point, it needs to be said that neuroscientific philosophizing is in general at least loose and uncertain, on the subject of behaviourism among others, and that things are said that may mislead. (It should come as no surprise that neurophysiologists do neurophysiology a lot better than they do philosophy.) In the following passage from an admirable book 'stimulation escape' refers to the fact that under certain conditions experimental animals stop pressing a bar and thereby giving themselves rewarding or pleasurable self-stimulation—when they have had too much of it. What was desired becomes 'aversive' or 'punishing'.

> . . . how do we know that the stimulation is really rewarding in the sense of being pleasurable, and do we know that stimulation escape really means the neural activity is aversive or punishing? Anthropomorphism (attributing our own feelings to animals) is a very risky and often misleading practice. Strictly speaking, reward and aversion should denote nothing other than their operational definitions in terms of bar presses to turn current on and off. . . . On the other hand, our own subjective feelings of reward and aversion must have something to do with brain self-stimulation and stimulation escape. People with chronic electrodes have reported pleasure and displeasure during stimulation of various sites. (Cotman and McGaugh, 1980, p. 599)

If the passage evidences some inclination to definitional behaviourism, in this case an inclination curiously limited to other animals, its burden is none the less otherwise. The favouring of 'operational' definitions of pleasure and pain does not conceal an acceptance of pleasure or pain as themselves being realities in connection with neural processes. What is said of human consciousness can hardly be said to be sceptical or unrealistic. So with very many other, indeed almost all, pieces of neuroscientific reflection on behaviourism and the like. (Carpenter, 1984, pp. 316–19; Kuffler, Nicholls, Martin, 1984, p. 4)

The first premiss of the argument for the Correlation Hypothesis assumes the falsehood not only of definitional behaviourism but also of Eliminative Materialism (p. 103) and any closely related Identity Theory: any which involves a denial of the existence of mentality realistically conceived. Can it be said that neuroscience espouses such a thing as Eliminative Materialism? The question, like the one above about behaviourism, is of importance but not fundamental. We do already have quite sufficient argument (pp. 73, 103 ff., 210) against both behaviourism and Eliminative Materialism, and are in no need of help from neuroscience in its philosophizing moments. Still, if it were the case that neuroscience—as distinct from several philosophers who attend to it—espoused either of the two doctrines, that would be uncomfortable. In fact, although it is again true that some slight qualification is needed, neuroscience does not espouse Eliminative Materialism or the like—which fact is often enough affirmed. (Kuffler, Nicholls, Martin, 1984, p. 4) The qualification is that neuroscience does indeed contain a certain amount of understandable bafflement (2.1, 2.2) about consciousness, which bafflement may issue in declarations a bit more bluff than serious. 'What is the poor scientist to do when confronted with the rumour of a phenomenon, which is all that consciousness is?' (Blakemore, 1977; cf. 1985) The author, one feels, if he is tempted to follow Pavlov by fining his students for using the word 'conscious', does not do so because there is no real and distinctive thing to which it applies.

Having got rid of the notions that neuroscience takes it that my wanting to get a book on Donne for my son *is* body movements in Dillon's Bookshop, or is something which has only neural properties, we are free to ask if standard neuroscience confirms that mental events realistically conceived are in intimate relation with neural events. There is no room for doubt. It does. It does so by findings of non-accidental connection in all the areas at which we have glanced: sensation and perception, learning and memory, pain and pleasure, the emotions, sleep, consciousness, and the higher functions. That all of these areas, and particularly the last two, are a matter not only of what is is proper to call findings, but also theory and speculation, often conflicting, does not affect that conclusion at all significantly. The breadth and weight of the evidence makes the conclusion irresistible.

Within the practice of neuroscience, there exists no dissent from psychoneural intimacy. That is, there is no dissent in ongoing standard neuroscience itself. There is some dissent by an occasional neuroscientist on holiday—which is to say actually doing philosophy. (5.5) Some indication of the acceptance by almost all neuroscientists of the proposition of psychoneural intimacy is to be had from a familiar galling response made to philosophers who propose it to them. That response, explained by the fact that they do not suspect psychoneural intimacy to be something about which a question is likely to be raised, is mistakenly to suppose that something more speculative is being proposed, something arcane or 'philosophical'.

Some may suppose that it is too much to say simply that neuroscience *confirms* psychoneural intimacy. That matter is to be decided by reflecting somewhat more closely on what it does provide. A look back over the data reported will show or suggest that it provides arguments of at least two types.

(i) In very many cases, a set *n* of neural facts is often, or very often, or in all experiment and observation, accompanied by a certain mental phenomenon *m*. It is not settled that *n* is the *whole* of a neural array in intimate connection with *m*. That is, it is possible, likely, or certain that sometimes precisely *n* occurs without *m*. There is no necessary connection, as required by the proposition of psychoneural intimacy, between precisely *n* and, on the other hand, *m*. The known connection between *n* and *m*, however, is the main premiss for an inductive or extrapolatory argument for intimate connection between *an array including* n, and *m*. There is an inference, so to speak, from a part to a whole of a neural intimate. There is a further inference from this conclusion about *m* and neural intimacy, and like conclusions about mental phenomena other than *m*, to mental phenomena generally. The first inference is of a dependable and familiar type. If a first thing is regularly associated with a second, and certain obvious alternative possibilities of explanation are excluded, there is reason to take the first as entering into something somehow sufficient for the second.

(ii) In very many cases, a set *n* of neural facts is taken as guaranteeing a mental phenomenon *m*. It is not settled, however, that *n* is no more than the neural intimate of *m*. That is, it is possible, likely, or certain that *m* occurs in the absence of some part or parts of *n*. There is no necessary connection, as required by psychoneural intimacy, between *m* and precisely *n*. The known connection between *n* and *m* is an argument for intimate connection between *a part of* n and *m*. There is an inference, so to speak, from more than a neural intimate to precisely a neural intimate. Such a conclusion about *m* and other mental phenomena issues in a conclusion about mental phenomena generally. The first inference is again of a dependable and familiar type.

There is no need to deny the fact that neuroscience does typically concern itself with less than, or more than, precisely the neural intimates of mental phenomena. The fact, certainly, is importantly owed to difficulties and ignorances of several kinds in fixing precise neural intimates. What is generally overlooked, and may lead to an underestimating of the argumentative value of the fact, is that it is also owed, far more importantly, to something else which is certainly not to be described as insuperable difficulty. It is something shared with most of science and also with common-sense inquiry. The primary concern here, which has to do with economy of effort and what is essential for progress, is indeed with somehow *necessary* conditions of phenomena, and effective although more than minimally sufficient conditions of phenomena.

Neuroscience, to speak differently, is not greatly concerned with the establishing of laws in the most ordinary sense. It is more taken up with doing so than, say, motor mechanics is taken up with establishing generalizations about motors, but it shares with motor mechanics an inclination to what we can label *presupposing questions* as distinct from *ultimate questions*. That is, it asks questions which derive from certain presuppositions, rather than questions in which nothing is taken for granted. That it does so in fact reflects the firm conviction of psychoneural intimacy. No one asks if *y* is a somehow necessary condition of *z*, or if *x* contains a somehow sufficient condition of *z*, if he or she believes that no minimally sufficient condition exists.

The success of neuroscience in giving answers to questions presupposing psychoneural intimacy does in fact provide greatly more evidence as to the answer to the ultimate question about psychoneural intimacy than does occasional direct concern with that question. This is so since the evidence, the basis of arguments of the two sorts, is of great bulk. It is not that there is but little evidence of somehow necessary conditions or of somehow more-than-sufficient conditions for mental phenomena. The evidence is of the first importance, secondly, because of its diversity. It involves, too, the mutual support of bodies of findings within the evidence. Further, it possesses other characteristics had by effective bodies of evidence in any part of scientific or good common-sense inquiry. (Hempel, 1966, Ch. 4) There is in fact quite sufficient reason to speak of neuroscience *confirming* psychoneural intimacy.

It is of course the burden of a great deal of contemporary and recent philosophy of science that theories and the like are never *established as true, proved, conclusively confirmed* or the like, but *tentatively accepted, taken as true for the purposes of further inquiry*, or the like. (Newton-Smith, 1981) We shall not enter into this large dispute, but rather take the common and persistent view that a great part of science can indeed be regarded as confirmed, and that the proposition of psychoneural intimacy is indeed such a part. It is confirmed by what can surely be called overwhelming evidence, entering into arguments of the two kinds and others.

The main role of neuroscience with respect to the argument for the Correlation Hypothesis is the confirmation of its first premiss. What remains to be said has to do with the third step of the argument, that Identity Theories generally, as distinct from just Eliminative Materialism, are unacceptable. As with the matters of definitional behaviourism and Eliminative Materialism, it would not be fatal, but uncomfortable, if neuroscience could be shown to oppose the rejection of Identity Theories. Neuroscientific attitudes to mental indispensability affect the question, but they are better treated in connection with the evidence for our second hypothesis. What is to be said here— allowed here—is that neuroscientific philosophizing does indeed include some declarations in support of Identity Theories.

In claiming that mind is the description of brain activity at a particular hierarchical level, I am adopting a particular version of what is known as the *identity hypothesis*. In some versions of the identity hypothesis, it is used as a statement of pure reductionism: the mind is reduced to being merely a less precise way of talking about the brain. This is not at all the version used here; in saying that the mind is the *same as* the brain at a different level of discourse, I mean that for any event it would be possible to provide two equally valid sets of descriptions, one in 'mind' language and the other in 'brain' language. (Rose, 1976, p. 31)

It is not wholly evident what a 'valid' description is, but presumably it is a true description. It is a description such that there do exist distinctive properties which it describes. This Identity Theory, then, as indeed is emphasized, is not Eliminative Materialism. It is in fact inexplicit, but is most akin to the Two-Level Identity Theory (pp. 99 ff.) or the Dualistic Identity Theory with Neural Causation (pp. 93 ff.). We shall not re-examine the worth of those theories. To glance at another piece of neuroscientific philosophizing, less restrained, its deceptive conclusion is as follows.

What, then, is the metaphysical position assumed by contemporary psychobiology? . . . In brief, I believe that psychobiologists implicitly assume that we live in a world that is: mechanistic and physicalistic, realistic, monistic, reductionist, empiricistic, methodologically behaviouristic, and is *not* idealistic, dualistic or pluralistic; and thus is not parallelistic or interactionist, rationalistic, emergentistic, vitalistic, operationalistic, or logically positivistic, or metaphysically behaviouristic. (Uttal, 1978, pp. 81–2)

The mind–brain 'monism', however, is so understood as to include a number of things, of which one is 'neuro-reductionism'. *That*, at bottom, is a matter of inquiry into the 'neuron correlates of mental processes'. (Uttal, 1978, pp. 45, 57–72) Neuro-reductionism, further, 'seeks to explicate those correlations between neural structures and mental or behavioural function in a way that allows the investigator to distinguish between the instances that are merely correlational and concomitant and the instances that represent a true expression of a structure–function identity.' (Uttal, 1978, p. 72)

It is very unclear, then, what this Identity Theory comes to. What mainly needs to be said is that it, and the previous 'identity hypothesis', are instances of what is rare in neuroscience. That is, it is rare to find even such indistinct support of Identity Theories. It is perhaps no more often found than is similar support for a view approximate to our own Correlation Hypothesis. By way of one example of the latter support, it is claimed by one neurophysiologist that a number of principles or postulates underlie neuroscience. The relevant three of these are as follows.

The physical correlates of our mental events are neurophysiological phenomena that develop in our brains.

Each specific mental event has as a correlate a specific spatiotemporal pattern of neuronal activity.

The laws of physics are applicable to all the material universe, including our bodies and nervous systems; furthermore, neither the neuronal nor the mental processes are subject to any determinism, causal or probabilistic, different from that of physics. (Rosenblueth, 1970, pp. 107, 108, 110)

These principles, in sum, and despite what it said of probabilism, which is very unusual, come close to the Correlation Hypothesis. What is also worth noting, and what is indicated by much of what has been reported of standard non-philosophical neuroscience, is that references in it to neural 'correlates' and the like are common.

5.3 EVIDENCE RELATING TO THE SECOND HYPOTHESIS

The second hypothesis of the present theory of determinism incorporates the proposition about psychoneural pairs, that a mental event and its neural correlate form a single effect and in a way a single cause. The hypothesis is that a psychoneural pair is in fact the effect of the initial elements of a causal sequence, which elements are neural and other bodily elements at about the time of birth of the person, and environmental elements then and thereafter. The hypothesis was arrived at by eliminating various other answers to the diachronic question of the explanation of mental events, some of them considered when that question was our official business (Ch. 3), some earlier in connection with the synchronic question of the nature of the psychoneural relation. (Ch. 2) We can put the main line of our reflections, looked at from the point of view of the diachronic question, into the following summary.

(i) An acceptable answer to the question of the explanation of mental events must satisfy the constraint of mental realism. It must not look elsewhere than to its subject. (2.1, 2.2)

(ii) An acceptable answer must satisfy the constraint of psychoneural intimacy. (p. 90)

(iii) It must satisfy the constraint of mental indispensability. (p. 91)

(iv) Identity Theories give answers to the synchronic question—we considered them in this way— but also entail answers to the diachronic question. Bound up with their failure to answer the synchronic question is a failure to give a defensible answer to the diachronic question. The most arguable of them fail because they do not secure mental indispensability. (2.3)

(v) Answers to the question must also respect a constraint of simplicity or parsimony. (p. 164)

(vi) They must use an adequate conception of causation. (p. 165)

(vii) They must also satisfy a further constraint, that of the neural or other bodily causation of neural events. It is that neural events are effects of standard causal sequences which, putting aside any environmental or pre-bodily part, are at every moment neural or otherwise bodily, whatever else may be true of them. (pp. 164, 170)

(viii) Further answers to the question of the explanation of mental events also fail—one of the most important being Neural Causation with Psychoneural Correlation, another the Psychoneural Relation as Macromicro. If they satisfy the constraint with respect to neural or other bodily causation, they do not preserve mental indispensability. (3.2)

(ix) Indeterminist answers to the explanatory question, while statable, are conceptually unsatisfactory, partly because of offending against the constraint of simplicity or parsimony. (3.4–3.9)

(x) The only defensible answer to the explanatory question is our second hypothesis, incorporating the Proposition of Psychoneural Pairs. The hypothesis and the incorporated proposition are owed most fundamentally to the proposition of the neural or other bodily causation of the neural, and the proposition of mental indispensability. (pp. 165 ff.)

Our present business is to look at neuroscience for evidence with respect to (vii) the important proposition of the neural or other bodily causation of the neural, for two reasons. First, as just set out, precisely it is fundamental to the argument for the second hypothesis. If true, secondly, it refutes Indeterminist pictures of the mind—it refutes them simply in virtue of the general character it shares with certain other causal propositions. That is, it does so by way of asserting that neural events are owed to *some* standard causal sequence. We shall also attend to a certain tendency in neuroscience to question (iii) the constraint or axiom of mental indispensability. Giving it up would affect a great deal of what has been said, as the above summary of our reflections indicates.

The proposition of the neural or other bodily causation of the neural pertains in part to what can be called neural events close to the environment. That they are caused by other bodily events is well established in neuroscience. There is only relatively little mystery about the *immediate* effect of the environment on an individual. That is, there is relatively little mystery about transduction, the processes whereby environmental energy of several forms is changed or transduced by bodily mechanisms into electrochemical properties which then give rise to action potentials in sensory neurons. Receptor cells, in the case of the visual environment, are the rods and cones of the eye, the former being sensitive to dim light, the latter to bright light and signalling colours. The cells contain light-sensitive pigments (rhodopsin in the rods) which absorb photons of light, and, by way of a certain chemical process, then alter the membrane of a cell.

If there is only relatively little mystery about receptor mechanisms and neural events close to the environment, that is not at all to say there is only relatively little mystery about the background of later neural events, say neural events in the cerebral cortex associated with such experiences as seeing a flag or hearing a trumpet, where these experiences are taken as involving the belief that one is seeing a flag or hearing a trumpet. These neural events obviously have greatly more in their background than the pulses first produced by the relevant photons or pressure variations. There is of course the linkage with such neural events close to the environment. Far more complex and obscure, there is the linkage with earlier neural events having to do with acquisition of concepts and so on—earlier neural events having to do with learning and memory.

Still, learning and memory are far from being *all* mystery. We looked at research into them in connection with the Correlation Hypothesis. The research is of yet more relevance to the second hypothesis and in particular to the proposition of the neural or other bodily causation of neural events. As implied in the earlier report, there have been extensive findings relevant to the somehow neural causation of the particular neural events associated with recognition, remembering, and so on. (pp. 274 ff.) Because of the defeating complexity of the human CNS, and for what are underdescribed as 'ethical' reasons, most of this research has been carried forward with simpler species than the human. In part the aim has indeed been to throw light on our own learning and memory. The dangers of extrapolating from a mollusc, frog, octopus, rat, or monkey to our own species is of course recognized.

One large fact of relevance is that the nervous system, in almost all species, is plastic, which is to say here—the term is variously used—that it is open to more or less permanent changes as a result of ongoing environmental stimuli. Axons sprout new branches, dendrites extend, levels of release of transmitter-substances alter. To speak in one general way, conditioning occurs. That is, there come into being new responses to stimuli. To speak differently, there is as much change and development in the brain as in the mind. That, of course, is exactly what must be true on the present and other tolerably empirical views of the mind.

Research into memory, as remarked in the earlier report, has not specified a single anatomical *locus*, but much is known of how memories are incorporated in several neural systems. Parts of the brain whose surgical removal results in various forms of amnesia include parts of the cerebral cortex and basal ganglia. Thus it is far from being true that there are no answers to the question of what neural states lie in the background of the neural correlates of remembering and the like. There are also relevant settled facts as to kinds of electrical activity, and synaptic changes. There is reason to think, in connection with neurochemical changes, that attempts to affect memory-transfers between members of simple species, say goldfish, by transplanting substances, will eventually be successful.

There is much more scientific evidence of other kinds that is pertinent to the neural or other bodily causation of neural correlates of remembering, recognizing, and the like—both when these enter into sensation and perception and when they do not. There is considerable knowledge of the relevant general development of the CNS in individuals of many species, which development is bound up with the given causal sequences. Genetic influences have been closely studied, as have details of the migration of neurons. With respect to genetics, strains of 'maze-bright' and 'maze-dull' rats have been bred. With respect to neuron-migration, it is now established, as already noted, that synaptic connections are modified by experience, that the brain cannot be regarded as 'hard-wired'. One simple relevant fact is that parts of the brain of trained rats are heavier than those of comparable untrained rats.

For all of that, and for all of what else might be said, it would be a remarkable understatement to say that there are gaps in the story told by neuroscience about neural and otherwise bodily causal sequences for those particular neural events associated with remembering or recognizing and the like, and with what they often enter into, perception and sensation. The situation, rather, is that there is greatly more gap than story, more mystery than clarity. Where does this leave us? Let us put off the question for a moment and glance again at the other parts of neuroscience surveyed earlier.

The survey in all of its parts inevitably implied at least something like the causal proposition now under consideration, for the reason that neural events in correlation with consciousness are partly or indeed typically characterized as being certain effects. It will be as well to make explicit what was implied. With respect to pain and pleasure, there is the fact that the causal history of certain peptides gives detail to the proposition of the neural and bodily causation of the neural events associated with lesser pain or the absence of pain. (p. 275) So with the emotions. Various known facts give content to the causal proposition in this connection. As for consciousness, all that was said suggests what is true, that there is very considerable knowledge of neural and bodily causal antecedents of associated neural events. One can mention the development and functioning of the hippocampus (p. 277) and of the corpus callosum (pp. 279 f.) Finally, with respect to higher functions, the development and functioning of the speech areas of the cortex are of obvious relevance (p. 280), as are facts of chemical function and such disorders as schizophrenia. (p. 281)

In all of this there is much knowledge and much ignorance. To repeat, where does this leave us? We began this chapter's summary of neuroscience with the neuron, and concluded that its causal functioning provides strong and clear support for the proposition that neural events are somehow standardly caused by prior neural and other bodily events. No more was said at that stage partly for the reason that (a) causation within the neuron or between any pair of neurons is held by some somehow to be consistent with (b) an absence of causation between neural systems or levels. The supposition is in fact remarkable. In fact, it is impossible or at any rate very hard indeed to see how anything like it can escape self-contradiction. As they stand, (a) and (b) *are* inconsistent. If one thinks of a temporal sequence of neural events finishing with the correlate of a thought or whatever, a sequence however complex and involving however many systems or levels, it is logically impossible to hold both that each event *is* somehow causally related to its immediate temporal predecessor or predecessors, and that causal connection does not hold between parts of the sequence conceived in terms of systems or levels. There is the simple fact of the transitivity of causation. (1.2, 1.4) *Whatever* further characterization is given of parts of the sequence one to another, it cannot be that they fail to be in causal connection, to be parts of a causal sequence.

One version of the remarkable proposition that causation does not hold between systems or hierarchical levels is, in part, that neural events at one level give rise to neural events at a higher level—there is 'emergence'—which higher events in turn affect events at the lower level, each of these developments being in some way indeterministic. No such doctrine, to my knowledge, is to be found in ongoing neuroscience itself—as distinct from rare pieces of philosophical reflection by neurophysiologists and others. (Sperry, 1952) As will certainly be expected, it is a doctrine readily disavowed by a vast majority of neuroscientists. (O'Keefe, 1985, pp. 60–1)

The answer to the question of where we are left by the existing but greatly incomplete evidence as to perception and sensation, memory and learning, and so on, is that it enables us to regard the proposition that neural events are somehow standardly caused by prior neural and other bodily events as yet more strongly supported than can be said on the basis of the functioning of the neuron. Suppose, despite what has been maintained, that there is some logical or conceptual possibility of accepting causality within and between neurons and yet denying it between parts of larger neural sequences, perhaps conceived in terms of hierarchies or whatever. The evidence which is available, with respect to perception and sensation and the rest, must undermine any such speculation. It *is* evidence as to precisely the larger neural sequences. It would not undermine such a speculation, perhaps, in the absence of what can be taken as settled, which is causation with respect to the particular links of the chains. It *does* undermine any speculation as to 'emergence' or whatever when conjoined with what is known of the neuron. If this is a matter of judgement, as certainly it is, it is a judgement in accord with clear and accepted criteria of evidence. (p. 285)

It needs remarking, no doubt, in line with what was said earlier in connection with the rarity of the the verb 'to cause' in neuroscience (p. 266), that there is no doubt that neuroscience everywhere uses the intuitive and standard conception of a causal circumstance and effect, such that the former necessitates the latter. It also everywhere uses the intuitive and standard conception of a causal sequence: a sequence of events each of which save those at the beginning and the final one is both an effect and an element in a causal circumstance. These facts are sometimes questioned by running them together with the falsehood that neuroscience conceives of causality somehow in terms of some atomistic materialism now abandoned. (Popper and Eccles, 1977) Nothing, it seems to me, needs to be said of that. The facts are wholly consistent with what is also true, that there is no explicit analysis of causal or other nomic connection in neuroscience. All that needs to be said of that is that neuroscience is not philosophy. If evidence is asked for the proposition that neuroscience uses standard ideas of causation, enough is provided by that most common of generalizations in it, that the brain is a machine, however uniquely complex a machine.

The conclusion to which we have come, precisely stated, is that standard neuroscience provides *very strong and clear evidence* that neural events are effects of causal sequences which, putting aside any environmental or pre-bodily part, are *at every moment neural or otherwise bodily, whatever else may be true of them.* That proposition, very intentionally, leaves open the possibility that the sequences in question are at some moments also mental. It is a proposition in conformity with the view to which we have come of psychoneural pairs as effects and causes within the given causal sequences.

The proposition taken as strongly supported is in fact a lesser or less radical proposition that some take neuroscience to establish. It is a lesser proposition than one which is quite often in a way implied by generalizations within neuroscience. What is quite often implied—sometimes by the generalization that the brain is a machine—is that neural events are effects of causal sequences which, putting aside any environmental part, are *wholly neural or bodily*. No mental fact enters into them.

As will be clear, the theory of determinism which we have developed requires the first proposition, that neural events result from causal sequences that are at every moment neural or bodily, but is inconsistent with the second proposition, that neural events result from wholly neural or bodily sequences. The second proposition, taken together with what goes with it, involves a denial of mental indispensability. What is likely to go with it is some account of the psychoneural relation which necessarily is other than the account which is partly in terms of psychoneural pairs. The account may be one of the Identity Theories, or the theory which asserts psychoneural nomic correlation in conjunction with the wholly neural causation of the neural, or the macromicro theory. These and any similar theories involve there being a full explanation of typical mental events and actions which does not have prior mental events as an ineliminable part. (pp. 91, 149 ff., 154 ff.)

Throughout this inquiry we have maintained mental indispensability as an axiom, and depended partly on it to reject the theories of mind just mentioned. Is it put into doubt by neuroscience? Are we to embrace not it, but the second proposition above together with an accompaniment, with the consequence that there *are* full but wholly neural or bodily explanations of typical mental events?

The reasons for saying no, or rather the reasons having to do with neuroscience, are two. The first and greater has to do with ongoing research itself. The second has to do with the occasional philosophizing within neuroscience.

The first is bound up with a general fact mentioned earlier in another connection: that neuroscience, like a great deal of science, is not much concerned to establish somehow sufficient conditions of phenomena, but rather concerned to establish theoretically useful or practically useful necessary ones. It is not overly concerned with establishing laws. It asks what were earlier labelled *presupposing* as against *ultimate* questions. (pp. 285 f.) More particularly, it is not much concerned to specify *causal circumstances* for mental events. It is indeed concerned with explaining the mental, as evidenced by behaviour, but it is typically concerned to specify *causes*, usually groups of causes, of mental events. It thereby assumes the existence of causal circumstances, but does not attend to them. Its concentration on causes is a natural consequence of methodological attitudes having to do with economy of effort that are common in science and in inquiry generally.

It is therefore not the case that neuroscience sets out directly to test the question of whether mental events have *wholly* neural or bodily explanations. It does not in fact ask if my mental event of recognizing the face of my brother is owed to a wholly neural or bodily causal sequence, but rather concerns itself with what we have taken to be the neural or bodily elements of that sequence.

I take it to be safe to say, further, that there exists no evidence within neuroscience that such a mental event as my recognizing my brother's face has a wholly neural or bodily explanation. Nor, I think, when the issue is made clear and kept clear, do typical neuroscientists maintain that there is such evidence. There is good reason for speaking of the *axiom* of mental indispensability.

It is possible, alas, to be distracted from it. As noted earlier (p. 269), there are usages in neuroscience that may suggest epiphenomenalism. Neural facts are sometimes said to *control, regulate, allow for, maintain,* or *elicit* mental facts. Neural facts are the *basis* or *substrates* of mental facts, that of which mental facts are a *function*. It is also true that neuroscience, in its rare philosophizing moments, can be said to show a tendency to epiphenomenalism. As with its tendencies to definitional behaviourism and Eliminative Materialism (pp. 282 f.), the philosophizing in question is typically uncertain and indeed indeterminate. Examples can be had from the first and third editions of an admirable general work.

It is a basic premise of physiological psychology that our minds are no more than the manifestations of functioning human brains. The mind does not *control* the brain, neither does the brain *control* the mind. Rather, the brain, in its operations, gives rise to the mind. Physiological psychologists have no answer to the mystery of self-awareness, that peculiarly private experience. If we are indeed machines, *why* are we conscious of our own existence? The fact that we cannot now answer this question does not mean that we must abandon the conviction that our behaviour, our feelings, and our thoughts are no more than functions of our physiology. After all, a denial of the physiological basis of mind does not explain anything either. (Carlson, 1977, p. 2)

Modern scientists take a monistic approach to studying the physiology of behaviour. That is, we believe that the physiological basis of behaviour will be completely understood once we know all the details of the working of the body, especially of the nervous system. . . . the reality we deal with is entirely physical in nature. . . . What we call 'mind' is a consequence of the functioning of the body and its interactions with the environment. The mind–body problem thus exists only as an abstraction.

What can a physiological psychologist say about human self-awareness? . . . We can . . . speculate about the usefulness of self-awareness: Consciousness and the ability to communicate seem to go hand in hand. . . . Perhaps the ability to make plans and to communicate these plans to others is what was selected for in the evolution of consciousness. (Carlson, 1985, pp. 4–5)

What runs through the passages, together with the official epiphenomenalism, is an evident tendency precisely against it. With respect to the first passage, we have it that minds are functions of brains—but also that brains do *not* control minds. In that passage and the second one, together with the official epiphenomenalism, it is implied or stated that consciousness or self-awareness, which certainly is realistically conceived, does itself have functions. It is easy to overlook the fact that what itself has a function does itself have an effect. If mental events are no more than side-effects of neural events, or, to speak more precisely, if they do not themselves enter into the explanation of subsequent events, they themselves can have no function whatever in the given clear sense.

A first conclusion to be drawn is that the so-called epiphenomenalism here, as in other neuroscientific writing, is not at all the real thing. The real thing, I suspect, is in fact the dubious property of some few philosophers caught up in philosophical doctrine. Of the pictures of the mind of which we know, the one most suggested by neuroscientific reflections of the kind in question is Neural Causation with Psychoneural Correlation. (3.2) What is in fact added to it, as is plain, and in good sense must be, and what destroys the picture, is mental indispensability.

A second conclusion to be drawn is that neuroscientific epiphenomenalism, so-called, or at any rate typical neuroscientific philosophizing-in-passing consists in large part in a set of somewhat unclarified attitudes which are *not* opposed to but *expressed in* the picture of the mind to which we have come. There is more agreement than disagreement between the two things.

Typical neuroscientific epiphenomenalism is partly opposition to Indeterminist Theories of the mind. Despite the atypical case of neuroscientific philosophy at which we shall look (5.5), there can be no doubt that neuroscience in general has no time for the idea of an originator and what goes with it. (Carlson, 1985, pp. 5–6) Neuroscientific epiphenomenalism, quite as importantly, is in part opposition to a conception of consciousness and mental events as non-physical. Perhaps this opposition to mental events as mysteriously outside the realm of the physical is fundamental to it. The picture of the mind to which we have come, while it properly distinguishes between the mental and the neural, makes both of them physical—both fall under a defensible definition of the physical. (p. 87) That view has consequences with respect to the possibility of scientific investigation of the mental, and in particular of the ancient and still intractable problem of the ultimate nature or mechanism of psychoneural nomic connection. (pp. 112, 170)

Neuroscientific epiphenomenalism, further, is opposition to mere psychoneural parallelism, and to any denial of psychoneural intimacy. It is thus opposition to Cartesian dualism and the like. In these various attitudes, and others, it is in fact one with the picture of the mind to which we have come. It is not too much to say that neuroscientific epiphenomenalism, which is certainly uncertain, is more consistent than inconsistent with the given picture.

5.4 EVIDENCE RELATING TO THE THIRD HYPOTHESIS

An action, conceived vaguely as a sequence of bodily events somehow owed to the mental, is in fact the effect of a causal sequence. One of the initial elements of this causal sequence and some of its subsequent elements are psychoneural pairs incorporating an active intention which represents the bodily sequence. The other initial elements of the causal sequence are other bodily events and direct environmental events. (p. 248) This, the Hypothesis on the Causation of Actions, raises less objection or calls for less support than its predecessors, fundamentally because it does not conflict with certain pictures we have of our existence, related to the Indeterminist Theories of the mind and rooted in deep attitudes. The hypothesis or something close to it does in fact provide the content of our standard conception of an action. We arrived at the hypothesis by way of reflection on that standard conception.

The main argument for it, not much attended to in our inquiry, is similar to that for the second hypothesis. (i) An acceptable answer to the question of the explanation of actions must satisfy the criterion of mental realism. (ii) It must also satisfy psychoneural intimacy, (iii) mental indispensability, (iv) personal indispensability, and (v) simplicity. An acceptable answer must (vi) use an adequate conception of causation, and, finally, satisfy (vii) the constraint of the somehow neural or otherwise bodily causation of neural events and also actions. (p. 164) As is clear, various possible answers to the question of the explanation of actions, derivable from various accounts of the mind, do not satisfy the criteria, while our third hypothesis does. (pp. 210 ff.)

It is of course (vi) the proposition that actions are effects of somehow neural or otherwise bodily sequences which is now relevant. That is the proposition for which neuroscience provides evidence. It is of course to be understood as consistent with mental indispensability, and so would be in conflict with any true tendency to epiphenomenalism in neuroscience.

One of several bodies of indirect or lateral neuroscientific evidence—we did not consider this sort of thing with the previous hypotheses but certainly might have—has to do with a part of the nervous system so far unmentioned. Our actions, as we know, do not include perspiration, dilation of the pupils of the eyes, digestion, adjustment of the heart rate, regulation of blood vessels, ordinary breathing, and the like. These processes, not subject to voluntary control, are owed to the Autonomic Nervous System (ANS). It operates more or less automatically, by way of feedback: roughly, the effects of the system affect the system itself, thereby producing different effects. The ANS is controlled by the hypothalamus, which receives neural input having to do with the state of the organs of the body, and in turn, by way of other neural connections, acts back upon them. The bulk of both input and output are outside of our conscious experience. The working of the ANS, and in particular its neurochemistry, are perhaps better understood than any other part of the nervous system.

Activities of the ANS, if all distinct from actions, are none the less of clear relevance to the proposition that actions are effects of standard causal sequences somehow neural or bodily in nature. That is so since the given activities, whose details we have left unconsidered, are indubitably causal in nature. Theories of them give indirect or lateral support to the theory, so to speak, of actions as causal—as effects of causal sequences. Certainly it counts in favour of a theory, heavily in favour, that it consorts with neighbouring and to some extent overlapping ones. (Cf. Newton-Smith, 1981, p. 228) If the ANS were not causal, that would be reason to doubt that actions are.

To come on to direct neuroscientific evidence for the causal proposition about actions, much of it has to do with the motor system, based in three parts of the brain. These are the cerebellum, basal ganglia, and, above all, that particular part of the cerebral cortex known as the motor cortex. There is great difficulty in closely delineating all of the motor system, as distinct from what might be called describing its bases, and, it must be said, there have in the recent past been doubts expressed as to the possibility of doing so. (Nauta & Feirtag, 1979; cf. Stein, 1985) It is exceedingly complex.

None the less, there is much knowledge of the motor cortex, a cortical representation having to do with movement and action. It is indicated by the motor homunculus given in Figure 27. It can be taken

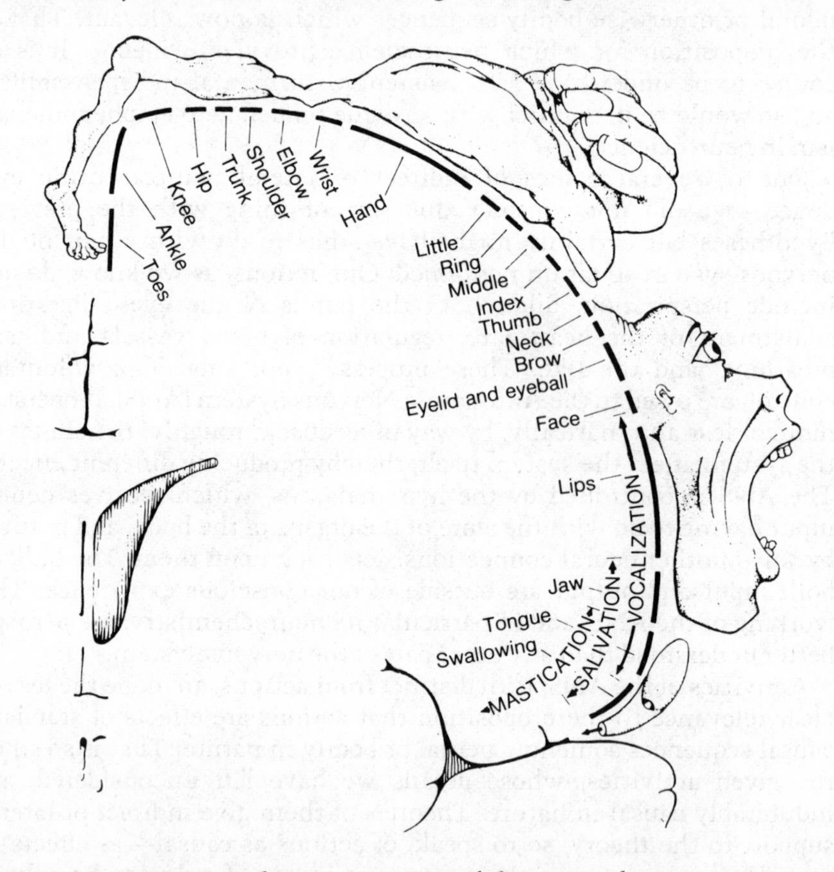

FIG. 27. Motor Homunculus: cross-section of the cortex showing motor areas for movement. From Penfield and Rasmussen, 1957. fig. 22, p. 57

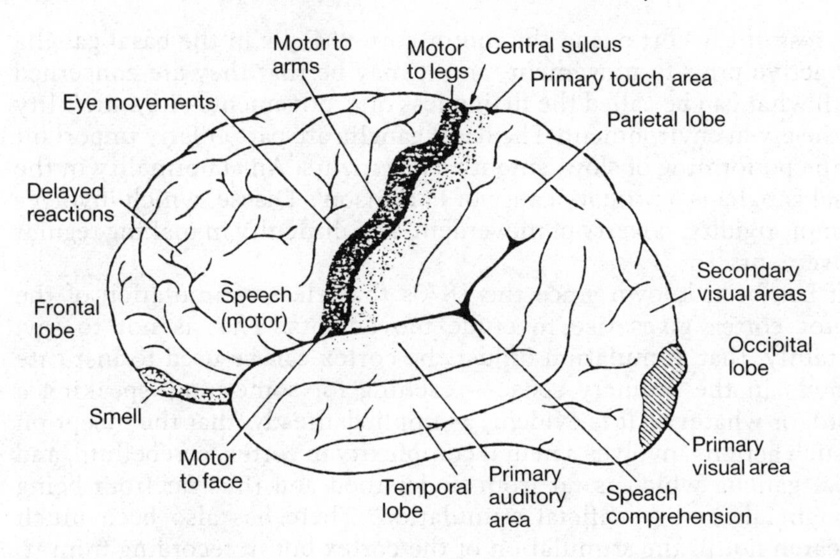

FIG. 28. External view of the cerebral cortex from the left. The motor area is black, and the primary sensory areas stippled. From Young, 1978, fig. 12.5, p 126

as relating movements, actions, or something involved in them, to cortical causes. This part of the brain, also called the motor strip (Figure 28), is of course different from the part represented by the sensory homunculus in Figure 23 (p. 271), but the predictable similarity of the two representations is apparent. The small of the back, say, is of relatively little importance to either sensation or action. Exactly what is represented by the motor homunculus is not a settled matter. 'Is it the conception of a movement, a plan for its execution, or merely a cortical map of the muscles to be employed? Are ideas, movements, or muscles mapped there?' (Stein, 1982, pp. 230–1)

The other two principal parts of the motor system—cerebellum and basal ganglia—are as essential to action. The cerebellum provides neural connections between the rest of the cerebral cortex and that part of it which is the motor cortex. In particular, it plays a fundamental role in passing information from the rest of the cerebral cortex to the motor cortex. In essence, it is said, its function is one of feed-forward control. As mentioned earlier, it regulates the rate, range, and force of movements. For this reason it is often spoken of as having the role of a computer in the motor system. It is taken to have a particular concern with rapid movements.

The basal ganglia, if less well understood, are as essential to action. They are like the cerebellum in that they make connections between the rest of the cortex and the motor cortex. Cells in the basal ganglia are active prior to movement, and it may be that they are concerned with what can be called the fittingness of a movement—its suitability to the given environment. The basal ganglia are particularly important in the performing of slow, smooth movements. An abnormality in the basal ganglia is a primary cause of Parkinson's Disese, which involves tremor, rigidity, poverty of movement, and difficulty in making regular movements.

It has been known since the 1870s that electrostimulation of the motor cortex gives rise to crude movements. That is not to say, certainly, that stimulation of just the cortex can be used to instigate actions in the ordinary sense—reaching for something, speaking a word, or whatever. It is evident, as implied already, that the inception of such actions involves a neural complexity in cortex, cerebellum, and basal ganglia which is far from understood and thus far from being brought about by artificial stimulation. There has also been much research not in the stimulation of the cortex but in recording from it, sometimes from single neurons, during the performance of actions.

A goodly amount is known, then, of the general inception and regulation of actions—quite enough to make possible good overviews (e.g. Greer, 1984), if not close delineations of function. Very little is known of what may be taken as of particular relevance to our hypothesis about actions, any neural counterpart of precisely the episode of willing or actively intending an action. Electrostimulation of human subjects does not give rise to movements that are in the ordinary way intended or desired. The subjects report no experience of intention or desire. However, there do exist findings that touch on that matter. Some, of which we shall see something more, have to do with recorded electrical activity at the time of willing. (p. 302) Some have to do with what can be called the triggering of programmes for action.

In a great deal of action or perhaps almost all, as is plain enough, we do not initiate each movement in the sense of intending precisely it. (p. 211) This is true, for example, of walking. It is a matter of a programme or somehow automatic system. Such programmes or systems are located at least in part in the cerebellum and basal ganglia, as already suggested, as well as in the motor cortex. The systems are not self-contained, but depend on feedback of various kinds through the Peripheral Nervous System, importantly from musculature. As for the triggering, which can be regarded as a true neural beginning of an action, it has been investigated in some of the lower animals. In the crayfish a specific neuron is a 'command element'. When stimulated it issues in a complex series of tail movements. There are related findings having to do with certain mammals. It is of importance that command neurons in our own species and perhaps others, while they can be affected by the environment, can also fire independently of it. Triggering, however, in our own and other higher species, remains obscure. It is possible that it itself involves a complex neural network. (Delong, 1971)

So much—very little indeed—for the fundamental structure of the motor system, consisting in motor cortex, cerebellum and basal ganglia. The remainder of the system, itself the subject of research of many kinds, brings in motor neurons to muscle, functions of the spinal cord, the Peripheral Nervous System generally, musculature, classes of movements, and so on. Enough has been said to convey what will have been anticipated, the judgement that the third hypothesis can be taken as *very strongly and clearly supported* by neuroscience despite very incomplete knowledge. As in the case of the second hypothesis, it is a judgement made in the context of what is known of the neuron itself. It is a judgement that for long has been routinely made in neuroscience. 'A basic working assumption of behavioural science is that the behaviour of organisms is like all other material events in that it is determined by physical events and can be predicted when all relevant conditions are known.' (Butter, 1968, p. 2)

5.5 SUPPOSEDLY CONTRARY NEUROSCIENTIFIC EVIDENCE

It is supposed by a philosophical–neurophysiological partnership (Popper and Eccles, 1977) that there is neuroscientific evidence against the Correlation Hypothesis and also the Hypothesis on the Causation of Psychoneural Pairs. It is evidence that establishes an indeterminist theory of the mind. The principal body of supposed evidence against the Correlation Hypothesis, or the interpretation of the evidence, involves the idea that a mental event may occur *before* there occurs 'neuronal adequacy' for it. The idea, as it seems, is that it is *after* a mental event, which is free-floating or ghostly, that its neural correlate occurs. To my mind this supposed refutation of the Correlation Hypothesis rests on demonstrable confusion. The demonstration, given elsewhere, will not be repeated here. (Honderich, 1984a, 1986; Libet, 1978, 1985a; Libet *et al.*, 1979; Popper and Eccles, 1977, pp. 364, 259, 257, 362)

The philosophical–neurophysiological partnership depends on a second body of supposed evidence to refute anything like the Hypothesis on the Causation of Psychoneural Pairs. This evidence has to do with the voluntary inception of action, and the research of another group of experimenters. (Kornhuber, 1974; Deeke *et al.*, 1969;

cf. Libet, 1985b) In this research, human subjects made small movements, such as flexing the right index finger, 'at will'. That is, in part, a subject chose when to make movements of that particular kind during a period when recording electrodes were attached to one of his finger muscles and also to his skull. The subject was given no instruction or signal of any kind as to when to move—which feature of the experimental situation is taken to be of great significance.

It was found that prior to the moment when electrical activity was first recorded in the muscle, there had been a buildup of electrical activity over a relatively wide area of the cerebral cortex. A slowly rising negative potential, named a readiness potential, was recorded through the skull from various parts of the cortex. Such a potential is of course a summation of patterns of neural activity, as in the case of much related research. (pp. 274, 289) Usually, the readiness potential began about 0·8 of a second before the action potential in the muscle. The readiness potential was followed by two different potentials, recorded as sharper EEG waves, beginning at about 0·09 of a second before the muscle activity. The last of these was not widespread but was concentrated on the relevant cells of the motor cortex, those long established as being in connection with movement of the relevant finger. This last potential occurred at about 0·05 second before the onset of electrical activity in the finger muscle. It takes about 0·05 of a second for the transmission of a message from cortex to muscle.

That, in general terms, is all of the relevant data. The philosophical–neurophysiological partnership, but not the original experimenters, take up the hypothesis that the readiness potential is in part the work of an originator, the self-conscious mind, and also that the self-conscious mind guides the electrical activity in its convergence on to the right motor cells, those for the movement of a particular finger. (Popper and Eccles, 1977, pp. 285–6) The principal comment made in support of this is as follows.

> It is impossible to carry out any scientific study on the decision-making propensity of a human being subjected to all the complexity of a 'real-life' situation even when that situation is ethically neutral—for example the decision to go home by train or bus, or which gramophone record to put on for the next playing. No doubt psychologists or philosophers could claim that in principle such decisions can be accounted for in a rigidly determined fashion by the present brain events and the stored memories. However the stringent conditions of Kornhuber's . . . experiment preclude or negate such explanatory claims. The trained subjects literally do make the movements in the absence of determining influences from the environment, and any random potentials generated in the relaxed brain would be virtually eliminated by the averages of 250 traces. Thus we can regard these experiments as providing a convincing

demonstration that voluntary movements can be freely initiated independently of any determining influences that are entirely within the neuronal machinery of the brain. . . .

The outstanding problem in the voluntary control of movement is of course the action across the interface between the self-conscious mind on the one hand and the modules of the cerebral cortex on the other. The existence of this influence is established by the empirical experiments of Kornhuber and associates. (Popper and Eccles, 1977, p. 294)

What is to be said of the idea that the given experimental data provide evidence against our second hypothesis and perhaps also the first hypothesis? The data amount to this: electrical activity of certain locations, durations and intensities, some of it comprising the readiness potential, takes place in the cerebral cortex prior to an action which the subject performs independently of any immediately prior instruction or environmental signal. That, however, is *perfectly consistent* with our two hypotheses, and with a denial of any indeterminist theory of the mind. It is possible and natural to suppose that mental episodes of inactively and actively intending the action are in nomic correlation with all or part of the 0·8 of a second of various cortical activity. It is possible and natural to suppose that the events of this neural activity are the effects of certain causal sequences.

Certainly our second hypothesis puts environmental elements into these causal sequences. But there is nothing whatever in the hypothesis which requires that all psychoneural pairs are owed to instruction or to cues from the immediate or more or less simultaneously obtaining environment. It is absurd to identify determinism in general, as the quoted passage seems to do, let alone our three hypotheses, with the idea that actions are always owed to 'determining influences from the environment'—i.e. *the more or less simultaneously obtaining environment*. That assigns to determinism the wonderful consequence, among others, of excluding the possibility of thinking about anything other than the current environment. The Hypothesis on the Causation of Psychoneural Pairs, along with all arguable views of the mind, does indeed assert that somewhere in the causal background of a neural correlate or whatever there are environmental influences. Nothing whatever in the given experimental data goes against that. There was in fact a great deal of environmental influence in the background of the actions of the subjects in the given experiments, both recent and distant background.

As for a further line in the quoted passage, there is nothing whatever in the given data, let alone a convincing demonstration, to the effect that 'voluntary movements can be freely initiated independently of any determining influences that are entirely within the neuronal machinery of the brain'. It very evidently does not follow, from the absence of current environmental cues, let alone from the exclusion from consideration of any random activity in the brain, that the initiation of voluntary movements is somehow independent of the brain.

The philosophical–neurophysiological partnership also write:

> we know now from studies by Kornhuber and others . . . that, when you are willing an action, you don't immediately trigger off the action; . . . the self-conscious mind is working up the neural machinery in wide ranging parts of the brain and gradually is moulding the patterns there, actively changing them. So eventually the patterned neuronal operation homes in on the correct pyramidal cells . . . in order to bring about the desired action. This whole process takes about 0·8 of a second and therefore you can think of the incredible complexity of the events going on. This is . . . an active influence of the self-conscious mind upon neuronal machinery.
>
> From these findings I develop the conjecture that there isn't any simple unitary relationship between neural events and the self-conscious mind. (Popper and Eccles, 1977, p. 476)

What is to be said in reply, first, is that it is no part of our three hypotheses that a mental event, a piece of willing or active intending, has some sort of instantaneous upshot, that in the given sense it 'will immediately trigger off the action'. Nor is it part of our hypotheses that there is a 'simple unitary relationship' between neural and mental events—where such a relationship would be other than what is suggested by the experimental findings. To assert psychoneural nomic correlation is not to assert anything whatever about particular locations, durations or complexities of the neural term. There is no conflict whatever between the Correlation Hypothesis and what is said to be true of electrical activity in the cortex.

5.6 QUANTUM THEORY

The theory of determinism of this book, consisting in the three hypotheses, is at least given very strong and clear support by standard neuroscience. Thus the theory is very strongly and clearly supported by the body of knowledge, theory, and informed speculation that is closest and most relevant to its subject-matter. That conclusion takes into account occasional neuroscientific claims as to mysterious emergence and the like and also the neuroscientific objections at which we have just looked. A retort can be heard, however. It is that there is a fact, a very large fact, that has been paid insufficient attention. Indeed there is, and it needs consideration before a final conclusion can be drawn as to the truth, falsehood, or whatever of our theory of determinism. It is the fact of Quantum Theory, that fundamental part of contemporary physics.

As we have already seen, Quantum Theory is one thing that has prompted philosophers into mistaken claims as to our standard conception of causation: the claim that we take effects to be preceded only by necessary conditions, and the claim that effects are no more than events made probable by their antecedents. (1.6) We shall not reopen the question of our standard conception of causation.

Quantum Theory, as we know, has also been relied upon by philosophers who advocate an indeterminist theory of the mind. (3.4 - 3.9) Such a theory, in sum, is that the brain is an indeterministic mechanism—a mechanism such that not all its events are standard effects or nomic correlates—but a mechanism subject to a kind of control by an originator. To speak differently, the idea may be that within a small permitted range of physical values, a neural event might have had a value other than it did, not because its neural and other bodily antecedents might have been different, but given those antecedents just as they were. This might be a synaptic event, on which the firing of a neuron depends. No value outside the range was nomically possible, but within the range, although each of the values was to some degree probable, none was necessitated by neural or other bodily antecedents. The event exactly as it was, according to this idea, was as it was because of some fact of *mentality*. That exactly it occurred, rather than another event, was the result of this mental control. (Eccles, 1953, p. 279; cf. Popper and Eccles, 1977, pp. 540–1) The originator or self-conscious mind selects one of the particular neural possibilities left open, although probabilified, by neural and bodily antecedents.

There is also reliance on Quantum Theory in connection with at least one greatly more restrained view of the mind.

> . . . in the sense used here, the brain is a deterministic system, so that if a brain state at any one time is specified, its subsequent states can also be predicted—within a certain degree of error. Thus I . . . maintain that any prediction that can be made about the brain is limited in the same way that the Uncertainty Principle limits prediction in physics: that is, the predictions are of a probabilistic and not an absolute nature. This uncertainty arises from the properties of the nervous system itself. (Rose, 1976, p. 37; cf. Burns, 1968)

This indeterminism—as indeed it is—is again a matter of synapses. Its proponent, however, as should be noted, is no friend to speculation about a self-conscious mind, and characterizes the philosophical–neurophysiological partnership and any like theorists as 'The Irrationalists'. (Rose, 1976, p. 361) His own summary account of mind and brain, as mentioned earlier, is an Identity Theory. (p. 286)

We thus have two reasons for considering Quantum Theory. First, it enters into indeterminist pictures of the mind, including the exceedingly rare neuroscientific versions. Secondly, it is relied on in criticism of determinist pictures by those who do not take up the challenge of providing an alternative. The determinism expounded in this book, certainly, will rightly or wrongly be regarded by many philosophers and scientists as at least put in doubt by Quantum Theory. They will suppose that the very strong and clear evidence provided by neuroscience is outweighed by the evidence of Quantum Theory. More particularly, they will take the second hypothesis as put in doubt, and, in consistency, the third. Also, of course, they will be required to doubt the proposition of the somehow neural and other bodily causation of neural events, which of course is in terms of standard causation.

It is to be noted, with respect to our two reasons for considering Quantum Theory, that if an interpretation of it were to put in doubt or refute our theory of determinism, it would not at all follow that Quantum Theory supported indeterminist pictures of the mind. For that to be true it would need to fit those pictures. It would need to be consistent with what else was asserted in such a picture. That matter will have our attention in due course. (5.6.7)

Quantum Theory is the successor at one level, the fundamental level of atoms and their constituents, to the great edifice of past physics, classical or Newtonian mechanics. In that view of physical reality, all of the universe is composed of elementary particles, or more precisely, mass-points, each having well-defined properties, such as a specific spatio-temporal location and momentum. Each particle, above all, is subject to fundamental laws of motion, and its relations with all other particles are indubitable instances of standard nomic connection. Thus, given a knowledge of positions and momenta of all particles in the universe at a time, it would be possible, in principle, to predict the future course of the physical universe and to retrodict its past course. The best known expression of this belief is of course that of Laplace. 'An intelligence knowing all the forces acting in nature at a given instant, as well as the momentary positions of all things in the universe, would be able to comprehend in one single formula the motions of the largest bodies as well as the lightest atoms . . . .' (Laplace, 1951 (1820), p. 4)

Classical mechanics, when conjoined with certain propositions about the mind, notably the Hypothesis of Psychoneural Nomic Correlation, evidently issues in a determinism of human decision and action, usually thought of as an epiphenomenal determinism. Each decision and action is a necessitated occurrence. Classical mechanics itself is almost universally regarded as itself a paradigm of theories of determinism, these being characterizations of specified larger or smaller classes of events as necessitated. A contrary view to the effect that classical mechanics is not a determinism (Popper, 1982b; cf. Suppes, 1970), has for good reason found few supporters, and been effectively criticized. (E. Nagel, 1979, p. 278f) Surprising and indeed compelling accounts of 'chaotic behaviour' in such simple systems as the spherical pendulum, or a weight on a string, accounts in terms of Newtonian dynamics, do indeed establish the existence of a certain unpredictability, but not the unpredictability (6.1) which is the mark of indeterminism. (Lighthill, 1986; Lichtenberg and Lieberman, 1983)

To say the least, Quantum Theory or Quantum Mechanics is very different. Interpretations of the several alternative bodies of mathematics in which the theory at bottom consists are regarded by many as making a determinist view of physical reality impossible. In this it is quite unlike that other revolutionary development in modern physics, Relativity Theory. The latter is sometimes taken to *require* a determinist view. (Maxwell, 1985) Quantum Theory, it is said, has brought about 'the downfall of causality'. That is, more precisely, the theory as interpreted is widely regarded as conflicting with a general view of physical reality as a matter of connections such that each event is necessitated, as distinct from made no more than probable. Quantum Theory, as is well enough known, does assert probabilistic nomic connections.

Another difference between classical mechanics and Quantum Theory bears on the principal one just mentioned, affects any exposition of the theory, and will be important in what follows. It is that there is doubt and disagreement, to say the very least, about the interpretation of the theory, which is to say about what its fundamental terms are to be taken to mean. Both classical and Quantum Mechanics do indeed consist in a formalism—mathematical formulae or equations—and an interpretation or model of the formalism. In the case of Quantum Theory there is general and great doubt and disgreement about the interpretation. That is to say, to repeat, that there is doubt and disagreement about how the *referents*, if any, of its mathematical expressions are to be conceived. There is general doubt and disagreement about how the formalism is to be understood in terms of physical reality, about what it can tell us of the physical world, at whatever level.

Interpretations speak of 'particles', of course, or of 'quantum objects' or the like, but how are the terms to be understood? The term 'particle', which we shall use, is certainly not to be understood in the classical sense. How are the more specific terms for subclasses of 'particles', such as 'photon' and 'electron', to be understood? Again, how are 'position' and 'momentum' to be understood? Certainly not as in classical mechanics. There is the same problem about the term 'wave', which is as essential to the interpretation of Quantum Theory. To say these cautionary things is *not* to engage in any punctiliousness peculiar to philosophers, but to report what is accepted by all physicists, and turns up early in one form or another in ordinary expositions of Quantum Theory. 'Students find quantum mechanics tough going .... Familiar concepts like speed, size, acceleration, momentum and energy take on weird features, or even become meaningless.' (P. C. W. Davies, 1984, p. ix)

Bound up with this fact about interpretation are others as indubitable and relevant to our concerns, but not unique to Quantum Theory. The formalism of the theory is far from being in any close connection with reality. It is very far indeed from being in the sort of connection, say, between the positive integers and sets of objects in the world—say the number 5 and five oranges. It is in an indirect and distant relation with many experimental results with which it is said to accord. The formalism, to speak differently, is connected to what is taken as relevant empirical evidence by way of long chains of deduction involving many subsidiary hypotheses. This is also its relation to something else of importance—its practical uses, which are many.

The formalism, which is in some sense fundamental, can be said to remain independent of its interpretations, and of evidence and practical uses, and not to depend for its own theoretical value on them. It is the practice of many quantum physicists, as it was of Werner Heisenberg and Paul Dirac, two founders of Quantum Theory, to persist with the mathematics and to be relatively unconcerned with interpretation and so on. Inevitably, however, expositions of the theory for non-mathematicians must fundamentally be in terms of interpretation and experimental results. That is not so much a cause for regret or insecurity as might be supposed, since it is of course exactly the interpretation of the theory—what it comes to in terms of physical reality—that is of relevance for us. As will be anticipated, what has been said of doubts and difficulties as to interpretation will enter into an objection to the general idea that Quantum Theory has brought about the downfall of causality.

The theory had its origin in Max Planck's departure from the classical assumption that physical quantities, such as momentum and energy, are in a sense continuous: they can take on any value. Taking some particular speed at which a car travels, the classical assumption is that there is not a single definite next lower speed separated from the first by some interval. Or, it is the case no matter how fast a car changes in speed, say from 50 to 60 m.p.h., that it passes through every intermediate speed. The assumption, although not without obscurity, remains natural and perhaps compelling. It was Planck's idea, however, in connection with heat radiation, that it comes in discrete units or packets: quanta. It is non-continuous. The energy only has values which are multiples of a certain basic amount. In Einstein's later hypothesis about light, in part a kind of revival of the particle theory as against the established wave theory of light, or rather a kind of marrying of such theories, light is not continuous but rather a matter of those quanta which are the 'particles' called 'photons'. Transitions between the quantal levels, or quantum jumps, were taken as having a random character. In Niels Bohr's subsequent work having to do with the atom, a development of the Einstein hypothesis about light, an 'electron' is taken as jumping between orbits, and as emitting a 'photon' in the process. But the interpretation of his formalism does not allow *any* location to the 'electron' between orbits, nor allow *any* precise location to the emission of the 'photon'. The 'electron' is a thing which for part of its history is *nowhere*, and the 'photon' turns up *nowhere*.

To put aside the history of Quantum Theory (Jammer, 1966, 1974), the theory itself may be expounded in terms of two logically related parts, one of them having to do with the *Wave Equations* of Erwin Schrödinger, the other with Werner Heisenberg's *Uncertainty Principle*. Let us consider them in turn.

Solutions of the Wave Equations, after certain mathematical transformations, are generally taken as giving a probability that a 'particle' will have a certain 'location' at a certain time. The equations can be, but generally are not, understood in such a way that they give rise to more than a probability as to 'location'. They do not give us a necessitated 'location'. The conclusion is inextricably bound up with and depends on what has long been referred to as *wave-particle duality*, which can be considered in terms of the well-known two-slit experiment.

Suppose that a stream of 'particles' and in particular 'photons' or 'electrons'—I shall now with reluctance now more or less give up the cautionary inverted commas—is directed from a source at a screen or barrier containing two slits, closely parallel and each about a millionth of a metre wide. Beyond the two-slit screen, some distance away, is a further screen which somehow registers the arrivals of individual particles that pass through one or both slits of the first screen. Figure 29 gives the set-up with both slits open. The two-slit experiment has to do with the difference between having one or both slits open.

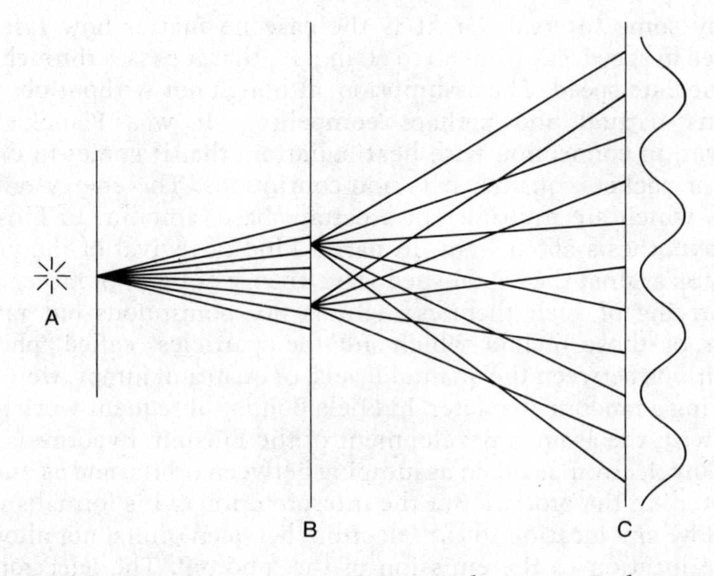

FIG. 29. Two-Slit Experiment. A: source; B: two-slit screen; C: detector-screen. The curve alongside C plots particle arrivals. After P. C. W. Davies, 1979, p. 60

(i) During a sequence of particle-emissions from the source when only one of the slits is open—say the left-hand slit from the perspective of the source—individual particles are registered as arriving on the detector-screen opposite that open slit. Their locations of arrival, in accordance with the usual understanding of the Wave Equations, are subject only to a certain probability. The record of their arrival, to follow a standard exposition, gives rise to a graph showing something like the solid as distinct from the broken curve in Figure 30. (cf. Feynman, 1965, Ch. 6)

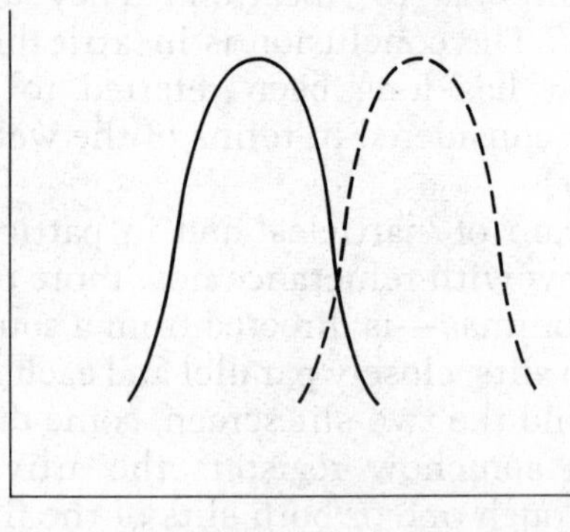

FIG. 30. Solid curve plots particle arrivals with only left slit open; broken curve for only right slit open. After P. C. W. Davies, 1979, p. 62

If one thinks of the particles as leaving marks or causing flashes on the detector-screen, these might be visualized in terms of a single vertical band or diffraction pattern, increasingly light in colour towards each side and increasingly dark towards a line down the middle of the band. The equations are taken as assigning a high probability to particles arriving on the line and lower probabilities to electrons arriving towards the sides of the band. They do not specify a more particular place of arrival within the band for any particle.

If only the other or right-hand slit were open, the result would be a similar distribution opposite that slit, as indicated by the broken line in Figure 30. That the graph, in either case, would not show a spike—as distinct from a hump—indicating arrivals in a *precise* target area, is not taken to issue in any conclusion about the paths of particles being affected by the nature of the slit or whatever—say bouncing off its sides in passing through—but in the conclusion of which we know, as to the absence of a standard causal circumstance for each arrival in its particular place.

In sum, what we have is that when one slit is open, it is as if a 'randomizing machine-gun' were being fired at the two-slit screen, and bullets passed through the open slit to strike the detector-screen. What is 'random', although within clear bounds, is the locations of striking or arrival.

(ii) The other case is where both slits are open. One's expectation, if uninformed, will be that the composite result will produce a graph of the kind shown in Figure 31, and *two* vertical bands of marks. The actual and mysterious result, to give it in the simplest possible form, instead produces a graph showing a curve like that in Figure 32—and

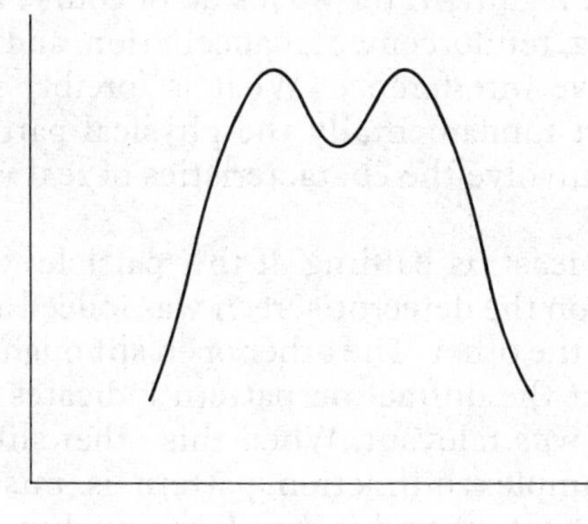

FIG. 31. Expected curve for particle arrivals with both slits open. After P. C. W. Davies, 1979, p. 62

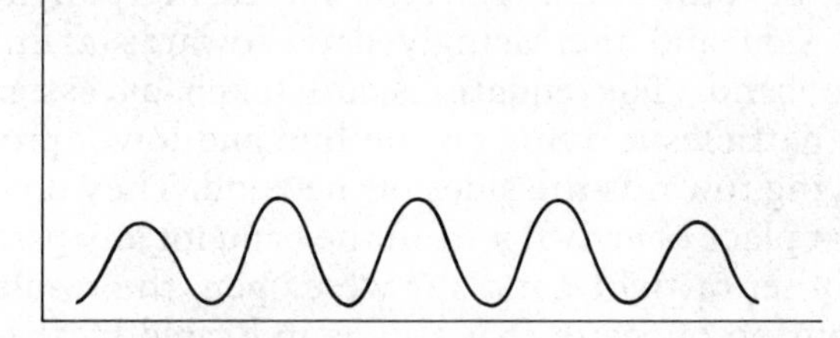

FIG. 32. Curve for observed particle arrivals with both slits open. After P. C. W. Davies, 1979, p. 63

also Figure 29. If one thinks again of the arrival of the particles as leaving marks, they are to be visualized in terms of a quite different diffraction pattern, a *considerable number* of vertical bands, each one again increasingly dark towards a centre line. (Cf. Powers, 1982, p. 132) With respect to each band, the Wave Equations again are taken as assigning a high probability to particles arriving on the centre line and lower probabilities to particles arriving towards the sides of the band. They do not specify a target, within the band, for any particle.

What is most notable about the second case, with both slits open, is the absolute inappositeness of the image of the machine-gun. What happens is wholly different in kind, not at all as if bullets were being fired. Rather, a wholly different image is required. It is as if a paddle wheel had propelled a wave of water at the two-slit screen, and the resulting waves emerging from the two slits had somehow merged to produce a complex result. Water-waves do of course act in such ways, involving spreading, reinforcement, cancellation, and so on. This is the known fact of wave interference. What is forcibly suggested by the second case is that fundamentally the physical particles in question somehow have or involve the characteristics of real waves rather than particles.

This, to say the least, is baffling. If the 'particle' whose impact we find at some point on the detector-screen was indeed a particle, it came through one slit or the other. The other open slit ought to be absolutely irrelevant to it. But the diffraction pattern indicates or indeed proves that the other slit *was* relevant. When this other slit is closed, as we have seen, the complex diffraction pattern is missing. But, if the 'particle' whose impact we find *is* therefore somehow to be regarded in terms of a wave, how *can* it be that it does indeed have a wholly localized impact? Such a fact is simply no fact about a real wave.

It needs to be added immediately that in this second case, when both slits are open, the mysterious result is not taken as in part the product of some interaction of particles. There is the same result if the particle-source emits particles *singly*, so that one particle has struck the detector-screen before the next is emitted. As several expositors remark, one can go further.

Suppose we build 10,000 similar pieces of apparatus and let one electron pass through each. We make a chart and mark a dot at the position that each electron arrives. Continuing in this way we build up a pattern of 10,000 dots, and find that the pattern is identical to [Figure 32] with all the peaks and troughs, even though each individual dot represents the result of a *completely separate* experiment occurring at a separate place and time. (P. C. W. Davies, 1979, p. 61)

As will be expected, there is more to be said of the Wave Equations and the two-slit experiment. Let us first turn briefly to the other part of Quantum Theory commonly expounded. It has to do with Heisenberg's Uncertainty Relations, often referred to as the Uncertainty Principle.

The burden of the Uncertainty Principle is that there exists a *necessary uncertainty* about, say, the position and momentum of a particle. Measures of these magnitudes, according to the Uncertainty Principle, are so related that it is somehow impossible, with respect to a particle at a time, to discover an exact value for both. More particularly, the Uncertainty Relations comprise a set of formulae, one of which can be expressed simply as $\triangle p \triangle q = h$. The variables p and q are taken as referring to what is called the momentum and position of a particle at a time. The symbol h is Planck's Constant, a specific numerical value, pertaining to quanta, discovered by Planck. $\triangle p$ can be described as a measure of the uncertainty to which we are subject in connection with the momentum. $\triangle q$, pertaining to the simultaneous position of the particle, can be similarly described.

The formula is therefore to the effect that the product of the two uncertainties is always equal to or greater than a certain value, Planck's Constant. It follows that if the uncertainty pertaining to position is small—that is, if certainty about it is large—then the uncertainty pertaining to momentum must be large. If uncertainty as to position is practically zero, then uncertainty as to momentum is in a sense practically infinite. If measurement specifies with great precision the location of an electron, then no measurement can specify with precision the simultaneous momentum of the electron. Alternatively, precise measurement of momentum makes it impossible in principle to have precise measurement of position. All that is possible, if either position or momentum is known, is to state a probability as to the other. If momentum is ignored, it is allowed to be possible in theory to determine position with absolute precision, and vice versa. Given all this, it is said, there is lacking a determinate relation, such as determinism asserts, between precise positions and momenta at a given time and precise positions and momenta at another time.

The usual one of a number of interpretations of the Uncertainty Principle, despite what has just been said, and despite a common ambiguity that has been preserved in my exposition, is that the uncertainty as to momentum if position is known, and vice versa, is not merely *our* uncertainty. Rather, it is an uncertainty or indeterminacy in nature. The uncertainty or indeterminacy in question is 'inherent' in physical reality. The trouble is not, as was first supposed, in our possibilities of investigation, and in particular in our instruments' disturbing what we investigate, which is a fact of classical physics as well, but in physical reality itself. (Cf. Squires, 1986, pp. 19 ff.)

This ontological view of the situation is in some intimate but obscure way related to several epistemological traditions of a positivistic kind, which seek to limit what can usefully, sensibly or meaningfully be said to what can be observed, measured, calculated and so on. Thus it is said, still, that when we determine the momentum of a particle, it is 'meaningless' to speak of its exact position at that time, and vice versa. '. . . it is meaningless to attribute, even in principle, a definite position simultaneously with a definite momentum. The concepts are such that we can have one, *or* the other, but not both.' (P. C. W. Davies, 1979, p. 56)

So much by way of a scant but typical introduction to interpretation of the Wave Equations and the Uncertainty Relations. It is certainly not the case, however, despite the existence of a certain orthodoxy to which we shall come, that there is settled and explicit agreement among those physicists who go beyond the formalism and give interpretations of Quantum Theory. This is connected, no doubt, with the clear fact that the endeavour to move from the formalism to a general characterization of physical reality quickly takes on philosophical difficulties. In fact, that is an understatement. As is widely accepted, and rightly so, the interpretation of Quantum Theory *is* a kind of philosophical enterprise. (It is one, for example, where seemingly serious use may be made of the speculation of the existence of God—speculation akin to that of Berkeley in his phenomenalist characterization of reality.)

What has been said so far, then, is not to be regarded as a précis of some canonical or established body of scientific belief, let alone knowledge. What has been said is to some extent open to judgement of a philosophical kind. The formalism is one thing, a matter for mathematicians, and what it is taken to imply is another thing, not a matter only for mathematicians. That there is no settled and explicit agreement about interpretation, in the first instance among physicists, is of course a point of some importance in any attempt to judge the upshot of Quantum Theory.

What *are* we to take as its upshot with respect to indeterminist pictures of the mind, and, above all with respect to our own determinist theory? Seven considerations or groups of considerations will be mentioned.

5.6.1 Success and Interpretation

Quantum Theory is a successful theory, which fact may be taken to have four parts. First, although there does exist theoretical opposition to Quantum Theory, to which we shall come, it has no serious rival in terms of acceptance by physicists generally. It has, with respect to microreality, supplanted classical mechanics, and there is no developed or indeed less developed theory of microreality which has anything like the same routine acceptance. This is true, at any rate, if Quantum Theory is taken as being *the formalism, certain experimental results, and practical uses*—that is, if it is not taken to include much by way of interpretation, or firm and reflective commitment to interpretation. (Cf. Hanson, 1967)

The second part of the theory's success is indeed that it is experimentally supported, and, what is closely connected and as important, strongly predictive. Its early history was of course the history of a theory that could do what classical mechanics could not do. This was the case in the very beginning, with Planck's quanta. Subsequently the theory has issued in predictions of atomic structure and spectra, various forces, the structures of solids, and so on. Thirdly, and in close connection, the theory has proved to be greatly fertile. It has continued to give rise to very considerable theoretical developments. Fourthly, and perhaps as important as anything else, the success of Quantum Theory includes the fact of its practical applications. These include nuclear reaction, the transistor, the laser, the electron miscroscope, and much other technology.

In short, to use the summary recommendation of Quantum Theory that is more familiar than any other, *it works*. The question of relevance to us, however, is that of what follows about its interpretation or implied general account of physical reality from that fact, the success of the theory. It might be said that much of contemporary philosophy of science has to do with this and counterpart questions—that is, questions about the relation between success and interpretation with scientific theories generally.

About all that can be said here of this large matter is that adherence to a ruling or common interpretation of a theory is very evidently strongly determined by the theory's success, in the given four-part sense, as distinct from what can too quickly be called *confirmation* or *proof*, and that just such success has in the history of science never turned out to be a guarantee of the ruling or indeed any interpretation of the theory. Any decent account of past science establishes both claims easily. It needs remembering that precisely *classical mechanics* enjoyed, and in part continues to enjoy, just such success. It needs remembering, too, that the history of science has been a history of revolutions, the very greatest of them associated with the names of Copernicus, Newton, Lavoisier and Einstein. What was overthrown on each occasion was, in brief, an *overwhelmingly successful* theory. Further, there have been instances of *contradictory* theories of the same subject-matter, both successful. To speak less historically, and to a slightly different point, the following is no outrageous proposition about theories and their interpretations or models, but rather a view of a widely accepted kind.

> The only point that can be affirmed with confidence is that a model for a theory is not the theory itself. In consequence, the adequacy of a theory as an instrument for systematic explanation and prediction cannot be taken without further scrutiny to establish the physical reality of every aspect of the substantive model in terms of which the theory may be interpreted. This is obvious when several models are known for the same theory, but it is equally true when only a single model is available. (E. Nagel, 1979, p. 116)

It may be no satisfaction to some contemporary physicists, but there is a fact perhaps more redoubtable than any in Quantum Theory, which is that Quantum Theory despite its success will be supplanted if human history continues. That proposition may call up a reply, of a kind rightly made to any scepticism. To advance the proposition, in the present context, is of course to imply at least an inclination to a determinist conception of reality. Can there be reason to take *it* as exempt from what has been said to be the perhaps depressing course of history? Part of an answer, tacitly allowed in very nearly all serious reflection, is that there *are* exempt propositions. No scientific revolution will ever overthrow the propositions that physical reality exists, that there are minds, and so on. I am inclined to add standard causal connection to the class of exempt propositions. It is, as it must be, a common inclination and more than that.

Still, it is perhaps best to retreat from that assertive position with respect to Quantum Theory and determinism—best to retreat to what was said before: that the success in the indicated sense of Quantum Theory must be greatly significant in an explanation of acceptance of the ruling interpretation of the theory, and that that success cannot guarantee the theory itself, let alone its interpretation.

5.6.2 The Copenhagen Interpretation

What is taken to refute determinism necessarily involves the Wave Equations and the two-slit experiment. The latter, to concentrate on it, is rightly said to be at least baffling. The principal difficulty that arises in thinking of it appears to be nothing less than self-contradiction or something close to it. We have it that the physical reality consists in particles, and indeed the impacts of particles on the detector-screen are recorded, perhaps by flashes on the screen. However, the mysterious result in the second case, when both slits in the other screen are open, suggests that the physical reality has essentially to do with, or importantly consists in, real waves of something. Roughly, it has to do with, or consists in, a periodic disturbance in some medium. Waves are necessarily *in* or *of* something. The waves in question cannot be waves in or of particles. Whatever other difficulties there are in that idea, there is the recalcitrant fact remarked on, that the mysterious effect can be produced by single particles coming one at a time. Whatever the medium, however, the mysterious result does appear to require waves.

One attempt to resolve the contradiction is the Copenhagen Interpretation of Quantum Theory, for long the orthodox or most accepted interpretation. Whether it is still that, and indeed whether it can be regarded as a single determinate interpretation, is in doubt. (Sudbery, 1986, pp. 210 f.; Squires, 1986) It is owed in the main to Bohr. His idea has to do with this: there are 'complementary descriptions' of the subject-matter, which is to say there are descriptions which are mutually exclusive without being self-contradictory. One description of the given physical reality is in terms of particles, the other in terms of waves. A key to the thought can be had from another claim of 'complementarity', having to do with physiology, made by Bohr's father. Mechanical explanations of behaviour, and teleological or purposive explanations, were claimed to be 'complementary'. As the claim is reported, 'If we dissect a body in order to discover how its mechanisms work, it becomes impossible to understand it as a living organism (we have killed it); on the other hand, if we consider it as a living organism we cannot gather the information necessary for a mechanistic analysis.' (Powers, 1982, p. 133)

However, whatever is to be said of that, it is not the essence of the Copenhagen Interpretation. The essence has to do, so to speak, with what is true before a description is applied, and what is true afterwards. The essence of the Copenhagen Interpretation, as has been widely accepted, is to the effect that observables exist only when an observation is actually being made, only when a measurement is actually being taken. (Putnam, 1965, p. 84)

This is taken to be true of what is called *position* in this interpretation. In classical mechanics, of course, and in ordinary understanding, a particle is taken to have position whether or not a measurement is being made. Given the Copenhagen Interpretation of Quantum Theory, what is true of a 'particle' at most times is that it *lacks* position-coordinates, that there is no more than a probability that it will have a position *if* a measurement of it is made. A 'particle' comes to have a location for the first time, for example, when it strikes a detector-screen. 'The quantum weirdness lies in the realization that as long as you are not actually detecting an electron, its behaviour is that of a wave of probability. The moment you look at an electron, it is a particle. But as soon as you are not looking at it, it behaves like a wave again.' (Pagels, 1983, p. 144)

So, something is not both of two seemingly contradictory things. It is one thing under a certain condition, and another thing under another condition. However, the quite peculiar condition under which it is, say, a particle with position, is a condition having to do with its being *investigated*—indeed, on some views, a condition having to do necessarily with our conscious activity. This interpretation, even among those who defend it, as indicated by the quotation above, is typically spoken of as crazy, weird, bizarre, involving a fusion of subject and object, marking 'the end of ordinary objectivity', bringing about the end of reality as ordinarily conceived, somehow not comprehensible, and so on—all for good reason. The central idea is related to but distinct from, and well beyond, the central proposition with respect to the Uncertainty Principle, having to do with indeterminacy of either position or momentum. (p. 313)

The Copenhagen Interpretation has been extensively discussed and disputed, but perhaps remains to the fore, if not dominant. What is to be said of it here? It seems mistaken to suppose, although often enough it is so supposed in one way or another, that what is claimed is repugnant to common sense, classical mechanics, and plain logic. What would be repugnant would be a claim to the effect that a particle in *an or the ordinary sense* changes its nature depending on our observations. There is a better conclusion, which indeed seems unavoidable.

It is that the 'particles' of which we have been speaking are *not* particles in anything at all like the classical or any ordinary sense of the word. To say the least, it is obscure what they are. It is the case with them that they come into existence only when a measurement of them is made, and that all that can be said of them before then is that there is a probability that they will be in a certain place *if* a certain measurement is then made. The important obscurity is not what sort of thing they are in the physical realm, but whether they in themselves are unquestionably in the physical realm at all. Things that are unquestionably in it are not, so to speak, *observed into it, investigated into it*, or *thought into it*.

The crucial question, given our own concerns, is this: *Are what are here called particles such that they are in fact in the category of those things which* can—*logically possibly or conceivably—enter into causal connections? That is, are what are called particles in fact within the physical realm as defined (pp. 87 ff.), which is the realm of all causes and effects and nomic correlates?* Not all things are in this realm, certainly. The positive integers, say, are not effects of causal circumstances. Nor are propositions, pure locations, times, and a good deal else. Of course this is no threat whatever to a general doctrine of determinism. To say the least, there is the possibility of arguing that determinism is quite untouched by the Copenhagen Interpretation of Quantum Theory, for the clear reason that the referents if any of such expressions as 'particle' are far from certainly in the relevant category. (Cf. E. Nagel, 1979, pp. 293 f.)

Those who unreflectively think otherwise, it can be argued, have not given enough attention to the great gap between classical and Quantum Mechanics, with the latter interpreted in the Copenhagen way.

> . . . a historical context appears to be the main reason quantum mechanics is generally thought of as the sort of theory by which, if determinism were true, one ought to be able to predict the exact location of the flash on the screen. Quantum mechanics appears as a more inclusive substitute for Newtonian mechanics. According to the latter, the flash is definitely caused by the arrival at the screen of a billiard ball-like particle. . . . quantum mechanics may be quite a different type of theory than Newtonian mechanics. The latter may be just the kind of theory that ought to predict the exact flash position if determinism is true. But, perhaps . . . quantum mechanics is not at all the type of theory that ought to be able to provide the same prediction if determinism is true. Then, of course, the failure of quantum mechanics to give the prediction is no argument for indeterminism. (Workman, 1959, pp. 256–7)

More specifically, it is quite impossible to have any acquaintance with Quantum Theory and to avoid a certain truth. It is that a veritable plethora of mentioned possible referents of the relevant terms of the theory, the terms either of the formalism or interpretations, are *not* clearly items which logically possibly could enter into causal relations, or, so to speak, the particular causal relations whose existence is doubted. The plethora of mentioned possible referents includes:

> observer-dependent facts, subjective ideas, contents of our consciousness of reality, epistemological concepts, ideal concepts, propositions, probabilities, possibilities, features of a calculation, mathematical objects or devices, statistical phenomena, measures and measurements, abstract particles, probability waves, waves in abstract mathematical space, waves of no real physical existence, abstract constructs of the imagination, theoretical entities without empirical reality, objects to which standard two-value logics do not apply.

No attempt can be made here to say more of this collection, and, for our purposes none is needed. What is at the very least the obscurity of the referents of the terms of Quantum Theory provides a premiss for the conclusion that it is at least doubtful that it does have an upshot for the theory of determinism of this book.

It is important that the premiss for this conclusion about Quantum Theory's possible irrelevance to determinism, whatever is said of the conclusion, is, as already suggested, far from being at odds with what has always been and now is said by physicists as distinct from philosophical and like commentators.

Einstein said of the fundamental equations of Quantum Theory that they are

> not to be interpreted as a mathematical description of how an event actually takes place in space and time, though, of course they have reference to such an event. Rather they are a mathematical description of what we can know about the system. They serve only to make statistical statements and predictions of the results of all measuresments which we can carry out upon the system. (1953, p. 259)

Schrödinger said that

> the object referred to by quantum mechanics . . . is not a material point in the old sense of the word. . . . it should neither be disputed nor passed over in tactful silence (as is done in certain quarters) that the concept of the material point undergoes a considerable change which as yet we fail thoroughly to understand. (1935, pp. 71–2)

Heisenberg took a similar view.

> There exists a body of exact mathematical laws, but these cannot be

interpreted as expressing simple relationships between objects existing in space and time. The observable predictions of this theory can be approximately described in such terms, but not uniquely. . . . (1930, p. 64)

. . . in the experiments about atomic events we have to do with things and facts, with phenomena that are just as real as any phenomena in daily life. But the atoms or the elementary particles are not as real; they form a world of potentialities or possibilities rather than one of things or facts. (Quoted in Pagels, 1983, p. 119)

According to Bohr,

the quantum postulate implies that any observation of atomic phenomena will involve an interaction with the agency of observation not to be neglected. Accordingly, an independent reality in the ordinary sense can neither be ascribed to the phenomena nor to the agencies of observation. (1934, p. 54)

Nor is it the case that physicists since the 1930s, including contemporary physicists, speak differently. We have it, for example, that Quantum Theory challenges

the basic belief, implicit in all science and indeed in almost the whole of human thinking, that there exists an objective reality, a reality that does not depend for its existence on being observed. . . . Many excellent books . . . convincingly demonstrate the power and success of the theory to make correct predictions of a wide range of observed phenomena. Normally . . . they omit to mention that, at its very heart, quantum mechanics is totally inexplicable. (Squires, 1986, pp. 2–3)

A recent exposition of Quantum Theory, partly in terms of three-valued logic (which adds *undecidable* to *true* and *false*) and partly in terms of another preferred interpretation, offers this judgement:

This logical interpretation of quantum mechanics can claim a measure of sense and reason. It is surely sensible to see the quantum peculiarities as products of weak description rather than as incomprehensible features of the world. It is surely more reasonable to suppose that our theory is inadequate than to argue that reality itself is bizarre . . . We should accept quantum mechanics as the most successful theory we have at present, while setting out to develop a new and better theory of reality. (Garden, 1983)

A recent exponent of the Copenhagen Interpretation is far from atypical in his equally pertinent declarations.

The Copenhagen interpretation . . . ended the classical idea of objectivity—the idea that the world has a definite state of existence independent of our observing it. (Pagels, 1983, p. 136)

Quantum reality is in part an observer-created reality. . . . with the quantum theory, human intention influences the structure of the physical world. This is radically different from the orientation of classical physics. (Pagels, 1983, p. 95)

The propositions asserted here are unaffected by what is rhetorically added to them, without explanation: 'The Copenhagen view is super-realistic—no fantasies or rationalizations about material reality are allowed.' (Pagels, 1983, p. 136)

To restate the main conclusion that can be drawn, along the latter terminological lines, it is at least arguable that Quantum Theory in the Copenhagen Interpretation does not unseat determinism since nomic connection is a matter of *reality*, the realm of so-called fantasies or rationalizations, and not 'super-reality', whatever that may be. It is no surprise, again, that a philosopher of science wholly committed to indeterminism, on various grounds, struggles against the Copenhagen Interpretation, 'subjectivism' as distinct from 'realism'. (Popper, 1982a) He is aware that the interpretation is no ally of indeterminism. The struggle is less than conclusive. (O'Hear, 1985)

There are other interpretations than the Copenhagen Interpretation. (Sudbery, 1986; Squires, 1986) They can be said to raise both similar and different difficulties when used as arguments against such a theory as the determinism of this book. A number of interpretations, for example, convert the fundamental propositions of Quantum Theory into propositions not about individual objects at all but about the universe as a whole. They cannot be considered here. Let us instead turn to something immediately relevant to such theories as the theory of this book.

5.6.3 Hidden Variables

It is inevitable that the idea will come to mind that the Quantum Theory as it is usually taken does not tell the whole story. What must come to mind is that there must be a possibility of a fully deterministic theory, different from and underlying Quantum Theory, which covers the relevant domain of facts. The domain is that of a deep level of physical reality. More particularly, there must be the possibility of a theory which covers all things in this domain which logically *can* be a matter of nomic connection, and *does* give them as being subject to nomic connection.

It is well known, and taken as something to be explained away by orthodox defenders of the Quantum Theory, that the physicist established as the greatest of the century held this view—as, by the way, did Schrödinger and Planck. Einstein was committed to and persisted in the view that the Quantum Theory is incomplete. Its indeterminism, he wrote, 'is so very contrary to my scientific instinct that I cannot forgo the search for a more complete conception.' (1950,

p. 89) It might be said that given the nature of science, which is very far indeed from excluding all appeals to authority, the fact of Einstein's conviction is a considerable one.

It hardly needs adding that it is not persuasive for a physicist or anyone else to offer *psychoanalytic* speculations about Einstein's supposed error in terms of his boyhood, speculations ludicrously irrelevant to science and to questions of truth and argument generally. (Pagels, 1983, pp. 20–23) Neither is it persuasive to maintain that despite *all* his writings and research to the contrary, Einstein adopted indeterminism at the end of his life. This is maintained (Popper, 1982b, p. 2) on the sole ground, if it can be called such, that a physicist who had conversations with him wrote in a letter that 'Einstein does not consider the concept of "determinism" to be as fundamental as it is frequently held to be . . . .'

It was for a number of decades maintained by many that the mathematician John von Neumann had provided a proof that there cannot be a determinism of 'hidden variables' or 'hidden parameters' underlying Quantum Theory. (1955, Ch.4) It is now widely accepted that the best that can be said for von Neumann's proof is that a deterministic theory would in part be logically incompatible with Quantum Theory. The proof has among its premisses what are taken to be indeterministic propositions of Quantum Theory.

There are now various deterministic hidden-variable theories—not all hidden-variable theories are deterministic. (Belinfante, 1973) One well-known deterministic alternative to Quantum Theory, or a preliminary form of such an alternative, was owed in the first instance to Louis de Broglie, another founder of Quantum Theory, and has been developed by David Bohm. It proceeds from the assumption that phenomena characterized at a certain level—that of Quantum Theory as it is now—can be regarded as indeterministic, but that this is consistent with their in fact being necessitated. The latter, of course, is a matter of the further theory.

The situation is taken to be of a kind familiar in a number of contexts. Insurance companies operate by way of statistical laws, which predict with sufficient accuracy the number of people who will die within given groups in a given time. The laws allow for no prediction as to a particular person, but they are consistent with what also exist, standard nomic explanations of individual deaths. Within physical science, there is the well-known counterpart of Brownian motion, the apparently random motion of spores and smoke particles subject to statistical law, but also to non-statistical law. It is statistically related to some things, but non-statistically related to others.

> . . . it seems evident that, at least on the face of the question, we ought to be free to consider the hypothesis that results of individual quantum-mechanical measurements are determined by a multitude of new kinds of factors, outside the context of what can enter into quantum theory. These factors would be represented mathematically by a further set of variables, describing the states of new kinds of entities existing in a deeper, sub-quantum-mechanical level and obeying qualitatively new types of individual laws. Such entities and their laws would then constitute a new side of nature, a side that is, for the present, 'hidden'. But then the atoms, first postulated to explain Brownian motion and large-scale regularities, were also originally hidden in a similar way. . . . (Bohm, 1980, p. 68)

The Bohm theory in fact gives a mathematical specification of possible new factors, sub-quantum-mechanical entities. Such a theory may to a very considerable extent give the same predictions as Quantum Theory with respect to observable results. It is a matter of considerable importance that to say, summarily, as it was above, that Quantum Theory *works*, is *not* to say that an enlarged and deterministic theory cannot do so. At the same time, this is consistent with something else, that an enlarged theory can to some extent diverge in its predictions from Quantum Theory, and so be tested against it. It is possible that a new theory will be superior to the present theory in certain respects. There is the possibility of experimental confirmation of a new theory over the present one.

Certainly there are objections to the possibility of a fully developed new theory. It is remarkable, however, that the principal objection has often been that it would not be Quantum Theory as it is, that it would in part be inconsistent with Quantum Theory. That in itself, as is plain, falls short of being an argument. Charitably regarded, it is tantamount to the conclusion for which the objection is to be an argument—i.e. the conclusion that Quantum Theory is the true last word on physical reality. (Feyerabend, 1962, p. 48f) There is also more recent objection owed to the physicist who defeated von Neumann's disproof of hidden variables. (Bell, 1966, 1971). The objection has not precluded a deterministic suggestion by the physicist in question (Bell, 1984) but, at the moment I write, it is embraced by some as conclusive—as was von Neumann's supposed disproof. The strength of the objection, one cannot but feel, is not yet settled. To succeed, it must do the very unlikely: prove a universal negative proposition—in this case that there are no causal circumstances for certain events.

Certainly a great majority of physicists discount the possibility that a hidden-variable theory will be confirmed. It is said by many, and allowed by Bohm, that his theory involves strange laws. It may be, despite the strangeness of Quantum Theory, that this will stand in the way of acceptance. Still, contemporary physics can properly be described as something less than stable. If there is something approximate to an orthodoxy, it is far from being untroubled. There has recently, perhaps, been a greater readiness to consider hidden-variable theories, which have recommendations to other than determinists. (Squires, 1986, Ch. 5) As tends to be forgotten by expositors in certain contexts, there is a general agreement that contemporary physics faces considerable problems. If physicists stand more or less together on central doctrine, their science is far indeed from being unproblematic or even a temporarily settled system.

There are other things to be said about the possible emergence of a new deterministic physics. More than the propositions and facts of physics itself are to be considered. One thing, already noted, is that the history of physics has before now seen the undermining of an orthodoxy *more* massive than that of Quantum Theory. This was, of course, the undermining of classical mechanics with respect to the microworld. More generally, as noted above, there is the unavoidable lesson of the history of science. If there can be dispute about its being a history of rising and falling *paradigms*, as has been forcefully argued (Kuhn, 1962), there can be no dispute that it consists in the rise and fall of theories.

A final fact of relevance here, which appears to need stating, is that the supposition of a new deterministic physics is not the supposition of a revival of classical mechanics, a turning back of the clock. Far too much has happened to make it conceivable that anything of the sort could occur. A new deterministic physics would no more be a revival in this sense than classical mechanics was itself a revival of Greek atomism. Consistent with that, of course, is a new theory's being like classical mechanics in being deterministic. There is a natural tendency in the writings of scientists to suppose that determinism is identical with classical mechanics, but that is far indeed from being the case.

5.6.4 Meaningless or Non-Existent 'Position'

Let us now give some consideration to the Uncertainty Principle. As remarked earlier, what is maintained is not merely that we must be uncertain as to the position of a particle if we ascertain its momentum, and vice versa. Rather, what is maintained is that if the momentum of a 'particle' is ascertained, (i) it *lacks* a position, or (ii) it is meaningless to speak of its having a position.

To look first at proposition (ii), surprisingly persistent among expositors of Quantum Theory, it is clearly owed to the attitude or theory of *operationalism*—which also enters into what is said of the

Wave Equations. This particular excess of empiricism, most defended in the philosophy of science of the 1930s, and in particular by the physicist P. W. Bridgman (1927, 1936) is typified by the claim that the statement that a liquid is acid means *nothing more* than that if a strip of blue litmus paper is put into it, the paper turns red. A statement as to the existence of a particle means *no more* than something about a visible line in a Wilson cloud chamber. The attitude or theory in question is akin to a number of past empiricisms within philosophy, notably Logical Positivism and radical behaviourism. Despite their good intentions, all of these empiricisms have for good reason been abandoned. Operationalism has the consequence, among others, that if we have two ways of measuring length, then the fact of the matter is that we have two different concepts of length. So with electrical charge. This cannot be correct.

> . . . the correct view is that when the physicist talks about electrical charge, he is talking quite simply about a certain magnitude that we can distinguish from others partly by its 'formal' properties (e.g. that it has both positive and negative values, whereas mass has only positive values), partly by the structure of the system of laws this magnitude obeys (as far as we can presently tell) and partly by its *effects*. All attempts to 'translate' *literally* statements about . . . electrical charge into statements about so-called observables (meter readings) have been dismal failures. . . . (Putnam, 1965, p. 76; cf. Hempel, 1966, p. 88f)

Of course if ascertaining the momentum of a particle makes it impossible to have an 'observation' of its position, and *if* operationalism were true, such that any statement about position would necessarily be a statement of the impossible 'observation', then it *would* be impossible to make any statement about position. Operationalism, however, is evidently not true, or even advantageous in inquiry.

The other proposition (i) is that if we ascertain the momentum of a 'particle' it *lacks* a position, and vice versa. It is not made wholly clear what proposition this is. Several things *are* clear, however. If 'position' is something whose existence is actually *dependent* on ascertaining it rather than ascertaining 'momentum', then 'position' is not position, and 'position' does not contribute to making whatever has it into something physical. The term 'position' is not being used in the sense it has in the determinist proposition that the position of a thing is necessitated by a causal circumstance. The point is essentially the same as one made before, in connection with the Copenhagen Interpretation.

There is a further point. It is anyway obscure how what is said here of the Uncertainty Principle can be thought to issue in an argument against determinism. The implied argument we are considering seems to reduce to this: since there is something which is not determined, determinism is false. What is that thing? Well, when momentum of a particle is ascertained, it is 'position'. But the fundamental proposition on the given supposition—the supposition that momentum is ascertained—is the proposition that *there is no position of the particle*. It must therefore appear, on further reflection, that the implied argument against determinism reduces to this: since something non-existent is not determined, determinism is false. It is difficult to resist the impression, therefore, that the supposed refutation of determinism depends on the further remarkable and in fact absurd premiss that determinism is the view that the position of a particle, *whether or not there exists such a thing*, is the effect of a causal circumstance. Something the same is to be said of proposition (ii) as against proposition (i).

5.6.5 Micro-Indeterminism and Macro-Determinism

An indeterminism found in Quantum Theory, as we know, is used in arguments for indeterminist theories of the mind. (3.4 – 3.9) This very necessarily involves the supposition that indeterminism at the subatomic level, or micro-indeterminism, can and does have repercussions above the subatomic level. That is, it gives rise to macro-indeterminism. In particular, micro-indeterminism gives rise to a neural indeterminism. We shall soon consider the attempt to link Quantum Theory and indeterminist theories of the mind. For now, let us simply ask what light is thrown on the suppositions of micro-indeterminism and of such macro-determinisms as the theory of this book by considering the relation between them.

It is quite often claimed or conjectured that any micro-indeterminism would be irrelevant to macro-determinism. The claim or conjecture, as noted, enters into certain theories of the mind. (pp. 2, 9) It has to do, in part, with the scale of magnitude of micro-events. The basic parameter of Quantum Theory (Planck's Constant) has the value 0·000 000 000 000 000 000 000 000 001 in terms of grams, centimetres and seconds. That is, the conditional proposition that if micro-indeterminism is the case, so too is macro-indeterminism, is doubted or denied. Or, more particularly, it is denied that any micro-indeterminism would be amplified into neural indeterminism. Thus the supposition noted earlier, that the firing of a neuron would not have occurred save for an undetermined micro-event (p. 180), is doubted or denied. 'The idea . . . of an amplifying mechanism which would allow the processes in the microphysical domain to influence the macrophysical organic event is quite conceivable. But we know of no such phenomena in the behaviour of organisms. . . .' (Mainz, 1955, p. 74; cf. Smart, 1963, p. 123; Mackay, 1964, pp. 398 f.; Rosenblueth, 1970, Ch. 8; Dennett, 1984, pp. 76–7, 136)

The conditional proposition, that if micro-indeterminism is the case so too is macro-indeterminism, is often rejected independently of consideration of the brain and mind, notably by those who are inclined to micro-indeterminism but unpersuaded that there is macro-indeterminism. If a macro-event occurs as an effect of a truly indeterministic micro-event, and hence the macro-event is also regarded as a matter of true chance, then there is the upshot that the macro-event cannot be regarded as the effect of a causal circumstance in the macroworld—which upshot, as remarked, may be taken as unacceptable. macro-events are taken to be predictable in principle on the basis of causation in the macroworld.

It is sometimes said in this connection, first, with respect to the conditional proposition, that an undetermined micro-event may be one of a specific and finite set of possible micro-events, and, further, that each member of the set would have had the same effect in the macroworld. In which case, the macro-event was fixed in so far as the microworld is concerned, despite its being the effect of a chance event. There is no reason, then, or so the argument goes, why the given macro-event should not also be an effect of a macroworld causal circumstance.

Secondly, the supposition that micro-indeterminism *would* produce macro-indeterminism is resisted by finding in it what is taken to be a mistaken assumption: that the elements or constituents of any complex are somehow fundamental, somehow prior to the complex. This may be taken to be the assumption that all 'true properties' of the complex must be properties of its elements. Hence, if a macro-complex has micro-elements, and the latter are indeterministic, the macro-complex must be indeterministic in so far as its 'true properties' are concerned. However, it is said, the fundamental metaphysical assumption, about what might be called the greater reality of elements as against complexes, cannot be defended.

Thirdly and differently, it is common to assert that macroworld effects of indeterminist micro-events are so small that they are in fact indetectable. There is the accepted fact that any experimental findings in any connection must be allowed to be subject to the possibility of a certain range of error. Any effects in the macroworld of an undetermined micro-event would be buried within this range.

It is noticeable that none of these three considerations satisfactorily defeats the conditional proposition that if micro-indeterminism exists, it does necessarily issue in macro-indeterminism. With respect to the first consideration, whatever else is to be said of it, it obviously is insufficiently wide. It cannot be accepted, in the absence of some considerable proof, that with respect to *every* possible set of micro-events somehow defined, each member of the set would have had the same macro-effect. As for the second consideration, it seems perfectly possible to separate the conditional proposition from the metaphysical assumption about the greater reality of elements as against complexes. Finally, with respect to the third consideration, it is in fact consistent with the conditional proposition. That is, the consideration allows the possibility of macro-effects, if undetected ones, of undetermined micro-events. There is a further point. Why should the undetected macroworld chance events not themselves have detected macro-effects? These too, given their supposed earlier cause, would be chance events.

None of this, to repeat, establishes that the conditional proposition is false. Is there more hope in a quite different view of the situation? All the above reflections tacitly rule out overdetermination. They assume, of a given macro-event, that there is no possibility of its being caused by an undetermined micro-event, and also caused by a certain causal sequence wholly within the macroworld. *Must* this assumption be made? Is it possible to feel, here if nowhere else, that a general overdetermination is somehow tolerable?

We shall not attempt to establish the truth or falsehood of the conditional proposition. Let us rather conclude what we safely can about it. We can conclude, first, that *if* it is false, the conclusion follows that any micro-indeterminism is irrelevant to our concerns. In particular, any micro-indeterminism is consistent with the Hypothesis on the Causation of Psychoneural Pairs and the other two hypotheses which make up the determinism that has been expounded.

More needs to be said of consequences of the second supposition, that the conditional proposition is true. It is, of course, a main part of the stock in trade of the several neuroscientific defenders of indeterminist pictures of the mind, used in defence of those pictures. It is also accepted by some Quantum physicists not at all or not much concerned with the mind and freedom. In the following illustrative passage, the image of 'the God that plays dice' is derived from Einstein's well-known summation of his own determinism: that God does *not* play dice.

Could the God that plays dice trigger a nuclear holocaust by a random error in a military computer? . . . the probability for such an event can be made extremely small. But this example raises the question of whether the quantum weirdness of the microscopic world can creep into our macroscopic world and influence us. Can quantum indeterminacy affect our lives? The answer is yes . . . . (Pagels, 1983, pp. 148–9)

The conditional proposition, if considerations against it can be contemplated, is in fact difficult to resist. However, to take it to be true, and to accept micro-indeterminism, is surely to have to deal with a certain proposition. It is that there may be no significant and uncontroversial evidence of macro-indeterminism, but that if micro-indeterminism is the case, and the conditional proposition is true, there *should* be evidence for it in a known fact of macro-indeterminism. The conviction that the macroworld is instead deterministic is a strong one. What may be offered against it is speculation in the entirely disputed area of human freedom, the area of human choices and decisions. Or, what may be offered against it is uncertain and unpersuasive speculation about, say, what may happen in a computer.

Surely, to repeat, if micro-indeterminism were true, we should by now have had solid evidence, if only a little bit? Over the course of history, even given propositions about improbability, *some* solid evidence should have accumulated? If all reality at a fundamental level is indeterministic, should we by now not have developed an acceptance of rare physical miracles, suspensions of nomic connection in the macroworld? If the particles of the human body are indeterministic, along with the particles of spoons, then, even if there is an immense probability that nothing surprising will happen, is it not arguable that by now there ought to have been occasions when all or enough of the particles of a body or spoon acted indeterministcally in concert, such that there occurred a levitation, or anyway a little spontaneous leap? (Cf. Weatherford, 1982, p. 256) Most of us are disinclined to suppose that there have been any. It may be said that the relevant calculations, into which we cannot enter, go against the objection. Still, there is some relevant speculation to which we now turn.

5.6.6 Radioactive Decay

Relatively simple claims as to micro-indeterminism are made often enough, usually having to do with radioactive decay. In such decay, an atomic nucleus is said to break down and emit a particle. The emission, it is flatly said, is not necessitated. That proposition, of course, is not a matter of simple observation, as seems sometimes almost to be suggested. Rather, it is something based entirely on what we have previously considered, the Copenhagen and other interpretations of Quantum Theory. Thus it is in no way independent of the problems we have been considering, but involves all of them. Consider the following passage.

> . . . a difficulty arises for determinism in the explanation of individual radioactive events. If the causes of decay are 'internal' to the nucleus concerned and if all nuclei of a given kind are precisely identical, then there can be no causal explanation of why one nucleus should decay now, whilst another one decays in ten years' time—deterministic laws must be invariable in their operation. Of course, one could try to get round this difficulty by denying the complete identity of all the nuclei concerned. However, this assumption is not to be discarded lightly: indeed it seems necessary for any workable atomic theory. If 'atoms' or 'atomic constituents' all had to be given 'personal profiles' then atomic physics would become a completely hopeless enterprise. Furthermore, one would have to explain why the individual characteristics of nuclei were always imprinted in the same proportions in a given type of material no matter when or how it was created; the 'randomness' would not be eliminated simply by pushing it one stage further back.
>
> The reason why the example of radioactive decay seems to provide a particularly strong argument for 'real indeterminacy' is that it does not depend on the interaction of an observer with the system being observed. (Powers, 1982, p. 143)

Part of a response to this is that the proposition that there *do* occur unnecessitated events of decay is derived from a premiss which is nothing other than Quantum Theory, whose interpretation *does* involve the great obscurities pertaining to an observer. They are not at all clarified by the truth that if the conclusion is true, and hence there exists the fact of unnecessitated events, the fact in question is not itself an observer-dependent fact. Problems having to do with an observer arise *before* we get to the conclusion. (Cf. Einstein, 1953, pp. 259–60.) Another part of a response to the passage is that it is impossible to avoid the conviction that the given argument for indeterminism, which rests on the proposition that all nuclei of a certain kind are *absolutely identical* in their constituents and operation, rests on no more than a greatly convenient but far from coercive assumption. Why should they be absolutely identical? The assumption is not made true by the consideration that physics depends on it, however essential the assumption may be taken to be. In fact the assumption is in effect denied by hidden-variable theories, and so it is at least disputable whether physics depends on it. As for the consideration in the last and to me unclear sentence of the first paragraph of the passage, it seems not to help. That a certain question arises seems to be no sufficient reason for thinking that it needs to be answered in terms of randomness.

5.6.7 Quantum Theory and Indeterminist Theories of the Mind

Finally, there is the question of whether an indeterminism found in Quantum Theory is in fact a friend to indeterminist theories of the mind. So far this question has not been examined, and it may have been assumed that the given indeterminist theories and the determinist theory of this book are not at all on a par as far as evidence from Quantum Theory is concerned. Indeed they are not, but for a reason other than may have been anticipated. The determinist theory does *better*. Quantum Theory is in a way *more* of a problem to the indeterminist theories. To introduce Quantum Theory into the indeterminist pictures of the mind, as the physicist A. S. Eddington may have seen (1939), is to produce something close enough to self-contradiction, as can be made clear.

Quantum Theory, if it is to suit indeterminist pictures of the mind, must be taken really to produce the proposition we have been doubting, that there are chance events. That is, Quantum Theory must produce the proposition that there are events lacking causal circumstances, or events which are effects of suitable causal sequences at least one of whose initial elements lacks a causal circumstance. What Quantum Theory must be taken to assert of a synaptic event, say, is that it is such a chance event. (Cf. 5.6.5.)

Except for a brief moment in recent philosophical history, it has not been supposed that a particular desired freedom and responsibility are given to us simply by such a proposition about chance. In fact, as we know, chance by itself is inimical to any freedom and responsibility. (pp. 184 f.) Therefore, as we have seen, indeterminist theories of the mind have as their fundamental proposition that an originator somehow *controls* decisions and actions. More particularly, as in the theory of the philosophical–neurophysiological partnership (Popper and Eccles, 1977), it *controls* relevant neural events. It brings it about, for example, with respect to a finite set of possible synaptic events, that one in particular becomes actual. The ideas of control, as we have seen, are thin and obscure. However, if we are to take the indeterminist pictures seriously, we must indeed suppose that there does occur something which is properly called the mental control or determination of synaptic events.

Proponents of indeterminist theories of the mind, therefore, when they look to Quantum Theory for help, tacitly take the view that it is a merely partial or incomplete theory. They take it to assert of a synaptic event, roughly, that merely in so far as *neural* antecedents are concerned, the event is one of chance. Proponents of the indeterminist theories take Quantum Theory as leaving it perfectly open whether or not the synaptic event is subject to another kind of determination, non-neural determination. Their ideas of control and determination are obscure, but their attitude to Quantum Theory, taking that attitude by itself, would be consistent with there being absolutely standard causal sequences of a *mental* kind for the events which Quantum Theory is regarded as making undetermined.

That attitude, certainly, is false to what is taken as the burden and purport of Quantum Theory by those, say, who take it to defeat a universal or LaPlacean determinism. Exposition and discussion of Quantum Theory in relation to universal determinism is hardly ever concerned with mentality or freedom. The subjects do not come up. But it is in fact bizarre to suppose that such expositions of Quantum Theory allow for the possibility that reality has within it a kind of unmentioned *determination* of precisely some of those events which are expounded as being a matter of chance.

One simple way of putting the point is that indeterminist theories of the mind, from the point of view of Quantum Theory as commonly interpreted, are precisely *hidden-variable* theories. The indeterminist theories assert that what may be taken to be a purely random event, from the point of view of the understanding of Quantum Theory most favourable to them, is in fact in some way absolutely and precisely determined by the hidden variable which is an originator. It would be entirely insufficient, to repeat, to rest on the fact that common expositions of Quantum Theory do not contain explicit denials of *mental* hidden variables. Such denials are missing since the question is not raised within physics and is alien to it. There can be no doubt that Quantum Theory is not commonly regarded as incomplete in the way it would be incomplete if it allowed the possibility of mental determination of quantum events.

Indeterminist pictures of the mind and our determinist picture are at first sight on a par with respect to Quantum Theory. It may be taken to provide an argument against each. At bottom, it is the same argument: that there are chance events. Such events are inconsistent both with determinism and with freedom and responsibility. It is possible, in order to defend the determinist picture of the mind, to advance certain considerations against the proposition that Quantum Theory establishes the existence of chance events, as we have seen. A defender of indeterminist pictures of the mind is in a less good position, in fact a hopeless one. In effect he must embrace and deny a certain interpretation of Quantum Theory at the same time. He embraces it to escape physical determination, and denies it by introducing mental determination.

Much was said earlier in this book against the indeterminist theories of the mind, and what has now been said completes the criticism. Such theories are none the less of a certain great importance. They express certain of our fundamental attitudes. They will, therefore, have another kind of attention in what follows. With respect to questions of clarity and truth, however, what is to be finally concluded about them is not only that they are thin and obscure, typically bound up with such confusions as teleology, and overwhelmed by neuroscience, but that they cannot get help from what in this century they have taken to be their greatest ally.

5.7 CONCLUSION, PARTLY TENTATIVE

What is evidentially most relevant to the question of the determinist theory of mind that has been expounded, to repeat, is neuroscience, a great and growing body of knowledge, theory, and informed speculation. The conclusion drawn from it was that it at least very strongly supports the determinist theory. To the idea that fundamentals of present-day neuroscience must be as likely to suffer falsification as Quantum Theory in the course of history, there is the reply among others that neuroscience stands a good deal closer to *data*, to the facts, than does Quantum Theory. It is, because it is less speculative and greatly less far from the sensory evidence, less vulnerable. Not all theories are equally theoretical—as seems sometimes to be implied in the outlook earlier named theorism. (p. 92) The account that has been given of neuroscience in this book will soon be outmoded. The required revision in the future, further, will not simply be a revision of details. To allow that plain truth, however, is not to allow that the fundamentals of present-day neuroscience which are of relevance to us are in any serious way vulnerable.

Of the other bodies of relevant evidence, a second consists in all other empirical sciences except for Quantum Theory. That they provide inductive evidence for a universal theory of determinism, having to do with all events without exception, and hence provide lateral support (pp. 297, 301) for our own theory, is not properly disputable. As will be clear, the universal theory of determinism is here taken, as it was earlier, to be a factual proposition, resting on evidence, as distinct from any sort of mere maxim or advice to the effect that we should always persist in looking for causal circumstances. The factual proposition is as good as irrefutable, since failing to find a causal circumstance for an event is always compatible with having missed it. Failing to have a causal theory for an event, likewise, is compatible with there being one. It does not follow that a general determinism is 'meaningless' or 'without content' or whatever. It *is* necessary to know what *would* refute it, as indeed we do: an event for which there exists no causal circumstance. (Cf. G. J. Warnock, 1961.)

The third body of evidence, which it would be irrational to disdain, is our extra-scientific experience of physical reality. It would be misleading to describe this vast experience as *pre*-scientific, since it evidently in ways constrains and guides ongoing science. Although the claim can be overstated or too simply stated (Boyle *et al.*, 1976, pp. 65, 66), it is clear enough that science is not wholly self-recommending. It is not the case that it can be argued for only circularly, by reference to itself, or only by taking one part as basis for the justification of another. Rather, there are external criteria, which criteria are given by our common extra-scientific experience, and which science typically meets. It is absurd to say, intending a certain equivalence, that 'science is another name for knowledge'. (Pagels, 1983, p. 348) Our extra-scientific experience is but weakly characterized as giving rise to 'common sense', so weakly that the characterization is useless. Leaving aside large questions here, there is the conclusion to be drawn that our extra-scientific experience of the physical world gives support to universal determinism and hence also to our own theory. Such a conclusion is no new one (Sidgwick, 1966 (1874), pp. 62–3) and remains unquestioned.

The fourth and remaining body of evidence is Quantum Theory. For the reasons lately given, it is not consistent with indeterminist pictures of the mind. That is not now our main question, however. I have argued, in sum, that Quantum Theory falls far short of disproving the determinist theory of this book. Again in sum, this is owed to what is certain, which is obscurity and doubt with respect to interpretations of Quantum Theory. Do those who enter the controversy about determinism waving the banner of Quantum Theory do greatly more than wave a banner?

It needs to be allowed, none the less, that the previous section of this book, of all its sections, is the one in which I and any reader must have least confidence. Not everything said in it satisfies me, and its subject-matter, of all the subject-matters of this book, is the one of which I have least understanding. It is a hard question, for example, despite all that has been said, what weight to give to the views of physicists on the interpretation of the formalism of Quantum Theory. The endeavour of interpretation, as remarked already, is in large part philosophical in nature. Still, it must be that some significant weight is to be given to the views of those who have a true grip of the formalism. I am not one, and have depended on secondary sources. More generally, it is undeniable that supporters of determinism after the rise of Quantum

Theory cannot be so sanguine as their predecessors.

The conclusion that this book's theory of determinism is true, therefore, is not to be drawn here. On the assumption that the best that can ever be said for any theory is that it is, say, *confirmed*, the conclusion is not to be drawn that the theory is confirmed.

We have not yet considered all that can be said against the theory. In the coming chapter we shall look at general objections having to do mostly with prediction and knowledge. These are philosophers' attempts to disprove or undermine determinism. We therefore cannot yet draw a final conclusion about the possibility that the theory is false. We *can* draw an interim conclusion, based on what we have considered so far.

On that basis, certainly, the conclusion is not to be drawn that the theory is false, or that it lacks very strong support, or that a general indeterminism is confirmed or better supported—let alone that indeterminist theories of the mind are tolerable. They are not. Again, given what we have, it is no more than a contentious misuse of language to describe this book's theory as a matter of 'faith'.

Hence, given what we have so far, the question of the consequences of our own and like theories of determinism for our lives, the question which has most exercised philosophers, is a question which arises from what may be false but may well be true, or indeed is much more likely to be true. It is very far from being a question made idle by arising from what is known to be false. In fact, of course, the question of consequences would not be made idle even by the falsehood of our own and like determinist theories. The question is raised by less strict theories of our existence, such as the near-determinism or naturalism mentioned earlier. (pp. 2, 9, 471) Dividing theories of our lives into three categories—determinist, near-determinist, and indeterminist—the first two kinds raise the problem of consequences. They are of differing likelihood but both are of vastly greater likelihood than the third.

6

Prediction and Knowledge

6.1 CONDITIONAL PREDICTABILITY

If the determinism we have been considering is true, then the future life of any person—each mental event and each action—is conditionally predictable, predictable in theory or in principle. The distinction between *conditional predictability* and simple *predictability* is not sharp, but must be drawn. If something is predictable, predictable in fact or in practice, it can be predicted on the basis of knowledge that exists, which is to say knowledge actually had by someone or at any rate recorded. If something is conditionally predictable, it could be predicted *if* certain knowledge existed, the having or recording of which knowledge is logically or conceptually possible, although perhaps factually or nomically impossible. (Cf. Ayer, 1980, 1984c.)

That determinism makes my future conditionally predictable comes more exactly to this, that there is a particular logical or conceptual possibility of knowledge of my future. That is, there does not exist what would be a logical or conceptual barrier to such knowledge. The logical possibility consists, exactly, in the existence as against the absence of certain facts: the facts of there being determinate things and nomic connections between them. Reality is determinate, which is to say, for our purposes, and at bottom, that it can be truly characterized in terms of individual properties. Reality is also nomic, which is to say there are connections between these items, statable by certain conditional statements.

The proposition of conditional predictability, which certainly can be said to follow from all doctrines of true determinism, has regularly been used by objectors to those doctrines. It has been used on other occasions than those—which are not rare enough—when the given conditional predictability is somehow elided with simple predictability, or when the distinction is not adequately made, or when other and irrelevant logical barriers are supposed.

In fact, the proposition of conditional predictability barely counts as an objection, since it is barely distinct from merely a denial of the determinism, however declamatory. What we have is very nearly a begging of the question at issue. This is so since the given conditional predictability is, so to speak, little more than an epistemological rendering of the ontological proposition of determinism. The denial of conditional predictability is no more independent of a denial of determinism than (i) a certain proposition that knowledge of the existence of a second daughter of mine is theoretically impossible is independent of (ii) the proposition that I have no second daughter. If it is proper enough to call conditional predictability something that follows from determinism, it is a peculiarly immediate consequence or entailment. (Cf. Berofsky, 1971, Chs. 1, 2.)

Those who may seem to introduce no more than the proposition of conditional predictability against determinism do often intend another claim. It is that determinism allows for *a particular kind* of conditional predictability. What objectors often have in mind is that determinism allows for the unconditional predictability of mental events and actions on the basis of only *bodily and environmental* conditions. (Popper, 1982b, p. 28) If the term 'physical' is used otherwise than we have used it, so as not to include the mental, the objection is that determinism allows for the conditional predictability of mental events and actions on the basis of physical facts alone. But the determinism to which we have come *does* respect the axiom of the indispensability of the mental and hence does *not* have the given consequence. The conditional predictability which is a consequence of the second hypothesis necessarily involves psychoneural pairs.

Still, it most certainly needs to be allowed that we share a *resistance* to conditional predictability, a resistance which does not depend on mistakenly running it together with epiphenomenalism. What explains this resistance? A suitable answer is needed. A part of the answer has to do with a certain factual inconceivability as it can be called—roughly, the utter hopelessness of anyone's coming to have the requisite immense knowledge in order actually to make the predictions which determinism allows to be logically possible. This somehow issues in resistance to the conditional proposition that *if* someone actually had the requisite knowledge, future choices and decisions could be known. The factual inconceivability of knowing the truth of the antecedent spreads itelf to the conditional as a whole. But the truth of the conditional is of course independent of the factual inconceivability of the antecedent.

If this is one part of our resistance to the idea of conditional predictability, it is certainly not all of it. It is not what is most important.

The subject of determinism falls mainly into three parts: the question of its conceptual viability, the question of its truth, and the question of its human consequences, above all its consequences for our feelings about our own lives. So far in this inquiry we have been concerned only with the first two questions. The theory we have constitutes an answer to the first of them, and we have latterly been considering the question of its truth. The third matter has to do with the attitudinal acceptability or tolerability of the theory. It would be footling to suppose that its consequences for our feelings are wholly agreeable, and mistaken to suppose, as it often has been supposed, that any disagreeability has to do only with an undercutting of moral responsibility. The essential fact is that a true determinism allows but one future—there is no more than one future course of events that is a possibility. (James, 1909, p. 150) A true determinism is therefore properly described as at least unsettling.

I shall not open the large subject of the given consequences of our theory prematurely, and so our consideration of conditional predictability must for the time being be left incomplete. Let us merely note that there is every likelihood that a resistance to the human consequences of our theory is transferring itself to that picture of the mind itself, and in particular to its epistemological formulation in terms of conditional predictability. This is the main explanation of our resistance to conditional predictability.

That is not to allow that in this resistance we have any reason of fact or logic which refutes the given picture of the mind. Such a reason, if it exists and is to be used in argument, needs to be produced. Declamation is not enough. Several philosophers have taken themselves to have found such a thing, and their claims will be among those as to the falsehood or whatever of determinism which we shall consider in what follows in this chapter.

6.2 NEWCOMB PROBLEMS

Imagine that yesterday the Being, as he may be called, predicted a choice of yours today and acted on his prediction. Your choice involves two boxes. B_1 definitely contains $1,000 and B_2 contains either $1,000,000 or nothing—you cannot see which. You can choose either B_2 by itself, or both B_2 and B_1. If, yesterday, the Being predicted you would choose B_2 alone, he put $1,000,000 into that box. If he predicted you would choose to take both boxes, he put nothing into it. He has done this many thousands of times before and never been mistaken. Very many people have chosen B_2 alone, and got $1,000,000. Very many people have chosen B_2 together with B_1, and got only $1,000. No one has ever got $1,001,000. In making your choice you are fully aware of this history of the Being's enormous predictive ability. Nevertheless, you also know that the Being did whatever he did with B_2 yesterday. The $1,000,000 is now in B_2 or it is not.

The problem, of course, is which choice to make, given the assumption that you want to maximize your gain. Evidently there appears to be a very strong argument for choosing B_2 by itself, for that being your *best choice*. But there also appears to be a very strong argument for your best choice not being B_2 alone, but both B_2 and B_1. There is now $1,000,000 in B_2 or there is nothing. Either way, it seems, if you think about it, choosing both boxes gains you $1,000. We thus appear to have a contradiction. By one argument your best choice is B_2 alone. By another argument your best choice is both B_2 and B_1. The problem was devised by a physicist, William Newcomb, and first expounded in print by a philosopher. (Nozick, 1969)

It has been argued, by way of a further consideration, that the best choice is indeed both boxes, and that this entails the falsehood of determinism. (Schlesinger, 1974, 1976) What is entailed is that we must give up the idea that there could conceivably be such a one as the Being. More particularly, it is impossible that there could in theory or principle be such prediction of a person's choice; conditional predictability of choices, however, is something that is entailed by determinism; therefore determinism is disproved.

However, there are also further considerations which seem to strengthen the argument for taking only B_2. It must be concluded in the end, I think, a bit disappointingly, that there is no provable solution to the problem, given the specification of the situation that we have. (Cargile, 1975; Levi, 1975; Mackie, 1985a; cf. D. Locke, 1978) As will have been apparent, the choice situation is not exactly specified, but left mysterious in several ways. Thus there is no single determinate Newcomb Problem, but a number of *Newcomb Problems*, depending on what else is read into the situation. These various problems, it can reasonably be claimed, have different provable solutions.

One central mystery in the situation described has to do with whether we are really to suppose, as at first we are likely to, that the presence or absence of $1,000,000 in B_2 is 'independent' of the actual event of a player's choosing that box or both. Is it instead possible that a player's choosing both boxes *causes* the $1,000,000 that is in B_2 to disappear? Are we to contemplate that there is 'backwards causation', such that the player's choice today 'causes' the right prediction yesterday? (Dummett, 1954) Differently, on the ordinary assumption that the prediction was not such an 'effect', how are we to regard it? Did the Being look the player over yesterday neurally and mentally, with extraordinary insight and knowledge, and somehow infer what he would choose? Or was the Being 'seeing' into the future, 'seeing' the player making his choice, where that involves no inference? Was he perhaps 'seeing' into an undetermined future, 'seeing' an undetermined choice?

On the last of these assumptions, to mention only it, a successful argument for taking both boxes would undermine not determinism but a supposition as to the given 'seeing'. That must lead us to a certain thought: that it may be possible, as indeed it is, to think of an exactly or sufficiently specified Newcomb problem that unquestionably *is* of relevance to our theory of determinism. Is it possible thereby to disprove the theory?

The thought experiment must actually incorporate or include the theory if we are to have anything pertinent to it. What we must think of, then, is a Being who predicts certain choices on the basis of just such neural and mental facts, and nomic connections, as enter into our hypotheses. The Being will be akin to the Ultimate Neuroscientist, but more than that. To remember in particular the second hypothesis—which involves environmental causes—the Being will also possess complete knowledge of the impinging of environments on persons. We suppose too of such a being that he has by the given means achieved a record of perfect predictions of players' choices as between B_2 by itself and B_2 and B_1 together. Can we allow that he ever makes mistakes? No. If we do, the *theory* on which he acts will not be put in doubt by a prediction-failure. We need also to make explicit what was no doubt assumed earlier. It is that the Being is wholly inflexible. There is no question of his ever departing from his practice of putting $1,000,000 in B_2 if on the given basis he predicts that only it will be chosen, and putting nothing in it if he predicts that the other choice will be made.

What our total imagining comes to, then, is that our theory of determinism is true of the player, and, to put it one way, a certain determinism is also true of the prediction-making and the subsequent activity of the Being. That is, with respect to the Being, there is nomic connection between features of the player and environment, on the one hand, and, on the other, both the Being's prediction and his activity. In sum, what we have can be described in terms of two states of the subject and the environment, S_1 and S_2. If subject and environment are in state S_1, the nomic consequence is a choice of only B_2, and its containing $1,000,000. If subject and environment are in state S_2, the nomic consequence is a choice of both B_2 and B_1, and B_2's containing nothing. It is nomically impossible that he chooses only B_2 and gets nothing, and it is nomically impossible that he chooses both boxes and gets $1,001,000.

His *best choice*, then, must of course be B_2 by itself. Is there a contrary argument for choosing both boxes? Since either the money is there or it is not, can it be that the player's best choice is to take both boxes, with all that follows from that, and in particular the inference that there cannot be conditional predictability?

That could be so only if taking both boxes and B_2's containing $1,000,000 *was a possibility*, and taking only B_2 and its containing nothing *was a possibility*. But we have it that neither thing is a possibility. It cannot conceivably be that a best choice is aimed at a benefit which it is impossible to have. If *that* were so, we might as well have said, at any point in our reflections up until now, that a player's best choice was to take *all* the Being's reserves of money, thereby putting an end to his diversion. That, if it were possible, would be *the* way for a player to maximize his gain.

Given the Newcomb Problem that we have constructed, the player's best choice is B_2 by itself, and there is no argument whatever for B_2 and B_1 together. Thus there is no possibility of an argument leading to the questioning of the assumption that there could be prediction of the kind in question. There is no argument against our theory of determinism.

Might it be that to speak of a best choice at all under conditions of determinism is somehow impossible? (Cf. Mackie, 1985a, pp. 150–1.) Might it be that a best choice is a rational choice, and that rationality does not accord well with determinism? We shall be considering that later. A second and related thought is that the Newcomb Problem we have been considering involves the player having a fixed future. Is there then point in talk of *choice* at all? That matter is also one which we shall in effect be considering.

6.3 THE TRANSITION TOWARDS ACTION AND OUR COMMONPLACE SCHEME OF EXPLANATION

Suppose I have the thought that Caroline will again propose her plan tomorrow, and I also think that because of this or that emotion, weakness, resistance, or habit of mine, I will go along with it. Or, I think that I will do so because of other facts about myself, or, extraordinarily, because of neural and mental facts about myself taken in conjunction with the three hypotheses of our theory of determinism. According to an independent and evocative doctrine of the mind, what I then do, to use an essential metaphor, is to *step back*. (Hampshire, 1959, 1963, 1965, 1972a, 1972b) There is in every such case, on every such occasion, the possibility of this interrogative and contemplative stepping-back or recessiveness, 'the recessiveness of I'. There is also the *necessity* of this stepping back. (1965, p. 90) I can and must detach myself from the grounds of my prediction. I detach myself from my emotions and beliefs, my perceived interests and arguments, and indeed from features of my personality and my will. (1965, p. 90) This is the recessiveness of a self, as distinct from all that goes with it.

The question which arises, if I think that I will succumb to my own inclinations tomorrow and go along with Caroline's plan, can be expressed as the question of whether I will do otherwise. However, its nature or a part of its nature is not simply that of a standard question about the future, say of whether it will rain or whether others will do certain things. This can be ascertained from my subsequent response to it. My response is not another prediction or speculation—perhaps a prediction contrary to the one which began the episode. My response to the question is the forming of an intention or the making of a decision. My response is the forming of what earlier in our own inquiry was distinguished as a forward-looking or pending or inactive intention, as distinct from an active intention. (p. 225) My activity is not that of a predictor or spectator, but an agent.

It is, further, a response tied to the thought or perception that something ought to happen, or has a recommendation, or is desirable, as distinct from a response tied only to a thought of a possible future fact, fully describable without evaluation. 'The question that I put to myself as "What shall I do?" almost turns into, and becomes very nearly identical with, the question "What should I do?"—as the question "What do I believe?" turns into the question "What ought to be believed?" . . .' (Hampshire, 1965, p. 75) This is as true when I accept the prediction in question. I do not merely believe the proposition that I shall do such and such. Rather, I *endorse* it.

These facts of what can be called the transition towards action consort with what is called *our commonplace scheme of explanation of conduct*. This is a scheme of explanation of our own actions. If I explain one of my own past actions, I do so, at bottom, by citing my intentions and decisions, of which I have a certain special knowledge. The most explicit objection to determinism, or rather to one kind of determinism, in the doctrine under consideration, depends on what is taken to be fundamental to our commonplace scheme: that it has 'a normative element'. (1965, p. 111) A thesis of determinism which, in addition to proposing a kind of explanation of all actions, also entails that this sort of explanation can 'replace' the commonplace scheme of explanation, a thesis of determinism which 'entails that the commonplace scheme of explanation of conduct is replaceable by a neutral vocabulary of natural law', is 'unacceptable'. (1965, p. 111)

It is not easy to be confident in attempting to make more definite the given line of objection. One supposition is as follows. A deterministic explanation purports to be complete. A determinist explanation of my buying flowers in Camden Town this evening purports to give a complete answer to the question of why I did so, why that event occurred. Certainly this is implicit in any determinism. Any view of actions which did not claim to offer, or to offer the possibility of, a complete explanation of actions—that could not be a determinism. However, it may be said, an essential or ineliminable part of any full explanation of action is in a way *normative*. In effect, any full explanation of action must have within it the commonplace scheme with its normativity, or something approximate to it. Any determinism which has no normative part, therefore, cannot possibly give a full explanation of actions. It cannot possibly replace the commonplace scheme.

The essential claim is that a determinism, if it is non-evaluative, must fail in its explanatory ambition. Is the theory of this book non-evaluative? Evidently it is no part of the theory to depend for its explanations on the acceptability of evaluative judgements, at any rate in any plain sense. An explanation of my purchase of the flowers, in terms of the theory, does not include the judgement that a drawing-room will be more pleasant if there are flowers in it. No matter of taste needs to be argued prior to the advancing of the explanation. On the other hand, there is no barrier to an explanation of my purchase, along the lines of our three hypotheses, which includes a reference to my evaluative feeling—of whatever worth it was, whatever degree or kind of taste—that flowers are a good thing. It may well be essential to include such a reference, in part a reference to the nomic correlate of a neural process, an element of a psychoneural pair, and in part a reference to a disposition. If the objection to our determinism is to succeed, it needs to establish that this is to give an inadequate place to evaluation. To consider whether this is so, let us consider the commonplace scheme of explanation of action.

Certainly, in the moment when I was moved by the consideration that the drawing-room would be more agreeable with flowers, my state of mind was different in kind from my state now when—let us say—I am somehow concerned only with ordinary truth or falsehood. When I was actually in the way of desiring, approving, inclining, intending or deciding, my state of mind was—let us say—of an intentional or decisional kind. But now, when I am in the commonplace way *explaining* my purchase by referring to such past episodes, it seems clear that my state of mind—to continue with the useful if not very enlightening descriptions—*is* one having to do with ordinary truth and falsehood. It is *not* of an intentional or decisional kind. What I do, in giving a commonplace explanation of my purchase, is precisely to refer to my past feeling, whose taste or worth is now irrelevant, that flowers in the drawing-room would be agreeable. My explanation is absolutely unaffected if it happens to be true that in the mean time I have been overcome by some aesthetic or social conviction to the effect that having cut flowers on view is weakly conventional, middle class, suburban, or whatever. The worth of my previous evaluative judgement, despite the reference to that judgement in my explanation, is of no import.

The conclusion must be that evaluation enters no more into my commonplace explanation of my actions than it enters into our determinist explanation of my actions. If we suppose, as we must, that the commonplace scheme satisfies the general criterion of explanation of actions, having to do with evaluation, then so too does the determinist explanation satisfy that criterion.

To return to a man's contemplative and interrogative stepping-back, and the subsequent intentional or decisional response, it is said that this 'opens new possibilities of things that he might now try to do, given his knowledge of the natural order of things'. (1963, p. 98) It is said that with respect to the initial conditions of a decision, 'we can always step back and *try* to alter the initial conditions'. (1963, p. 103) Again, 'no matter what experimental knowledge of the previously unknown causes that determine a man's beliefs is accumulated, that which a man believes, and also that which he aims at and sets himself to achieve, will remain up to him to decide in the light of argument'. (1972a, p. 3) It is said, finally, that the fact of stepping back makes a place for 'freedom of mind'. (1972a, p. 20)

These reflections may well suggest that the given doctrine of the mind is a traditional indeterminist one. This is not its official intention. On the contrary, it is allowed that *some* theory of determinism may be true. What is mainly insisted on is the reality of the described phenomena that enter into the transition towards action. None the less, whatever the intent of their defender, they may be seen as obstacles to determinism.

The first of them is fundamentally the fact of some awareness of alternative possibilities. Perhaps one can fall into thinking that the reality of such episodes is somehow impugned by determinism, but it is in fact plain that nothing in the three hypotheses puts into question the existence of the given phenomenon. Why should something whose content is of the form 'Will *a* or *not-a* happen?', although somehow evaluative, not be a nomic correlate and an element of a psychoneural causal pair? As for the forming of an intention or deciding, there is the same consideration. If determinism entailed, simply, that there were no intendings or decidings, that would of course destroy it. Determinism does not entail that.

Nor is it the case that determinism must somehow attempt to transform intending or deciding into discovering, attempt to lead us to experience or to see forming an intention or deciding as discovering. It is a further theme of the doctrine of the mind we have been considering that there is an irreducible distinction between 'that which I discover, and observe, about my desires and interests, emotions and thoughts, and about my actions, at a particular time, and that which, stepping back from these observed facts, I decide'. (1965, p. 104) It is said, very truly, that it is inconceivable that we should dispense with this distinction. But why should it be supposed, if it is, that determinism at all threatens this distinction? That determinism claims the possibility of certain discoveries about my deciding—in a word, that it is nomic—does not turn my decidings into discoverings. That determinism is like many other theories in making us spectators of agents, including spectators of ourselves after the events of our deciding—this fact in no way makes agents into spectators. More precisely, it does not make our deciding and intending into anything else, and in particular not into contemplation or belief.

Another consideration about the given phenomena of intending and deciding has to do with the fact that one cannot or can hardly engage in forming an intention or deciding and at the same time also contemplate one's activity. (p. 81) It seems that we cannot *decide* or *be deciding*, and at the same time be spectator of our activity. Hence we cannot then contemplate it as being a matter of nomic connections. This is an instance of a larger fact: that in a certain sense self-consciousness is impossible. If I am *am* contemplating my future, or for that matter now contemplating a lady-bird, or in the moment of desiring a thing, or feeling the leather of the table-top, I cannot *now* be reflecting on my experience. A true double-act is impossible. This puts intendings and decidings, along with other mental episodes, into a category in one way unlike other events, processes, and so on. I *can* examine the nomic functioning of a motor or a corkscrew while that is going on, or the causal course of a chemical process, or the nomic sequence of my own arm movements. The impossibility of witnessing my decidings in the given way, and thus the impossibility of attending to them as nomic, is perhaps a significant source of resistance to determinism.

However, to come to a predictable judgement on this fact, it cannot enter into anything so substantial as an argument against determinism. That there is an obstacle to seeing the character of a thing at a time is no ground for concluding that it lacks that character. The point is close to one traditionally made against the traditional claim that we can and do introspect that the working of the will is free. The traditional reply is that our failure to perceive causes is no guarantee that they do not exist.

6.4 DETERMINISM WITHOUT PREDICTABILITY

According to the Correlation Hypothesis any change in my consciousness, as a matter of nomic connection, is accompanied by a neural change. That is fundamental to an audacious argument now to be set out. (Cf. MacKay, 1960, 1964, 1966, 1967) Another premiss is that I myself cannot at a time *t* have a complete conception of my own state or states of consciousness at *t*. This is so, it is said, because that conception would have to include a conception of my having that same conception. Logically, no information system can wholly represent itself as it then is.

Suppose an attempt is to be made to predict whether I will choose a *Times* or a *Guardian* tomorrow, *and also* to give me good reason beforehand to believe the prediction. More precisely, an attempt is to be made at t_1 to predict a neural process of mine at t_3, the correlate of my choice then, and to give me good reason to believe the prediction sometime before then, say at t_2. The predictor has total knowledge of my CNS at t_1, and also of environmental effects on me between t_1 and t_2. He also has a complete knowledge of relevant nomic connections. In this first part of his endeavour to arrive at his prediction, we are told, he must arrive at a view of what my overall neural states will be up to t_2. Then he must try to press on in the same way to a conclusion about the neural facts at t_3, of which my choice will be a nomic correlate.

His task, however, is also to give me good reason to accept his prediction. This must involve him in giving me an adequate understanding at t_2 of my neural state at t_2. (*Whenever* he tries to convince me of his prediction, his reasoning will include a description of, among other states, my state at that time.) This he cannot do, since I logically cannot have such a picture. This conclusion depends on the premiss of which we know, and also on the Correlation Hypothesis. The argument is as follows.

Suppose the predictor says to me at t_2 that my complete neural state is now *n*. I entertain that thought. Can it be true? No, since I cannot have a complete picture of my consciousness now, and every fact about my consciousness now has a nomic correlate of a neural kind. Therefore I *cannot*, whatever the predictor says, come to have a complete grip on my neural state at t_2. I needed such a grip in order to have good reason to believe what was supposed to come after, the prediction as to what I will choose at t_3. Therefore the predictor must necessarily fail in his second task: to give me good reason beforehand to believe a prediction as to whether I will choose a *Times* or a *Guardian*.

That, it may be said, does not matter much to determinism, but worse follows. For the same kind of reason, the predictor must also fail in his first task, predicting my choice. To repeat, he must arrive at a complete view of my neural state at t_2. Here, as in my own case, there is an insuperable logical obstacle—essentially because of the fact, owed to his communicating with me, that *his* thoughts affect *my* brain. There is the peculiarity that his attempt to arrive at a true and full account of a subject-matter alters the subject-matter.

That too may be taken as less than fatal to determinism, but it is not the worst of it. Suppose a predictor is not in communication with me, and thus *does* arrive at a prediction that I will take a *Times*. He does so in the way outlined above, save that he keeps all his facts to himself. He can be certain of his prediction, it is allowed, but more is said. *If* his reflections were to have been communicated to me, I could not have been certain of his prediction, because of the self-conception impossibility. Thus, it may be said, his certainty is only a *conditional certainty*. This is so since it depends on who contemplates the prediction. It is an odd sort of certainty—a sort that involves concealment of a prediction from someone else, and more particularly concealment of its grounds, if it is not to be accompanied by that person's justified want of certainty about the prediction.

Can it be, none the less, that the future *true state of affairs* is given by the view of the predictor? Well, it may be said, his prediction is not universally true. It is not true for me. I could not know that I will take a *Times*. Thus it cannot be that the prediction that I will do so captures the future true state of affairs—since a true state of affairs *is* such that it can in theory be known by anyone.

We thus arrive at the conclusion of our reflections on communicated and uncommunicated predictions. It is that prediction of a person's choices and actions is in principle impossible. It is said to be consistent with, and indeed in part based on, determinism. The conclusion if accepted is at least likely to seem to be some kind of barrier to determinism. We have a denial of what has been taken to be an implication of determinism, that choices and actions are conditionally predictable.

Very much can be questioned (Thorp, 1980, pp. 78 f.; Landsberg and Evans, 1970), but let us look only at the last part of the argument. Is the incommunicative predictor's certainty flawed because if I did have what I do not have, awareness of his reasoning, I would not be certain of the prediction? The ground for saying so comes to this: his prediction would *cease to be well supported* under certain conditions. It would cease to be well supported if I were informed of his reasoning, above all in so far as it involves a complete conception of my neural state at t_2. That is true in a trivial way, and false in the only way that matters.

It is true that if I were informed, a premiss of his reasoning would be false. That is a premiss which excludes the possibility of an effect of my being informed of my neural state. This is trivial. It is true of any derived proposition whatever that one can have no more than a conditional certainty about it in this sense. That is, it would cease to be well supported if a premiss taken as true were instead false.

To repeat, the ground for saying the predictor's certainty is flawed or conditional is that his prediction would cease to be well supported if I were informed. That can be taken to mean something else. Unreflectively taking it to mean this other thing is mainly what gives any persuasion to the argument against predictability. One takes the claim that a prediction would cease to be well supported if I were informed to mean that *the predictor's reasoning is open to doubt*. It is not that it has a false premiss, but that it wants validity.

To think so is a mistake, as can be shown clearly. The incommunicative predictor's reasoning, for our purposes, can be identified as reasoning that does not take into account any effect on me of any awareness of it. *Ex hypothesi*, the predictor can rightly forget about such an effect. There is none. He proceeds from a knowledge of initial neural conditions, environmental conditions and nomic connections to a prediction that I will take a *Times*. Would that reasoning become doubtful or somehow invalid if I were to learn of it, or any part of it?

The answer, which gives us a refutation of the argument against predictability, is obviously no. That reasoning, if I were to be aware of it, would be of no use to the predictor, since one of its premisses would then be false. However, if it *is* rock-solid reasoning as things stand, it *would be* rock-solid if I were to hear of it. I could be, and should be, precisely as certain of it as the predictor. Of course, the reasoning in question, if I were to hear of it, would be pointless or without application, since it would then involve a false premiss. By way of analogy, one's reasoning that if it was raining earlier, the chairs on the balcony will now be wet, is made no worse or better by one's discovering that it wasn't raining. The reasoning is of course in the given way pointless.

The conclusion must be that no scintilla of doubt attaches to the predictor's reasoning or prediction in virtue of the fact that it would be pointless if conditions were otherwise than *ex hypothesi* they are, and hence a premiss of the reasoning were false. Taking it in the only way that matters, it is false that the predictor's reasoning would cease to be well supported if I were informed of it. His certainty is not flawed certainty in the only sense that matters.

Consider now the idea that the incommunicative predictor does not have a hold on the future true state of affairs. In fact he does have an irrefutable basis for predicting that I will choose a *Times*, does indeed have a hold on the future true state of affairs in any relevant sense. This can be seen by a simple analogy. Suppose I have been struck on the head by a maddened indeterminist and am wholly unconscious. If we apply the given ideas to this situation, the onlookers do not and cannot have a hold on the true state of affairs. This is so since—we have it—such a thing can in theory be known by anyone, and there evidently is a theoretical barrier to my knowing it. If I entertain the proposition that I am unconscious, I do not know it, since it is false. There is no need to press on further. What is clear is that it does not follow in any significant way, from the fact that in a certain sense a proposition cannot in theory be known by everyone, that it does not represent a true state of affairs. Here too we have a refutation.

The argument against conditional predictability, to repeat, is alleged to be consistent with determinism and in fact to depend on it. As was also said, it is likely to be taken as some kind of barrier to determinism. In fact, if the argument were successful, it would destroy determinism. This is so, as it is now yet safer to say, for the reason that determinism *entails* conditional predictability. (Cf. Berofsky, 1971, Chs. 1–4)

It is therefore also proper to conclude, I think, against what is alleged about the consistency of determinism and unpredictability, that the audacious argument has in fact been no less than an attempt to prove the truth of a self-contradiction.

6.5 GÖDEL SENTENCES

Determinism, it is intriguingly said, is refuted by the consequence that it makes us out to be machines incapable of doing things that in fact we do. (Lucas, 1961, 1970) The refutation, it is said, can be made precise by way of Gödel's Incompleteness Theorem, to the effect that each formal language which is consistent, and adequate for the expression of ordinary arithmetic, contains a sentence that is not provable within the language, but which we can none the less see to be true. For our purposes, which require no close approach to Gödel's work, we take it that the sentence is the self-referring one of a given language *L* which says of itself

This sentence is unprovable within the formal language *L*.

This we can call the language's Gödel Sentence.

What is to be said of its truth and of its provability within the language? To answer in the recommended way, there are four possibilities, as in Table 1.

TABLE 1 Possibilities as to truth and provability of a language's Gödel sentence

	True	Provable
1	yes	yes
2	no	yes
3	no	no
4	yes	no

(1) As in the first line of matrix, might the sentence be true, and also provable in *L*? No, since if it were provable in *L*, it would not be true but false, since what it asserts of itself is that is *un*provable in *L*.

(2) Might the sentence be untrue, and also provable in *L*? To think so would be to fall into self-contradiction, since, it is said, what is provable in *L* is to be understood as what is *provable as true*. A system in which the sentence was provable would be inconsistent.

(3) Might it be untrue, and unprovable in *L*? No, since if it were unprovable it would not be untrue, but true—since it says of itself that it is unprovable.

(4) The remaining possibility can therefore be seen to be what is right: the sentence is true, but unprovable in *L*.

In general, all consistent formal languages which contain simple arithmetic are *incomplete* in the sense that each contains a Gödel Sentence which is true but unprovable in the given language. The argument just given for this result, certainly, is no more than a slight impression of Gödel's work. (Gödel, 1934; Hanson, 1961; Nagel and Newman, 1958; van Heijenoort, 1967; Lacey and Joseph, 1968; Findlay, 1942)

We now have the first part of the supposed refutation. The second is a characterization of a kind of determinism. This is 'physical determinism', close to Neural Causation with Psychoneural Correlation. (3.2) The characterization, with as much or as little success, can also be put on our own theory of determinism, and on any account 'in which all the relevant variables', say those having to do with a person's thoughts or actions, 'are determined by the values of some or all the variables at some previous time'. (Lucas, 1970, p. 93)

Such a determinism, to come now to the characterization, places a limitation on the perceptual capabilities of a person: only a finite number of kinds of states of the environment can be perceived. '. . . on the view of the...determinist, there are only a finite number of circumstances—possible environments—that are discriminably different. ... ...very small differences, at least in the environment, are not perceived by the human organism, and make no difference to its responses.' (Lucas, 1970, p. 131) It said to follow that our theory limits to a finite number the kinds of causal sequences issuing in psychoneural pairs, and hence the kinds of inferences to conclusions. It also follows, perhaps by way of the idea of limited perceptual capability, that a person can have only a finite number of kinds of beliefs. These points, it is to be noted, do not have to do with *particular and actual* perceptions, sequences, conclusions, and beliefs, which of course are also finite in number, but with the *kinds* of these things which might enter into even an immortal life—or, to say the same thing differently, with *possible* particular perceptions and the like.

A third part of the argument has to do with a consequence of the characterization that has been put on our determinist theory of a person. It is said to follow that all of a particular person's thinking, say Trenholme's thinking, can be regarded as an instantiation of one particular formal language or system. If such a determinist theory is true, '. . .each human being's reasoning . . . may be viewed as a proof-sequence in some logistic calculus. . . . The conclusion which, according to the determinist, a particular man can produce as true, will correspond to the theorems that can be proved in the corresponding logistic calculus.' (Lucas, 1970, p. 132)

This gives us all we are thought to need in order to proceed to the conclusion. As we know, Trenholme's particular formal language has its Gödel Sentence. Trenholme, given the determinist view, can in effect be conceived, in so far as his thinking is concerned, as no more than an instantiation of the given language. So conceived, Trenholme cannot produce the given Gödel Sentence as true. He is unable to see a certain thing. But he *can* see it. The fact of the matter is that Trenholme, being rational, may indeed follow an argument like the one above, involving the four possibilities, and indeed see that the given Gödel Sentence is true. (Lucas, 1970, p. 133) Therefore, contrary to the assumption, determinism is not the case with Trenholme.

In sum, what follows from the assumption that determinism is true of us, and a characterization of that fact, is that we are calculating machines of a certain kind; each of us instantiates a certain formal system; such machines cannot arrive at the truth of certain propositions, Gödel Sentences; but we can do so; therefore determinism is untrue.

The first part of this distinctive attempt at refutation is no matter of agreement among logicians. (Chihara, 1972), and the whole of the attempt has been regarded as inconsequential by one advocate of Free Will. (Kenny *et al.*, 1972) With respect to the second and third parts, which move at a certain speed, questions and objections do certainly come to mind.

(i) What is the reason for thinking that if my actual perceptions of my environment are determined, it follows that my kinds of perceptions are finite in number? How would the proposition, whose truth or falsehood can be left unconsidered, that my kinds of perceptions are infinite in number, be inconsistent with the proposition that each of my actual perceptions is a part of a psychoneural pair which is the effect of a certain causal sequence? For that matter, how would the proposition that my possible perceptions are infinite be inconsistent with supposing that each *possible* perception is to be conceived as such an effect? No explicit answer is given. We have no reason to take the proposition about infinite kinds to be inconsistent with nomic connection.

(ii) Even allowing that determinism carries with it the given proposition about a finite number of possible beliefs etc., how does it follow that a person—say Trenholme—can then be regarded as the instantiation of a single formal system of the given kind? It has sometimes been noted at this crucial point, in similarly sceptical discussions of the supposed refutation of determinism, that *one* formal system's Gödel Sentence can indeed be proved within another system. The possibility then arises that Trenholme's perception of the truth of a given Gödel Sentence is owed to the fact that he instantiates two formal languages. His knowledge of the truth of the Gödel Sentence of the first language is owed to the fact, allowed to be consistent with determinism, that it *is* provable by him in the second language. The argument has been pursued further and ingeniously (Dennett, 1972), but let us not follow it.

In place of asking why Trenholme should be regarded in terms of just one formal system, it is proper to ask why the propositions of finitude, thought to be entailed by a determinism, establish that Trenholme can be regarded *at all* as being an instantiation, of however many formal languages, in so far as his reasoning or truth-acquisition is concerned.

(iii) There is a related objection. What is it about determinism that is inconsistent with the reasoning, in a sense external to a formal system, which is taken to sum up Gödel's proof—the reasoning given above having to do with the four possibilities of truth and provability of a Gödel Sentence? Further, just what is the connection that seems to be implied between being open to *infinite* kinds of thoughts, which presumably is true given an indeterminist picture of the mind, and being able to judge that a Gödel Sentence is true?

(iv) It is strongly suggested that determinism needs to give *some* answer to the controverted question of how our reasoning is to be conceived, on what model or analogy, and so must be assigned the formal-systems answer. (Lucas, 1970, p. 139) In fact, determinism need not give an answer. (p. 5) It is not at all necessary, in order to confirm the proposition that our reasoning is a matter of nomic connections, to have an adequate model of that reasoning. If it were, then an indeterminist theory of the mind would be in the same difficulty, or rather, given its obscurity, a worse one.

(v) It is necessary to Gödel's proof, of a certain true sentence's being unprovable within a language, that the language be consistent. If the conclusion is to be drawn that determinism turns us into instantiations of formal systems, which instantiations lack a certain capability, it needs to be established that the instantiations are consistent. In a plain sense, we often enough are not consistent. Any sense in which we *are* fully consistent can with reason be regarded as elusive. (Cf. Dennett, 1984, p. 29.)

(vi) There is nothing to be gained from an argument so far unmentioned. Here, a determinist conception of a person or his reasoning is taken by definition to exclude the possibility of his seeing the truth of the Gödel Sentence assigned to him. Or, as the point is expressed in terms of computers, anything capable of seeing the truth of its assigned Gödel Sentence is by definition 'not a computer within the meaning of the act'. (Lucas, 1970, p. 138) A determinist is left with the ready rejoinder that determinism as he defines it is adequately defined for his given purposes, and does not have anything to do with Gödel Sentences.

Replies to some of these questions (Lucas, 1970, pp. 133–8) should be but cannot be considered here. They take matters further. They do not do not persuade me that determinism is false because, in the proposed way, it makes us less than in fact we are, into mere calculators of a certain kind.

6.6 FAILURE OF NEUROSCIENCE

It is the advance of neuroscience, unbroken progress in providing evidence for nomic connections specified by the three hypotheses of the theory of determinism, which gives to that theory its greatest empirical support. This support, consisting in nomic explanations and part-explanations, itself rests crucially on ever-increasing success in prediction. More kinds of neural and mental events are in fact successfully predicted. The first or philosophical member of the philosophical–neurophysiological partnership noticed earlier (pp. 194, 301) grants that there has been this steady growth in the predictive power of neuroscience. He further grants that there will continue to be such growth in the future, and that, in a certain sense, as elsewhere in science, there is no limit to it. (Popper, 1982a, pp. 14, 15) That would not impede him from the hypothesis that the theory of determinism of this book is completely baseless, utterly unbelievable, wild, and such that no good arguments have ever been offered for it.

In fact he does not consider anything very close to the theory, but rather what is named by him '"Scientific" Determinism'. This is partly defined as a determinism which is alleged by its proponents to have scientific support. Our theory would certainly evoke the same condemnations. In considering his objection, or at least what *may* be his objection, in relation to our own theory, rather than to 'Scientific' Determinism and in terms of a certain machinery of conceptions and method, we not only attend to what is of first relevance to us, but also have a better opportunity of clarity.

To sketch the objection first, it has to do with the character of neuroscience and the character of the theory of determinism. The existing practice of making and confirming predictions within neuroscience enables it to achieve its own goal consistently. If, however, neuroscience is contemplated as a source of evidence for the theory of determinism, the existing neuroscientific predictive practice will not do. Rather, it is said, an alternative predictive practice in accordance with a more stringent condition would be required. This condition, it is said, could not be satisfied. It is not close to being satisfied now. Therefore neuroscience, for all its own success, goes no way *at all* towards supporting the theory of determinism.

More particularly, to start with universal determinism (pp. 4, 306), it does not merely entail that *general kinds* of events are conditionally predictable. It entails, as we may say, that each future event is conditionally predictable *with any desired degree of precision*. Each minute aspect, each smallness of magnitude, is predictable. If some minute aspect of a future event were not predictable, it would not be true that all events without exception were predictable. If it were not possible in principle to predict event e_1 with any desired degree of prediction, to catch hold of some minute aspect of it, it would not be possible, so to speak, to predict the contained event e_2, the minute aspect. Universal determinism, let us therefore say, is *complete*, or entails *complete* conditional predictability.

Is our theory of determinism in this sense complete? It has to do with mental events, and these are conceived in such a way that any discriminable content of consciousness counts as such an event. (2.2) The Correlation Hypothesis claims of each discriminable item within consciousness that it is the nomic correlate of a cerain neural event. Of each such mental event, further, the Hypothesis on the Causation of Psychoneural Pairs claims that it is the effect of a certain causal sequence. So too the correlated neural event. More particularly, anything about a neural event which is in correlation with a discriminable mental event is an effect of a certain causal sequence. Thus, there is the theoretical possibility of predicting every mental event. So too the correlated neural event.

Are the two hypotheses then *complete* in the above sense? If we take the view that there is nothing within consciousness that is not logically possibly discriminable, as is inevitable, that is a reason for saying that our theory is complete in so far as consciousness is concerned. If we consider a neural event as a correlate—that is, consider an event in those respects in which it is in nomic correlation with a mental event—then whatever respect is in question, it is by the second or causal hypothesis a certain effect. It is natural enough, then, also to take it that the second hypothesis entails complete neural predictability. Similar remarks are to be made about the Hypothesis on the Causation of Actions.

Turning now to neuroscience, we are told that it consists in theories of a certain character. Like all scientific theories, they are human inventions. Further, they are general, and lack the 'uniqueness' of reality. They are 'nets' and thus, however fine their mesh, they fail to catch all of reality. To speak more literally, they share in the merely *approximate* character of all scientific knowledge. Given that it is the nature and goal of neuroscience to achieve only approximate knowledge, its making and confirming of predictions is satisfactory.

Let us now consider the attempt to regard neuroscience as evidence for determinism. More particularly, let us consider the idea that increasingly successful prediction in neuroscience gives reason to believe that mental and neural events, and actions, are conditionally predictable—conditionally predictable with any desired degree of precision, or completely conditionally predictable. To come to the crux of the objection we are considering, the record of increasing predictive success in neuroscience gives no evidence whatever for the proposition about complete predictability in principle. Predictive success, in order to do so, would have to satisfy what is called the Principle of Accountability.

The principle can best be approached by way of such facts as these, that it is open to a neuroscientist to attempt to save what he takes to be a nomic explanation of a type of event, despite predicted instances not occurring, by arguing that he had insufficient knowledge of antecedent or initial conditions. Suppose that he has advanced an explanation of a type *m* of mental event. It is, he has maintained, the effect of neural events of type *n*. When he has isolated instances of what he has taken to be *n*-type events, however, they have not as predicted been followed by *m*-type events. He may now conclude there is no nomic connection of exactly the kind he has supposed. However, it is also open to him to persist in maintaining that there *is* precisely such a connection. This he does by way of the claim that the isolated neural events were in fact *not* events of precisely type *n*.

Clearly we may find this unsatisfactory. We may wish in such a situation to make use of a principle to the effect that certain failures in prediction *are* taken as falsifying the supposed explanation. We may reasonably decide to require some specification of the degree of knowledge of initial conditions, in this case of neural events, which, given failure in prediction, *would* license the conclusion that the supposed explanation is false. To put it differently, we may require specification of the lesser degree of knowledge of initial conditions which, given failure in prediction, would allow the conclusion that the explanation need not be abandoned, that the failure can be accounted for.

Let us leave aside what principle of this kind, if one can be stated, is in fact used in neuroscience. What different principle would have to be satisfied if neuroscientific predictions were to count as evidence for the theory of determinism, and in particular for complete conditional predictability? It is the *Principle of Accountability*, so named because it requires of us that we account for any failures in prediction. That is, we are to account for them by having shown, in advance of a prediction, that we had an insufficiently precise knowledge of initial conditions. An unsuccessful prediction which cannot be accounted for in this way must refute or count against the supposed explanation or theory. More precisely, and crucially, the Principle of Accountability in one of its two forms, a less stringent one, requires a certain *calculation*. If we seek to use predictions derived from a particular neuroscientific theory or explanation in order to confirm the theory of determinism,

> . . . we must be able to determine in advance, from the prediction task (which must state, among other things, the degree of precision required of the prediction), in conjunction with the theory, how precise the initial conditions or 'data' have to be in order for us to carry out this particular prediction task. . . . [we must satisfy] the principle (i.e. the principle of accountability) that we can calculate from our prediction task (in conjunction with our theories, of course), the *requisite degree of precision of the initial conditions*.) (Popper, 1982a, p. 12)

Let us consider, as we are invited to, an explanation of the firing of a neuron, to be tested by the making of a precise prediction as to a particular firing. It is the burden of what is maintained about it by the objector that the Principle of Accountability cannot be satisfied. We *cannot calculate* what degree of knowledge of initial conditions would suffice to make a failed prediction count as a refutation of the explanation or theory. Further, 'our knowledge may steadily increase without approaching that very special kind of knowledge which satisfies the principle of accountability'. (1982a, p. 18)

What we thus have, at bottom, is the claim that neuroscientific predictions cannot satisfy a certain condition which applies if they fail. What may seem to be left possible is the rejoinder that successful predictions within neuroscience *are* evidence for determinism, even overwhelming evidence. It is clear, however, and often stated, that we are to conclude that *no* neuroscientific prediction gives any support to such a theory as our theory of determinism. It is not supposed that we need to discard only what might be regarded as a slight part of the neuroscientific evidence: that is, those proposed explanations, still retained, whose predictions have not been successful. What is no doubt intended, and certainly is needed for the objection, can best be set out by way of another principle.

To return to the proposed causal explanation of mental events of type *m*, let us suppose, differently from before, that the related predictions turn out right consistently. Whenever what is taken to be an *n* event is isolated, it is followed by an *m* event. The predictions do not satisfy the Principle of Accountability, but this is of no direct relevance, since that principle pertains only to failing predictions. Still, it might be said, there evidently is a counterpart-principle which is relevant. The Principle of Accountability, in effect, requires that we be able to calculate the minimum knowledge of initial conditions needed for a prediction whose falsity will defeat an explanation. The counterpart-principle, in effect, is that we are able to calculate the minimum degree of knowledge of initial conditions needed for a prediction whose truth will confirm an explanation. Just as we must account for failures we must vindicate our successes. There is as much need for the *Principle of Vindication*. The neuroscientist who claims to explain m-type events by n-type events must of course have sufficient knowledge of what he takes to be n-type events—knowledge sufficient to justify him in taking them to be n-type events. We need the Principle of Vindication, also having to do with calculation, if neuroscience is to support determinism. The end of this line of thought, of course, in conformity with the objection, is that the Principle of Vindication is not satisfied by neuroscience, and hence none of its successful predictions counts as evidence for the theory of determinism.

It has taken some time to expound the objection. Not much more is required, however, in order to assess it. What we need to do is to get it into a succinct form.

(i) Neuroscience, on the basis of evidence of precision p, succeeds in predictions of precision p.

(ii) The theory of determinism entails that if we had evidence of greater or absolute precision p', we could succeed in predictions of precision p'.

(iii) Neuroscience does not and could not have evidence of precision p'. More particularly, it could not satisfy the evidential requirements of the Principles of Accountability and Vindication.

Therefore (iv) the entailment of the theory of determinism, that if we had evidence of precision p', or evidence satisfying certain requirements, we could succeed in predictions of precision p', is wholly unsupported by neuroscience.

The argument is wholly without force. It is plain that the conclusion (iv) does not follow from the premisses. The fundamental weakness of the argument is the mistake of thinking that the falsehood or worse of the antecedent of a conditional, short of something like self-contradiction in the antecedent, falsifies the conditional. The conditional mentioned in the conclusion is *not* falsified by what goes before. If this were the 'philosophically most fundamental' argument against determinism, determinism would be home and dry. (Cf. Popper, 1982a, p. 55)

That the argument is without force is plain from its succinct form, and nothing changes if we enlarge it. We can enlarge the first premise by using the conception of neuroscientific theories as merely approximations or nets, and perhaps the conception of neuroscientific practice as governed by some relatively undemanding pair of principles about accountability and vindication. We can enlarge the third premiss, about evidence of precision p', by further specification of the Principles of Accountability and Vindication, and also calculation and completeness. None of that helps. From (i), (ii) and (iii), in whatever form, (iv) does not follow.

Stray thoughts may confuse the issue. (a) One is that the determinist needs to suppose that evidence of precision p supports predictability of precision p' just as it supports predictions of precision p. In fact no simple inference is intended from evidence of precision p to predictability of precision p'. There is, rather, an inference from a multitude of conjunctions of evidence and prediction both of precision p, and constant improvement through the history of science, to another conjunction, about conceivable evidence and predictions both of precision p'. (b) It is certainly not the case that we have been given any argument for what would be embarrassing to all parties, that it is self-contradictory or incoherent to to speak of complete knowledge of initial conditions. (c) The conditional proposition that *if* we had complete knowledge we could predict completely in no way entails a prediction that we shall ever have complete knowledge. (d) There is no argument from the proposition that some fish are not caught in the net of a theory to the conclusion that those fish are indeterministic.

There is only one important question, a question pertaining to the fact that we have a great and increasing corpus of actual evidence and predictions of precision p. What reason has been supplied for the claim that this fact does not support—indeed does not support *at all*—the proposition that if we had evidence of greater precision p' we would be able to make predictions of precision p'? No reason has been given.

6.7 DETERMINISM AND KNOWLEDGE

Nothing looked at so far in this chapter is as challenging to determinism as the long tradition of objection now to be considered. That it is typically inexplicit and rhetorical does not destroy one's feeling that it carries within it a forceful consideration, or at any rate a large fact. The tradition can be taken as beginning in the third century B.C. with the opinion of Epicurus, recorded in an extant fragment.

> The man who says that all things come to pass by necessity cannot criticize one who denies that all things come to pass by necessity; for he admits that this too happens of necessity. (1926, p. 113)

The thought is to the effect that the determinist's response of objection to the indeterminist is somehow undermined by his own avowed determinism, which applies to his response. His response, purporting to be a criticism of indeterminism, cannot in fact have the standing of criticism. If so, of course, the determinist's affirmation of his own doctrine is also undermined by that doctrine. To claim the truth of determinism is somehow self-defeating. Further, there is the proposition that if it is claimed that criticism and affirmation *are* possible, then that entails that determinism is false.

The given tradition of objection has very likely been an unbroken one, and has many recent expressions. (Taylor, 1966, pp. 168 ff.; Lucas, 1967, pp. 171–2; Malcolm, 1968; Chomsky, 1971; MacIntyre, 1971, pp. 244 ff.; Snyder, 1972; Boyle, Grisez, Tollefsen, 1976; Popper and Eccles, 1976, pp. 75 ff.; Denyer, 1981; Popper, 1982b, pp. 81 ff.)

There has also been something like a tradition of rejoinder to the objection, and it too has recent expressions. (Ayer, 1963, pp. 266 f; Smart, 1963, pp. 126 f.; Flew, 1965; Glover, 1970, p. 41; Cowan, 1969; Wiggins, 1970, pp. 132 f., 1973, pp. 34 f.; Grunbaum, 1971; Lyons, 1975; P. S. Churchland, 1981; O'Hear, 1984; Dennett, 1984, Ch. 5)

There are also those who change their minds about the objection, sometimes more than once. (Haldane, 1932, p. 157, 1954; Honderich, 1969, pp. 123 ff., 1970, pp. 210 ff., 1988, pp. 361 ff., pp. 521 ff.)

Very clearly the opinion of Epicurus needs clarification. Various things might be meant by saying (i) that determinism is inconsistent with criticism and claims as to truth, and (ii) thereby is self-defeating, and (iii) that if criticism and affirmation *are* possible, determinism is false. We need something more precise than the common staples of the Epicurean tradition of objection. Let us concentrate for the most part on various ideas suggested by the first opinion, that determinism is inconsistent with criticism and claims of truth, from which the other two things follow. We shall look at eight ideas or interpretations.

6.7.1 Determinism and Truth Itself

Although it has often enough been said, of such a determinism as the theory to which we have come, that it threatens truth itself, that cannot be right. Suppose that the truth of a statement consists in correspondence with fact. Or, more analytically, suppose that its truth consists essentially in this: the statement itself or a referring statement related to it in a certain way—related, for example, as 'That is a troglodyte' is related to 'Troglodytes do exist'—refers to a thing and assigns it to a descriptive category, and the thing is indeed in the category. The referring and assigning are a matter of semantic conventions. (Austin, 1950; Honderich, 1968; cf. P. F. Strawson, 1950) Evidently, if this is the nature of truth, the truth of a statement is unaffected by the fact that my believing the statement is an element of a psychoneural pair which is the effect of such a causal sequence as specified by the second hypothesis of the theory of determinism. The truth or falsity of the statement—a matter wholly of certain semantic conventions and the way the world is—is in general independent of the causation of anyone's belief. Truth or falsity would be as independent of Free Will or the belief's having been originated, if that were the explanation of it—that is, the answer to the question of why it occurred.

The given account of truth is to my mind fundamentally correct, but the present point does not depend on that opinion. All arguable accounts of truth (Tarski, 1944; Dummett, 1978) make *it*, as distinct from knowledge of it, independent of the explanation of the occurrence of beliefs. Suppose that the truth of a statement consists in its logical relations to other statements, perhaps its derivability from them. Theories to that effect, bringing together truth and provability, have been forcefully elaborated. That the given logical relations hold is a fact unaffected by the explanation, in the given sense, of the occurrence of a belief in the statement.

As just implied, indeed partly stated, the situation with validity and with evidential support taken on their own is of course the same. If a statement is in fact entailed by given premisses, or is supported, confirmed, or made probable by a body of evidence, those facts are invulnerable. They can in no way be threatened by a deterministic answer to the question of why there occurs a belief that the statement is a logical consequence or somehow warranted conclusion. For good reason it has never been suggested, in accounts of validity and evidence—or of truth—that among their criteria is one which requires that beliefs in the given statements be uncaused.

Let us put this first idea aside. The Epicurean tradition of objection to determinism, rather, properly has to do with *knowledge* of truth, truths of fact and of logic. It has to do with those beliefs, which is to say certain mental events, which have among others the property of having factually or logically true contents. The objection must be that determinism is inconsistent with one of the necessary conditions of knowledge other than the condition that what is known must be true.

6.7.2 Reason

What is the condition of knowledge which, according to the objection, is inconsistent with determinism? One vague but seemingly indestructible idea is that those of our beliefs which are in fact pieces of knowledge are necessarily owed only to Reason. Other closely related and clearer ideas, of which there is no shortage, are that our beliefs which are pieces of knowledge are owed only to truth, validity, the laws of logic, reasons, logical necessity, sufficient reason, facts of entailment, evidential support, criteria of reasonable trustworthiness, norms, or logically relevant considerations. A scientist mentioned above, before his change of mind, had one of these ideas in mind in writing that 'If my opinions are the result of the chemical processes going on in my brain, they are determined by the laws of chemistry, not logic.' (Haldane, 1932, p. 157) Knowledge, as this may be taken to imply, must be owed only to such as the laws of logic. Determinism, by intruding another kind of determination of our opinions, precludes them from being knowledge.

This claim as to the necessary genesis of knowledge can be made clearer by the distinction, as it can be expressed, between *the reason* or *a reason* for believing something, and *my reason* or *the reason I had* for believing it. The reason or a reason for believing such-and-such is always a good reason, and would in some sense exist if no one ever thought of it or believed the thing in question. That it was raining would be *the* or *a* reason for the truth of the proposition that the balcony is wet in a world without consciousness. A logical law would be the reason for its not being both raining and not raining. Reasons in this sense are best regarded as certain propositions—abstract objects or possible contents of thought. Certainly they raise questions about the mode of their existence, which we need not pursue. By contrast, such reasons as *my reason* for concluding something a moment ago, or *the reason I had*—which may have been a good, bad, or indifferent reason for my conclusion—are evidently best regarded as themselves also *being* beliefs, mental events.

Is it a condition of a belief's being knowledge that the belief is owed only to Reason, in the sense of *the* reason for it, or *a* reason for it—certain propositions? Is it a condition of the belief's being knowledge that it has no other provenance, that it is such a product? If so, it seems, there could be *no* knowledge, since propositions are in no relevant sense efficacious. In no relevant sense do they produce, determine, or explain anything, or have consequences. They are not causes, or like causes. They stand in relations with other propositions, certainly, but they do not themselves produce, give rise to, or evoke the beliefs which are our pieces of knowledge, or do this with any other non-abstract entities or events. This is owed at least to the fact that propositions are not spatio-temporal.

Far from its being the case, then, that a piece of knowledge must be owed only to propositions, no piece of knowledge *could* be owed only to propositions. If the Epicurean tradition had in it no more than the notion that knowledge is owed only to Reason in this sense, it would have in it nothing worth consideration. It is clear that an Epicurean objector to determinism must in fact have in mind, however obscurely, that knowledge has what can be called some suitable *real* as against *abstract* genesis—into which, no doubt, propositions somehow enter—which suitable real genesis cannot coexist with a belief's being an effect, or at any rate the effect of such a causal sequence as specified in the second hypothesis of our determinism.

Can the objector achieve anything by taking the possible view that *the reason* or *a reason* for a belief is not an abstract entity, a proposition, but rather a state of affairs, where such a thing may somehow be in or a part of the world, the spatio-temporal order? Can it be claimed that a belief of mine constitutes knowledge only if it is owed *only* to a reason in this sense? The claim is absurd or wholly obscure. We cannot entertain the idea that *the state of affairs* of its being raining gives rise to my belief *directly*, where that means not by way of my sensory experience, not by way of *my reason* for believing that it is raining. We cannot think any such thing, either, of the state of affairs, if it can be called such, that it is either raining or not raining—we cannot leave out my reason for my belief. Any other claim that is intended is wholly obscure.

6.7.3 Epiphenomenalism

'If my opinions are the result of the chemical processes going on in my brain, they are determined by the laws of chemistry, not logic'—that sentence, to quote it again, if it suggests the hopeless idea as to the wholly abstract genesis of knowledge, also suggests another. It suggests that determinism's supposed inconsistency with knowledge consists not merely in the fact that it involves the causation of beliefs, but that it involves their causation according to 'the laws of chemistry'. There is the strong suggestion that determinism's inconsistency with knowledge is owed to the fact that determinism makes our beliefs into effects of wholly non-mental antecedents. Knowledge requires a real genesis of a mental kind. Whatever the particular intention of the sentence, however, it has often been claimed explicitly by others that determinism is or entails epiphenomenalism, and that epiphenomenalism is inconsistent with knowledge. For several reasons there is as little hope for this idea—which, incidentally, is distinct from the simpler objections having to do with epiphenomenalism noted at the start of this chapter. (p. 338)

It is notable that the philosophical controversy over the analysis of knowledge—a controversy carried on almost entirely without reference to determinism—has not included the contention that mental indispensability (pp. 91 ff.) or the falsehood of epiphenomenalism is a condition of knowledge, and rightly so. The familiar and indubitable requirement that if I am to know *p* I must have something so far unmentioned, *good reason* or the like for my belief, in fact seems not to include the requirement that either my belief which constitutes my knowledge or the beliefs which constitute my good reason for that belief are to be conceived in other than an epiphenomenal way. Secondly and wholly decisively, however, as we have several times seen, there is no general connection between determinism and epiphenomenalism. Our own theory indubitably preserves the indispensability of the mental.

It does seem clear, as has long been supposed, that if I am to know *p*, it must indeed be the case that I have *good reason* or the like for my belief. In general, what is it to have good reason? In terms of the distinction used above, it is for the antecedent of my piece of knowledge, *my reason* for it, to have as its content *the reason* or *a reason* for the proposition which I know. Putting aside anything about epiphenomenalism, does determinism conflict with this requirement? It is roughly the requirement that I have suitable experience or believe warranted propositions such that the different particular proposition I am said to know is entailed by or somehow supported by the given experience or propositions. It *is* this requirement, somehow, that can be taken as threatened by a determinism. Still, if we are now in the right neighbourhood, it remains possible to go wrong, or to fail to get to the heart of the matter. To my mind, indeed, the Epicurean tradition generally has failed to get to the heart of the matter. It has not succeeded in making explicit the objection and thus its strength.

6.7.4 Reasons and Causes

One less than formidable proposal, made much of recently but perhaps with less history behind it than some others, is that determinism is inconsistent with our having good reasons for our beliefs because of the nature of causal and all nomic connection, and the nature of the connection between good reasons and the beliefs they support. Determinism, it is said, makes my reasons for my beliefs into causes of them, or at any rate puts my reasons into some nomic connection with my beliefs. Causes and effects, it is said, like all things in nomic connection, are *logically or conceptually independent*. Good reasons and the beliefs they support, however, are *not* logically or conceptually independent. Therefore, to claim that antecedents of my beliefs are causes or other nomic antecedents of them is to debar those antecedents from being good reasons.

The theory of determinism of this book does indeed regard our good reasons as causes of beliefs which they support. Our good reasons are elements in psychoneural pairs which occur in causal sequences which have the beliefs as elements in other psychoneural pairs which are effects of the sequences. In what sense, however, must items in causal or other nomic connection be logically or conceptually independent? This question is not explicitly answered by the objectors. What is typically said, vaguely, as was noticed earlier, is that since Hume it has been accepted that a cause and its effect cannot be in logical or conceptual connection. Quite as important, in what sense is it supposed to be true that good reasons for beliefs *are* logically or conceptually connected with those beliefs?

What is true, obviously, to take the latter question first, is that the proposition which is the content of my reason (say the conjunction of *If p then q* and *p*) may strictly entail the proposition which is the content of my belief (*q*). Or there may be some lesser conceptual connection between the two propositions. But that is obviously not to say that the *cause* in question, my first belief, entails the *effect*, my second belief. Or, to speak more properly, that is obviously not to say that any statement of the occurrence of the cause entails a statement of the occurrence of the effect. Nor is it to say, what is different, that the causal relation *is* the entailment relation. It is plainly logically possible in every case when I have a thought which is my reason for a second thought, even if the first proposition involved strictly entails the second, that I have the first thought without having the second. My mind could wander from the subject or I could fall asleep.

To approach the matter by way of the first question, that of in what sense it is true that causes and effects cannot be in logical or conceptual connection, Hume did indeed make clear, for the first time, that causal connection does not consist, to speak loosely, in a cause entailing its effect. But, as we have just seen, to allow entailment between the proposition in a reason and the proposition in a belief is far from asserting entailment between reason and belief, or asserting that a causal relation between reason and belief *is* the entailment.

Can Hume or anyone else be thought to have established something else relevant that might be spoken of in terms of the absence of logical or conceptual connection between cause and effect, or the non-identity of causal and logical or conceptual relation?

Might it be that if two things are cause and effect there can be *no* statement of the occurrence of the first which entails a statement of the existence of the second? No—it is certainly not true of cause and effect that no statement of the occurrence of the cause can entail a statement of the occurrence of the effect. The statement specifically that *there occurred the cause of the goblet's breaking*—that *there occurred that which caused the goblet's breaking*—does indeed entail the statement that there occurred the effect of the goblet's breaking. Every case of cause and effect allows for such an entailment. Hence the causal connection between reasons and beliefs is such that there is always such an entailment, but that simply does not debar a reason from being a cause of a belief. (Cf. Davidson, 1980h, pp. 13–14.)

It is true that any cause and effect of which we can speak usefully are such that there is a statement of the occurrence of the cause which describes it as other than merely and no more than the cause of the effect. If we lack such a statement of the occurrence of the cause, we shall, in brief, have no more to say than that the cause of *e* caused *e*. Are my reasons such that there is no statement of their occurrence which describes them as other than just the causes of the beliefs they support? Obviously not. Even if it were the case, absolutely extraordinarily, that I could say of something *only* that it was *the belief that caused me to have a second belief*, I should in those very words have something minimally useful. I would not be describing something as, merely, *that which caused me to have a belief*.

6.7.5 Completeness

If *p* entails *q*, and *p* is true, my believing both things is a good reason for my believing *q*, which I claim to know. My reason, it may be said, is *complete*. What can be meant by that, presumably, is that I need no further reason for my belief. Is it the case that our grounds for all our pieces of knowledge are in this way complete? Let us, without attempting to be more precise about the idea of completeness, suppose that the answer is yes. Such an answer has been taken, in the Epicurean tradition of objection, as a premiss for the further contention that there is *no room for a further explanation* of the piece of the knowledge. Determinism, it is objected, attempts to intrude a further explanation. It attempts to intrude a causal explanation of the belief. The attempt, it is maintained, must fail. It may be objected, similarly, that a *causal* explanation of a belief is in a sense complete, and that therefore it excludes the possibility of an explanation by way of good reasons. The first line of thought, then, is that knowledge excludes determinism. The second line of thought, on which we are concentrating, is that determinism excludes knowledge.

Much might be said of this, but not a lot is needed. The objections appear to consist in no more than *non sequiturs*, and nothing is done to show they do not.

We can indeed suppose, as in the first line of thought, that it is a condition of knowledge that our reasons for our beliefs are in a sense complete. That is not to say, and has not been shown to entail, that no other or further explanation of our beliefs is possible. That one kind of explanation, so-called, needs nothing added to it, does not entail that no other explanation, whether or not of the same kind, is possible. We have no ground for thinking that our having a complete logical basis for knowledge—what might be called a complete logical explanation—entails the absence of a causal explanation of it.

Nor is it the case, with respect to the second line of thought, that the completeness of a causal explanation has been shown to entail the absence of a logical explanation. It has not been shown p must be untrue, or cannot entail q, or that I cannot believe p and that it entails q, if my belief in q is taken to have a causal explanation. If something lies behind these arguments, we are given no idea of what it is.

6.7.6 *Incompleteness*

Not content with the objection from the completeness, in a sense, of good reasons, an objector may offer another from the premiss of their being in another way *incomplete*.

> . . . if I have given . . . a rational explanation of a belief of mine, then there is always room for a further 'but'. There may be some further argument . . . and this argument may not be any further fact about the physical world . . . . When we are invoking rational explanation, therefore, the logic of the discourse requires us to keep open the possibility of . . . the same facts supporting some different belief. . . . Truth is a perpetual possibility of being wrong.

However, a deterministic explanation of a belief 'leaves no room for our having held any other'. It rules out the possibility just mentioned, of the same facts supporting a different belief. (Lucas, 1967, p. 171)

It is not clear what character is here assigned to good reasons for a belief. They must, it seems, be in some sense inconclusive or defeasible, and hence the belief in which they issue in some sense tentative. This is presumably consistent with the belief being a piece of knowledge, and indeed being knowledge of a necessary truth. It is difficult to see clearly that such inconclusiveness and tentativeness must always be the case. Let us assume, however, that good reasons are always inconclusive or defeasible, and the resulting beliefs always tentative. It is true, certainly, that a deterministic explanation of my belief—whatever character it has—does in a sense exclude the possibility of my having believed otherwise. But why, from this, is it supposed to follow that my reasons cannot be somehow inconclusive and my belief somehow tentative? Certainly it does not follow from the hypotheses of the theory of determinism that I cannot be in *uncertain* states of mind—unsure, puzzled, baffled, or whatever.

Perhaps more to the intended point, it does not follow that my reasons must indubitably entail just exactly what I believe.

It is likely that this objection from the incompleteness of reasons is a misdescription of some substantial consideration. The objection as it stands seems uncompelling.

6.7.7 *Belief For No Reason*

It is said that a particular counterfactual consideration establishes that determinism conflicts with our having good reasons and hence with our having knowledge. It is said that if I come to think, about one of my beliefs for which I take myself to have good reasons, that it is the effect of such a causal sequence as specified in the second hypothesis, I shall also think that I would also have the belief in the absence of the given reasons. Or, what comes to the same thing, I would have the belief even if my reasons were not good ones. How can I be confident, then, that my reasons *are* good ones? Determinism must in fact put into doubt all my claims to knowledge.

A first thing to be said of this is that we in fact have no basis for the counterfactual proposition that I would have my belief even if my reasons were not good. Suppose I take myself to know that it is very likely that the cruiser General *Belgrano* was sunk during the Falklands War as an act of political ambition rather than military necessity. The belief is owed, in my mind, to a number of what I take to be good reasons: for example, my belief that the ship was far less of a threat when it was ordered to be sunk than when it was first sighted and reported, thirty hours before. Suppose that the latter proposition were false. Would I none the less have the belief that the *Belgrano* was sunk as an act of political ambition? There is no ground for saying so. If the proposition about its being a lesser threat when it was sunk were false, it is natural to suppose that certain claims about its being a lesser threat then would not have been made, and there would not have occurred the causal sequence that issued in my belief. It is natural to suppose that I would not have the belief.

Still, it might be said, I have no guarantee that I would not have had my belief even if the supporting proposition were false. To that, there must be the rejoinder that it is simply a pre-theoretical fact that we are mistaken in some of what we take ourselves to know. It can be no objection to determinism that it has the consequence, or at any rate allows for the possibility, that we are sometimes mistaken. Can the objector concede that we cannot reasonably ask for such a guarantee of our knowledge, but claim that we can and do require something else, and that determinism does not give this other thing to us? Well, we need to be told what the thing is. It might be called a *sufficient consonance or linkage* between causal antecedents of a belief and good reasons for it, but that takes us nowhere.

There is something else to be said of this objection to determinism, or at any rate a different expression of what has just been said. The objection purports to put in doubt all our claims to knowledge, by means of the proposition that a causal genesis for a belief does not have something called a sufficient consonance or linkage with good reasons for the belief. Consider again what I take myself to know, that the *Belgrano* was sunk because of political ambition rather than military necessity. One good reason, to repeat once more, is that it was far less of a threat when it was sunk than when it might first have been sunk, thirty hours before. How is that good reason affected by the proposition that my belief was an effect of a certain causal sequence? The reason, it seems, is untouched by the proposition. Certainly we are given no idea as to how, if it is, the reason is affected by the proposition. The same is the case, of course, with all my reasons for anything I take myself to know.

The objector, then, fails to make clear the demand for a sufficient linkage or consonance between causal antecedents and good reasons for a belief, and fails to make clear how seemingly good reasons are adversely affected by determinism. What this comes to, more generally, is that this objector in the Epicurean tradition does somehow suppose that good reasons have a character which they cannot have if determinism is true, but he does not, in arguing in the given way from the counterfactual consideration, make clear that supposed character. The inefficacy of the objection, however, like the inefficacy of its predecessors, is consistent with something else—the rooted idea that there none the less is something substantial in the Epicurean tradition. Indeed there is.

6.7.8 The True Burden

We can approach what is fundamental in the Epicurean tradition by way of certain rejoinders to the tradition. Against the general idea that causation prevents knowledge, it can rightly be pointed out that we do not decide on or choose our beliefs. This is true of my belief now as to the windy weather outside my window, and as true of any other belief, on whatever subject. It is true, certainly, of belief in determinism or in its denial. Nor do we decide on or choose our reasons for our beliefs. Our reasons are themselves beliefs or experiences, and thus as little open to decision or choice. In sum, belief is passive or involuntary, which consorts well with causation. The same is true of want of belief, the common state of neither believing nor disbelieving a proposition.

Also, given what can be called the aim of belief, which is truth, our involuntariness in belief is perfectly satisfactory. What we want, with respect to beliefs, is for them to fit the facts. What better situation, then, than that our beliefs should, so to speak, be imposed upon us by the facts? It is understandable, it is said, that we should want to be free in our decisions and actions, since our desire there is not to fit them to the facts, but to have or make the facts fit them. It is absurd to want to be what fortunately we are not, free in our believing. (Wiggins, 1970)

It may be maintained, further, that our very conception of knowledge, when fully analysed, has a feature which is in some connection with this fact. It is not sufficient for my knowing the weather is windy, it may be said, that it is in fact windy, that I believe it, and that I have good reason for so believing. There are instances which satisfy these three criteria but of which it can be maintained that they are not cases of knowledge. (Gettier, 1963) What needs to be added, it may be said, is a fourth criterion, that the belief is an *effect* of the state of affairs which is believed to exist. The state of affairs is a part of a causal circumstance for the belief. (A. I. Goldman, 1967; Benaceraff, 1973; Nozick, 1981, Ch. 3)

Belief certainly is involuntary, and that can be seen as satisfactory. To point this out—whether or not the further claim about a fourth criterion of knowledge is correct—is to make an apposite rejoinder to some of what is said in the Epicurean tradition, where there has seemed to occur the absurd argument that determinism undercuts knowledge because it does not leave us free to decide on or choose our beliefs.

Consider another proposition, however. It is that I *can* decide or choose *to try to come to have* a belief about something, and often succeed. I do not at the moment have a belief as to whether Colin McGinn is in his room. I *can* decide to go next door and see. If I cannot decide to believe *p* as against *not-p*, or vice versa, I can be said to decide to try to come to have one of those beliefs, where I now have neither. Further, I can act on my decision. Often, although obviously not always, to do this is to perform an ordinary action, to bring about a certain bodily event. To speak more ordinarily, there is the idea that my life has in it many questions whose answers I can decide to get, and which often I am able to get.

This idea about each of us is fundamental to our conceptions of ourselves as knowers. We do not only believe reasons, and believe what they are reasons for. We do, as we say, attend, concentrate, question, inquire, investigate, try to prove, deliberate, speculate, think up possibilities, consider alternatives, choose lines of thought to pursue, and so on. All of these acts and activities may be subsumed under the general description already given: we can decide to try to come to have a belief, perform fitting actions, and often succeed. To speak differently, our conception of ourselves as knowers is in part a conception of ourselves as *agents*, the performers of mental acts and of ordinary actions. We so act as to come, involuntarily, to have a particular belief.

To see this is to see the possibility of something else. The mental acts and bodily actions which are essential elements in the backgrounds of at least many of our beliefs may be more or less free, or free in different ways. Exactly what connections such degrees and kinds of freedom have with knowledge is evidently a considerable subject. Perfectly clearly, however, there is a possible response to determinism here. It is the true foundation of the Epicurean tradition of objection to determinism, the heart of the matter.

It is that determinism destroys a freedom of act and action that is essential to knowledge. By way of a kind of preliminary metaphor, it can be said that for each question we entertain there are sectors of evidence—the places of fact which determine the true answer. If my acts and actions are subject to a determinism, I cannot be confident that I shall investigate all such sectors. The nomic connections which govern my existence may exclude me from some of these.

Of course, a further thought will now occur to anyone familiar with something in traditional reflection on our freedom. (Ch. 8) It is the thought that we do have *a* freedom, including a freedom in inquiry, which is said with reason to be in no way threatened by determinism. Is *that* freedom not sufficient for knowledge?

Given this, what remains to be done here is to draw not a final but a preliminary conclusion. Still, it is a conclusion of importance.

The long-running philosophical controversy about our freedom, as remarked at the beginning of this chapter, has had three principal parts, and some others. The first principal part has been concerned with whether or not any deterministic picture of our existence is clear and in general conceptually satisfactory. A second principal part has had to do with the question of truth. That is a question on which empirical and philosophical considerations bear. The third principal part of the controversy, and perhaps its largest, has concerned certain *consequences* of determinism, in the main its consequences for morality. Latterly, this part of the controversy has also had to do with our non-moral attitudes to ourselves and others. These consequences are judged in relation to the question of what freedom of mental act and of bodily action is required for certain things. The questions here are all of one form. They ask what is the consequence for *x* of this or that freedom of act or action, a freedom allowed or disallowed by a determinism. *x* may be moral blame, or gratitude, or whatever.

A fourth and lesser part of the philosophical controversy about our freedom as it has been carried on, a part always separated from the three just distinguished, has been the part we have latterly been considering, having to do with the Epicurean tradition. Is a determinism consistent with truth? Is determinism inconsistent with our being governed by Reason, or by the laws of logic or the like? Is its supposed epiphenomenalism consistent with knowledge? Is determinism consistent with the supposed conceptual connection of reasons and conclusion? Is it consistent with the completeness, or with the incompleteness, of reasons? Is determinism defeated by the consideration that a caused belief is one we would have had even in the absence of what we take to be good reasons?

None of these questions, or others that might have been included, is an adequate expression of the challenge of the Epicurean tradition. That challenge, as we have now seen, is about the freedom of mental acts and actions, as much so as any question about, say, determinism and morality. The challenge is properly expressed by way of this question: What consequence for knowledge is carried by the freedom of decision and action, or the lack of it, that is entailed by a determinism? The preliminary but important conclusion, then, is that what has been considered separately as a fourth part of the controversy about our freedom is in fact within one of the principal parts, the third. In general, indeed almost always, this has not been seen, although we shall notice two exceptions later. (p. 470) It has sometimes unreflectively been denied. (Wiggins, 1970)

The importance of the conclusion is not, so to speak, taxonomical. There is no great gain, in itself, in discerning the likeness or unlikeness of disputes, in locating questions appositely. The importance of the conclusion that the issue of determinism and knowledge is within the most traditional philosophical dispute about determinism—the third of the principal parts of the controversy—is that the right location of the issue can make its resolution more likely. It can in fact lead to a resolution of the issue. That resolution will depend, of course, on a settling of the traditional philosophical dispute about consequences. That is the endeavour of the third and final part of this book, to which we are about to turn.

6.8 CONCLUSION ON THE TRUTH OF THE THEORY OF DETERMINISM

At the end of our consideration of neuroscience and Quantum Theory, one final conclusion was drawn. It was that this book's theory of determinism cannot be taken as established, as true. If the most that can ever be said for any theory is something less than that—that it is in some sense confirmed or to some large degree probabilified—then the theory of determinism cannot be taken as in that way confirmed or to that degree probabilified.

With respect to the question of whether the theory could be judged false, an interim conclusion was drawn. It was said that a consideration of Quantum Theory and of course neuroscience could not establish that the theory of determinism was false, or that it lacked very strong support. On the contrary, it was said, taking into account only neuroscience and Quantum Theory, the theory of determinism *could* be judged to be very strongly supported. However, there remained to be considered certain philosophical objections to the truth or acceptability of the theory—the objections of which we now know.

Those objections, although the first and last will have further consideration in another form, certainly do not weaken the claim to truth of the theory of determinism.

My final conclusion about the theory—it would be better to say my final official conclusion—is that *it is very strongly supported*, certainly as well supported as a general theory of indeterminism of the non-mental world, and, further, that it is greatly superior to indeterminist theories of the mind. Attempts have been made, certainly, to specify more precise and graded conceptions of the acceptability of theories—more precise than 'very strongly supported', 'weakly supported', and the like. There seems little to be gained by these attempts to rigorize what is in the end and fundamentally a matter of informed judgement. (Brown, 1987; P. F. Strawson, 1952, pp. 233 ff.) We shall not pursue the matter.

It is my own inescapable conviction, which will be no news, that the determinist theory of this book *is* true, or at any rate has the strength of, say, the theory of evolution. This conviction, I grant, goes beyond the sum of the evidence for and against. Such a conviction, however, is no idiosyncrasy. It is a conviction about our existence similar to those of very many thinkers in various disciplines, including a number of sciences. Such convictions are had by most thinkers of a certain discriminable cast of mind. They are convictions that stand in connection with the fact that philosophers generally have not begun to give up the question of the consequences of determinism as a question whose traditional presupposition—determinism as against near-determinism (pp. 2, 9, 336)—has been established as false. Quantum Theory has had no such effect in its seventy years.

The convictions in question are fundamental to a long and dominant scientific and philosophical tradition, which has in it, among others, the greatest leaders of science and the most acute of philosophers.

Hume, foremost among the latter, necessarily depended, in large part, on one of the supporting grounds for determinism other than neuroscience, the ground of our ordinary experience of the non-mental world.

> It is universally allowed that matter, in all its operations, is actuated by a necessary force, and that every natural effect is so precisely determined by the energy of its cause that no other effect, in such particular circumstances, could possibly have resulted from it. (1902 (1748), p. 82)

There is more reason now for being convinced of a determinist theory of the mind. To revert to the principal ground, it derives from the history of scientific inquiry since Hume's time into the Central Nervous System, above all the brain. We are not at all in Hume's position.

In abbreviation of all of this, but mainly of what was said above of determinism as very strongly supported, I shall speak of the *likely truth* of the theory of determinism. As will be gathered, that is not to imply that *no* probability whatever attaches to its denial.

PART 3

The Consequences of Determinism

7

Two Families of Attitudes, and Dismay and Intransigence

7.1 SEVEN QUESTIONS

Of what consequence is the likely truth of the expounded theory of determinism, or of theories akin to it, including a number of near-determinisms, to our thoughts and feelings about our individual prospects, thoughts and feelings we have in contemplating our futures? Of what consequence, secondly, is this determinism to our feelings about other persons who affect our lives well or badly? That is, of what consequence is it to our appreciative and our resentful feelings, these being non-moral feelings? Thirdly, how does it stand to our claims of knowledge, our confidence of laying hold of truth? Of what consequence is it, fourthly, to our holding ourselves and others morally responsible for things, or crediting ourselves and others with moral responsibility—which is to say of what consequence to certain feelings on our part of moral approval and disapproval? Of what consequence is it, fifthly, to the general moral standing or worth of persons, their standing over time or indeed in their whole lives, and, sixthly, to judgements of the rightness or wrongness of particular actions and of kinds of actions? Finally, how does determinism stand to certain large and encompassing institutions, practices, and habits, including punishment by the state? These go a considerable way to making our societies what they are.

All these questions are pressing, but the most pressing, or at any rate one which is as pressing as any other, is the first, that of consequences of the theory of determinism for an individual's engaged outlook on his or her own coming life, his or her own future. It is not a question which has been discussed, at least explicitly, in connection with any determinism. On the contrary, there has in the truly vast literature on determinism and freedom been a persistent concentration on the upshot of determinism for moral responsibility, for our holding others and ourselves responsible for particular actions, and crediting others and ourselves with such responsibility. What is arguably most significant, then, has been overlooked or avoided.

This is a yet larger fact than the failure to locate the question of the Epicurean tradition, our third question, in the right place. Arguably it is with respect to our conception of our own future lives that a determinism is most challenging to us, least tolerable. Determinism can be a black thing, as many have attested. It can, as in the case of Mill, and as he said, weigh on our existence like an incubus. (1924 (1873), p. 118) It did so for a time on mine. That it is our conception of our own prospects which perhaps is most sensitive to a determinism is in accord with a considerable fact, no doubt in some ways a disgreeable one. It is the fact whose nature is disputable, whose rationality and tolerability will continue to have attention in various ways (Parfit, 1984, Pt.1), the fact that we are not in the first instance moral agents, concerned in an impartial way with ourselves and others. We are first of all self-interested, concerned with ourselves and with those to whom we are connected.

To leave out life-hopes and knowledge in considering the consequences of determinism, let alone to leave out everything but moral responsibility and perhaps punishment, as has been common until recently, is not only to fail to treat of all of a subject-matter but also to fail to treat of what seems to be its most pressing part. Further, it is to fail to have all the help one can have in dealing with the question of the consequences of determinism. The fact of the matter is that certain general truths are clearer with some consequences than with others, partly because they are less obscured by theory—moral theory above all—and by controversy. Life-hopes are of particular importance in this connection. Having seen what is true there, it is easier to see what is true elsewhere, and in general.

It is my contention that reflection on the challenge of determinism in the seven areas—our outlooks on our own futures, non-moral feelings about other persons, knowledge, responsibility, the other two moral matters, and social institutions and the like—will bring into view or evoke two families of attitudes, attitudes in a fairly ordinary sense, and, in the first instance, two general responses to determinism, which responses are further evidence of the attitudes. For example, to think of determinism in connection with our own futures brings into view two attitudes, which in fact are two sorts of hope. Thinking of determinism in connection with moral disapproval brings into view two attitudes which are two sorts of moral disapproval. Thinking of the likely truth of determinism in each of the seven connections does or or at any rate can issue in two attitudes—and initially the two general responses.

It is not that some of us take one of the attitudes and make one of these responses, and others of us take the other attitude and make the other response, but that *each* of us is, at the very least, capable of both pairs. Each of us has, or at the very least is capable of taking, both attitudes, and each of us makes, or is capable of making, both responses.

One of the responses we may call the response of dismay, the other the response of intransigence. Both, at least initially, are in a sense second-order responses. Whatever they may give rise to, they are not themselves outlooks on one's future life, feelings towards others, kinds of confidence about knowledge, moral judgements and feelings about responsibility or the standing of persons or kinds of actions, and commitments and the like to social institutions. Rather, the two responses are to determinism, and more particularly *to the challenge or other relation of determinism to the attitudes*—to the outlooks, feelings, claims, and so on.

It is best to speak of two attitudes in the case of *each* of life-hopes, personal feelings, and so on. Thus there are fourteen discriminable attitudes in all. It is *possible* to speak in the same way of responses—that is, to speak of two of them in connection with each of life-hopes, personal feelings, and so on. In the case of responses, however, it is perhaps best to speak generally, as I have and most often will, of two types.

Is it the case that there are further consequences of determinism? It is possible to think of an eighth: its consequence for our *sense of ourselves*. It is brought to mind by a uniquely strong inquiry into personal identity, one which raises the possibility of an alteration in our senses of ourselves. (Parfit, 1984, Pt. 3) This eighth consequence is perhaps also implied by thoughts that determinism diminishes us or makes us puppets. (Nozick, 1981, p. 291) Aided by a certain spirituality, F. H. Bradley, may have regarded this consequence of determinism as somehow primary (p. 465), and philosophers of very different temperaments have given it distinctive consideration. (Dennett, 1984; Glover, 1987; G. Strawson, 1986) The concern in question, I think, is best expressed in terms of what turns up in what has already been distinguished—determinism's consequences for us as possessors of hopes; consequences for us in certain of our personal feelings which are reflexive or self-directed, and particularly pride (p. 403); consequences for our own moral responsibility and particularly our own moral credit; and consequences for things that enter into the other listed considerations. If one could succeed in distinguishing determinism's consequence for our *sense of ourselves* where this latter thing was otherwise—that is, not to be resolved into the given concerns—it would, I think, not be a consequential thing.

The seven questions with which we began—'Of what consequence is the likely truth of determinism to our thoughts and feelings about our futures?' and the others—can be expressed in several different more or less summary ways. *What is the consequence of determinism for families of attitudes which are integral to our lives?* Or *Is one or the other of the responses of dismay or intransigence, or a third instead, the somehow satisfactory one?* There is no short way to good answers to these questions. We begin on the way towards answers with the subject of our own attitudes to our own future lives.

7.2 LIFE-HOPES AND DISMAY

To contemplate one's own future in a general way, if one is ordinarily lucky or in an ordinarily tolerable state of life, is to feel a life-hope. Such a thing is reasonably so-called since it is, among one's hopes, the dominant one whose realization is taken at the time as what would make one's life or a coming part of it fulfilled, happy, satisfactory, or anyway of worth. It is the hope with which one is most persistently taken up. To lack such a thing absolutely is barely to have a life at all in what might be called a full human sense.

Like any hope, it is an attitude, which is to say a feeling in a wide sense. A typical attitude is open to different general characterizations, of which one is this: *an evaluative thought of something, feelingful and bound up with desire.* It involves feeling in a narrower and elusive but traditional sense, where feeling is somehow akin to sensation but unlocalized. It may involve desires of several kinds. It involves less of what can be called excitement or bodily commotion than typical emotions. A hope in particular is an attitude which has a single desire or want to the fore, so to speak, and involves a kind of uncertainty as to its satisfaction. A hope, in line with the general characterization of an attitude, is perhaps to be seen as *a desire for something, feelingful and incorporating evaluation, which desire is taken as not certain to be realized.* Hopes, evidently, are typically dispositions which recurrently issue into consciousness.

To contemplate one's own future in a general way, again, if one is ordinarily lucky, is to have one of the succession of life-hopes which mark so well the stages of an individual's life, and enter importantly into their definition. Perhaps a great majority of us are lucky in this way, since frustrations and defeats ordinarily do not destroy an individual's spirit, but are followed by new hopes. They need not be lesser ones. The fulfilment of hopes also gives rise to new ones. No one rests very long, let alone forever, in what is among the finest of states, the state of ambition realized or aspiration satisfied. One's name on a list, certificate, door, or book is not an enduring satisfaction, if there are any of those, but transient.

Often enough a life-hope has quite sharp definition: the achieving of some standing, position, or rank, the doing or the finishing of a thing, the possession of something, the securing of some state of mind, the making or altering of some relationship with another person, the development or succouring of someone, succeeding in a struggle, getting reparation, getting even or returning an injury, avoiding disaster, delaying death. Or, certainly, one's life-hope at any time may be a desire or want for some combination, somehow weighted, of such quite well-defined things. Despite the considerable fact of amoral self-interest alluded to above, of course, our life-hopes are sometimes properly described as having something other than an egoistic character. Often enough, they centre on the good of another person.

If life-hopes, whether or not combinations, and whether or not egoistic, are often well-defined, it is something like as common that they are not. What is hoped is that things will somehow turn out decently, that life if it changes will go on as well as it does now or no worse, that what is to come will be more or less all right. That is, the hopes have little more specificity than is suggested by those very descriptions—'life somehow turning out decently' and so on—and many like them. The descriptions are not replaceable by more precise ones.

The contents of all of these hopes, including the less well-defined ones, are at least in part states of affairs, where states of affairs do not include their own geneses, causes, or explanations. One may hope for the state of affairs in which enough food for oneself and one's family is assured, a patch of ground possessed, a job or a car got, a mortgage paid off, a shop opened, a political party or movement forwarded, children got on their way, a past lived down, a reconcilation achieved, a book written, a defeat put into the past, or a sickness survived. It barely needs adding that the depth or fervency of the hope does not have much to do with the relative worth of the possession, condition of life, judged from some external point of view. One can have a great hope for, say, something less than one's name on the door.

However, as most of their descriptions imply, life-hopes are at least typically for more than states of affairs. Our hopes in their contents are not only for the thing to be had, but for the *achievement* of it. They are not only hopes that some state of affairs should come into existence, persist, or change, by whatever means, say good fortune or someone else's action. My hopes are typically hopes that I will make something happen, bring something into existence, keep it, or change it—that I will succeed through my own actions in securing certain states of affairs. Most often what we aspire to do is to succeed in changing or maintaining our lives by way of our own strength, determination, judgement, wariness, guile, and by way of the various endeavours which flow from our own capabilities. In short, my hopes are typically *for* certain actions as well as states of affairs. It may be that such a feeling is more rooted or pervasive in some cultures as against others. It seems unlikely to be missing from any. It is persuasive to think of it as selected by evolution.

Even in so far as hopes are just for states of affairs as understood, as distinct from achievement, they are bound up with our future actions. They rest on and draw in beliefs about our own actions, what we will do in the future. We believe, for the most part, that we depend on ourselves. If what I hope for is the state of having the marks of success, and I am not greatly desirous of having earned them, I am none the less likely to believe that I will come to have them only through my own actions. Even the hope of meeting the right person, or dying in the right company, or the hope of surviving in the unpredictability of battle, is in this way taken to be in a way dependent on one's own actions. Certainly, despite some moods and moments, we are not fatalists in a certain traditional sense. That is, we do not believe the unbelievable thing that all or even much of what happens to us at future times will happen irrespective of what occurs between now and then, including our own behaviour.

In sum, then, our life-hopes of whatever kind have in them, or at least depend on, ideas or beliefs about our own actions to come. Actions figure as both ends and means. More particularly, our hopes have in them or depend on beliefs, ideas, or whatever somehow to the effect that in some significant way we do and will *initiate* our actions, much of the time or at least some of the time. They will not happen to us. Still more particularly, and fundamentally, we somehow believe or conjecture that *we stand to our actions in such an initiating way that we have at least some chance of fulfilling our hopes*. This initiation is integral to the idea of achievement, and as relevant, although in a different way, to actions as means. There are other expressions of the unspecific beliefs, ideas or whatever as to initiation, which initiation is as important to actions as ends as it is to actions as means. It is (I say) I who will act, and so may come to have what I want. It is I who will give rise to my actions, or bring them about, and so can have some optimism. It is I who in some way will conduct myself into my future.

There is no fatalism in this—fatalism of a second conceivable kind. We do not believe, as already remarked, that our actions will come into existence irrespective of antecedents intrinsic to ourselves (p. 91), and we do not believe either the vaguer and more ambiguous thing that these antecedents of our actions—our decisions, intentions and the like—come into existence irrespective of *us*. This fatalism too is unbelievable.

The large question must arise of whether the given beliefs and the like, somehow to the effect that we at least sometimes do and will initiate our actions and so have at least a chance of fulfilling our hopes, are out of place since or if the theory of determinism is true. An answer to the question is to be sought, evidently, by attempting to make our beliefs or whatever more clear, specific, and of course literal. They have been gestured at and labelled, rather than specified or clarified, by talk of initiation. What is important here, and wholly fundamental to the argument to come, is that there is more than one way of attempting to do so, more than one direction in which to go.

To follow one of these ways or directions, which we shall in this section (7.2), is to come upon certain distinctive thoughts about the initiation of our future actions and hence something bound up with those thoughts, a distinctive kind of life-hope. It is to come upon a kind of hope, perhaps to be an actress, which is one member of a certain family of attitudes. To follow the other line of reflection, which we shall later (7.3), is to come upon different thoughts about the initiation of our future actions, and a different kind of life-hope. This second kind of hope, in a sense, can have the same content—*to be an actress*. However, it is a member of a different family of attitudes. Neither of these lines of reflection is a matter of invention or advocacy. In each case we find out something about ourselves.

To proceed on the first line, let us bring to mind two related groups of natural and more contentful thoughts about the initiation of our future actions. The first group of these thoughts has to do with a particular conception of the future, and the second with the constraints of environment and our own natures.

We may think and say that our actions are such that our futures are in part *open, alterable*, or *unfixed*. Questions about them are not yet answered. More particularly, it is not that the questions have already got answers, stored up, although answers as yet unknown to us. There are no answers yet—they do not exist. Our futures are not laid out for us, waiting to be discovered, but in some crucial way or degree are to be made or formed by us. In part they will be our products. Our futures are not absolutely bound to the present and past, such that they can involve no break from the present and past. They do not merely *run on* from the present and past, containing no real opportunity.

There is really only one troublesome fatalist, the logical fatalist, who argues from the premiss that it is *true now* that the future consists in certain events, however unpredictable in practice most of them are, to the conclusion that *it is fixed and unalterable now* that the future consists in only those events. (van Inwagen, 1983; Sorabji, 1980) We resist the premiss, or the inference, since we take the conclusion to be false or are fundamentally disposed against it. That this fatalism is resisted by virtually all philosophers and others who consider it, that the problem of fatalism is very nearly *defined as* the problem of showing the conclusion of the inference to be false, is entirely indicative. The fact confirms the existence of the attitude to the future, and particularly to our future actions, which is our present subject. It amounts to a proof, not that one is needed, of the existence of the attitude.

A second and related group of thoughts about our initiation of future actions takes our futures to be not wholly the products of, specifically, our environments and our own characters, bents, desires, weaknesses, temptations—the corpus of our dispositions. Our past life-hopes, unless we are very lucky indeed, have often enough been frustrated or partly frustrated, in our view, because of our own dispositions rather than anything else. Often enough, and rightly, we put down our past failures to ourselves. We are sustained, however, by the idea that we have the possibility of overcoming ourselves in the future—by which we mean the possibility of overcoming our characters, fears, and the rest. I am not inevitably the creature of my dispositions, but rather I have the chance—even if only a small chance—of being their master. I will not forever succumb to certain fears or illusions. I will not forever be rattled and made inept by certain persons or in certain circumstances. Similar things are to be said about environmental constraints. In such ideas there is a further indication of what may be taken, when we think about ourselves in the first way, as that which our hopes contain and on which they are founded.

It is evident that in these two related groups of thoughts, about an unfixed future and about escape from environment and dispositions, we take our actions to be owed to some determinate centre of ourselves, somehow intrinsic and fundamental. Or rather, we have a certain elusive image of such a centre. All the thoughts noticed so far may be taken to indicate this, directly or indirectly. It is plainest when I say I will overcome myself, but not much less plain when I say I will make, form or produce my future. What is the determinate centre in question? It is no body, or machine, or system, or programme, or mental event, or succession of mental events. Above all, to take the most likely candidate, it is no disposition or sum of dispositions.

The old and familiar answer that is hard or impossible to resist is that it is *my self*. It stands in some wholly obscure relation to one of the two elements in each mental event as earlier defined, the subject as distinct from the content. (pp. 80, 194 ff.) It is particularly important that I take my self to be something that has the chance of mastering my dispositions, and hence to be distinct from them. As it seems, this self has the chance of mastering any one of them, and so is distinct from any of them.

There is a temptation, into which philosophers have fallen before now in other contexts of inquiry, and of which we have already seen something (p. 196), to declare the centre in question to be just a *person*, with the implication that we are all clear about what a person is. In an ordinary notion of a person, however, a body or part of one is at least a constituent of a person. But clearly, or anyway arguably, neither my body nor a part of it is a constituent of the initiating centre of which I conceive. My body, in fact, is typically conceived as *subject to* the centre in question.

Our thinking and speaking with respect to our selves, further and fundamentally, does not imply that they are merely items in standard causal sequences, caused causes of actions. My thinking and speaking about my self does not imply that it is merely a *link*. Talk of initiating, making, giving rise to, bringing about, producing, and so on implies, rather, an activity that is not itself a product. Above all, such an activity is implied by the idea that it secures to us open futures. More than that, our various expressions do positively suggest that an action, in being owed to an initiating self, is not owed to a standard causal sequence, or rather, to one which has the initiating self within it as an effect. The expressions do positively assert the self, an active source of the action.

Finally, the self is not presented to us as superfluous or otiose with respect to the initiation of actions. That is, actions are not such that although they are owed to a self, they also have some other explanation sufficient in itself. What we have, in the conception of an unfixed future and also of an escape from environment and dispositions, includes a denial of an alternative explanation of action. The expressions, to speak differently, are informed by something other than, but related to, the convictions of the indispensability of the mental and of the person. (pp. 91, 146) They are informed by an idea of the indispensability of the self.

Is there room for doubt about all this? Is there room for doubt that in our thinking and speaking of the initiation of our actions we may claim or presuppose the existence of an uncaused and unsuperfluous self? Can it be claimed that in fact we do not ever have such an idea? Something like this *has* been claimed of what is called the self in contexts other than our present one, having to do with life-hopes. It has been claimed, in particular, in the context of the philosophy of mind, as a result of our failure to introspect a self, and also, very differently, in connection with ideas of moral responsibility—as we shall see. (Ch. 8) When life-hopes are the subject, the claim that we do not have such an idea of self is wholly unpersuasive.

What else but an uncaused and unsuperfluous self *could* rescue us from the situation in which our actions are simple products of our environments and the manifold of our dispositions? Still more clearly, what else but an uncaused and unsuperfluous self *could* provide for us an open future? Such a future is not merely a future which is somehow in our control, our doing, which future is none the less one that is already settled and not open to change. What we propose to escape is fixity, not just this or that kind or character of fixity. No doubt a fixed future which in some sense will be our own doing is better than a fixed future not in that sense our own doing, but both of them run against our aspiration. (Cf. Dennett, 1984, Ch. 3)

Again, it seems clear enough that the aspiration in question is not merely for a future that is in practice unpredictable, with nothing said as to it being fixed or unfixed. We are not satisfied with the idea that all the events which constitute our coming lives are in a category with the event which occurred when the ball on the roulette wheel came to rest unpredictably but inevitably on red 7. What we want are futures that are not settled.

This claim is not properly a matter of argument, but of every person's experience. Still, is there some other argument against it, perhaps drawn from the present context, rather than the philosophy of mind or moral philosophy? Might it be argued that there is no reasonableness in resting our hopes on an unfixed future, and hence, by way of the sanguine premiss that we *are* in general reasonable, that in fact we do not desire unfixed futures? That is, since there is no good reason to rest our hopes on a certain thing, it is no part of the obscure ground of our hopes. More fully, it may be argued that there is little or nothing to choose between two conceivable situations now in contemplation, both of relevance to the subject of life-hopes. One is the situation where future things are already fixed but not yet known, are unpredictable in fact, and the other is the situation where future things are not known, also unpredictable in fact, but are fixed. There is no reasonableness, it will be said, in going for the second option. Our real state of knowledge makes no difference between them.

One reply must be that it all depends on a person's experience up to the present. If things have gone well for a person, there is more to hope for in what follows on the assumption that the entire run of his or her life is fixed. This involves a large but decent inductive inference, an extrapolation from past and present to future. If things have not gone well, or not so well as was hoped, it is at least not unreasonable to have greater hopes on the assumption that the whole of one's life is not fixed, but is connected with the activity of a self. The alternative assumption, that of fixity, when conjoined with disappointment or defeat up until now, issues by way of an inductive inference in the prospect of more of the same.

Are we generally of the view that our lives have gone well, or of the view that they have not gone well, at least not so well as was hoped? I take it, although a qualification or two might be added, that generally we are more of the second and sadder view than the first. In reply to the objection, then, it can be said there is better reason for hope in the denial of fixity than in any acceptance of it. Given the sanguine premiss of our reasonableness, there is reason to think that we do *not* tend to the idea of a fixed personal future.

The direction of the reflections in this section will have become clear before now. To characterize them again, we first took it that our life-hopes depend on pre-philosophical and pre-theoretical beliefs, ideas or whatever to the effect that somehow we will initiate our future actions. The question was whether these beliefs and the like are out of place if or since the theory of determinism is true. There is one clarification of them which finds them to be bound up with a conception of an open, unfixed, or alterable future, a future such that in it we escape environment and dispositions. This clarification of beliefs and the like as to initiation can also be said with some reason to be pre-philosophical and pre-theoretical. The same can be said of what follows, that we have the idea or whatever of a determinate centre, a self, which is uncaused in its activity and which is not superfluous. There is one final thing which needs to be made explicit, or more explicit. The self is not, so to speak, an agency or mechanism of chance. The futures owed to it, if unfixed, are not a matter of *chance*. True chance, at any rate on a large scale, would be about as damaging to hope as fixity. To take it that I, in or as my self, give rise to my actions, is not to make them a matter of randomness. Rather, they are things which are subject to the self.

Indeed the direction of all this will have been clear enough. What we arrive at is the conclusion that we share, when we think of our life-hopes in one way, a picture of ourselves related to what earlier in this inquiry were considered closely, *indeterminist theories of the mind*. In particular, the common idea of one's self is in a way akin to the idea of an originator, that which originates our intentions, decisions and the like, and our actions. (3.4–3.7)

The conclusion is not at all well expressed as being that the given beliefs and so on about initiation can be taken to include some indeterminist theory of the mind. It is rather the case that the given beliefs and ideas include what is best called an *image*, an image to which the indeterminist speculations stand as theories—attempted developments, improvements, or clarifications. There can be no doubt that what is included in our thoughts is not so explicit or developed as the indeterminist theories of the mind. It would be absurd to claim that in considering our futures we are all possessors of an articulated conception. What is not absurd, but seemingly undeniable, is that we have an image or idea that stands to the indeterminist theories as many of our ordinary ideas stand to more or less developed conceptions and theories in science, other parts of philosophy than the philosophy of mind, and also politics.

When we think in this first way of the initiation of our actions, then, we partly think of it as effectively indeterminist. To do so, as will perhaps already have been clear, is to have *a certain sort of life-hope*. This is partly so since, as we have seen, ideas of our actions—ideas of them either as ends or as means—give much of the content to life-hopes of whatever sort. To have the present sort of life-hope, whether about being an actress, surviving in battle, or whatever, is to have a hope best characterized in ways of which we know: a hope for an unfixed future, a hope for a future in which we are not creatures of our environments and our dispositional natures. A life-hope of this sort, as anticipated earlier, is a member of one of two families of attitudes. The other members have to do with the other parts of our existence which are affected by determinism.

It has not been argued that our initiation-ideas with respect to life-hopes of this kind amount *only* to a kind of antecedent image of indeterminist theories of the mind. Our ideas, when we think in this way, do not have to do only with *origination* of actions. My hopes of this kind also depend, to be brief, on such beliefs as that I will not be in jail, or under persistent threat, or greatly ignorant of what can be done. They also have to do, that is, with what is underdescribed as the absence of obstacles or external forces. They have to do with willingness or self-directedness, what will be called *voluntariness*. (p. 397) To speak only of this part of our beliefs or ideas, however, the part having to do with *origination*, there is no doubt that it is out of place since or if the determinist theory is true. It is out of place in the plain sense that it is logically inconsistent with the determinist theory.

No question whatever arises about the fact of this inconsistency. It would be perfectly pointless to set out to give a proof of the logical inconsistency of this part of our beliefs or ideas and the determinist theory. This is so since *no sense* can possibly be attached to the idea of an unfixed future, or a future which somehow escapes or is released from environment and dispositions, other than that of a future not subject to the nomic connections asserted in a determinist theory, but somehow subject to the person. That is the clear and inescapable *definition* of an unfixed or released future, or, at worst, an ineliminable part of the definition. Similar remarks are to be made, as they were earlier, of the image of the self, as distinct from its activity. Here too there is palpable inconsistency with determinism. (Cf. 3.4–3.8.) There is no more need or place for proof of the given inconsistencies than in the case of the inconsistency between the assertion, say, that Napoleon was for a time married and that he was at that time a bachelor. There is no question of whether or not we are in a situation which must be resolved partly by way of the proposition that not both *p* and *not-p* are true. We are in precisely such a situation.

What we come to, then, is that the likely truth of the theory of determinism forces us to abandon certain images or ideas which indubitably are part of our life-hopes of a certain kind. The given ideas or images, it seems, must go. Thus, it may seem, we are deprived of the hopes. These desires must collapse. What is important to us, in our contemplations of our future lives, is denied to us. Nothing significant remains of these hopes. This is *the response of dismay*, the response of dismay as it pertains to our life-hopes. We shall also encounter it in other connections. Here it is a dismay which is owed to coming to believe of life-hopes we have that they *depend on a necessary condition which cannot be satisfied since or if determinism is true.* Further, there is no adequate replacement for these hopes.

It is worth remarking in passing that it is of course essential, here and in all of what follows, to have in mind a firm and strong conception of determinism, or of near-determinism. A vague or uncertain one, or indeed one unattended to in the course of reflection, is near to useless in considering the consequences of determinism. Although what will be said of consequences of determinism does *not* depend on exactly the determinism that has been expounded, but has a considerably more general application, it will come as no surprise to be advised to keep just that in mind. It is preferable to have in mind the three hypotheses in their literalness. (p. 247) To do that, however, will no doubt bring to mind more—something like William James's understandably coloured account of a more general determinism.

It professes that those parts of the universe already laid down appoint and decree what other parts shall be. The future has no ambiguous possibilities hidden in its womb: the part we call the present is compatible with only one totality. Any other future complement than the one fixed from eternity is impossible. The whole is in each and every part, and welds it with the rest into an absolute unity, an iron block, in which there can be no equivocation or shadow of turning. (1909, p. 150)

The existence of dismay in connection with life-hopes will be of importance in itself to the argument to come. What is yet more important is the existence of the particular attitude to the future, a kind of desire or want, whose foreclosure by determinism issues in the dismay. This attitude, whose existence is indubitable and which contains a part inconsistent with determinism, is certainly no transient phenomenon which arises in thinking about determinism. It is *natural*. That is to say, at least, that we do have it, and have it spontaneously, and that it involves no internal incoherence or any such thing.

On the first of these three points, I shall not struggle to establish propositions that are not needed, that every human being without exception feels the desire—we may call it just the desire for an unfixed future—or of course that having it is some kind of condition of being human. That the desire, or at the very least the capability of it, is a general fact about us, in an ordinary if not wholly strict sense, is indubitable. That the desire is spontaneous in the sense of not being put upon us or evoked in us by philosophers, priests, or other such personnel is as clear. As for the third point of naturalness, it seems as clear that the desire, whatever else is to be said of it, and whatever has been manufactured out of it by speculative philosophers, is unconfused and cannot be said to contain anything that could be called an internal mistake.

A word or two more can usefully be said of these points, given one recent account, distinctive and admirable, of the human consequences of one particular theory near to a determinism, in fact called 'naturalism'. (Dennett, 1984) The account does not contemplate life-hopes, or advance or consider the general line of argument on which we are engaged. (Cf. p. 471) The account, none the less, may give rise to a suggestion which is certainly false, that the only desire in the neighbourhood we are now considering is one which is had by some few persons adversely affected by certain benighted philosophers—philosophers who conceive of determinism or argue about it in terms of one or another too simple or too extravagent idea or metaphor. They may conceive of or argue about determinism in terms of an exceedingly simple causal model of persons, a 'Sphexish' model suitable to the seemingly mindless digger wasp *Sphex ichneumoneus*, but wholly unsuitable to our own causal complexity. Or, they operate by way of such extravagant metaphors as those of the Invisible Jailer, the Nefarious Neurosurgeon, the Cosmic Child Whose Dolls We Are, or the Malevolent Mind-Reader.

On the contrary, to repeat, the attitude with which we have been concerned is not something owed to philosophers, and it is not, to speak quickly, a no doubt rare desire to be other than a *simple* causal mechanism, or a desire to be free from the grip of the Invisible Jailer or whatever. These desires *can* indeed be satisfied despite determinism. It is rather a kind of desire, which is clear and distinct and has nothing to do with simplicity or some other agent, for an unfixed future.

The response of dismay, incidentally, is also natural, in at least the sense that it too is a general and spontaneous fact and is unconfused and without internal mistake. More remains to be said of it (p. 349, Ch. 9), but not anything that will conflict with that judgement.

7.3 LIFE-HOPES AND INTRANSIGENCE

It is certainly possible, and not necessarily on another day or in another mood, to accept the theory of determinism as true, and come to feel differently about its bearing on life-hopes. This different response in fact involves life-hopes different in character or kind from those we have been considering. This is not a matter of a different starting-point of inquiry, although I shall say a word more about that starting-point in a moment. That is, we can begin again from thoughts to the effect that our life-hopes do indeed rest on our *initiating* our actions, at least sometimes, on our being related to our actions in such an initiating way that we have an opportunity of fulfilling our hopes.

To begin with, let us look not at a case of good hopes for the future, but at the case of someone whose hopes are small and declining. A man without a job, for the first time in his life, may hope in this weak way that he will be able to get one, to have again the existence he has always had before. His life-hope for the future centres on a picture which has in it a number of elements—say his independence of others and of society, his satisfaction in having his children perceive him as their able supporter, purposeful activity in place of empty hours and unsustaining activity, the enjoyment of things now denied to him, an escape from condescension. Certainly it is a hope involving both a state of affairs and achievement. If he does have such a hope, however, it is small and declining, indeed a dying hope. Now, as these words are written, such hopes for many men are just that.

Why is his hope as it is? To enter into his situation is very easily and naturally to come to a particular answer. It is that his hope has declined because he believes or is near to believing that the world of his life will not change, but will continue to frustrate his hope. It is that belief or near-belief of his, that the world of his life will continue to deny him what above all he wants, that is the explanation of his near hopelessness. More particularly, it is his perception that his future activity in simply passing time—the drudgery of watching television and of whatever else goes in place of work—will flow from certain desires and intentions of his, but certain reluctant or entirely second-best desires and intentions, as distinct from desires and intentions into which he enters fully, desires and intentions which he embraces.

Reluctant desires and intentions, we can say, are those which operate in situations to which the agent is somehow opposed. *Embraced* desires and intentions satisfy the condition that they operate in situations which the agent at least accepts. These two types of situations—there is more than one sub-type of each—can also be called *frustrating or obstructing* situations and *satisfying or enabling* situations.

We can then take it that the beliefs or ideas which enter into and on which we ground hopes may consist in something so far unconsidered. To say that life-hopes falter when a man lacks the thought that he will initiate his actions is to say, in part, that his hopes falter when he lacks the thought that his actions will flow from certain desires and intentions which he embraces—no doubt, to return to an earlier distinction, both active and inactive intentions. (4.2) His actions will flow instead from reluctant desires and intentions. In explaining full hopes by saying that a man believes that he can initiate the actions which will shape his life, we can take it that he believes just that his actions will flow from his embraced desires and intentions.

One may feel a resistance to the idea just proposed. The formulation of our grounding thought with respect to life-hopes, that we can initiate our actions, does in a direct way convey or suggest an unfixed future, escape from environment and dispositions, and a self. That we can initiate our actions does not in so direct a way convey the present idea, that a person's actions may flow from his own embraced desires and intentions. It may be said rightly that the present idea, that hopes are fully supported by way of the idea of embraced desires and intentions, conflicts with the previous idea, that they are supported by the idea of what can somehow override the corpus of one's desires and intentions—and it may be maintained that it is only the latter supporting idea that can be found in our thought and talk about initiation.

It would in fact be something like rash to attempt to rule out the proposal that our thoughts about the initiation of action can be taken as having to do fundamentally with embraced desires and intentions. That would be to depend on a particular formulation of those thoughts in too confident or exclusive a way. It is reasonable to think that we should not be wedded to one formulation. There are others. Here is one: We shall often or sometimes be in such a position that actions which will secure our life-hopes will be within our doing. Or: We shall often or sometimes be in such a position that we will not be defeated in the desire which is our life-hope.

Evidently there is no single canonical pre-theoretical expression of our grounding thoughts with respect to our hopes. There can be no doubt that philosophy, in so far as it advances, advances through precision, clear distinctions, a resistance to loose talk. That truth needs to be brought into compromise with another. It is the truth that in attempting to deal with our experience, it is necessary to resist the temptation of thinking that it is to be indubitably captured in just these words or just those, by exactly this perception rather than any other. In the beginning, at any rate, a certain looseness is necessary. It is worth adding, in recollection of Aristotle, that different subject-matters allow of different degrees of precision. This third part of our inquiry into determinism (Chs. 7–10) will allow of less precision than the earlier parts.

We have so far seen one sort of satisfying or enabling situation which may be expected to turn up in one's future and so give a basis for a life-hope. The *world of a man's coming life*, as it was said, may be one which allows him hope. The *way of a woman's world*, in her anticipation, may constitute a situation which allows her hope. By contrast, her hopes may falter or die when the way of her future world is not one to which she is disposed, but rather the related frustrating or obstructing situation. 'The world of a man's coming life', and 'the way of a woman's world'—certainly these are descriptions of a general and relatively vague kind. They are none the less at least useful, if not essential. In speaking of the world of a person's coming life, we speak of such things as the customary or ordinary responses and activities of other people, the state of a society and its economy, the availability of things, a person's corpus of feelings, and a good deal more.

The world of a person's coming life, or the way of his or her coming world, is surely the *principal* kind of satisfying or frustrating situation with respect to very many of our desires. It is such situations that are often most pertinent to our life-hopes. It has been an oversight of philosophers, in inquiries distantly related to our own, to have paid them no attention. Their large importance can be gauged from the large extent to which we are social beings, involved in and dependent on societies. That is not to say there are no other sorts of satisfying and frustrating situations, no other sorts of determinants of embraced or reluctant desires. Let us glance at them in something like a descending order of pervasiveness, beginning with those most likely to enter our lives.

Suppose it is central to a man's life-hope that he direct his capabilities and energies effectively at a single goal. He seeks to arrange his obligations, various endeavours, his time, to some extent his personal relations, perhaps his expenditures, and no doubt his self-indulgences and distractions, so as to further his life-hope. He may want and need, differently, to have a realistic view of himself, or to escape the grip of what can illusorily present itself to him as truth, rationality, principle, decorum, or taste. The state of affairs which enters into his hope can be something of the sort mentioned earlier. He wants above all to own a little land, to become a salesman or a craftsman, to bring to fruition a line of research, to finish a novel, to become a judge, or to have an acceptance in his family or among acquaintances.

It may be that the way of his world, in the sense we know, has not been against him in his ambition, and will not be, or at any rate nothing so much as the state of the economy and so on is against the man with no job. However, something else may be against him. He may, looking to the future, not have anything like a decently confident anticipation that he will be able to direct his energies effectively towards his goal. This will be a matter, evidently, of desires which delay him in his progress, or take him off his track. There is certainly nothing incoherent in supposing that the desire which is a life-hope may be frustrated by other desires. Lesser desires, indeed desires for what is also disdained, can be stronger desires. One can do what is typical of many lives—and part of the lives of all people—which is to succumb to the temptation of immediate satisfaction.

It is evident, then, that a second kind of enabling or obstructing circumstance with respect to one's life-hope is the circumstance in which the hoped-for state of affairs is or is not achieved as a result of the effective dominance or want of effective dominance of the desire which is one's life-hope. If our hopes require the idea that our world will not frustrate us, they also require the idea that we will not frustrate ourselves. The desires and intentions that move us will be those that are constitutive of or essential to our fundamental hopes.

There are related cases of self-frustration which are stark and extreme, but of less relevance to the general run of persons. The alcoholic or the heroin addict may be deprived of the expectation that he will be able to act on the desire, which itself may be a life-hope, to give up drink or heroin. He faces a future in which desires which he has for drink, or heroin, are desires he wants not to have. (Cf. Frankfurt, 1969, 1971.) He will do what in a fundamental way he wants not to do. Others of what are sometimes called behaviour-disorders are similarly stark and extreme: sexual abnormalities, kleptomania and the like, and compulsions generally.

A third sub-type of enabling or frustrating circumstance has to do with independence of other particular individuals, or of the ascendancy of other particular individuals over us. My hopes may depend on my belief that the desires on which I shall act will not in some way be formed for me by someone who has a dominance over me. I shall not be as a child to another person. Relatedly, and more starkly, there are threats and what have been called coercive offers. It is possible to take too narrow a view of threats in particular. Not all of them, by any means, are close to the philosopher's staple example, that of a man who has a gun held to his head. A person's prospects may be greater or lesser, in ordinary life, and seen to be such by him or her, as a result of respectable threats, threats without guns.

Fourthly and finally, there is the kind of frustrating circumstance which amounts to personal bodily constraint. A bodily disability, want of health, an injury, prison or other confinement—these too must touch or drag down or destroy a person's hopes. As before, there is the contrasting situation, where a person can have the belief that his desires will not be impeded in such a way.

A richer and more precise account might be attempted of beliefs and ideas as to these four kinds of satisfying and frustrating circumstances. It is not needed. What seems evidently true of each of the four reasons for hope and contents of it—anticipations of satisfying or enabling circumstances of the four kinds—whatever fuller account is given, is that each is logically consistent with determinism. The truth, if it is one, that my actions and activities will not be owed to serious constraints of my social world, that I will not in this respect act out of reluctant desires, is evidently not in conflict with the truth of determinism. So with truths about actions and activities not issuing from such things as my own weakness, but from my dominant desire, the desire which is my life-hope. So with truths about independence of others and freedom from threat, and truths about personal bodily constraint.

To speak generally, there is no logical inconsistency between a determinism and the summary proposition of *voluntariness* or willingness that *a man's or a woman's actions can issue from his or her own embraced desires, be done in satisfactory or enabling rather than frustrating or obstructing circumstances.* Nor would there be inconsistency if we enlarged the account by adding one or two other satisfying or enabling circumstances of somewhat lesser importance to life-hopes—say a circumstance having to do with anticipated knowledge rather than ignorance.

As in the case of the logical *inconsistency* between determinism and the earlier foundation and content of life-hopes, having to do with one's self, there is no room for doubt about the fact of logic—in the present case, the fact of logical *consistency* between determinism and the ground and content having to do with embraced desires. To speak of our own theory of determinism, nothing whatever in the specified nomic connections and in particular the specified causal sequences is in conflict with the proposition about embraced desires. In order to have inconsistency, we should need to have some primitive conception of causation, one involving a kind of animism. That is, we should have to suppose that a caused desire, in virtue of being caused, was an unembraced or reluctant desire. Time does not need to be spent on that. We *have* in the beginning (Ch. 1) spent time in arriving at a clear and at least defensible account of causation. It settles the matter. The characterization of causation in terms of certain conditional statements leaves no room for doubt, having to do with causation itself, as to the consistency between the theory of determinism and the given grounds for life-hopes.

As with the earlier ground and content of life-hopes, having to do with one's self, there is no call for anything that could reasonably be called a proof of the fact of logic. It may well be possible, with our fact of consistency, as in the case of many indubitable consistencies, to give a further analysis of the conceptions involved, and so to provide a further account of the consistency. To do so would not be to settle what was until then unsettled. Indeed, an analysis of plain consistencies which purported to put their consistency into doubt would in fact be self-defeating, such as to unseat the analysis.

To come to the end of these reflections, what we have is that it is entirely possible to rest a life-hope on the sufficient belief, a quite clear belief, that one will in the future be acting out of embraced desires, acting in satisfactory rather than frustrating situations. Or rather, to remember again that one's future actions are likely to be ends as well as means, it is possible to have as both content and basis of a life-hope the prospect of acting out of embraced desires. Evidently this is an attitude fundamentally different from and in conflict with the one considered in the last section, involving the idea of a self. It is a life-hope different in kind from the life-hopes involving a self.

This attitude, whatever else is to be said of it, is a natural one in the same sense as the other. (p. 392) It is as good as universal. It is certainly not factitious, the work of philosophers or the like. It involves no internal conceptual shortcoming. Further, there is the evident reasonableness of the desire. That is, it is for and based upon a clear, large fact, that for the most part we do and shall act out of embraced desires.

Given the naturalness and reasonableness, it is hard to resist the idea, at least for a time, that in so far as the initiation of actions is concerned, our life-hopes need involve and rest on nothing more than embraced desires, satisfactory situations.

That we have or can have such substantial life-hopes is the principal source of a certain response, different from dismay, which we may make in the matter of determinism and life-hopes. It is *the response of intransigence*. It also has a lesser but significant source of a different kind. It is in part a response to the response of dismay and its associated desire, involving a self, an unfixed future, escape from environment and dispositions. Intransigence and perhaps its associated desire are in part a *dismissal* or *rejection* of dismay and its associated desire.

That latter desire, if certainly we feel it, and if it is natural, is crucially a kind of image—of one's self. An image, we may feel, is unsatisfactory, something that falls well short of giving us the reassurance of an explicit idea or argument. Its unsatisfactoriness is something for all of us, certainly, and not just for philosophers or other conceptually demanding persons. No one, in serious matters, is likely to remain content with the insubstantial. Further, although this is of much less importance with respect to the wide fact—the fact that the given desire and the response of dismay are as good as universal—the image is not much improved on in indeterminist theories of the mind, which are near to chimerical. As for dismay itself, it is by definition no satisfactory thing, but a thing which gives rise to resistance or opposition.

The response of intransigence, at its strongest, is that a determinism is no threat whatever to our hopes of the kind that matter. A determinism leaves our important hopes as they were. Nothing changes. We have no need whatever to succumb to the response of dismay in connection with our hopes. The truth of determinism is indeed inconsistent with something. But it is inconsistent with what does not matter. There is that which *does* matter with respect to our hopes, which is action out of embraced desires, and determinism is fully consistent with that. The response of intransigence, when it is not at its strongest, is that determinism does not much touch our hopes. Things remain fundamentally as they are. Here and elsewhere, as will be clear, brevity excludes the giving of a full and non-schematic account of a phenomenon in its diversity. To speak of the response of dismay, or of intransigence, is to speak of what has variations, degrees, colourings—all of which must be passed by. It may indeed be that really to convey the diverse reality of the phenomena we should have to attempt to use something close to the methods of imaginative literature. (Cf. p. 403.)

The desire we have been considering, a second kind of life-hope, is like the first in being a member of a family of attitudes. We have attempted no full or general picture of the character of either family, but it was remarked in passing (p. 390) that the first family of attitudes has to do fundamentally with *origination*. To use again a second label, it also has to do with *voluntariness*. The second family has to do *only* with *voluntariness*. Attitudes of the first family issue in dismay, and attitudes of the second family do or may issue in intransigence. We now proceed with the other members of both families, and with further instances of dismay and intransigence.

7.4 PERSONAL FEELINGS

Few parts of our experience are as satisfactory as our experience of the love of others for us, and their affection, loyalty, trust, goodwill, and support. Much the same is true of what is somewhat different, our experience of certain more judgemental attitudes of others to us: their admiration, their approval, their inclinations to commendation, compliment, and the like. To speak a bit less lamely than one does in calling the experiences merely satisfactory, it can be said that love and the like and admiration and the like enrich our lives, give delight to them, and sustain them.

What is in question, to describe it more generally, is our having the good feelings and the good judgement of others, as evidenced in their particular actions on our behalf, or in their continued endeavours, or perhaps in their conduct of the whole of their lives. To have nothing of the love of others, their acceptance of us and their sustenance in adversity, and to get no honour, esteem, or marks from them, is to have merely a grey time in this world. To have a life from which one's reciprocally warm feelings were removed, and one's pride or satisfaction in having the good opinion of others, would be to exist in less than a full human way.

What we want, of course, is not properly described as our feelings of reciprocal warmth, pride, and so on, if those are somehow understood as separate from our *true* beliefs about others' actions and lives. That is, we do not value the conceivable situation where we have certain feelings, but for no good reason. What we want and value is not merely something internal to ourselves, but something which essentially includes the relevant doings and manifestations of feeling by others.

The worth of it is more than indicated by the fact that we are not at all inactive in connection with the good feelings and good judgements of others. We do in fact seek out their good feelings and judgements. It is not too much to say that we pursue them, and that sometimes we pursue them indefatigably. Life-hopes, certainly, are importantly hopes for the good feelings and good judgements of others with respect to us. If argument were needed for the importance to us of these attitudes of others to us, our behaviour would provide it.

As for the character of our reactions to those actions of others which evidence good feelings or judgements with respect to us, it is clear that they are not rightly conceived as moral reactions, although it is obvious that the two kinds of reaction shade into one another, and that we are quick in all circumstances to find moral support for ourselves. It is possible enough to be grateful to someone for a good turn which one feels—in so far as one can manage impartiality on the subject—to be something he ought not to have done. I may not feel less warmly to the good reviewer of my book as a consequence of the thought that the review was not a work of monumental impartiality. No doubt I shall be inclined to avert my eyes from the partiality or whatever. It would no doubt be difficult or impossible to have a recognizable feeling of gratitude as a result of having been benefited, with the agent's intention, by an act that was monstrous or on the way to it. Part of the reason may be that one would feel not only benefited but also injured—the latter by being a kind of accomplice. I shall leave unconsidered this question of the relation between morality and our personal responses to the good feelings and judgements of others, although more will be said of relevance to it. It is plain enough that the personal reactions or attitudes in question are discriminable from moral reactions and can persist in a kind of opposition to many moral responses.

So much, in introduction, for the place in our lives of experience of the good feelings and good judgements of others—in sum, the place of our *appreciative feelings* towards others for their good feelings and judgements. As will be clear, no close or taxonomic account has been attempted, and at least one side of the subject-matter has been left untouched—our appreciation of others for actions which benefit those close to us, notably our children.

There is a counterpart of the appreciative feelings. It is that part of our experience which can be put under the heading of the *resentful feelings*. The heading covers a good deal. The feelings in question are personal responses to others on account of actions which evidence dislike and so on of us or bad judgements on us. The spectrum of our responding feelings can be taken to run across hatred, vengefulness, grievance, reproach, umbrage, hurt and pique. Perhaps more so than the appreciative feelings, they are desirous. They may have in them desires to return the hurt.

The resentful or accusatory feelings with respect to others are not a satisfaction in life. They are sometimes nurtured, and sometimes fed on, but it is impossible to think that anyone who was spared them would sensibly choose to bring them into his or her life. That is to say, importantly, that no one would choose to bring into his or her life that which gives rise to his or her resentful feelings: judgements and feelings of others against oneself. Also, the feelings themselves, considered intrinsically, are at the very least a burden. Still, there is something else. In our lives as they are, some others will in fact reject us, go back on their loyalties, disdain our works, or withhold their approval. Do we wish, in this circumstance, not to have the resentful feelings?

It is possible, in some moment of high rationality, to wish oneself free of the resentful feelings, where that is a wish not about the actions and attitudes of others, but a wish about oneself. It is to wish oneself free of the feelings despite the existence of grounds for them. Still, there presumably is some sense in the idea of a proper indignation, or indeed the idea of a decent rage. If keeping a feeling of affront alive or feeding on resentment may be transparently a bad thing, for the person engaged in the activity, it strikes one as unsatisfactory that a woman betrayed or ill-used should be short of feelings in response. One may feel this too of classes of people, indeed nations, who have been victimized or viciously treated. This has to do, not with a recognizable moral judgement against betrayal, ill-use, victimization, or viciousness, but some less clear impulse having to do with the propriety and humanity of resentful feelings, or at least ways, degrees, and durations of them.

To move to a more tractable matter, there can be no doubt of the *rootedness* as distinct from the desirability or defensibility of the resentful feelings in our lives. Despite what has just been wondered about their desirability or defensibility, we have in fact often had urged upon us the policy of escaping them. Never, it may seem, was a policy less likely of settled or general success. Far from escaping them, we must struggle even to make a start. They seem to be part of the stuff of our lives, and, despite some small victories, it may indeed be difficult to imagine ourselves free of them. No doubt good men resolve to have good cause for their resentful feelings, but few good men resolve with any confidence not to have them at all. Given this, any proposal about the ending of them which has something effective behind it, say an arguable claim of fact or logic, as distinct from religious or conventional moral exhortations, which have lost any force they may once have had, is unsettling. One may feel asked to subtract a part of one's nature.

Both the appreciative personal feelings and the resentful personal feelings are feelings directed towards others. There are related groups of feelings which are directed by ourselves towards ourselves. When these reflexive feelings are related to the resentful feelings towards others, they have to do with our acts of self-hurt, and perhaps self-disdain. In the other case, they have to do with self-help and the like. I can certainly reproach myself for an act by which I damaged my own prospects, and feel something uncommonly like self-commendation on other occasions—certainly there does occur self-praise, and it has something like the basis of our sincere praise of others. None the less, it seems that self-directed feelings of the two kinds play a lesser part in our experience than the two kinds of feelings about others. (Cf. p. 381) What is in question in connection with the self-directed feelings, as with the feelings about others, is not morality. When I reproach myself for the self-hurting act, I do not think in a moral way that I have wronged myself.

What are our grounds of belief and the like, or presumed grounds, for our appreciative and resentful feelings? Suppose I have in the past wounded a friend and patron by my unkind talk of him. Since then I have made good recompense, and taken care not to cause hurt again. He is now in a position to help me again. A third party, a past adversary and competitor of mine, with a score or two to settle, and aware of my hope, contrives to have my friend and patron understand that I have persisted in my wounding ways. There is some truth in his report.

Why, despite that, when I come to know of my adversary's activity, am I vengeful or bitter? That is, on what beliefs or ideas—no doubt ready assumptions on my part—does my vengefulness rest? Or, as it may be as correct to ask, what beliefs or ideas enter into my attitude? One thing is not in doubt. It is that they are beliefs or ideas which can be expressed by saying that my competitor *could have done otherwise* than damage me as he did. Even given our opposed interests, and our past history of competition and strong feelings, he could have refrained from doing what he did. But what precisely are the beliefs or ideas in question? As with life-hopes, there are two possibilities.

The first set has in it, first, the assumption that my adversary's act of reporting my talk was not something done without a realization of the nature and likely effect of the act. He knew what he was doing, and hence the effective distinction between that and what he might have done instead. The act was certainly not done in ignorance, or merely thoughtlessly, or inadvertently. His report, I am sure, was far from being the activity of someone who in fact makes trouble because he lacks a sense of the uncertainty of human relationships, or the susceptibilities of a hearer. In general, I assume that the act was done out of clear knowledge, not as a result of any relevant kind of ignorance, and hence in an awareness of the alternative.

Secondly, I assume that my competitor was not unwilling in his action—in the sense that he was driven or compelled to act, by some despised or unsupported part of himself, against another more significant or even fundamental part of himself, perhaps an impulse to truth, fair dealing, or considerateness. Nor was the action the somehow uncertain upshot of a conflict within him. The action, then, was not subject to constraint in one of its several internal forms. His desire in doing what he did was not in that way a reluctant one.

Thirdly, I assume that the act in question was not, so to speak, unnatural to the agent. Such an act as he performed, although done in knowledge and not internally forced upon its doer, is sometimes an act out of character, or at odds with the doer's personality, or clearly not within his ordinary way of carrying on. I do not suppose, in the present situation, that *this* act was such. It was far from being true, I suppose, that my adversary in his malicious report was not his usual self. He was, in my view, exactly his usual self—a person who also in the past has not been restrained by honourableness or whatever.

Fourthly, I do of course assume that he is, despite what I see as his shortcomings, as normal as the next man. He is a man, no more immature in any radical sense than the rest of us, and not of a personality that is radically out of kilter. He is a long way from being a deranged character who sees the world as consisting in a plot against him, or a person who quite fails to have any empathetic comprehension of the feelings of others. Nor is he so extraordinary as to be the victim of all of his passing desires, a leaf in their breeze.

Fifthly, although the question of the influence of others on him may arise, I of course take it that his action cannot be described as owed to someone else. His judgement was not distorted by someone able to manipulate him. He was certainly not the puppet of another person, or acting under threat, or in another way not his own man.

Taken together, these five considerations, to which some others might be added, unquestionably enable me to say things which fortify or indeed increase my feeling. The vicious action, I say, was wholly his. It was no accident or misfortune, not something which befell or happened to him. It came from his nature, his established and persisting nature. He was in no way external to it, a bystander or a victim of it.

This account of ordinary assumptions we may make, assumptions which very certainly can underpin our resentful personal feelings, or be incorporated in them as the only assumptions about the initiation of action, has been derived in good part from the unprecedented essay which first widened the consideration of consequences of determinism, and which will be further considered below. (P. F. Strawson, 1962) As will be apparent enough, the assumptions are of the same character as, and in good part overlap with, one of the two sets of grounding beliefs for life-hopes considered earlier. (pp. 393 ff.) Those were beliefs having to do with a man's actions flowing from his own embraced desires, and hence beliefs about actions not frustrated by the way of his world, certain desires of his own, the dominance of others, or bodily constraints. The first and last of these possible considerations is not so germane to personal feelings as to life-hopes. Also, when we think of our life-hopes, we are not likely to have in mind, in connection with our life-hopes, that we ourselves are relatively normal individuals. More generally and more importantly, what is fundamentally a very similar set of considerations making for what has been called voluntariness or willingness (p. 397) differs significantly and takes on different casts of feeling in different contexts. There is a good deal to be gained by reflecting these differences in our inquiry, rather than relying on a general and uniform account for all contexts. For one thing, we defeat, in ourselves as well as others, a resistance to the general which derives from no more than forgetting that the general proposition can be replaced by a set of particular and persuasive propositions.

The five assumptions I make in the imagined case of my patron and my competitor, which are quite sufficient to support my feeling, do not exclude a determinism. They can be and are properly understood in such a way that they do not. In particular, nothing in them fights with the three hypotheses of the determinism that has been expounded. It does not at all follow from the three hypotheses that all actions are done out of any kind of ignorance, that they are somehow forced upon somehow unwilling agents by themselves, that they are acts unnatural to the agent, or done by abnormal agents, or that they are somehow owed to someone other than the agent. The three hypotheses do not deprive our resentful feelings of the given assumptions.

Evidently we can transfer these conclusions to counterpart beliefs which underpin our appreciative personal feelings. It follows in no way about a generous act of giving, or a piece of shrewd support, from the fact that the desires and intentions behind it were items in certain causal sequences, that the act was done obliviously or out of internal compulsion, was an act out of character, or was owed to abnormality or external constraint. If there are some differences between the grounds or elements of resentful and appreciative feelings, as indeed there are, they are fundamentally alike and both are consistent with determinism.

What we may do, then, when we bring together the likely truth of determinism with the fact of personal feeling, some of it entrenched and enriching, some at least entrenched, is to see that we evidently do have the possibility of persisting in this feeling. We can maintain that certain considerations are quite sufficient for this feeling, and that these considerations are in no way threatened by determinism. To speak differently, and more precisely, appreciative and resentful feeling of a certain fundamental type can persist. It is feeling for or against another person, bound up with a set of facts about another person, a set which has certain good or bad desires and intentions to the fore. Such feeling is a member of the family of attitudes having to do with only voluntariness or willingness (p. 397), the family whose other member of which we know is an attitude to the future, an attitude of hope, resting on or having in it only an idea of embraced desires.

This is the response of intransigence. It comes roughly to this: we *do*, in considerations having to do with the absence of ignorance, overmastering desire, unnaturalness, and so on, have real and effective grounds for personal feeling; they *are* consistent with a determinism; any other grounds are as nothing beside them; if these other imagistic and insubstantial grounds are inconsistent with determinism, the fact need not unsettle us, need not disturb us in our human ways—things can remain as they are.

In this intransigence, whatever else is to be said of it, we have the natural idea that the five grounds of personal feeling that have been noticed are the worthwhile sum of what is conveyed by saying that the persons who do us good or ill generally *could have done otherwise*. As remarked before, that those words catch hold of whatever beliefs or assumptions support our feelings is not in question—no more in question that that the grounds of our life-hopes are somehow caught by the words that we do in some significant way *initiate* our actions. The natural idea, to speak a bit differently, is that what is conveyed by saying someone could have done otherwise than he did is properly to be taken as this: he could have done otherwise in so far as considerations of ignorance, his own desires, his not being himself, abnormality, and dominance by others were concerned. That is, none of those possible limiting facts was a fact about his action.

In this intransigence we have one way, but not the only way, of responding in the matter of determinism and the personal feelings. We can respond differently, as a result of taking or recalling another attitude. We can be dislodged from the position we have taken up. We can again begin with the plain thought, that personal feeling is based on a proposition somehow to the effect that others generally *could have done otherwise*, and come to rest in another place.

We can be taken by the perception that persons who could have done otherwise have true of them *more* than that they acted with knowledge, without internal constraint, naturally or in character, in a normal maturity, and without being subject to threat or the like. It is wholly possible to take the five considerations to be a part but not always all of what is rightly expressed by saying that they could have done otherwise. More important, it is possible to think that the rest is what is fundamental.

Consider again the example of my adversary and the harm he does me. He reports my unkind talk, in my view, out of desires to pay back old scores and to try to deprive me of the support of my patron and friend. My vengeful feeling, we have so far supposed, can rest on the fact that these specific desires did not involve certain things—ignorance and the others. Hence his action was his own. My vengefulness, as we have said, is directed at a certain considerable corpus of dispositions, with the mentioned pair of desires having a prominent place in the corpus.

It is all too possible to raise, and to be captured by, certain questions. Should my vengefulness be directed at only the corpus of dispositions? Can it properly be? It is natural to reply that despite what has been said so far, my feeling is rightly directed against precisely *a person*. If I have strong feelings about what I take to be a pair of vicious desires, within a certain corpus of dispositions, I also have strong feelings about *my adversary*. I indubitably have feelings about *him*. It is not that I have feelings of which it can be said only that they are directed at a human being. Can we contemplate the idea that the person in question is to be taken as just identical with the corpus of desires and other dispositions? Well, to speak in a quick but defensible way, this is but a *collection* of things. It is, to me, a collection of things that is disagreeable or worse, but that is not the point at issue. A collection of dispositions must seem to lack just the *unity* which is the mark or essence of a person.

There is another stronger if related line of thought. My vengefulness may be objected to, perhaps objected to as unreasonable. What will be offered against me, above all, is *explanation* of my adversary's action. It was the consequence of those desires, which desires themselves have a certain history. It was owed, as well, to other things, which things also have their explanation. I shall naturally say, in reply to this sort of objection, that my adversary could have restrained himself. He could have overcome those things, his dispositions. That, as it seems, distinguishes him from the corpus of dispositions. I shall need to say, in order to defend my vengefulness, that *he could have done otherwise* where that means something other than we supposed a little way back. (p. 403)

Saying he could have restrained his passions does not seem to carry only the thought, for what it is worth, that no single desire of his was so strong as to overbear all his other desires and inclinations. It seems to carry the thought that he, as distinct from any one of his dispositions, might have overborne the desires which issued in his action. My adversary, in my vengeful view of him, was not just a set of directed pressures, so to speak, in this case pressures not opposed to one another, and all of them directed towards the act. Something else was on hand, different from a pressure, but capable of resisting pressures. If the metaphor is baffling, it is not therefore misleading.

The general idea to which we have come is one encountered earlier, in connection with grounds for life-hopes—the idea of a self. We can also get to the idea by other routes related to those we have. If I suppose that my vengeful feeling is owed to my perception of a certain corpus of dispositions, the question will indeed arise of their explanation. I cannot take it, in answer, that they have *no* explanation. Nor, it must seem, if they do have an explanation, can I resist transferring my vengeful feeling to *it*, to whatever has given rise to them, shaped them, or released them. Furthermore, to be secure in my vengeful feeling, it must be that *this* source or whatever is not itself the product or whatever of something else. I must find a proper object of feeling which does not allow for further regress. I come again to what we gesture at in speaking of a self.

Two or three related lines of reflection, then, issue in the image of an uncaused and unsuperfluous entity, the image which gives rise to the originator and its activity of origination in indeterminist theories of the mind. In the image we have what must seem to be a necessary condition or indeed element of personal feelings of appreciation and resentment. It is a necessary condition of personal feelings of a certain type or character. It does not need remarking at this point, except out of pique, that there remains an occasional philosopher who can succeed wonderfully in mistaking it for just a requirement of randomness. It is not that. (pp. 184 ff.; Watson, 1982, p. 13; cf. Honderich, 1973a, p. 208)

There is a general way of distinguishing between the ground we now have for appreciative and resentful feelings, involving origination, and the conflicting one considered before, involving only voluntariness, which ground gave rise to the response of intransigence. The earlier ground, it was said, can be expressed in this way: typically when a person acts, in so far as ignorance, internal constraint, his not being himself, abnormality, and dominance by others are concerned, he could have acted otherwise. None of these possible limiting facts was a fact about his action. The ground we have lately been considering for our appreciative and resentful feelings can be expressed in this way: typically when a person acts, in so far as the possible limiting facts just mentioned are concerned, *and also* in so far as his dispositions and indeed all his properties are concerned, he could have acted otherwise. *He*—he who is not a property of himself—could have risen above all his desires, temptations, weaknesses, and so on. Given his entire state just as it was, he could have acted otherwise.

When we have in mind this ground for our personal feelings, a ground involving both voluntariness and origination, we have in mind what is inconsistent with necessitated intentions and actions—inconsistent with the theory of determinism. It is not true, if determinism is the case, that a person could ever have acted otherwise in the given sense. No doubt can exist about the logical relation between the theory of determinism and the given ground for personal feelings. The two are indeed inconsistent, since the theory of determinism is to the effect that given a man's dispositions and other properties at the time of his forming of an intention, the latter thing was necessitated, and, further, that the subsequent action was also necessitated.

As dismay was a natural response for each of us with respect to determinism and our life-hopes, so too is it a natural response to determinism and personal feelings. Appreciative and resentful feelings, the first kind entrenched and of great value to us, and the second kind at any rate entrenched, are out of place given the truth of determinism. Or rather, a determinate category of both appreciative and resentful feelings are out of place, and, to speak of the appreciative feelings, no other category can be satisfying—there can be no satisfaction in feelings involving only voluntariness, which by comparison are of little worth. The feelings that are out of place depend essentially on or involve a proposition or propositional image which is false. That it is unclear exactly how the feelings depend on or involve the proposition —that question has been left untouched—does not affect the issue. (p. 478)

In sum, with respect to personal feelings, we have two inconsistent possibilities. We can take them to be attitudes having to do with only the voluntariness of actions, and respond intransigently to determinism, and we can also take them to be attitudes having to do not only with voluntariness but also origination, and we may then fall into dismay with respect to determinism. It is better to say, simply, that we *have* the two conflicting attitudes, and make the two responses. We can in fact move from either attitude and its associated response to the other attitude and its associated response.

7.5 KNOWLEDGE

Our life-hopes and our feelings towards others are fundamental parts of our existence. They are parts of an existence which also has something different in it, wholly pervasive attributes, notably the attribute of being informed by knowledge. A danger to our knowledge is therefore a wide or encompassing danger. Further, since our knowledge is directive not only of our hopes and personal feelings, but of all our desires and intentions, anything that puts into question our knowledge also puts into question the rationality of what is also pervasive, which is desire and intention. It makes suspect for us what we take to be the truths on which our desire and intention depend. It is not too much to say that what threatens our knowledge threatens our existence as we conceive it.

The Epicurean tradition of objection to determinism, as we have recently seen, does not often make clear its own strength. (6.7) Those in the tradition have most often been concerned somehow to trace knowledge to the laws of logic or the like, or to claim the inconsistency of such suppositions as that reasons could be causes. What lies under such objections, or is uncertainly conveyed by them, it was argued, is the proposition that mental acts and ordinary bodily actions are prerequisites of knowledge, or, more precisely, that a certain freedom in these acts and actions is such a prerequisite. The Epicurean tradition, then, if it can rightly be seen as offering a direct objection to determinism, perhaps by way of the argument that determinism applies to and undermines itself, is fundamentally an answer to one of a specifiable class of questions about the consequences of determinism, with which we are now concerned.

The questions of this class, although they have not until now been described in this way, are those which arise from the seeming conflict between determinism and *our freedom in mental acts and bodily actions*. They include the questions at which we have just looked about life-hopes and personal feelings. What is the consequence of determinism for assumptions of our own freedom made in connection with life-hopes? What is the consequence of determinism for assumptions of others' freedom made in connection with personal feelings? That the issue of determinism and knowledge comes here, as we saw, cannot be denied by way of the truth that belief is involuntary. As might have been remarked in our earlier discussion, it is an ordinary fact that questions of *responsibility* arise about belief and knowledge. The maxim that ignorance of the law is no excuse points directly to the further truth that belief and knowledge do involve more than what is involuntary. So too does the ordinary fact that we blame people for being careless about finding things out.

Our knowledge is with reason divided into three sorts, firstly knowledge of propositions or that such-and-such is the case, secondly the kind of knowledge which consists most importantly in sense-experience, and thirdly what can be called practical knowledge or simply ability, such as knowing how to tie one's shoelaces. (Pears, 1972) It is the first sort that must be most important here.

It is plain, as argued already, that our propositional knowledge is in essential ways owed to our ordinary actions. This is no elusive claim to the effect that believing or knowing in themselves have an 'active' character, that they themselves are somehow intrinsically active, or the more arguable claim that we construct 'theories' which guide our sense perception. Nor is it any philosophical speculation about connections between the meaning of concepts and our activities or operations. (p. 325) It is a claim which is trivially but clearly exemplified by the simple truth, to revert to an example, that my knowledge that Colin McGinn is next door is owed in part to my having gone there to see. Many of my pieces of ordinary knowledge do have as necessary conditions intended movements of mine which I can easily specify. More generally, the vast corpus of my particular knowledge about the world is evidently the result, in part, of my past actions—actions, again, in the most ordinary sense. This truth is not affected by further ones, that no particular piece of my knowledge has but a single action as necessary condition in its background, and that I cannot begin to specify all of my actions which were necessary conditions of any particular piece of knowledge. Evidently an individual's ordinary actions will have to have a role in any model or theory of the accumulation of his knowledge.

Also, as anticipated above, there is a like proposition that is more important, having to do with mental acts. It clearly depends on there being a distinction of a satisfactory kind between active mental events—mental acts—and passive mental events. We did specify a distinction in connection with the Hypothesis on the Causation of Psychoneural Pairs. (p. 174) Mental acts, it was said, are purposive or goal-directed, which is at least in part the fact that they are subject to an ongoing desire. They can be characterized further by way of such ideas as those of interim goals and plans. Despite the existence of this proposal as to a difference between mental acts and other mental events, the attempt has sometimes been made, occasionally by determinists, to deny any distinction. Also, it has happened that *all* of what goes on in our minds has been taken to consist in mental acts somehow conceived. (Geach, 1957) These contentions, passed by earlier without notice, need consideration. Let us pause to give it.

That there is a prima-facie difference between active and passive mental events is evident enough. We do ordinarily speak in many ways of our beliefs and also our desires as not being a matter of activity but of passivity. We cannot but think or feel such-and-such, or escape this or that sensation. However, as remarked earlier, we also speak of what we evidently take to be somehow active: attending, questioning, inquiring, speculating, judging, trying to prove, making up our minds. (pp. 107, 174) We make a difference between deliberating and vacillating, and also make like contrasts in other cases, for example between (i) the claim that someone solved a problem and (ii) the claim that the solution came to him. It is not difficult to be inclined to the argument that the latter difference is evident in the fact that the two claims can indeed be so understood that the second does not entail the first. (Taylor, 1966)

Perhaps the denial of a difference between mental acts and other mental events has been partly owed to the fact that the difference does not involve a sharp boundary, that mental acts do shade into other mental events. This is no good reason for the denial. For what it is worth, there is a somewhat related difference between ordinary actions and other bodily events. Some bodily events are *in a way* or *in part* actions—falling down can be an example. Sometimes the denial has not been owed to direct reflection on the matter but has been a consequence, correctly or incorrectly drawn, of a general theory of the mind.

Perhaps the main source of a denial of a difference between mental acts and other mental events has been the absence of an account of mental acts which makes them one with or sufficiently like bodily actions. It is clear that we cannot give an account of them much like the account given earlier in this inquiry of ordinary actions. (4.1–4.3) One difficulty has to do with intention. It does seem that *some* mental acts are preceded or accompanied by active intentions, or things very like them. This may be so of the acts which enter into considering a proof or carrying forward an inquiry, or indeed remembering an address or simply deciding on something or choosing between alternatives. However, there also appear to be mental acts which are *not* preceded or accompanied by active intentions—it seems this can be true of those mental acts which *are* the active intentions involved in ordinary actions, or the like things that sometimes precede or accompany other mental acts.

The absence of or even the impossibility of a unitary general analysis of both mental acts and bodily actions is perhaps no better reason than the others for a denial of the difference that has been made between them and other mental events, in terms of purposiveness or goal-directedness. Furthermore, more can be added to that account. It brings mental acts and bodily actions more closely together. Mental acts do in fact share discernible attributes with actions. They may involve trying or struggling. They may be deliberate or careless. We may indulge in them or refrain from them. We may take credit for them or feel remorse about them. As already implied by saying that questions of responsibility arise in connection with knowledge and belief, others may blame us or hold us responsible for mental acts, or for not having performed them. Any attempt to argue that there are the same or very like facts about *all* mental events—say seeing something, being subject to a desire, or having a thought—must necessarily be a lame attempt.

Therefore, in addition to the fact that our propositional knowledge is owed in essential ways to our ordinary actions, there *is* also the more extensive fact that it is owed to what can properly be distinguished as our mental acts. Pieces of knowledge owed to discriminable ordinary actions are necessarily also owed to the discriminable mental events involved in those actions. Other pieces of knowledge are not owed to discriminable bodily actions but are owed to discriminable mental acts. To find the solution to a puzzle, or to get the answer to a question, often enough involves mental acts whose existence is clear to us, but does not involve ordinary actions which we can specify.

If ordinary actions and also mental acts are necessary conditions of our pieces of particular knowledge, it is also true that they are necessary to what can be distinguished from such pieces of knowledge, and also from our general knowledge in a certain ordinary sense. Acts and actions are conditions of what can be called our *general conceptual scheme*. A first part of this is made up of the arrays of categories upon which we depend to organize our existence. A second consists in what can be inadequately described as logical conceptions and generalizations. It includes the conceptions of truth and probability, the laws of logic, and the canons of empirical or scientific inquiry. It also includes informal counterparts and ramifications of these, which govern our ordinary way of carrying on as believers and knowers. In a third part our general conceptual scheme can be taken to include several evaluational systems, one of them having to do with morality. A good deal of our evaluational systems cannot be given the name of knowledge, but evidently much can. It involves descriptive categories. Further, by way of a single example, morality involves a principle of transitivity: if *a* is worse than *b*, and *b* worse than *c*, then *a* is worse than *c*.

Our general conceptual scheme may appear to us to have the character of a great datum, something different in kind from a product, and far from a contrivance. Particular parts of that scheme, say the laws of logic and arithmetical truths, have a yet more independent standing. Be that as it may, we cannot suppose that our conceptual scheme is the work of anything other than ourselves. Most of us, despite occasional defences of what is called the independent reality of propositions, numbers, and so on, cannot suppose that our conceptual scheme has any reality apart from us, that it could outlive consciousness. Our conceptual scheme was in fact the slow creation of our species. To think of that process, inevitably, is in part to think of the ordinary actions and the mental acts of our predecessors and ourselves.

So much for introduction. Suppose now (i) I am the returning officer for an election, obliged to determine and declare the winner, and lack returns as to the voting in one crucial constituency, or (ii) that my aim is to complete a calculation, but I lack a value for one variable, or (iii) that I am faced with a moral problem and cannot satisfactorily conceive the effect of a possible action on a lover, or (iv) that I am trying to decide on the acceptability of a philosophical theory, and cannot persuade myself of the coherence or incoherence of some transition of argument in it. In each case there is a specifiable bodily action or mental act, or a number of them, that I cannot perform. In each case, then, there is evidence which I lack. The result is a want of belief and knowledge. What reasons I have for a concluding proposition cannot count as good ones.

Suppose now that I get the final piece of evidence in each case. But suppose also that I come to accept or at any rate contemplate that my resulting belief, and what I would ordinarily take to be my knowledge, is an effect of a certain causal sequence, as specified by the theory of determinism. Suppose, above all, that I accept or contemplate that each act and action of mine and of others which contributed to that result was an event of which the theory of determinism is true. These will include, further back, myriad acts and actions which entered into the development of our conceptual scheme, and so contributed indirectly to the result. In each case, what occurred was a necessary event: an effect of certain causal sequences. To say the same thing differently about each act and action, no other event could have occurred in its place.

Consider in particular my fourth enterprise, to decide on the acceptability of a particular philosophical theory. I may now feel that my final situation, including a view of the coherence of a part of it, and also including my contemplation of determinism, is in a way like my situation *before* I became clear about the coherence question. That is, I can suppose that there may exist facts such that different acts and actions, logically possible acts and actions never performed, would have issued indirectly or directly in their discovery, and that these facts would have provided further reasons for or against my conclusion about the theory. In particular I can suppose that these facts would provide reasons such that the reasons I actually have for my conclusion are not good ones, or are not sufficient.

For some of us, it may require an effort of will even to contemplate the truth of a determinism, or a near-determinism. Perhaps it may be an impediment, too, that the present contemplation of it, in connection with the given example, must have to do not with a particular act or action, but with an indeterminate number of them, of which we have no specific conceptions. Be that as it may, to succeed in contemplating or to accept the truth of a determinism as applied to our knowledge-claims is to do something which can evoke the response of dismay. To have a picture of ourselves and of all the makers of our maps of belief, so to speak, as having been fixed to following only certain of the logically possible paths of inquiry, is to have a picture which can very naturally raise doubt in us as to the terrain. They and we have not *explored* reality, but have been guided, however voluntarily, on one tour. By way of this thought and like ones we may succumb, at least for a time, and indeed recurrently, to an agnosticism or scepticism. Or, at any rate, we can feel that such an agnosticism or scepticism should be the consequence of an actual acceptance of determinism. Our knowledge-claims should be abandoned. Or, at least, we should lose confidence. What we took to be a ground for them is illusory.

What contributes essentially to this dismay, evidently, is the attitude that a wider and in some sense a complete spectrum of actually possible acts and actions in the past would have been *necessary* to present knowledge: a spectrum wider or richer than the one fixed by the nomic connections of a determinism. The idea of such a spectrum of possibilities is tightly attached to a certain image of the initiation of acts and actions. It is again the image of origination, the germ of indeterminist pictures of the mind. Like the image of movement across all of the spectrum, it is an image which cannot coexist with a determinism.

In short, we have or can have the attitude that what is needed for knowledge, true knowledge, is not only voluntariness—centrally, here, the absence of certain impediments, of which more below—but also origination. That this is inconsistent with determinism issues in the response of dismay. To speak differently, we have an attitude to propositions, beliefs, and the like, an attitude of confidence, which is vulnerable to an acceptance of determinism, and issues in dismay. Similar remarks are to be made about a related attitude to ourselves and others as *inquirers* and *knowers*.

This response in connection with knowledge is undoubtedly most natural in connection with certain things. These include claims to knowledge with respect to philosophical, scientific, or other theories, and also our resolutions of moral problems. This is so since both, as we ordinarily suppose, issue in the end from exercises of judgement. Dismay is most natural here for the reason that it is mental acts that have our attention. It is much the same fact, evidently, that verdicts on philosophical or scientific theories, and the outcomes of our moral deliberation, have neither the necessity of syllogistic conclusions nor the necessity of the deliverances of sense-experience.

Is it the case that the given attitude having to do with origination, and the response of dismay, are not only most natural here, but that they are possible only here? Elsewhere, as in thinking of (i) the example of the election or (ii) the calculation, involving a variable, we may give less attention to mental acts. They have less prominence. So with other cases, including the fundamental case of perceptual knowledge. Ordinarily I do not *judge* that I see a tree or a colour. None the less, it would surely be odd if our claims to knowledge, of all kinds, did not share a common vulnerability, to one or another extent. This would be odd if knowledge of all kinds is owed in part, as surely is true, to our own acts and actions and those of others, however antecedent these are to our concluding beliefs.

In fact it is possible to take a determinism to raise doubt about *all* of what we ordinarily take ourselves to know. This can come about by way of its relevance to what was called our general conceptual scheme. sense-experience is categorized by us. As for syllogistic conclusions and the like, impregnable as they seem to be, it is not impossible to feel that we might have had an alternative and in some sense a superior system of necessary truths. We can and do overcome psychological barriers to doubt in this connection, or at any rate we can perceive reason for the attempt.

The response of dismay, then, with respect to knowledge, is that a determinism undermines it. Given a determinism, the whole growth of human knowledge, including the development of our conceptual scheme, and what each of us claims to know now, is a matter of one vast fixed sequence, one system of nomic connections. Questions which might have risen in the growth of our knowledge in fact could not arise. Possibilities could not be considered. Ways of regarding reality were barred. Our good reasons therefore cease to be such, and all belief is or should be subject to doubt. They do not much matter, but there are also the further consequences claimed in the Epicurean tradition. The determinist's defence of his own doctrine as a piece of knowledge is undermined, and we accept that any affirmation of the possibility of knowledge would entail that determinism is false. (p. 361)

If it is possible to make this response owed to a certain attitude, or at least to contemplate it, it is perhaps impossible to persist in it. Hume's famous remark, about what it was possible for him to believe in his study—in the course of his philosophical reflections—and what was the case outside, in practical life, is of relevance. Dismay with respect to knowledge, further, is rightly seen by us as in a way more consequential, because more general, than dismay about either life-hopes or personal feelings. At any rate, each of is capable of moving to a different attitude to the acts and actions which are conditions of what we ordinarily take to be our knowledge, and thus each of us is capable of a different response.

I must allow, certainly, that some kind of freedom in act and action is necessary to knowledge. More particularly, I accept that in order to know something, I must have a certain confidence that requisite questions have been asked. I must have a confidence, perhaps, that requisite sense-experience has been had, or requisite theoretical possibilities considered. To lack this freedom, whatever it is, to acquire good reasons, is to be barred from belief. As to what this freedom is, an answer different from the one we have is pressed upon me by my empiricism, a common and untheoretical empiricism, and my attention to our ordinary existence as knowers. The answer is that the requisite freedom is voluntariness. We can resist our own demand for more.

To recall life-hopes and personal feelings, we can take them to be fully secured by voluntariness—by my actions being done out of embraced desires, by the actions of others not being subject to certain constraints and disabilities. Or, to speak more carefully, certain types of life-hopes and personal feelings, satisfying to us, can be secured by voluntariness. We can take the same position with respect to knowledge. We can take the attitude that knowledge requires the satisfaction of certain *desires for information*, desires involving the proposition that certain acts or actions will issue in belief and knowledge. My wanting to have the return of votes from the crucial constituency, or to know the value for the variable, or the fact about the effect of a possible action on a lover, or the fact with respect to the coherence of the transition in the philosophical theory—each of these desires is informed by the proposition that the answer to a certain question will rightly settle an issue, give me what I want, which is one or the other of the two possibily true beliefs.

It is an uncontested fact that when I cannot act on my desires for information in certain ways—when I am constrained in certain ways—I cannot have the knowledge which is my goal. Is it not reasonable to think that knowledge, in so far as acts and actions are concerned, is no more than a matter of being able to act on existing desires for information? Is it not reasonable to think that when I can act on my existing information-desires, and they themselves have arisen out of previous acts and actions owed to information-desires, I have *all* of what is with reason required for knowledge, in so far as acts and actions are concerned? No doubt there is no discerning, now, the beginning or all of the growth of the knowledge which informs my present information-desires, including the beginning and growth of our conceptual scheme. More than that, there is a kind of problem. There were, presumably, early acts not so informed, or not informed at all. But facts about our present state—that we do act out of a great history of information-desires—are not to be abused on the ground that we cannot see back to the beginning of the history.

Is there not a pointlessness in the other view, the one requiring origination? Our knowledge rests, assuredly, on truth and good reasons. To have knowledge is necessarily to have a conception of truth, and to have good reasons. Let us again suppose that the truth of propositions is correspondence with fact, or, as we can say, with reality. Good reasons, we can say, are a matter of that reality, and also of logical connections: in general, that propositions entail or provide evidence for other propositions. The reflection that gives rise to the response of dismay depends on the conception and supposition of a reality *other than or in addition to* the one we know, the known subject-matter of the growth of our knowledge. Dismay can also be taken to depend on a supposition as to *other reasons or further reasons* than good reasons of the kind we know. This is a supposition pertaining to the further reality and also to something other than the logical connections we have.

No doubt it must remain unpersuasive to declare, as philosophers have, that such conceptions are in fact no conceptions at all, that they want content, significance, or sense. We *can* properly be said to speculate intelligibly about the existence of facts that have been and will forever be beyond our experience and definition, facts associated with logically possible acts and actions which have never and will never be performed, logically possible paths of inquiry forever closed to us. We can give *some* sense to talk of unknown and and in fact unknowable logical connections. To do this is not to do the impossible, actually to get outside of the only conceptual scheme we have, but to make some use of ultimately general conceptions in it, notably the conception of a fact or circumstance. The philosophies which have made such endeavours are not unintelligible. Kant does not elude all sense when he maintains that there does exist that of which we cannot have experience, the noumenal world which lies behind the phenomenal world. Nor does Spinoza in maintaining that God or Nature has an infinite number of attributes, of which we know but two, Thought and Extension.

Still, to repeat, there is surely a pointlessness in any doctrine of a reality beyond. This is so in part since it involves so thin a conception of that reality. To whatever extent that it is put beyond our reach, it is indeed to that extent beyond our reach. The reality from which we are excluded by a determinism is as greatly beyond our reach as Kant's noumena or Spinoza's unknown attributes. What is the value of this faint picture? What are we to do with it? Certainly it is pointless in the sense of being useless to us, not only in any practical way but also theoretically, as a basis or criterion, or as stimulation to further knowledge. As for the satisfaction of knowledge-in-itself, as it may be called, and the dissatisfaction of ignorance-in-itself, these might be taken to depend on concreteness and particularity. There can surely be no great sense of loss with respect to what is near to I-know-not-what.

If we must allow the conceivability of other ideas than the one we have of reality, there is not only the pointfulness but the necessity of persisting with our own. We cannot, by the very hypotheses of our determinism, escape it. Furthermore, if it is possible to conceive of comparing or assessing alternative global knowledges—ours as it is against what logically might have been—it cannot be that our knowledge is anything like useless as an approximation of reality. We depend upon *it*, after all, for the thin conception we have of what might have been. We cannot conceivably take it to be insignificant. If we do this, we must discard what it allows us by way of intimation of a better.

These reflections could be added to—for example, by way of a treatment of what was noticed earlier, about a dependency of knowledge on causation (p. 371), but let us go no further. What we have can indeed enable us to avoid dismay, at least for a time, about determinism and knowledge. We can come to rest, although uncertainly and without guarantee, in a conflicting and arguable view. We may take the attitude that what is required for the only knowledge that we can possess, the only knowledge that can matter to us, is a voluntariness. This confidence consists in the large fact that we are not wholly frustrated, and often not frustrated at all, in our desires for information. In large part we are not subject to a set of constraints which it is fairly easy to describe, and which are akin to those of which we know, pertaining to life-hopes and personal feelings. They do have a character peculiar to the matter of knowledge. Further, we can continue to attempt to reduce the clear constraints under which we do find ourselves: ignorances, apprehensions, restricting orthodoxies, self-frustration, practical obstacles to inquiry, barriers and coercion and manipulation by those not concerned with truth or desirous of concealing it. A determinism is consistent with the voluntariness we have and which we can struggle to increase.

Our second response with respect to determinism and knowledge, then, may be a certain intransigence, to the effect that knowledge is untouched by determinism. It is not affected, let alone near to being undermined. Our existence as knowers is untroubled by determinism. We can rest easy. The main burden of the Epicurean tradition is quite mistaken, as are its related propositions to the effect that the determinist's defence of his own doctrine as a piece of knowledge is undermined, and that since we do have knowledge, determinism is false. (p. 361)

7.6 MORALITY

I may regard a particular action, say that of the Member of Parliament for Hampstead and Highgate in voting for a bill to reduce the resources of the National Health Service and to give advantages to those who have private medical insurance, as *right or wrong*. I may have the more or less reflective attitude, at least when doing moral philosophy, that right actions generally are those that fall into a certain category or categories. There are various possibilities. I may take it that right actions are all those not proscribed by the criminal law, or perhaps by my sense of some customary morality. I may take it, very differently, that they are those in accord with the Principle of Equality, that we are to take effective steps to try to make well off those who are badly off, with effective steps and the conditions of being well off and badly off defined in certain ways. I may take it that right actions are those in accord with some version of the Principle of Utility, or those somehow called for by some collection of duties—perhaps with duties of loyalty to one's family and certain others to the fore. I may take it, again differently, that right actions are those which are done out of a sense of principle, or benevolent intention, or truth to oneself, or some other fact having to do with the initiation of action, some fact about the agent in his action as distinct from concomitants and consequences of his action. (MacIntyre, 1967)

The particular judgement or conviction about the action of the

Member of Parliament, if I am consistent, will of course be consistent with my general view about the rightness of actions. The particular judgement and others like it may be consequences, or sources, of my general view—or in some other relation of mutual dependency. (Rawls, 1972)

I may also have a view of the MP himself, most importantly of his character, a view perhaps but not necessarily based in part on his particular action. I have an attitude to him *as a person*, as having a certain *moral standing*. He is a man whose actions generally are, or generally are not, in line with the correct principle for right actions according to my lights. Partly because of this, if not only, he may be for me a man of honour, or of moral tenacity, as his action may help to show, or an opportunist, or someone with a very selective eye for injustice or need, as again his action may help to show. I am likely to have, as well, some attitude or view about the general question of what all men and women ought to be like. All of us, I suppose, should strive to have certain dispositions rather than others. Again, if I am consistent, the particular judgement about the MP and the general view of what people ought to be like will be related.

It is to be noticed that to have a view of the MP himself, as what we can call in a traditional way a good or bad moral agent, is not the same as *morally to approve or disapprove of him for the given action*. We may think well of him in general as, say, a person of humanity or decency, perhaps to be revered, and yet disapprove of him for his vote, taking it to be out of character, or a lapse, or owed to misjudgement. We may have a generally low view of his character, moral perception, or sensitivity, but feel what might be called moral admiration for him in so far as this particular and uncharacteristic action is concerned.

In speaking of this third matter, moral approval and disapproval, I have in mind our feelings or attitudes, and not any resulting acts of commendation or condemnation, praise or blame, which will be considered later. (Ch. 10) They are feelings having to do with responsibility. As already indicated (p. 379), in speaking of our *holding* a person responsible for something, I have in mind exactly our feeling of moral disapproval, and in speaking of *crediting* someone with responsibility for something, I have in mind our feeling of moral approval. To hold a woman responsible for an act, or to credit her with responsibility for it, is not to perform a purely intellectual operation, but to have a feeling about her with respect to the act. It is not barely to ascribe moral responsibility to her, where that is to have beliefs or whatever about the initiation of the act, perhaps to judge truly or falsely that she is morally responsible. Such beliefs, images or judgements, which are of the ordinary factual kind, enter into or support our feelings, in a way analagous to the way in which such things enter into or support life-hopes and personal feelings. (pp. 384 f., 403, 409) Something more will be said of the matter in due course. (pp. 449 ff.)

So—we have moral feelings and make moral judgements about the rightness or wrongness of particular actions, and also about types or classes of actions. Secondly, we have feelings and make judgements about people with respect to their characters and personalities, to which their actions over periods of time, perhaps their whole lives, are of primary importance. Here too we are concerned with individuals and types. Thirdly, we have feelings about particular people with respect to particular actions. That is, we hold an individual morally responsible for a particular action or group of actions, or credit him or her with responsibility for it. (Cf. Parfit, 1984, pp. 98 ff.)

The three parts of morality now roughly and perhaps too firmly distinguished, having to do with right actions, general moral standing, and moral responsibility, evidently stand in certain relationships. These, to my mind, have not traditionally, if ever, been made clear. A clearer understanding of the relationships is fundamental to an economical and effective consideration of the consequences of determinism for morality. To be clearer about them is to see that the challenge of determinism to morality is wider and more fundamental than has been supposed, and also more easily specified. It will not be possible, however, or necessary, to arrive at much more than a schematic understanding of morality. Questions will be passed by.

To put aside for a time the approving and disapproving of persons for particular actions, and to persist with the subjects of feelings and judgements about the rightness of actions themselves, and about what men and women ought in general to be like, it is a fact that a good deal of moral philosophy is addressed to the general question of what actions are right. In answer, argument is given for a general moral principle, one of those mentioned above or another, some specification of the class of right actions. The question of what it is to be in good moral standing, the question of what men and women ought to be like, is dealt with, if at all, very much on the side. One may get the impression that the question is somehow peripheral to morality. So with the moral philosophy of Mill, and Utilitarianism generally.

In some other moral philosophy, perhaps quite as much, the concentration is the other way on. The question addressed is that of what sort of persons we should be. The good man gets all the attention, along with the bad man. We are told what sort of character or what virtues are had by good men and women. There may be the implication that the question of what actions are right is somehow peripheral in morality. So with the ethics of Aristotle and successors of several kinds.

As already implied in connection with the example of the MP, it is possible—it certainly need not be self-contradictory—to say that a man did the right thing, and that in the doing of it he gave an indication of being a morally questionable or indeed morally appalling character. (The point is different, of course, from the point that one can morally approve of a bad man for an act, or disapprove of a good one.) The MP voted as I feel he ought to have, but did so, perhaps, out of a disposition to self-serving careerism or to toadying. He does not actually have the principle which in my view made his action right, and perhaps has no respect for it. As evidently, a man may be taken to do the wrong thing, and in doing it give some evidence that he is morally admirable. The standard case here is one where the agent acts out of feelings or commitments of which I wholly approve, and judges the nature of the effects of his action to the best of his ability, but in fact misjudges them. Good intentions, but an innocent or condonable mistake as to the situation.

This fact, that right actions and good moral standing of agents do not always go together, or wrong actions and bad characters, is one of three facts which may consort to produce a certain upshot of mistake, very common indeed. The second fact is one noticed a moment ago, the concentration for whatever reason of much moral philosophy on either the question about right actions or the question about the good man. The third is the fact of our ordinary way of carrying on, illustrated so far in this discussion, in which feelings and judgements about actions, and feelings and judgements about men and women, are certainly discriminable parts. We do ordinarily separate the two matters.

The mistake is the idea that if one puts aside the issue of morally approving or disapproving agents for their actions—crediting them with responsibility or holding them responsible—and of course commending or condemning them, there remain *two* fundamental moral questions rather than one, two logically distinct questions. There is the question of the rightness of actions and the question of the goodness of agents. In fact it is true, or nearer the truth, despite our proceedings so far, to say that there is but one question where two are perceived. This one question is certainly open to different expressions.

It can be expressed in this way: (i) *How ought the world to be?* Or, better, since in morality we are concerned with the world in so far as it is in our control, there is this expression: (ii) *How ought the world to be, in so far as we can affect it?* What that comes to can also be put in this way: (iii) *What actions are right?*—so long as the words are not taken in a certain narrow way, of which more in a moment. It can also be put as (iv) *What sort of people ought we to be?*—so long as those words are not taken in a certain narrow way.

It is possible to answer this fundamental question in a way which directs attention to actions and their concomitants and consequences. The Principle of Equality, other answers having to do with fairness or justice, and versions of the Principle of Utility, are evidently of this kind—consequentialist answers, as we shall call them. (Cf. Parfit, 1984, Pt. 1.) In these cases we are given instructions which have to do with the satisfaction of desires in acting, and also, more important, such satisfaction as a consequence of acting. We are to secure certain totals or distributions of satisfaction, or certain totals *and* distributions.

However, it is possible to attempt to answer the fundamental question in ways which concentrate on the agent. Agent-related answers, as we can call them, to a greater or lesser degree identify right actions as those which have a certain initiation, often an initiation in and true to a certain kind of person. Do the act, we may be told, which flows from a pure good will, from being a certain sort of person. Do the act which is true to a universal principle about how men in themselves ought ideally to be. Certainly there is nothing inapposite or irrelevant about saying, in answer to the fundamental question, that right actions are those which come from the Kantian pure good will, or are in accord with some other universal principle about the genesis of actions in sympathy or love, or are actions which issue from or preserve the integrity of the agent. Nor, however, is there anything inapposite or irrelevant about saying, in answer to the fundamental question, that we ought to be the sort of people whose actions are in accordance with the egalitarian principle or some other consequentialist principle.

In the two sorts of moral philosophy noted above, consequentialist and agent-related, the relevant subject-matter must in fact be taken as specified by something close to our single question, best expressed as that of how the world ought to be, in so far as we can affect it. However, the question actually addressed in each of the two sorts of moral philosophy is typically so formed as to presuppose, or by custom to suggest, a certain range of answers to the single question. In the first kind there is addressed the question 'What actions are right?' taken in a certain narrow way. That is, what is presupposed or suggested is that all the arguable answers to the single question are answers having to do with certain concomitants and consequences of actions. Answers in terms of the nature of agents are in effect put on one side. In the other sort of moral philosophy there is posed the question 'What sort of persons ought we to be?', again taken in a narrow way. As that is usually understood, it presupposes or suggests a range of answers that has to do only with agents, with their initiation of actions. As will have been noticed, this range of answers has to do not wholly with what we have called the general moral standing of agents, but also with moral approval and disapproval or the initiation of actions, but let us leave that complication alone.

Both ranges of answers are in fact about how the world should be in so far as we can affect it. No doubt the selection of a question carrying a presupposition is natural enough, given someone's conviction that the answer to the question of how the world ought to be falls into a certain range. There is the unhappy result, however, that the subject-matter is characterized in such a way as to close off relevant moral issues. A range of answers is excluded from consideration without argument. Argument *is* needed. To exclude the other range of answers is to make a fundamental moral claim. It is to make the claim that either consequentialist or agent-related moralities are mistaken.

There is a related point. It sometimes seems to be supposed, curiously, by those who give agent-related answers to the fundamental question, that those who give consequentialist answers are making some *philosophical* mistake, or, more precisely, that they are failing to perceive the somehow necessary nature of morality. (B. Williams, 1973) That is, something is supposed other than the fact that the dispute between agent-related and consequentialist doctrines is a ground-level moral dispute. That is exactly what it is, a dispute as to the right answer to the question of how the world ought to be in so far as we can affect it.

That there is a single question rather than two logically independent ones is best indicated and perhaps proved by the fact that each sort of moral philosophy, despite appearances, gives and must give a recommendation both as to actions and as to what sort of persons we ought to be. (Cf. B. Williams, 1981, p. 53.) At any rate, both go a good way towards dealing with both matters. If a kind of Utilitarianism is pressed upon us in connection with actions, there is a certain upshot with respect to persons. The good man or woman is at bottom this kind of Utilitarian. If being true to oneself, or preserving one's integrity—whatever that may finally mean—were given as the whole recommendation with respect to being a person of good moral standing, then the matter of actions must be resolved in a certain way. Right actions are those in which the agent is true to himself or herself.

More generally, and whatever the form of question which is chosen, it evidently is impossible to give an adequate and tolerably complete view of the good agent without giving a view as to what actions are right. It is similarly impossible to give a view of right actions without giving a view of the good agent. This is fundamentally what makes it true, or more true, to speak of one rather than two questions.

None of this gives us the conclusion, however, that we cannot do either of the different things of concentrating on persons or concentrating on actions. The questions at the start of this section and at the start of the chapter can stand, and we shall persist with them. I can, in considering a certain episode involving a person and what he or she did, direct myself either to the person or to the act—both matters being sufficiently distinct from that of moral responsibility. Given my moral outlook, one of these things will be more natural or apposite. That is certainly not to say, to repeat, that there are *logically independent* questions. The analogy is a weak one, and in some ways misleading, but there *is* an analogy with concentrating on the effects with respect to a sequence of physical events, or concentrating on the causal circumstances. I can ask 'What are the effects?' or 'What are the causal circumstances?' It is clear that to do so is to direct attention to a part of what is a single subject-matter. It is reasonable to say that the fundamental question is what causal relations exist with respect to the sequence which is the subject-matter.

So much for a single fundamental question of morality, its various expressions, and two possible concentrations. The other large question, not less fundamental, has to do with moral approval and disapproval of persons for particular actions, crediting them with responsibility and holding them responsible. Images, ideas, beliefs, and judgements as to initiation of action enter into these feelings, and the feelings may result in acts of commending or condemning others for particular acts. To approve of a man for an action, to have such a moral feeling, is not to do something which is to be identified either with judging him to be in general a good man, as already noted, or with judging an action to be right. It is in fact something different from both, although there are relations between the three things. Let us consider what in a way are the most fundamental of the relations.

Suppose I have the view that right actions are those which are in accord with some principle of fair distribution of satisfactions and dissatisfactions, perhaps the Principle of Equality. (pp. 420, 591) I judge a particular action, say a politician's, which is in accord with the Principle of Equality, to have been right. It take it, secondly, that the politician acted out of a settled disposition to act on just the principle in question, and, of course, since he does have that settled disposition, that he does generally act on the principle. I thus take him as a man to be morally all right. Thirdly, I may of course approve of him for the particular action in question, this approval involving a cerain belief about the initiation of the action, a belief to the effect, as we can loosely say, that he was responsible for it.

Does my judgement as to the first matter, the rightness of the particular action, involve only the fact that it was in accord with my favoured principle of fairness? Does it also somehow involve the possibility of *assigning responsibility* to the agent—holding him or her responsible, or crediting him or her with responsibility? Certainly judging the particular action to be right does not presuppose that the agent *is* in fact assigned responsibility for it. The matter of assigning responsibility need not arise. Nor, of course, is an action by definition something for which an agent is in fact assigned responsibility. We have not so defined actions. An action, not to be precise, is a movement or stillness somehow owed to the mental. (p. 247) To say an action is owed to the mental is not to say that it is something for which the agent is to be assigned responsibility. Nor would we get that upshot if we were to define an action, as is reasonable, as something owed to intention. That an action is intentional is not the same fact as the one about responsibility. Indeed there are many perfectly intentional actions for which the agent is not hold responsible, and not such that he or she is credited with responsibility.

Is it true none the less, by way of some argument, that to judge an action to be right or wrong is to presuppose that the agent *could* be assigned responsibility for it? The principle that 'ought' implies 'can' is first of all a principle about responsibility, and about acts of praise and blame. There, what it comes to is that to blame a man, to say that he ought to have done otherwise than he did, requires that he could have done otherwise. (Cf. Frankfurt, 1969; van Inwagen, 1983, pp. 162 f.) What of the rightness of actions? Suppose, to repeat, that I am committed to the Principle of Equality. An action is done, and it serves the end specified by that principle. It contributes effectively to making well off those who are badly off. However, it was an action for which the agent cannot be assigned responsibility. He was, perhaps, absolutely compelled to perform it. Is the action right?

Difficulties attend answering the question, and more so to answering it briefly, but the answer is surely no. The action is neither right nor wrong. One can come to this answer by keeping in mind that the fundamental question in morality, other than that of responsibility and praise and blame is properly expressed as this: How ought the world to be *in so far as we can affect it*? Morality is concerned with the world in so far as it is in our control. An action for which the agent cannot be assigned responsibility, and which he therefore cannot be taken to control, is an event which falls outside of morality's concern or province. Certainly if such an action by a man was one he was compelled to perform by the action of another man, that prior action may be within the province of morality, but that is nothing to the point.

To speak more generally, the very existence of the question of right actions depends on the existence of actions for which responsibility can be assigned. If there were no responsible actions, there would be no judgements, as we have them, as to the rightness of actions. The subject-matter as we know it would not exist. In brief, the rightness of actions depends on the possibility of holding people responsible and crediting them with responsibility.

To return to the action mentioned a moment ago, in accord with the Principle of Equality, but a compelled action and hence not one for which the agent can be assigned responsibility, I may certainly take it that it was a good thing that it happened. It was a good thing in just the sense that it was a good thing for Ghana that the first rainy season in 1986 was as it was, and so ensured a full harvest. I may say that it was an action which would have been right, rather than neither right nor wrong, if it had been an action for which the agent could be assigned responsibility. None of this affects the conclusion that the principle that 'ought' implies 'can' has to do with judgements of actions as right or wrong as well as moral approval and disapproval of agents.

We are examining the relations between right actions, good men, and moral responsibility for actions. One such relation, we have it, is that the possibility of assigning responsibility to someone for an action is presupposed by taking an action to be right. We have it that right actions involve a certain initiation. The rightness of an action presupposes something about the agent.

To revert for a time to a separate matter, one may be tempted to think at this point that *any* answer to the question of what actions are right—at bottom the question of how the world ought to be—must bring in something about the goodness or badness of the intention and dispositions from which it derives. That is, to recall the two sorts of moral philosophy, any answer must bring in something from agent-related as distinct from consequentialist moral philosophy. Very roughly indeed, an action cannot be right unless it comes from a morally acceptable intention. There is no need to succumb to the temptation. We must not run together the fact that a right action is one which presupposes that the agent can be assigned responsibility, and the disputable and disputed proposition that to be right it must come from a morally acceptable intention or a certain kind of agent. Also, the proposition that an action's being right does entail that men ought to have certain intentions is not the proposition that to be right an action must come from a certain intention.

Some illustration of the main point will be useful. For Kant, right actions are those done out of the a priori command of reason, expressed in the moral law somehow conceived. They are done, that is to say, out of the pure good will, which is distinct from feeling for oneself or others. Are we to understand that an action's rightness does not merely presuppose but *is* in part its being an action for which the agent can be assigned responsibility? (Cf. Nozick, 1981, pp. 310 f.) There are notorious obscurities and difficulties in Kant's moral philosophy, and in particular at this very point, but it may be that an action's rightness is so conceived. That is, the requirement that a right action is such a responsible one is a moral requirement. Part of what makes it right, as distinct from what is presupposed in taking it to be right or wrong, is that it was an action for which the agent can be assigned responsibility. One may wish to resist the judgement that such a fact is itself to be valued, as one may wish to resist the idea that the colour mauve could itself be valued in a moral way, putting aside any connection it is taken to have with other things. Perhaps there is no conceptual obstacle, however, to what may be the Kantian idea. It will hardly need adding, however, and this is the main point, that when right actions are conceived differently, responsibility-assignments are not in the given way internal to judgements about actions.

To turn now to the relation between responsibility-assignments and the matter of the general moral standing of persons, it will rightly be expected that we have the same situation. The question of what men and women ought to be like, what dispositions they ought to have, involves the matter of responsibility-assignments as much as the connected question of right actions. This is so, in part, since the question of what dispositions we ought to strive to have is itself a question of the rightness or wrongness of certain actions: those that contribute to the formation of dispositions. The actions need to be responsible ones in the given sense. That is, if there were no actions for which agents could be assigned responsibility, there could be no exhortations as to what we ought to be like.

Will it be said against this that in judging someone to be of an examplary moral character, we appraise his dispositions, as distinct from any of his actions which formed or contributed to his dispositions? There is some inclination to agree. To praise a man for his honourableness may be to praise him for his character itself. It is likely we should not do so if we supposed he had not contributed at all to the formation of his character, but none the less we do direct our admiration to his character. However, this fact does not disconnect judgements as to good men and women from responsibility-assignments. At the very least, *something* of the connection mentioned above persists. To favour someone for his character would be a very different enterprise if we took it that his previous actions had *nothing* to do with it.

In any case, there is another point. Suppose that such character-judgements had absolutely nothing to do with the actions which issue in a man's character. They would none the less be judgements about a man's dispositions *to responsible action*. Dispositions being such, that is the very nature of judgements about them. In line with the fact that morality's fundamental concern is with how the world ought to be, in so far as we can affect it, our judgements about dispositions necessarily have to do with the controllable. That is, they would have to do with judgements as to dispositions to perform actions over which we have control, actions for which we are held responsible or for which we are credited with responsibility. The subject of good, humane or estimable men and women thus presupposes the subject of responsibility in this second way.

The account of the structure of morality now completed is indeed schematic, and hence in several ways unsatisfactory. There are perhaps yet closer connections between the three matters than have been indicated. What we have gives a certain fundamentality to moral approval and disapproval or responsibility, a fundamentality which is hardly disputable. What we have also suggests, although the subject has been left untouched, that our attitudes to actions, and our attitudes with respect to the standing of persons, owe their characters in part to our attitudes in connection with responsibility. There is more to be said than that our attitudes with respect to actions and standing *presuppose* the possibility of attitudes with respect to responsibility. The latter attitudes contribute to the nature of the former. If, as will be anticipated, and to look forward, we have a *pair* of attitudes with respect to responsibility, so too we have a pair of attitudes with respect to each of right actions and general personal standing.

7.7 MORAL APPROVAL AND DISAPPROVAL

Suppose a man foresees the divorce which will end his embittered marriage, and, out of certain desires, forms a certain intention and acts on it. One desire, as we believe, is to try to capture the affection and loyalty of his son, and in so doing to deprive his wife of them. He desires too, we believe, to reduce the money that will go to his wife in the division of their assets. He forms the intention to give a large gift of money to his son, which he does. The situation might be an extraordinary one, such that it would be possible to condone or excuse his action, but, if the situation is at all ordinary, we shall certainly disapprove of him morally, hold him morally responsible for the action and effects of it. We shall have disapproving feelings about him, which feelings may have expression in blame or condemnation, spoken judgements to the effect that he has behaved badly or appallingly. Such feelings or attitudes, as we have just seen, although their nature has not been much explored, stand in certain relations to and shape or at least colour our views of the rightness of actions and the moral worth of persons.

Such feelings or attitudes of disapproval, and also feelings of approval, are disapproval and approval of a person for or in relation to a particular action. The action of the husband, given that disapproval is in question, must be a wrongful one. Thus, while morally disapproving or approving of persons for particular actions is connected with the rightness of actions in the general way we have seen—in part, the existence of right and wrong actions presupposes the possibility of assigning responsibility to persons—the two things are also connected in the simple way that moral approval and disapproval depend on or are otherwise related to a view of what actions are wrong and what right. This remains true despite the fact that few of us, if any, operate in our lives by way of an explicit, general principle of right actions such as that of Equality or Utility. Something related might be said—it would not be simple—about the connection of approval and disapproval with our conceptions of good men and women.

In the example, to repeat, we take the husband's action to be wrong. The large remainder of what is involved in our disapproval of him importantly includes beliefs, ideas, or whatever about the initiation of his action. What do these come to? There can be no doubt, as in the case of the personal feelings of appreciation and accusation, that they can be expressed this way: *he could have done otherwise than he did*. However, the words do not in themselves reveal what is behind them. As readers will expect, given what has been found with life-hopes, personal feelings, and knowledge, the words can give expression to two different sets of thoughts as to the initiation of action, and enter into two attitudes we have—which attitudes issue in two responses to determinism.

It was remarked at the beginning of this part of our inquiry that a typical attitude may be regarded as an evaluative thought of something, feelingful and bound up with desire, the mentioned feeling being feeling in a narrow sense, somehow similar to or perhaps involving sensation. (p. 382) Although our fundamental concern has been with the thought within an attitude, and will continue to be so, it will be enlightening to attend paricularly to the *nature* of the feeling and desire involved in moral disapproval, partly because all attitudes are in some way fusions of their elements. We might have done so with life-hopes, personal feelings, and, to a lesser extent, our confidence of knowledge.

Certainly, when we disapprove of the husband, we do have feelings about precisely his initiation of his action. It is not that we take his action to have been wrong, and thus have feelings about just it, no doubt pertaining mainly to the effects on his wife, and that for the rest we merely have bare beliefs or ideas about the initiation of the action. Certainly we have feelings about the initiation itself. Again, it is not as if moral disapproval were a matter of taking an action to be wrong, and, for the rest, only wanting to affect the future—wanting to prevent more such actions, or, perhaps more likely, wanting to bring about an act of restitution. Disapproval, to whatever extent it is forward-looking in this way, is not only this.

Consider first certain immediate feelings—*feelings of repugnance*—which perhaps are always involved in moral disapproval, of whatever kind. They need to be clearly distinguished from something else. With respect to the example, these feelings of repugnance have to do with the man's pair of desires, to capture his son's loyalties and affections away from his wife, and to deprive her of a fair share of the monies of the marriage. We have such feelings, too, about his forming of the intention, or deciding, to make the large gift to his son. We are likely to characterize the desires and the decision as shameful, low, or terrible. They have such a character for us, presumably, mainly because of the repugnant state of affairs to which they pertain: the man's wife being in the several ways injured. To regard the desires and intention as shameful, low, or terrible is also to have a feeling about the man. In fact, the two things run together. Whatever his general standing as a moral agent may be, he is now repugnant in having the desires and intention. He is in this episode such a man as does not reject them, knowing their import. The desires and intention are consonant with the whole of a corpus of dispositions.

The feelings of repugnance in question, if connected with intention and action, are not connected in the way of certain feelings to which we are coming. The feelings of repugnance are distantly related to those we have about things or instruments which have a grim or terrible purpose. These feelings are thus not passive, in the way of some aesthetic feelings, but have in them tendencies to action by which, perhaps, they are best identified. They have in them tendencies to withdrawal from, or avoidance of, the husband, or indeed tendencies to more positive action, that of trying to bring an end to, or to change, his persisting desires.

These, to repeat, are not our only feelings in disapproving of him, and they may not be the feelings which are to the fore. The feelings which may be to the fore, and can seem of greater importance, are conveyed to us by the most common expressions of disapproval. To say, if I do, that he has behaved terribly or abominably, that he should not get away with it, is at bottom not to be repelled by his desires or his decision, although I am that, but to be in another way of feeling. To say his wife's subsequent sad state is to be put down to him, or laid at his door, or to say that he is to be held responsible for it, in the most common sense of these expressions as well, is not to be repelled or appalled by his desires or decisions, but to be engaged in more aggressive feeling likely to issue in blame or condemnation. This feeling is other than repugnance.

What is most important for our concerns is the character of this feeling that may be directed at the agent in his act. It is not a tendency of the kind mentioned above, to withdraw from him or avoid him, perhaps to try to reform him, but a tendency to *act against* him. It is, further, a tendency in which we feel ourselves to have justification, the justification of moral principle. This gives to it, clearly, a certain strength and character. We are in fact allowed to be more overt and less restrained here than with the personal feelings (p. 401) where we lack moral sanction.

There can be little doubt that Mill states or overstates a fact when he writes, although not explicitly about moral disapproval, that 'we do not call anything wrong unless we mean to imply that a person ought to be punished in some way or other for doing it'. (1972b (1863), p. 45) Moreover, despite others of his views, having to do with the preventive or deterrent value of punishment, it is clear that he has in mind that in calling something wrong what we mean to imply is not only a need for preventive punishment, but something quite different, what we can call retribution for the act.

It is true, certainly, with respect to moral disapproval, that typically we stop short of the deed, because we take it that punishing a man for an action is not our business, or, say, because we take it that it would not be prudent, or we have some residual doubt about the situation. It is as true that a feeling of disapproval can coexist with a readiness to excuse or half-justify an action, with feelings of pity and with empathetic understanding, and, as already remarked, with the view that the agent in question is in general of decent or good moral standing. All of that qualifies the idea that disapproval may involve retributive feelings, but it does not at all refute it.

There can be no doubt that these feelings which we may have do consist in *retributive desires*. Our feelings consist in desires, which we take to be justified, that the person of whom we disapprove should suffer at least some discomfiture. He should in some degree have distress for the wrong he has done, if perhaps only the distress of knowing that he has the disapproval of others, a disapproval which is not self-concerned or idiosyncratic but is based on moral convictions or principles. What we want is at least that what he has done should be brought home to him. In other cases, perhaps such a case as has been imagined, our desires in disapproving of a man are stronger, and do issue in resolutions to act and of course in actions itself. The spectrum of retributive feelings of disapproval is thus a wide one, but all the feelings are essentially desires that the disapproved person should suffer at least some discomfiture, tendencies to act against him.

More will be said of the exacting of retribution in another context. (Ch. 10) Enough has been said here to raise the question of what beliefs, ideas, or the like are involved in moral approval and disapproval when it includes, as certainly it often does, not only feelings of repugnance and desires to affect future behaviour, but also retributive desires. Put differently, what is our particular attitude of moral disapproval? We shall approach an answer by way of a route different from before. With life-hopes, personal feelings, and knowledge, our procedure was in part to elicit and characterize an attitude, which attitude could then be seen to be inconsistent with determinism. Here let us proceed by first assuming determinism, and asking what can be concluded by way of that assumption with respect to the question of the existence of an attitude.

Suppose then that we do the unusual thing of contemplating that the theory of determinism was true of the vicious husband's action. We contemplate that his relevant thoughts and feelings, and his decision and action, were necessitated events, in the ways specified by the three hypotheses. Suppose we were to move on from merely contemplating this, and came to *believe* it. What effect would this settled acceptance of determinism have on our retributive desire that in some way he be subjected to some distress? There seems little need to linger about the answer. Is it not impossible to avoid the judgement that if we came to *believe* that his action, decision, and all things related to them were necessitated events, our retributive desire for his distress—to say nothing of anything else—would falter, weaken, and collapse? Surely our desire would not survive our coming to have the belief about necessitation.

It is proper to say, despite misleading suggestions (p. 122), that our state of feeling would be related to our states of feeling towards machines. If we do in fact manage vindictive personalizing feelings about machines, something close to retributive desires, we cannot persist in them. If we engage in kinds of fantasy or spuriously accusatory rages about things that do not work—a car or a corkscrew—these are but transitory and unsustainable passages of experience.

Will a sceptic say that we cannot judge now that settled belief in determinism would issue in the collapse of our retributive desires? There is a relevant fact here, which will also be of importance at a later stage. (Ch. 9) It gives strength to such scepticism, but not rational support. It is the fact that we, or at any rate very nearly all of us, do not believe any determinism, any determinism that deserves the name. Or, almost all of us within one great tradition of culture do not believe such a thing. More than that, we share a resistance to believing it. This certainly stands in the way of speculation, but it is a difficulty that can be overcome, and needs to be overcome in the interests of truth. In fact we *can* see from present evidence that the retributive desires which may and very often do enter into moral disapproval would in fact not persist if we came really to accept a determinism. It is not that the fact that disbelief in determinism is rooted in nearly all of us, and is one thing fundamental to our view of the human world, actually makes any effects of actually coming to believe it unpredictable. We can have evidence as to those effects in the lesser but clear effects on our feelings which attend the lesser thing—just contemplating that a determinism is true. This can be shown.

To return to the example, and to add something very relevant, suppose that one is at the pitch of one's moral feeling of the given kind about the vicious husband. The facts of the matter have just now become clear, and one's reaction to them is new and strong. Why is it that one is irritated, or more than irritated, by someone who introduces, however tentatively or commonsensically, the idea of a determinism? If we do not imagine this person as an irritating or Olympian figure pleased to bring truth to us from above, a philosophical determinist given to tutorial ways, and do not in any other respect introduce irrelevancies, it may seem that the answer is clear enough.

It can hardly be that we take him to be introducing a falsehood. If we do not take determinism to be true, we do not take it to be false either. It can hardly be, either, that our irritation has to do only with being faced by a relevant proposition of whose truth-value we are uncertain. No doubt our irritation *would* have to do with the intrusion of theory into our lives at an unsuitable moment, but the thought is unavoidable that we are mainly irritated by the possibility of *an obstacle to our own retributive desire*. We are irritated by being faced with an unwelcome excuse, in a way a total excuse, for the man against whom we have retributive feelings.

To accept that such desires would in fact not survive one's coming to believe a determinism is to accept that there exists a kind of our moral disapproval—that we now enter into a kind of moral disapproval—which is inconsistent with determinism. Further, if certain of our desires would falter, weaken, and collapse with the acceptance of a determinism, the reason can only be that one conception of the initiation of actions, whatever it comes to and however it enters into moral approval and disapproval, is inconsistent with determinism. The safely predictable effect on our desires provides us with an indirect argument to those related conclusions.

There can be no doubt as to the sort of conception of action-initiation that is suggested by this indirect argument. It can be indicated by saying, for example, that in taking a man to be morally responsible for an action in the given way, an essential part of what we take to be true is that the action had *an individual explanation*. (Honderich, 1976) It was in a particular sense his doing. That means, in part, that it is not to be explained by his dispositions. It is not explained in such a way that it would follow that another person of like dispositions would in the same situation perform a like action. The explanation, then, is not of a general kind. So to conceive the initiation of his action, certainly, is to have but a vague conception. Its clearest content, in the given respect, is the image of a self, which in fact is not greatly more than a kind of denial of a nomic explanation of action. If we struggle to clarify the idea, we come again to indeterminist pictures of the mind, and above all what is said of an originator and origination.

What we have so far is that an argument beginning from determinism shows that we engage in a certain kind of moral disapproval. That disapproval, involving an image of the self, is inconsistent with determinism. The response of dismay, in connection with moral responsibility, is that since or if a determinism is true, moral responsibility is an illusion. It is such since the truth of the propositional image of the self enters into and is a necessary condition of it. Since or if a determinism is true, our feelings of disapproval and approval for ourselves and others are indefensible.

This dismay, at bottom, has to do with both an apprehension and also a moral uncertainty related to guilt. Is this moral approval and disapproval now not deep-rooted in our lives—on the way to being as deep-rooted as the personal feelings themselves? *Can* it be given up? We must have the apprehension that it cannot. The thought that others cannot in this way be held responsible for their actions also carries the implication that we, in our ongoing feelings, are involved in a falsehood which has certain consequences. Certain of our dispositions to action which cause distress to others rest on falsehood. Here there is cause for moral uncertainty if not guilt.

There is, too, a third and perhaps more direct way in which we may be affected by the idea that moral responsibility of a fundamental kind is illusory. If it is such, *I* am not in this way morally responsible for my actions, and hence not a subject for approval or disapproval. I get no credit of a certain kind, and I am diminished by not being open to judgement. The occasional moral pride I feel rests on illusion, as does what is more common, the lesser feeling that by my life, at least for the most part, I have not earned moral disdain. It must seem, too, that certain feelings of regret, remorse, and guilt are without sense. To think these things is to feel deprived of standing.

Furthermore, to recall what was shown earlier, there is the fact that what is false can be taken as a presupposition of the rest of morality. Only those actions are right or wrong which are actions for which an agent can be held responsible or credited with responsibility. Men and women can be of this or that moral worth only if they can be assigned responsibility for their actions. Is this presupposition not the presupposition of a moral responsibility involving the retributive feelings? In short, to respond with dismay to the issue of determinism and moral responsibility is to be dismayed about the possibility of all of morality. To answer in a certain way the fourth question with which we began, that of the relevance of determinism to moral responsibility, is to be committed to a related answer to the fifth and sixth questions, about the worth of persons and the rightness of actions. (p. 379)

If this moral responsibility is an illusion, there is also another kind of consequence for the connected matters of the worth of persons and the rightness of actions. It was implied earlier that it is at least arguable to take right actions to be those which have certain concomitants and consequences. This consequentialism carries the corollary that good men are of a certain kind. However, as was said, it is also possible to take right actions to be those which have a certain inception in the agent, and again to take good men to be of a related kind. One agent-relative principle of right actions is the Kantian one that we are to do the action which flows from a pure good will. As was asked (p. 429), can it be that this principle in part *defines* right actions as responsible ones, as distinct from presupposing that right actions, somehow defined, are responsible ones? If so, doubt about moral responsibility will have a direct effect on the principle.

Whatever is to be said of the Kantian principle, there is another principle of right actions which definitely *does* bring in responsibility in something other than the presupposing way. It is the principle that we are always to act so that each man gets what he *deserves* for what he has done. For a man to deserve something for an action *is* in part for him to have been responsible for the action. The idea of desert, as distinct from the idea of right action, is itself in part the idea of responsibility. What follows, it seems, is that what we can call the

Principle of Desert must be given up if moral responsibility is an illusion. There is that consequence, and it is a further possible source of dismay. We shall consider it further in connection with the seventh of the questions mentioned at the beginning, that one having to do with the relation of a determinism to large social institutions and practices. (Ch. 10) A main one of these is punishment.

It is clear, then, that our moral disapproval and approval can take the form of an attitude which involves thoughts as to the initiation of action which do not have to do only with voluntariness or willingness —centrally a matter of unconstrained and unconflicting desires on the part of the agent—but also and fundamentally with origination. To have this attitude is inevitably to be dismayed by determinism, and dismayed as well with respect to the matters of right actions and the general moral standing of persons.

However it can also be made clear that we are the possessors of another attitude, and thus have the possibility of the response of intransigence with respect to moral responsibility and the other matters. The attitude and the response can be brought into view, or rather a part of it can, by seeing more of the size of a certain fact than we did when passing by it earlier. Here too, incidentally, we shall proceed by way of a kind of argument or reflection different from the one employed in connection with life-hopes, personal feelings, and knowledge—and also different from that one just employed in connection with the first of our two attitudes having to do with moral responsibility. To some extent, as already implied, arguments of all these types could be advanced in the case of each of life-hopes and the other six things affected by determinism.

The fact passed by earlier is that our moral feelings, judgements, and activity—our lives in so far as they have to do with the three parts of morality that have been distinguished—are not only reactive but may also be purposive. It is not just that we react in feeling and behaviour to actions and people, but also that we may do so to a certain end. This fact was in sight when it was allowed that moral disapproval does involve wanting to affect the future—to prevent more actions of a certain kind, or to bring about acts of restitution. (p. 432) Also, it was allowed that feelings of repugnance may include tendencies to change or end a person's repugnant desires. (p. 432) It might have been allowed, too, in connection with Mill's claim that our calling things wrong implies a need for punishment, that the punishment in question is not to be conceived entirely in a retributive way but also in a preventive way. (p. 433)

Nothing said so far about moral approval and disapproval, right acts or moral standing provides anything that could be called *an explan-*

ation of the content of ordinary morality. That is, no attempt has been made to specify what purpose we have in morality, considered in terms of its content. To try to give at least a sketch of such a thing is another way, different from the way of reflection on examples, or the assumption of determinism, to bring into view attitudes we have, or to succeed in not overlooking them. The enterprise is peculiarly in place with morality, as against life-hopes, personal feelings, and knowledge, since ordinary morality is a construction of ours in a way that they are not.

In trying to sketch such an explanation of ordinary morality, it is preferable not to fall into speculation about purposes other than our ordinary purposes now, purposes which are open to being discerned without the aid of imagination. (There is something to be said in some contexts for at least a variant of Francis Bacon's alarming opinion that in doing science the imagination is to be hung with weights.) It is therefore preferable to eschew speculation about the original rise of morality, its primitive emergence, or such speculation about all of us as turns up in psychoanalytic theory. The latter theory, whatever insights it includes, is in the view of a great majority of philosophers not such as to satisfy general criteria of conceptual adequacy, evidence, and explanatoriness. What is to be considered, then, is the plain question of why we now have morality, or better, why we give the support we do to it. However, there are presupposed questions which need some prior attention. What, in terms of its content, *is* ordinary morality? *Can* we be said to support it?

As implied a moment ago, the characterization of ordinary morality so far given has for the most part not been concerned with its content. What has been sketched is its several structural parts—responsibility, right acts, moral standing—and their relations, and certain of our moral feelings. That is, not a great deal has been said of the claims in which ordinary morality consists—claims as to what obligations we have, what our ideals should be, and so on. It was noted that different and opposed principles of right action are advocated in moral philosophy. It can be said of almost all of these—including the Principles of Equality and Utility and opposed principles which locate the rightness of actions in intentions or other sources—that they do not give us much of the content of ordinary morality. As remarked above, few of us run our lives by such general principles, and those institutions of society which are in part shaped by morality can hardly be said to be shaped by the general principles. As for several less well-specified general moralities—perhaps the outlook of an established church—their want of specification stands in the way of there being a clear question of what adherents they have, and what institutions they shape.

However, almost all of the philosophical and like moralities can be regarded as having things in common. What they share, unsurprisingly, is also what can also be discerned as the content of ordinary morality. Ordinary morality has been effectively characterized as consisting in four requirements, those of non-maleficence, positive beneficence in some degree, fairness, and non-deception, the latter including a requirement about the keeping of promises and agreements. (Warnock, 1971; cf. Foot, 1978) These are specifications of what actions are right and what men and women are good. We are required, first, to restrain our passions in the pursuit of our own interests, and so not do positive harm to others. We are, secondly, to help others or to seek their good, in certain ways and to certain extents. Thirdly, we are in a sense not to discriminate, but to extend our concern to all the members of relevant classes of individuals rather than only to those who are somehow closer to us. Fourthly, we are for the most part to be truthful. What needs to be added to this, and complicates the matter considerably, is that ordinary morality allows or indeed requires a considerable qualification of all these requirements—in the direction of benefiting those who are close to us or to whom we are in other special relationships, above all our families. To them we are taken to have special obligations. (Cf. Parfit, 1984, pp. 95 ff.)

If the philosophical moralities can be said to have these requirements in common, they do of course give them different shapes and weights. They conceive them differently, and fit them differently into some system. In the case of moralities of justice, of which the morality of the Principle of Equality is one, dominance is given to the third requirement, fairness. Each of the first and second requirements has also been given dominance in philosophical moralities, and truth has also been given at least a precedence in some systems.

Given that these four features provide us with a tolerable characterization of the content of ordinary morality, can we in fact be said to support it? This large question can be given only a fragment of an answer. Certainly our support for morality is in fundamental ways self-interested and thus inconsistent. It is also uncertain. With respect to self-interest and inconsistency, there can be no doubt that ascendant classes within societies construe and manipulate this morality to their advantage, sometimes managing remarkably well to conceal the fact from both others and themselves. Certainly they go beyond what is allowed for by special obligations. It cannot conceivably be said that morality is followed consistently. Marx's importance, it can be supposed, properly rests largely on his determination to record this large fact—the use of morality, law, and a good deal else by ascendant classes—more so than on his philosophical elaboration of the fact in a theory of history which makes morality a part of a society's superstructure, somehow determined by a supporting structure of what are called economic relations and forces. (G. A. Cohen, 1971; cf. Honderich, 1982d) To move quickly from social classes to individuals, it is clear that almost all of us fall into *akrasia*, seeing the better course and following the worse. We are in general not resolute about following the better course as we see it, but uncertain.

Despite these several considerations, however, and others tending in the same direction, we do give *a* support to morality. We are persistently inclined in our lives to kinds of defence of non-maleficence, beneficence, fairness, and truth. Social institutions support these requirements in theory, and to some degree in fact. To be self-interestedly and uncertainly in their support is not to give them no support at all.

To come on to the main question, of why we support morality, it would be strange to suppose that we do not see something to be gained by our support, that our support for it has no advantage to us, that it does not serve an ongoing purpose we have. It would be almost as strange to suppose that all that can be said for our support, or even what can mainly be said for our support, is that it serves the end of satisfying what what have lately been considered, our retributive desires.

Our purpose emerges from our awareness that for several reasons, a principal one being the limited sympathies of individuals as agents, *things are likely to go badly*. (Warnock, 1971) Things are likely to go badly, more precisely, for those other individuals whom agents affect and who are beyond the ranges of sympathy of the agents. As a consequence of the limited sympathy or worse of the husband in the example, things go badly for his wife. We ourselves, in some degree detached from the struggle, have his wife within the range of our sympathy. More generally, our general sympathy with those others whose interests do not directly affect our own, which sympathy is as much a fact about us as our retributive feelings, issues in an attempt to constrain the conduct of agents with respect to their competitors, opponents, and adversaries. The constraints we attempt to place on their conduct are of course constraints having to do with maleficence, non-beneficence, unfairness, and deception. It is not to be forgotten, of course, that each of us, in different circumstances, is agent, victim, and sympathizer. If our third role goes a considerable way to explaining morality, our second must be as important. The purpose of avoiding or limiting the distress of individuals—including ourselves—is the principal explanation of morality.

If we now reconsider moral disapproval, with these general facts in mind, it is possible to come to something different from before. We can very reasonably think and feel that we may disapprove of others in order to serve the purpose of morality, whatever else may be true of our feelings. Better, we can *recall* that we do so. Our disapproving, at least to an important extent, may be a matter of desires pertaining to the future. Those people we select for disapproval, or of course approval, and perhaps for blame or praise, are those whose selection serves our purpose.

Clearly we do not disapprove of all persons who have performed acts which have given rise to unfortunate states of affairs. Some of those who fail to help others, in ways that are generally required, are excused disapproval. So with some of those whose acts go against fairness and truth, and perhaps, depending on how it is conceived, non-maleficence. We disapprove, roughly speaking, of those whose behaviour can be altered by condemnation or blame, and, unless it is altered, will continue to conflict with the requirements of morality and hence our moral purpose. We may have in mind the altering of likely future behaviour somehow related to the act that has been done—perhaps the behaviour of not putting right what has been done, not making restitution for it. We may have in mind altering or preventing likely future behaviour of the same kind as the wrongful action that has been done.

A fuller picture of moral disapproval, in so far as it involves our purpose in morality, can be had by reflecting on groups of persons of whom we do not disapprove, despite the states of affairs to which they give rise or contribute. As might be expected, what turns up in this context is fundamentally like what turned up in connection with one way of thinking and feeling about determinism and life-hopes, and determinism and personal feelings. As might also be expected, and was mentioned earlier, there are certain differences, at least differences of emphasis. The simplest difference is that certain considerations are naturally to the fore with personal feelings and morality, and less so with life-hopes.

We do not morally disapprove of, hold morally responsible, a man who performed an action which we take to have gone against, say, the requirement of fairness, if we believe the action was owed in an essential part to his *want of knowledge* of its effects. He predicted them badly, but to the best of his ability. That is, his desires and intention in the action were perhaps in conformity with fairness, but his possible judgement that his action would conform to fairness was mistaken in some non-culpable way. No doubt we will seek to improve his information or his judgement—our endeavours of this kind are also determined by our purpose in morality—but we take it that his desires and intention, as evidenced in his judgement about the action, were themselves not such as to make it likely that he will again frustrate the purpose of morality. Moral disapproval, which has to do precisely with desires and intentions, is out of place.

We shall not hold responsible, for the same general reason, a man whose action was subject to *external constraint*. We do not think the worse of the Pennines rock-climber who sits where he is, and so fails to get help for his injured companion, if a rock-slide has trapped both of them. His desires and intentions are very likely also acceptable to us, both in themselves and in their relevance to future conduct. We suppose that he does have a disposition to beneficence, such that he would get help but for being trapped.

The point is not the simple one that what he actually does is in part the effect of something external to him, the rock-slide. All of our actions are affected by external facts. What is fundamental to our not disapproving of the climber is not that he was affected in his intentions and action by the rock-slide, but that his desires, intention, and action do not in any way repel us or put us off, and do not suggest any future frustration of the purpose of morality. To say we do not think the worse of him because his action was subject to physical constraint is to say we do not think the worse of him because a fact of the world stood in the way of his acting on his own acceptable disposition to give help. The external fact is not what makes disapproval out of the question, but is consistent with our supposition as to dispositions he has, which dispositions make disapproval out of the question.

Thirdly, we do not disapprove of a man whose action of failing to give help, say, was owed to an effective threat by someone else. My sitting where I am, and not telephoning the police gets me no disapprobation if a man with a gun is sitting across from me. My desires and intentions in so doing do not bode ill for the future, regarded in terms of the aims of morality. The point again is not simply that I am affected by another person's possible action, but that the fact defeats any presumption that my dispositions are not in accord with the purpose of morality. As for the instrinsic nature of my desires and intentions, two things can be said, of which the first is that they are reasonable. The second, which will get more emphasis in the case to which we now go on, is that I may be reluctant to do what in fact I do, which is to sit still.

Suppose, fourthly, that a man desires both to go on drinking as much wine as he does, and also to stop, since his drinking is dragging down his life and thus affecting others. Moreover, with respect to this conflict of desires, he has a third strong desire, which is that his desire to reduce his drinking should be the effective one. It, rather than the desire to go on as he has been, should actually govern his behaviour. Perhaps he has no desire even to have the desire to drink so much. Or, perhaps more likely, he greatly prefers the desire to reduce his drinking to be the one that issues in action. Or, he unequivocally judges that it is morally right, given the effect on his family, that the desire to reduce his drinking have sway over the other. The judgement is sincere, in no way suspect. Suppose, however, despite his feelings to the contrary, that what actually happens is that he goes on drinking as before. (Cf. Frankfurt, 1971; Slote, 1980.)

Our attitude to him will be one governed by the firmness of our belief about his second-order desire, preference, or judgement. If we have a settled belief that his second-order desire is strong, and unopposed by a second-order desire going the other way, or that his second-order preference is indeed great, or that his moral judgement is an unequivocal and committed one, we shall at least be impeded in any disapproval of him. We may be puzzled about what it is that explains his going on drinking excessively, but we shall have to suppose that there *is* something more, perhaps a bodily fact, a bodily cause of alcoholism, which is of relevance to the situation. We shall not be fully disposed to hold his behaviour against him.

One reason will be that he will not respond to our doing so. That is, given that he already has a commitment to only moderate drinking, which commitment is unfortunately inefficacious, it must seem that blame would have no use. That is on the assumption, of course, that the facts are indeed as they have been described. If we suppose, differently, that some strengthening of his second-order desire, preference, or judgement would have the desirable effect of reducing his drinking, we shall be differently disposed. A second reason for our not being disposed to blame him is that certain of his desires and intentions are tolerable or indeed impressive. Conjoined with something else, the upshot is that he is running down his life, and hurting others. None the less, in virtue of his second-order disposition, and the struggle that his life involves, he himself is unlikely to have our revulsion. Certainly we shall try to inhibit any such tendencies we feel.

Fifthly, and briefly, we do not disapprove of the insane for their conduct. At any rate we do not do so if we have certain beliefs about their desires and intentions. Sixthly, we do not disapprove of the very young. It can be said that they are the victims of their desires, often desires which are owed in some direct way to their environments. They have not come to that maturity in which one actually has an awareness of an attitude towards one's desires, as distinct from their objects. They do not have second-order desires, preferences, or judgements.

Given this brisk survey, we can draw a certain conclusion, or better, perceive a certain fact. As must be expected from the discernible purpose of morality, we do have a certain attitude of moral disapproval, and of moral approval, different from the one previously considered. We do feel moral disapproval for or hold responsible those whose behaviour can be altered, and, unless it is altered, will continue to conflict with the requirements of morality and hence our moral purpose. We do feel moral approval for those whose behaviour can be confirmed, and so will continue to accord with the requirements of morality and hence our moral purpose.

We do have an attitude of moral approval and disapproval, which, in so far as the initiation of an action is concerned, is centred on a certain set of facts, those we have surveyed. To order them differently, they are the facts that the agent is mature and sane, acts out of a certain knowledge, does not act out of a certain conflict of desires, and is not subject to external constraint by either his situation or another person. If we return to the example of the vicious husband, it is certainly possible to take an attitude to him which is informed by just such ideas.

They could be elaborated in various ways. More could be said which would draw on our earlier inquiry into the nature of all actions. (Ch. 4) For example, more could be said of the relation between the antecedents of the action and the action itself—say the degree to which the action is represented by its intentions. More could be said, certainly, in enlargement of reflection on desires for desires. Here we could attempt richer accounts of personality, character, and mental life. In place of speaking of desired desires, we might speak instead of the unified, unconflicted, and directed person. We could consider the question, in developing such a picture of the person, of a higher order of desires than the second. None of this is essential the conclusion to which we have come—that we are possessors of attitudes of moral disapproval and approval which importantly have to do with our purpose in morality.

As at like points before now in our inquiry into the consequences of determinism, no question can arise over the logical consistency of a determinism and the given ideas as to the initiation of action. An action whose initiation has the given set of properties is perfectly consistent, in particular, with the three hypotheses. It can be said, incidentally, with the aid of certain definitions which themselves give expression to the attitude we are contemplating, that not only moral responsibility but also 'free will' is logically consistent with the three hypotheses. A man's will in an action is defined as the desire that actually moved him to act as he did. His will was free, it can be said, if it was the desire which he himself desired to be effective, and so on. (Frankfurt, 1971)

The response of intransigence with respect to a determinism and moral responsibility rests in part on the reassuring fact of consistency. It rests as well on the intrinsic recommendation of the given ideas as to the initiation of action: they are the ideas that accord with the purpose of morality. We can feel that the truth of a determinism is consistent with what principally matters in connection with moral responsibility. A determinism is to be allowed to be inconsistent with something else, a certain image, that of the self, and hence with retributive feelings which are central to the definition of a different attitude. We lose nothing of value, so far as our rational purpose in morality is concerned, if we abandon the image, those feelings, and the given attitude. They are irrelevant to our purpose. So we can feel—while we persist in intransigence with respect to moral responsibility.

What remains to be said is something about a conceivable misconstruction of the attitude that we now have in view, an attitude which has to do only with voluntariness, and not also with origination. The misconstruction consists in mistakenly taking the given attitude for something else, which can be called *the objective attitude.* It is an attitude which we may take up to a mixed group of individuals whom we believe to be wholly immature, perhaps permanently so, or to suffer from a severe disorder of mind, personality, or feeling. They may be deluded, deranged, or entirely untouched by the distress of others and hence bereft of the usual restraint owed to sympathy and a moral sense.

Such people are regarded by us as individuals to be managed, treated, controlled, trained, guarded against, detained for their own protection, or imprisoned for the safety of others. In general, our attitudes to such individuals are attitudes of a detached and at least a manipulative or managing kind. Our attitudes are informed by *no more than the desire to take efficacious steps in dealing with problems, challenges, and dangers*. In place of resentment at an insult or a piece of offensiveness, we may be taken up instead with devising the most efficacious policy, treatment, or the like.

The objective attitude, as specified, *does* share something with the attitudes of moral approval and disapproval with which we have lately been concerned. That similarity may perhaps lead to the idea that the attitudes with which we have been concerned are not true moral attitudes at all. It may be objected that to have what is in fact an attitude of moral disapproval to someone is not merely to be concerned to affect his behaviour, to take the most efficacious steps to avoiding certain future possibilities. It may be objected that latterly in our inquiry we set out to discriminate a certain attitude *of moral disapproval*, and come to something that is nothing of the sort, but no more than the objective attitude.

The proper reply is that while the attitudes that have lately been discriminated do indeed have a character that derives in part from our purpose in morality, that is not all that is to be said of them. The given attitude of moral disapproval is indeed one, as was said initially (p. 438), that *importantly* is a matter of desires pertaining to the future. *In part* it is that. The given attitude, none the less, if it shares something with the objective attitude, is evidently also different from it—as must be expected from the simple fact that the given attitude is specified as *not* having to do with those who are either insane or very immature. The line of thought we are considering depends on taking the objective attitude as specified: informed by no more than the desire to take efficacious steps in dealing with problems, challenges and dangers. That is *not* a full or proper characterization of the attitudes having to do with moral responsibility that we have been considering.

A full and proper characterization will include what has not much concerned us, but must be included: what were earlier called feelings of repugnance. (p. 432) These were noticed not in connection with the attitudes that have lately been discriminated, which have to do only with voluntariness, but in connection with the earlier attitudes considered, which have to do with both voluntariness and origination. However, feelings of repugnance are as much a part of the attitude of disapproval lately discussed. Their being such distinguishes what we have from the objective attitude as defined, and goes a good way to making evident that what we have is in fact a recognizable attitude of *moral disapproval*. If it is now supposed, differently, that these feelings are to be assumed to be part of the *objective attitude*, the argument we have been considering loses its force. It becomes uncertain that the objective attitude, as now defined, is not recognizably moral.

More can be said to distinguish the discriminated attitude of moral disapproval from the objective attitude as originally or lately defined. It is a part of the discriminated attitude, although there has been no need to mention the fact until now, to give a kind of consideration to the person in question, the person of whom we disapprove. That is, his or her views are to be considered. This part of the discriminated attitude, which cannot be clarified quickly, has no real counterpart in the objective attitude as defined. Further, it is part of the discriminated attitude, although unmentioned until now, to present *reasons* to others. It is not as if morally disapproving of someone, in the given way, were no more than doing whatever would get him to act in a certain way. It is, rather, a matter of *argument*.

The objective attitude was first characterized in the distinguished essay mentioned earlier (p. 404) which first widened consideration of the consequences of determinism. (P. F. Strawson, 1962; cf. 1985, Ch. 2) The central theme of that essay, which does not reflect the main contentions of our own inquiry, cannot easily be related to them, but it does suggest a further contention.

It is that the mistaken view of the attitude that has lately been discriminated—the mistaken view of it as being tantamount to the objective attitude—has a further consequence. The mistake, and only the mistake, causes us to think of and to move to the other attitude, the one that issues in dismay. To put the matter differently, it may be contended that it is only those who have a quite insufficient grasp of the attitude of moral approval and disapproval lately discriminated, and so take it not to be a recognizable moral attitude at all, who are moved in the direction of the attitude involving both voluntariness and origination, and issuing in dismay. The contention would therefore leave us with the proposition that those who have a full and proper understanding of the attitude lately discriminated will have no other inclination. They will not also be inclined to require both voluntariness and origination in connection with moral approval and disapproval.

The contention is mistaken. Our present concern is the attitude having to do only with voluntariness and issuing in intransigence. It is not the case, however, that when that attitude is rightly seen or recalled, we thereafter have no other inclination. We *can*, by considerations of the kind set out before (pp. 432 ff.), find ourselves thinking and feeling differently. We do not get to the attitude having to do with both voluntariness and origination by having a mistaken understanding of the attitude having to do only with voluntariness. Rather, we find good grounds in ourselves.

7.8 CONCLUSION

What has been said of the first four of the listed areas of consequence with respect to determinism—life-hopes, personal feelings, knowledge, and moral responsibility—suggests what we can take to be the case, without further inquiry, with respect to the fifth and sixth matters, right actions and the moral standing of persons. Because of the connection of these matters with moral responsibility, as already suggested in part, we here have two discriminable attitudes and may make two responses. Or, to say the very least, each of us is capable of two attitudes and the two responses. The attitudes and responses are not so definite, but the situation is not so different from what we have as to require us to pause over it.

In sum, then, we may focus on a conception of the initiation of action which has in it an image of origination in addition to ideas of voluntariness. In so doing we take it that origination is a necessary condition of what is in question—life-hopes, personal feelings, knowledge, or the parts of morality. On the other hand, we may focus on a different conception of the initiation of action, which has to do only with voluntariness, and have attitudes involving only it. If we give our attention to origination as well as voluntariness, we fall into dismay in all cases. If we attend only to the voluntariness of actions, and give it a certain regard, which we can, we may and are likely to make the response of intransigence.

We have not much delayed, in our survey, to reflect on the general nature of an attitude or feeling, or to enter into an ongoing philosophical discussion of attitudes and like things. It is, although attitudes are not exactly emotions, the discussion having to do with the analysis of emotion. (Leighton, 1985; Kraut, 1986; Solomon, 1986) We have, rather, proceeded on the basis of a quite standard view, roughly to the effect that an attitude is no relatively simple or unitary thing—not, say, a sensation-like feeling or a judgement—but rather is an evaluative and feelingful thought bound up with desires. So too is a response such a complex or fusion, although we have not looked into that at all.

If our concern had been the philosophical analysis of an attitude or feeling, we would have considered much more of the relations of the elements within them, or rather, their somewhat different relations in different attitudes. We would also have considered the several confusing ways we have of thinking and speaking of them. For example, as has been apparent, to be describable as 'having an attitude of moral disapproval', taken as *including* the thought of the voluntariness of an action, is also to be in a state which falls under the description of 'taking the attitude that the voluntariness of the action justifies disapproval'—which latter description makes the thought *external* to the disapproval in the given sense. Further, if our enterprise had been the analysis of attitudes, we should have given more attention to physiological states or changes. (Lyons, 1980; Alston, 1967; Budd, 1985)

None of this, of which something more will need to be said in another connection (pp. 478 ff.), has been necessary to the enterprise in hand. It has been, to repeat, the enterprise of establishing the broad fact that each of us has or tends to have each of two inconsistent attitudes with respect to consequences of determinism, and each of us in the first instance may make two inconsistent responses. Each of the families of attitudes is natural. In itself each attitude involves us in no incoherence or any like thing. It is certainly not something which is put upon us by theory—certainly not by philosophers—but something we find in ourselves. So with the responses.

Clearly this situation calls for further reflection. In particular, it is no satisfactory situation and appears to face us with a kind or kinds of choice. Despite what has been said of the defensibility of an attitude or a response in itself, there must be a question about attitudes, and, above all, the question of whether dismay or intransigence or some third response is, as we can say, our best course. There is also the further matter, so far left untouched, of determinism's seventh area of consequence—social institutions, practices, and habits.

Before coming to those final and culminating stages of our inquiry, there is one other issue to be considered. Indeed it is rather more than what is properly called an issue. It is *the traditional philosophical dispute as to the consequences of determinism*. If we have not yet resolved the problem of the consequences of determinism, we have quite enough on hand to deal with the traditional dispute. What is still to be said, in the final stages of our inquiry, will be of the order of what has been said already—it will have to do with attitudes and responses. If we have not yet seen all of the possibilities of our situation, we have seen that it is, so to speak, fundamentally a situation of attitudes and responses. That general fact is established, and is enough in order to deal with the traditional philosophical dispute as to the consequences of determinism.

8

Compatibilism and Incompatibilism

8.1 A LONG BATTLE

The claim that we have two attitudes in the case of each of the consequence-areas of determinism, and the claim that in the first instance we do or may make two responses, and a further claim to be advanced in due course about a choice to be made from these two responses and another—these three claims will be thrown into sharper definition and their distinctiveness clarified if we look at the long battle, the traditional dispute, about the consequences of determinism. It is this traditional dispute about consequences between two parties, rather than the subjects of the conceptual adequacy and the truth of determinism, that has most engaged philosophers, at any rate philosophers within the main current of philosophy in the English language. It is a dispute carried forward as zealously now as in the past.

There is also a second reason, other than the making of a clear distinction between doctrines, for looking at the long battle. The two parties to it, Compatibilists and Incompatibilists, or exponents of freedom as voluntariness and exponents of freedom as voluntariness plus origination, propound the principal alternatives to the contentions of Chapter 7 about the consequences of determinism, which contentions will be defended further and added to in what follows, centrally by way of an answer to the question of the choice among responses. Any adequate defence of these contentions depends in part on an argued rejection of their entrenched alternatives, alternatives which have always been thought and sometimes declared to exhaust the possibilities. (G. Strawson, 1986, p. 6) That very definitely they do not is a principal contention of this examination of the consequences of determinism.

Our clarificatory purpose in particular will be securely served by allowing Compatibilists and Incompatibilists to speak for themselves. They themselves, that is, will establish the distinction in question. Also, the quoting of considerable passages from their works will be as quick and effective a way of exposition as any other. In fact we shall do little more than glance at the long battle. In the seventeenth century it was already a controversy of which Hobbes could say that it had given rise to 'vast and involuble volumes . . . which fill not only our libraries but the world with their noise and disturbance.' (1839–45, (*c.* 1650), Vol. iv, p. 234) It has since persisted, if a bit more quietly. By one reckoning it is something like four centuries old. (Kenny, 1975, p. 123) It may well be older. By another reckoning it began, along with the dispute about the truth of determinism, with the Stoic philosophers of the several centuries before and after Christ. (Sorabji, 1980, Ch. 4, Ch. 14; Huby, 1967) The fact of its longevity need not be dispiriting. It need not make one abandon hope of a solution. As we shall also see, to claim to have the solution, as I do, may not be entirely vainglorious.

The materials of the dispute were certainly in existence before the dispute—or before its clear beginning. Among the many distinctions made by Aristotle between causes, disposition, origins, originating principles and the like, one is of importance to us. The following passage is from the *Eudemian Ethics*.

> . . . if anything existent may have the opposite to its actual qualities, so of necessity may its principles. For what results from the necessary is necessary; but the results of the contingent might be the opposite of what they are; what depends on men themselves forms a great portion of contingent matters, and men themselves are the sources of such contingent results. (1915 (*c.*350 BC), 1222^b)

There is obscurity in this, and more in the surrounding passages, and certainly ground for objection, all of which may be owed in part to the historical transmission of the original text and by difficulties of translation. Aristotle's thought, or one of his thoughts, appears to run as follows. Each existent thing has a character which is in some way shared by whatever it is from which the thing comes. A thing which *necessarily* happens must have a like source—it too must occur necessarily. A thing which does not necessarily happen, and thus which might have been otherwise, or might not have happened, must also have a like source, such that this source might have been otherwise or might not have existed. At least some of a man's actions are such that they might not have occurred. They must depend on something of a certain kind in him. Perhaps it can be speculated that this thing, for Aristotle, is *a somehow sufficient cause that may or may not operate.* (Cf. Kenny, 1979, Ch. 7)

In the *Metaphysics* (1048^a3 f.), at any rate, according to one translation, we have the enlargement that

> . . . it is necessary in the case of non-rational powers that when an agent and a patient are brought together the action and effect take place, whereas in the case of rational powers this is not necessary; for every one of the non-rational

powers can have but a single effect, whereas the rational can have contrary effects, so that if they were under the same necessity as are the irrational they would have contrary effects at the same time. But this is impossible. It is necessary, accordingly, that something else be decisive in rational action; I mean wanting or deliberate choice. (Kenny, 1975, p. 124)

What we have added here is roughly that a man has a *rational power*, which power can give rise to contrary effects, his doing of one thing or another—say his voting for or against Pericles. A rational power is therefore different in kind from a non-rational power or ordinary disposition (p. 139), which issues in a single effect. If we ask how a rational power gives rise to what it actually does give rise to, say voting for Pericles, we cannot suppose this is a matter of necessitation, since, by some sort of parity of reasoning, we should have to conclude that it also necessitates voting against Pericles. But voting for and against Pericles is impossible. Hence rational powers not only can give rise to contrary actions, but these actions are not necessitated. Such a power consists in deliberative choice.

In Book Six of the *Nicomachean Ethics* (1139^{a}30 f.) more is said of rational power or deliberative choice.

The origin of conduct—its efficient, not its final, cause—is choice; and the origin of choice is appetition plus means–end reasoning. So without understanding and reasoning on the one hand, and moral character on the other, there is no such thing as choice; for without reasoning and character there is neither well-doing nor its opposite. Reasoning, in itself, moves nothing; only means–end reasoning concerned with conduct. . . . for good conduct is the end in view, and that is the object of the appetition. Therefore choice is either appetitive intelligence or ratiocinative appetite. (Kenny, 1975, p. 16)

Something of what is intended here, again roughly, is that the efficient cause of a man's action, which efficient cause is choice, has essentially to do with desire or the like and also with reasoning about means to an end, the end being the goal or final cause of the action. The reasoning in itself would not issue in action. Desire or the like for the end is also necessary. Choice is thus desirous means–end reasoning. (Cf. Kenny, 1979, Ch. 2.)

In the ideas of these three passages—ideas of a somehow sufficient but non-necessitating source or principle of action, of a rational power which can give rise to either of two upshots, and of choice as desirous reasoning—we clearly have an antecedent, perhaps the prime antecedent, of indeterminist pictures of the mind, and in particular the idea of an originator. In certain essential respects it is about as contentful as those pictures, which is not to say very contentful.

Earlier in the *Nicomachean Ethics*, in Book Three (1109^{b}30 f.),

Aristotle is directly concerned with morality, and in this connection offers his well-known account of what can be called voluntary action.

> . . . moral excellence or virtue has to do with feelings and actions. These may be voluntary or involuntary. It is only to the former that we assign praise or blame, though when the involuntary are concerned we may find ourselves ready to condone and on occasion to pity. It is clearly, then, incumbent on the student of moral philosophy to determine the limits of the voluntary and involuntary. . . . Actions are commonly regarded as involuntary when they are performed (a) under compulsion, (b) as the result of ignorance. An act, it is thought, is done under compulsion when it originates in some external cause of such a nature that the agent or person subject to the compulsion contributes nothing to it. Such a situation is created, for example, when a sea captain is carried out of his course by a contrary wind or by men who have got him in their power. But the case is not always so clear. . . . An involuntary act being one performed under compulsion or as the result of ignorance, a voluntary act would seem to be one of which the origin or efficient cause lies in the agent, he knowing the particular circumstances in which he is acting. (1953 (*c*.350 BC), p. 77, p. 81)

Again there are problems of interpretation, some of them having to do with the fact that the relevant Greek terms are only inadequately rendered into the English terms 'voluntary' and 'involuntary', or indeed into any other English terms. What we can take away from Aristotle's discussion, of which the given passage preserves no more than the main theme, is that there are certain things, having to do with ignorance and compulsion, which are commonly regarded as standing in the way of a man's action being voluntary, and hence his having praise or blame for it. Certain conditions which must be satisfied if a man's action is to be voluntary—in one understanding they are necessary conditions but do not comprise a sufficient condition—are that the cause of his action lay within him, and he acted in a knowledge of what he was doing.

Aristotle in another place in the *Nicomachean Ethics* appears to bring together the first three passages, having to do with origination, with the fourth, having to do with voluntariness, into something like a logically necessary and sufficient condition, in so far as the initiation of action is concerned, for moral approval and disapproval, for crediting a person with responsibility for something or holding a person responsible for something.

> It is when the act is voluntary that the moral issue presents itself. We blame the doer and, with that, his deed becomes an unjust act. . . . By a voluntary action, let me repeat, I mean one which (a) it was in the agent's power to do or not to do, (b) he performs not in ignorance but with full knowledge of the person affected by his action, the instrument he is using, the object he seeks to

attain, (c) in no particular is determined by accident or *force majeure*. (1953 (*c*.350 BC), p. 159)

There are, it is to be admitted, some difficulties in taking this fifth passage as an amalgamation of the first three and the fourth. Here we clearly have 'voluntary' used in a wider sense than in the fourth passage, and as we have been using it in our own inquiry. But it seems a secure assumption that Aristotle takes us to have it that moral responsibility for an action to depend both on its deriving from rational power or desirous means–end reasoning, and its being voluntary in the narrow sense, where voluntariness is only a matter of the absence of compulsion and ignorance.

Aristotle can hardly be said to consider seriously any clarified thesis of determinism (Sorabji, 1980; Huby, 1967), although in one passage in the *Nicomachean Ethics* he reports on a deterministic argument which he then quickly and carelessly rejects. (1953 (*c*.350 BC), p. 92) Clearly, however, he has an idea of determinism.

My principal interest in him has to do with two matters, two speculations, connected with what is plainly true, that he took rational power to be an indubitable fact. The first and most important speculation is in part that he claims or would have claimed that all men or all clear-headed men have a certain conception of the initiation of action, which conception has to do with rational power as well as voluntariness. Further, all men see that it is rational power as well as voluntariness that is a necessary condition of moral approval and disapproval. All men agree, as a matter of a single, settled conception having to do with the initiation of action, that this approval and disapproval would be impossible if men lacked choice or rational power as conceived. They would agree that the mere voluntariness of an action, voluntariness in the narrow sense, having to do with ignorance and compulsion, would not be sufficient to preserve the agent's moral responsibility for it. They have some sort of settled belief to this effect. Further, their settled belief as to rational power is, and is taken by them as, a belief inconsistent with determinism.

Given this speculative understanding of Aristotle, what stands in the way of his being described as our first Incompatibilist is not greatly more than the implication that the dispute existed within the Greek philosophy of his time. As remarked earlier, it seems not to have. That is, there may have been no exponents of freedom as no more than voluntariness—a freedom consistent with determinism.

A second and certainly more secure speculation, well supported by the general tenor of Aristotle's work, is that he would take the fact of a man's rational power to be itself a fact which confers a status upon him. It separates him from the brute world of non-rational powers. The fact of rational power, then, is not merely one more fact about the world, or no more than a fact of great actual consequences, but *the* fact which gives dignity to human existence. A man would be another and a lesser thing, a thing of far smaller significance, if he lacked choice of the given kind, the power which distinguishes him from all of the non-rational, and lies at the foundation of all excellence.

There is little need to speculate about the views of the determinist Hobbes. In *Of Liberty and Necessity* he writes:

For . . . what *liberty* is; there can no other proof be offered but every man's own experience, by reflection on himself, and remembering what he useth in his mind, that is, what he himself meaneth when he saith an action . . . is *free.* Now he that reflecteth so on himself, cannot but be satisfied . . . that a *free agent* is he *that can do if he will*, and *forbear if he will*; and that *liberty* is *the absence of external impediments.* But to those that out of custom speak not what they conceive, but what they heard, and are not able, or will not take the pains to consider what they think when they hear such words, no argument can be sufficient, because *experience* and *matter of fact* are not verified by other men's arguments, but by every man's own *sense* and *memory*. (1839–45 (*c.*1650), pp. 275–6)

A free action, as every man knows from reflection on his own thought and use of language, and in particular his use of the term 'free', is no more than one which is willed by him, which is to say desired by him, one which is not subject to external impediments. It is an action of which no more is required than that it be in a plain sense voluntary or willing. That is to say that nothing is required by way of an origin in rational power or the like. More fully, as Hobbes writes in *Leviathan*,

LIBERTY, or FREEDOM, signifieth, properly, the absence of opposition; by opposition I mean external impediments of motion; and may be applied no less to irrational, and inanimate creatures, than to rational. For whatsoever is so tied, or environed, as it cannot move but within a certain space, which space is determined by the opposition of some external body, we say it hath not liberty to go further. And so of all living creatures, whilst they are imprisoned, or restrained, with walls, or chains; and of the water whilst it is kept in by banks, or vessels, that otherwise would spread itself into a larger space, we use to say, they are not at liberty, to move in such manner, as without those external impediments they would . . .

And according to this proper, and generally received meaning of the word, a FREEMAN, *is he, that in those things, which by his strength and wit he is able to do, is not hindered to do what he has a will to.* But when the words *free*, and *liberty*, are applied to any thing but *bodies*, they are abused; for that which is not subject to motion, is not subject to impediment . . . from the use of the word *free-will*, no liberty can be inferred of the will, desire, or inclination, but the liberty of the man; which consisteth in this, that he finds no stop, in doing what he has the will, desire, or inclination to do . . . .

Liberty, and *necessity* are consistent: as in the water, that hath not only *liberty*, but a *necessity* of descending by the channel; so likewise in the actions which men voluntarily do: which, because they proceed from their will, proceed from *liberty*; and yet because every act of man's will, and every desire, and inclination proceedeth from some cause, and that from another cause, in a continual chain, whose first link is in the hand of God the first of all causes, proceed from *necessity*. So that to him that could see the connexion of those causes, the *necessity* of all men's voluntary actions, would appear manifest. (1839–45 (*c*.1650), p. 196)

Finally, a man's having this voluntariness in his action, a voluntariness quite consistent with necessity, is entirely sufficient for his having moral commendation or condemnation, and hence what we have called moral approval and disapproval. It is in no way required that the action not be subject to necessity.

... *praise* and *dispraise* ... depend not at all on the necessity of the action praised or dispraised. For what is it else to *praise*, but to say a thing is good? Good, I say, for me, or for somebody else, or for the state and commonwealth? And what is it to say an action is good, but to say it is as I would wish? or as another would have it, or according to the will of the state? that is to say, according to the law. Does my Lord think that no action can please me, or him, or the commonwealth, that should proceed from *necessity*? Things may be therefore *necessary* and yet *praiseworthy*, as also *necessary* and yet *dispraised*, and neither of them both in vain, because *praise* and *dispraise* ... do by example make and conform the will to good. ... (1839–45 (*c*.1650), pp. 255–6)

These passages are of interest to our inquiry in a number of ways.

(i) Unquestionably Hobbes is convinced and declares that the solution to the problem of Free Will—or more particularly the problem of the consequences of determinism—is to be settled by recourse to a single conception shared by virtually all men. There is that which 'we use to say', that which is 'every man's own experience', that which 'he himselfe meaneth when he saith an action is free'.

(ii) He appears to take it in the last passage quoted that virtually all men share a true proposition that it is this freedom and no other that is necessary to moral approval and disapproval. They have a clear belief to this effect, perhaps true in the ordinary sense.

(iii) Only those of us who are led astray by custom and argument of a philosophical or theological kind, custom and argument in what can be called the Aristotelian tradition, and do not attend to our own beliefs and usages, are unclear that this free action consists in nothing more than voluntary action as conceived.

(iv) Thus, since necessity or determinism is fully consistent with voluntariness, an acceptance of determinism has no effect whatever on our approval and disapproval. The question of moral responsibility is unaffected by an acceptance of determinism.

(v) Less precisely, as the second passage indicates and as other passages demonstrate (1839–45 (*c*.1659)b, pp. 274–5), Hobbes can be said to be *captured* by a determinism. It consists not merely in truth, but in truth of great sway, truth which must carry all before it. His commitment to a determinism is not unlike Aristotle's commitment to rational power or origination.

(vi) Finally, and importantly, Hobbes takes the claims of his philosophical opponents in the Aristotelian tradition to be the product of no more than confusion, a confusion produced by philosophical and theological custom. They have not done all that is necessary, which is to attend to their own thoughts and words.

His adversary, Bishop Bramhall, was taken to be opposed to him on all counts. In fact there is an agreement between them, and between them and others, which is fundamental to the argument now being pursued in this inquiry. However, Bramhall is absolutely persuaded that anyone worth attending to has a determinate conception of liberty not far from that of Aristotle—according to our speculation about Aristotle—and hence he is wholly disdainful of the definition of liberty as no more than voluntariness.

> . . . true Liberty consists in the elective power of the rational will. . . . Reason is the root, the fountain, the original of true Liberty which judgeth and representeth to the will, whether this or that be convenient, whether this or that be more convenient. Judge then what a pretty kind of liberty it is which is maintained by T.H., such a liberty as in little children before they have the use of reason, before they can consult or deliberate of anything. Is not this a Childish Liberty? and such a liberty as in brute beasts, as bees and spiders . . . ? Is not this a ridiculous liberty? Lastly (which is worse than all these) such a liberty as a river hath to descend down the channel. . . . Such is T.H.'s liberty. . . . T.H. appeals to every man's experience. I am contented. Let everyone reflect upon himself . . . . (1676, pp. 651–2)

Bramhall like Aristotle also takes man's rational power, the supposed fact of origination, as distinct from voluntariness, to be that which gives significance and dignity to human existence. In Bramhall's case, this is intimately connected with religious belief, and in particular doctrines of righteousness, sin, and a divinely ordered scheme of things.

Hume, although he is sometimes cited as the founder of the tradition of Compatibilism, of freedom as voluntariness, does not advance greatly beyond Hobbes in fundamental conceptions, as can be seen from the following famous passages. Still, they have shaped, indeed governed, the tradition of Compatibilism, and are as important as any to be considered.

It is true, if men attempt the discussion of questions which lie entirely beyond the reach of human capacity, such as those concerning the origin of worlds, or the economy of the intellectual system or region of spirits, they may long beat the air in their fruitless contests, and never arrive at any determinate conclusion. But if the question regard any subject of common life and experience, nothing, one would think, could preserve the dispute so long undecided but some ambiguous expressions, which keep the antagonists still at a distance, and hinder them from grappling with each other.

This has been the case in the long disputed question concerning liberty and necessity; and to so remarkable a degree that, if I am not much mistaken, we shall find, that all mankind, both learned and ignorant, have always been of the same opinion with regard to this subject, and that a few intelligible definitions would immediately have put an end to the whole controversy. . . .

I hope, therefore, to make it appear that all men have ever agreed in the doctrine both of necessity and of liberty, according to any reasonable sense, which can be put on these terms; and that the whole controversy has hitherto turned merely upon words. (1963 (1748), p. 81)

After making use of his definition of causation as constant conjunction in his argument for determinism, Hume comes on to his definition of freedom.

. . . to proceed in this reconciling project with regard to the question of liberty and necessity; the most contentious question of metaphysics, the most contentious science; it will not require many words to prove, that all mankind have ever agreed in the doctrine of liberty as well as in that of necessity, and that the whole dispute . . . has been hitherto merely verbal. For what is meant by liberty, when applied to voluntary actions? We cannot surely mean that actions have so little connexion with motives, inclinations, and circumstances, that one does not follow with a certain degree of uniformity from the other, and that one affords no inference by which we can conclude the existence of the other. For these are plain and acknowledged matters of fact. By liberty, then, we can only mean *a power of acting or not acting, according to the determinations of the will*: that is, if we choose to remain at rest, we may; if we choose to move, we also may. Now this hypothetical liberty is universally allowed to belong to every one who is not a prisoner and in chains. Here, then, is no subject of dispute. (1963 (1748), p. 95)

He then maintains that his liberty, voluntariness, which is wholly consistent with necessity, is in effect universally believed to be all that is essential to moral approval and disapproval, and to commendation and condemnation. Indeed, for us to add something else would be destructive. We see that necessity, the very denial of origination, *is* required for praise or blame.

. . . necessity . . . has universally, though tacitly, in the schools, in the pulpit, and in common life, been allowed to belong to the will of man, and no one has ever pretended to deny that we can draw inferences concerning human actions, and that those inferences are founded on the experienced union of like actions with like motives, inclinations and circumstances. . . . The only proper object of hatred or vengeance is a person or creature, endowed with thought and consciousness; and when any criminal or injurious actions excite that passion, it is only by their relation to the person, or connexion with him. Actions are, by their very nature, temporary and perishing; and where they proceed not from some cause in the character and disposition of the person who performed them, they can neither redound to his honour, if good; nor infamy, if evil. . . . as they proceeded from nothing in him that is durable and constant, and leave nothing of that nature behind them, it is impossible he can, upon their account, become the object of punishment or vengeance. According to the principle, therefore, which denies necessity, and consequently causes, a man is as pure and untainted, after having committed the most horrid crime, as at the first moment of his birth, nor is his character anywise concerned in his actions, since they are not derived from it, and the wickedness of the one can never be used as a proof of the depravity of the other. (1902 (1748), pp. 97–8)

For Hume as for Hobbes, (i) it is plain that we all share a single firm conception with respect to freedom which must be decisive in settling the problem of liberty and necessity. All men have ever agreed, and could not mean otherwise by their words. (ii) It is said or at the very least implied, further, that we have a belief to the effect that this freedom alone is the prerequisite of disapproval, hatred, blame, and vengeance. (iii) To add a denial of necessity to this freedom—to add origination to voluntariness—would in fact make moral disapproval and the like impossible. (iv) The acceptance of a determinism leaves moral responsibility precisely where it was, given the compatibility of determinism and voluntariness. That nothing changes can be no surprise for the additional reason just mentioned, that moral responsibility *presupposes* a determinism. (v) Importantly, the whole problem of liberty and necessity is an intellectual or theoretical one. It is to be settled by putting a few intelligible definitions in place of ambiguous expressions, by attending to what we mean.

Kant accepts that all that of which we have experience, including all the springs of our own actions, are subject to a determinism. It is, of course, a foundation of his philosophy that the very existence of our experience is in a way dependent on universal causality. Every aspect of my performance of an action, then, and all of what preceded it, considered in terms of ordinary reality, or in its 'empirical character', conforms to 'the laws of causal determination'. (1950 (1781), p. 468)

However, the existence of moral obligations is no less indubitable a fact, if a fact about another order, another realm. That I am sometimes culpable for an action, and to be blamed for it, is such a fact. It inescapably follows that I performed it freely. As for what this comes to,'. . . we must understand . . . by the term freedom . . . a faculty of the spontaneous origination of a state; the causality of which, therefore, is not subordinated to another cause determining it in time.' (1943 (1781), p. 300) This faculty is one of Reason.

The seeming contradiction is escaped, in Kant's view, by the idea that the faculty of origination can also be regarded as outside the realm of experience. It is not in its 'empirical character', but in its 'intelligible character', that 'the faculty of spontaneous origination' can be truly so described. (1950 (1781), p. 300) Very evidently this way of dealing with the unavoidable idea that 'the freedom ascribed to the will seems to stand in contradiction to natural necessity' (1959 (1785), p. 75) bears no relation whatever to the Compatibilism of Hobbes and Hume. The latter makes no use of an idea of two realms.

The doctrine is most plainly expressed by Kant in the following passage from the *Critique of Pure Reason*, which is free of his usual technicalities.

> . . . let us take a voluntary action, for example, a malicious lie by which a certain confusion has been created in society. First of all, we endeavour to discover the motives to which it has been due and then, secondly, in the light of these, we proceed to determine how far the action and its consequences can be imputed to the offender. As regards the first question, we trace the empirical character of the action to its sources, finding these in defective education, bad company, in part also in the viciousness of a natural disposition insensitive to shame, in levity and thoughtlessness. . . . We proceed in this inquiry just as we should in ascertaining for a given natural effect the series of its determining causes. But although we believe that the action is thus determined, we none the less blame the agent, not indeed on account of his unhappy disposition, nor on account of the circumstances that have influenced him, nor even on account of his previous way of life; for we presuppose that we can leave out of consideration what this way of life may have been, that we can regard the past series of conditions as not having occurred and the act as being completely unconditioned by any preceding state, just as if the agent in and by himself began in this action an entirely new series of consequences. Our blame is based on a law of reason whereby we regard reason as a cause that irrespective of all the above-mentioned empirical conditions could have determined, and ought to have determined, the agent to act otherwise. This causality of reason we do not regard as only a co-operating agency, but as complete in itself, even when the sensuous impulses do not favour but are directly opposed to it; the action is ascribed to the agent's intelligible character; in the moment when he utters the lie, the guilt is entirely his. Reason, irrespective of all empirical conditions of the act, is completely free, and the lie is entirely due to its default. (1950 (1781), p. 477)

As for the idea that the freedom required for morality might consist in voluntariness alone, it is dismissed by Kant in the following passage from the *Critique of Practical Reason*.

Suppose I say of a man who has committed a theft that this act by the natural law of causality is a necessary result of the determining ground existing in the preceding time and that it was therefore impossible that it could have not been done. . . . how can he be called free at this point of time with reference to to this action, when in this moment and in this action he stands under inexorable natural necessity? It is a wretched subterfuge to seek an escape in the supposition that the kind of determining grounds of his causality according to natural law agrees with a comparative concept of freedom. According to this concept, what is sometimes called 'free effect' is that of which the determining natural cause is internal to the acting thing. For example, that which a projectile performs when it is in free motion is called by the name 'freedom' because it is not pushed by anything external while it is in flight. . . . So one might call the actions of a man free because because they are actions caused by ideas we have produced by our own powers, whereby desires are evoked on occasion of circumstances and thus because they are actions brought about at our own pleasure; in this sense they are called free even though they are necessary because their determining grounds have preceded them in time. With this manner of argument many allow themelves to be put off and believe that with a little quibbling they have found the solution to the difficult problem which centuries have sought in vain and which could hardly be expected to be found so completely on the surface. (1949 (1788), p. 99)

These passages do not give greatly more of Kant's view than a general impression. They overlook distinctions and doctrines expounded elsewhere. Nor shall I give more than an impression of why it is hard to demur when the view is described as 'a hopeless failure'—a description put on it in a full account of Kant's philosophy. (R. Walker, 1978, p. 148; cf. Boyle, Grisez, Tollefsen, 1976, pp. 112–15) There is no possibility of thinking that the contradiction between a determinism and origination is in fact avoided by trying to put Reason, the faculty of the will, character or whatever into two realms. Further, if the contradiction *could* be argued to be avoided by this division, then the originating will is wholly subtracted from the world of choices, decisions, and moral responsibility. That which actually does originate is quite removed from our actual subject-matter, the one to which we must attend.

As for the argument that moral responsibility must be regarded as such a kind of fact as licenses an inference to the existence of what it is taken to presuppose, origination, and in some way a denial of determinism, it is an argument that has to some extent persisted. It has lately been refurbished in a considerable defence of Free Will. (van Inwagen, 1983) For reasons that may already be apparent, and will become clearer, having to do essentially with the presumed *fact* of moral responsibility, it seems to me an argument without force. In a sentence, it is an argument which moves from desires and the like as to the nature of reality to a factual conclusion about the nature of reality.

Our present concern, however, is neither the attempt to avoid the inconsistency of determinism and origination, nor the argument for origination, but the consequences of determinism. Kant's view is very relevant to this. My purpose has been to illustrate a number of things.

(i) Kant takes it that a determinism in itself, with no more said, would be wholly destructive of morality. If a determinism could not be supplemented, so to speak, by its denial, nothing of moral responsibility and what presupposes it could exist. (ii) This conclusion is the result, in part, of Kant's absolute conviction that there is a truth, indeed 'a law of reason', to the effect that the freedom that is fundamental to morality involves origination, and that the attempt to argue that only voluntariness is required by morality is wholly mistaken, a departure from truth and usage, indeed a 'wretched subterfuge'. (iii) We are to understand, obviously, that this conviction about the necessity of origination to morality is not peculiar to Kant, but rather that all men or at least all clear-thinking men have his beliefs. (iv) Kant's philosophical opponents—no doubt he had Hume in mind—in conceiving man to lack the power to begin an entirely new series of consequences, have deprived man of his splendour, his capability of truly being a moral agent.

It would be mistaken and indeed absurd to suggest that the traditions of Incompatibilism and Compatibilism are wholly uniform —wholly uniform in the respects in which we are interested. There is none the less a considerable uniformity, quite enough for conclusions to which we are coming. Mill, to come on to him, illustrates both facts. He is to some extent different from Hobbes and Hume, but is well within their tradition. In his *Examination of Sir William Hamilton's Philosophy* he attends to the question of the good man, as distinct from the related question of moral responsibility. His question is whether goodness depends on an assumption of Free Will.

My position is, that a human being who loves, disinterestedly and consistently, his fellow creatures and whatever tends to their good, who hates with a vigorous hatred what causes them evil, and whose actions correspond to in character with these feelings, is naturally, necessarily, and reasonably an object to be loved, admired, sympathized with, and in all ways cherished and encouraged by mankind; . . . and this whether the will be free or not, and even independently of any theory of the difference between right and wrong; whether right means productive of happiness, and wrong productive of misery, or right and wrong are intrinsic qualities of the actions themselves, provided only we recognize that there is a difference, and that the difference is highly important. What I maintain is, that this is a sufficient distinction between moral good and evil: sufficient for the ends of society and sufficient for the individual conscience:- that we need no other distinction; that if there be any other distinction, we can dispense with it; and that, supposing acts in themselves good or evil to be as unconditionally determined from the beginning of things as if they were phaenomena of dead matter, still, if the determination from the beginning of things has been that they shall take place through my love of good and hatred of evil, I am a proper object of esteem and affection. . . . (1979a (1865), pp. 456–7; cf. 1961 (1843), p. 547 f.)

Mill adds a note which sums up and rejects the response of one of his many critics, Patrick Alexander (1866), and further indicates his own view, to the effect that morality in general does not depend on Free Will.

Mr. Alexander draws a woeful picture of the pass which mankind would come to, if belief in so-called Necessity became general. All 'our current moralities' would come to be regarded 'as a form of superstition', all 'moral ideas as illusions', by which 'it is plain we get rid of them as motives': consequently the internal sanction of conscious would no longer exist. 'The external sanctions remain, but not quite as they were. That important section of them which rests on the moral approval or disapproval of our fellow-men has, of course, evaporated': and 'in virtue of a deadly moral indifference', the remaining external sanctions 'might come to be much more languidly enforced than as now they are', and the progressive degradation would in a sufficient time 'succeed in reproducing the real original gorilla'. A formidable prospect: but Mr. Alexander must not suppose that other people's feelings, about the matters of highest importance to them, are bound up with a certain speculative dogma, and even a certain form of words, because, it seems, his are. (1979a (1865), p. 457)

Does Mill allow in the first passage that among ordinary men, as distinct from such servants of doctrine as Mr. Alexander, some are not of his own convictions—that human goodness depends wholly on certain human qualities, and does not require any denial of determinism, and, as is certainly implicit in the passage, that voluntariness is sufficient for moral approval and disapproval? Does he allow that some ordinary men *do* take the goodness and badness of their fellows to be bound up with origination, with the question of 'whether the will be free or not'? Perhaps Mill can be taken to allow that goodness and badness *are* taken by some to presuppose something in conflict with the idea that all actions are 'as unconditionally determined from the beginning of things as if they were phaenomena of dead matter'—despite his assertion that this presupposition is in fact not needed either in individual morality or for the ends of society. It is to be granted, at any rate, that Mill is not so definite as Hobbes or Hume in conveying that the generality of men have some sort of belief, as distinct from anything else, as to the sufficiency of voluntariness for moral standing and of course moral responsibility. He does indeed speak of feelings. It could be difficult to deny, however, that he somehow has in mind a kind of belief, or also has in mind a kind of belief.

F. H. Bradley, while claiming to disavow both the tradition of origination with voluntariness and the tradition of voluntariness by itself, is in fact within the first one. In 'The Vulgar Notion of Responsibility in Connection with the Theories of Freewill and Responsibility', he writes:

> . . . if, at forty, our supposed plain man could be shown the calculation, made by another before his birth, of every event in his life, rationally deduced from the elements of his being, from his original natural endowment, and the complication of circumstances which in any way bore on him. . . then . . . he would be most seriously perplexed, and in a manner outraged. . . . one sees directly the ground of our man's dislike for rational prediction; for such prediction is, in a word, the construction of himself out of what is not himself. . . . If, from given data and from universal rules, another man can work out the generation of him like a sum in arithmetic, where is his self gone to? It is invaded by another, broken up into selfless elements, put together again, mastered and handled, just as a poor dead thing is mastered by man. And this being so, our man feels dimly that, if another can thus unmake and remake him, he himself might just as well have been anybody else from the first, since nothing remains which is specifically his. The sanctum of his individuality is outraged and profaned; and with that profanation ends the existence that once seemed impenetrably sure. To explain the origin of a man is utterly to annihilate him. (1927, pp. 15, 20)

Bradley subsequently considers Mill's idea, shared with predecessors, that unphilosophical persons may for a time be unsettled by a determinism because they take causation to consist in 'necessity', where that is something more than the reality of causation, which reality is no more than constant conjunction. Bradley replies to Mill on behalf of the vulgar.

> When you speak to us plainly, you have to say that you really understand a man to be free in no other sense than a falling stone, or than running water. In the one case there is as little necessity as in the other, and just as much freedom. And we believe that this is your meaning. But we know that, if these things are so, a man has no more of what we call freedom than a candle or a coprolite, and of that you will never succeed in convincing us. You must persuade us either that the coprolite is responsible, or that we are not responsible; and with all due respect to you, we are going to believe neither. (1927, p. 25)

Our single shared conception of freedom, then, if Bradley is right, involves a conception of a self, which conception cannot exist in conjunction with the predictability entailed by a determinism. His own developed view of the nature and distinctness of this self, and its relation to acts, is derived in good part from Hegel. Certainly it is a view different in kind from Aristotle's, and also Kant's, but not so different as to put it outside of the tradition of origination.

My intention is not to pursue these matters, but to illustrate another philosopher's certainty that we possess a single shared conception of freedom, one which is inconsistent with predictability and determinism. Further, Bradley takes it that we somehow believe this freedom to be necessary to responsibility. Also to be noticed is his particular fervency, if it is only that, in what can be called the humanist commitment of some philosophers opposed to determinism. Man, if a determinism is true, falls to the level of a coprolite.

Jean-Paul Sartre, with respect to our main interest, if not much else, seems to be of Bradley's mind.

> It is strange that philosophers have been able to argue endlessly about determinism and free-will . . . without ever attempting first to make explicit the structures contained in the very idea of *action* . . . No factual state whatever . . . is capable by itself of motivating any act whatsoever. For an act is a projection of the for-itself towards what is not, and what is can in no way determine by itself what is not. . . . This implies for consciousness the permanent possibility of affecting a rupture with its own past, of wrenching itself away from its past so as to be able to consider it in the light of a non-being and so as to be able to confer on it the meaning which *it has* in terms of the project of a meaning which *does not have*. . . . I am condemned to exist forever beyond my essence, beyond the causes and motives of my act. I am condemned to be free . . . man being condemned to be free carries the weight of the whole world on his shoulders; he is responsible for the world and for himself as a way of being. We are taking the word 'responsibility' in its ordinary sense as 'consciousness (of) being the incontestable author of an event or of an object'. (Sartre, 1956 (1943), pp. 433 ff.)

For Sartre we have an idea of action—certainly the idea is not to be taken as something peculiar to him—as necessarily in a way free, as something owed to a kind of origination. Such action, as nothing less would, carries with it a lot of responsibility. That much seems clear, whatever else is to be said. (Cf. M. Warnock, 1965, 1973; Danto, 1975; Caws, 1979; Olafson, 1967.)

G. E. Moore focused the attention of philosophers on what is meant by saying that a man could have done otherwise than he did, or can act otherwise than he does, as distinct from saying that he acted or is acting freely. In well-known passages in his *Ethics*, Moore takes it, rightly, although he does not spell out the connection, that the rightness or wrongness of an act depends on its being true that the agent could have done or can do otherwise.

We certainly have *not* got Free Will, in the ordinary sense of the word, if we never really *could*, in any sense at all, have done anything else than what we did do. . . . But, on the other hand, the mere fact (if it is a fact) that we sometimes can, in *some* sense, do what we don't do, does not necessarily entitle us to say that we *have* Free Will. . . . Whether we have or not will depend on the precise sense in which is true that we can. (1912, p. 203)

What, for instance, is the sense in which I could have walked a mile in twenty minutes this morning, though I did not? There is one suggestion, which is very obvious: namely, that what I mean is simply after all that I could, *if* I had chosen; or (to avoid a possible complication) perhaps we had better say 'that I *should, if* I had chosen'. In other words, the suggestion is that we often use the phrase 'I *could*' simply and solely as a short way of saying 'I *should*, if I had chosen'. (1912, p. 211)

There is, therefore, much reason to think that when we say that we *could* have done a thing which we did not do, we *often* mean merely that we *should* have done it, *if* we had chosen. And if so, then it is quite certain that, in *this* sense, we often really *could* have done what we did not do, and that this fact is in no way inconsistent with the principle that everything has a cause. And for my part I must confess that I cannot feel certain that this may not be *all* that we usually mean and understand by the assertion that we have Free Will. . . . (1912, p. 217)

Moore adds that many people will require for Free Will not only that we could have done otherwise than we did if we had so chosen, but also (a) that we could have chosen otherwise than we did. He then notes that by the latter requirement we may mean (b) we should have so chosen, if we had chosen to make the choice. (1912, pp. 218–9) The sentence (a), if it is analysed into (b), is of course taken to be consistent with determinism, since it is then taken to assert only that the given acts and choices could or should have been otherwise if their causal antecedents—certain choices—had been different.

Moore is inclined to think that the several sentences as analysed are what is meant by saying we have Free Will. However, he ends by saying that he can find no conclusive argument against, or for, the view that saying we have Free Will involves saying we can could have done otherwise in some other sense.

J. L. Austin in his paper 'Ifs and Cans' produced what has often been taken as a refutation of all or some of Moore's account of 'could' and 'can'. (1961) His refutation depends in part on noting that 'I can if I choose' does not involve the idea that the choosing is a cause of the action, since, if this were true, it would follow if I cannot act, I do not choose to. Effects are in a sense necessary conditions of their causes. (Cf. 1.2, 1.3.) Moreover, it does follow from 'I can if I choose' that I can, whether or not I choose to, which sort of entailment would certainly not hold if my choosing were a cause. Causes, he supposes, are necessary conditions of effects. Austin also offers strong argument against the idea that 'I could have done otherwise' means, say, 'If I had the opportunity and the motive to do otherwise, I would have done otherwise'. (Nowell-Smith, 1954)

Austin says hardly anything of what he takes to be the proper understanding of such sentences as 'I could have done otherwise' where they are 'absolute'—indicative rather than conditional. That is, he says little of such claims as 'I could have ruined you this morning (although I didn't)' as against 'I could have ruined you this morning if I had had one more vote'. There is the implication that it is claims of the former indicative sort that are involved in what we think about our freedom, and that they are inconsistent with determinism. He ends by saying:

> It has been alleged by very serious philosophers (not only the two I have mentioned) that the things we ordinarily say about what we do and could have done may actually be consistent with determinism. It is hard to evade all attempt to decide whether this allegation is true—hard even for those who, like myself, are inclined to think that determinism itself is still a name for nothing clear, that has been argued for only incoherently. At least I should like to claim that the arguments considered tonight fail to show that it is true, and indeed in failing go some way to show that it is not. Determinism, whatever it may be, may yet be the case, but at least it appears not consistent with what we ordinarily say and presumably think. (1961, p. 197)

Moore and Austin are in agreement about something, and thus in agreement, despite various differences, with all the philosophers we have noticed. This agreement of theirs is of more significance to the present inquiry than their disagreements.

Moore supposes that there is a single use of 'He could have done otherwise' and the like that settles the issue of the compatibility of freedom and determinism, whatever other uses there may be. That is, we have a single understanding, some understanding or other, of 'He could have done otherwise', which settles the issue of whether the freedom somehow required for moral responsibility and the rest of morality is consistent with a determinism. Moore takes it, evidently, that this is our ordinary or standard use. Further, he is inclined to think that he has specified this crucial use of 'He could have done otherwise', despite not having a conclusive argument for his inclination. If he has some doubt about his inclination, he has no doubt that there is a single thing we mean by 'He could have done otherwise', which thing settles the principal issue.

Austin, despite his still greater caution in all matters of language, shares the idea that we do have a single way of thinking and talking in connection with freedom, which way of thinking and talking is what can settle the issue. Certainly its analysis remains difficult, even after errors of analysis have been put aside. But to get the right analysis of our single common belief would be to settle the issue. It is notable that this inclination of his is not much affected by his feeling that determinism had not been made clear.

To come on to disagreement between Moore and Austin, it was noted above about Hobbes, Hume, and Mill that they suppose that the settled conception in question is a conception as to voluntariness by itself. Moore, although uncertainly, is of their mind. In the cases of Aristotle, Kant, Bramhall, and Bradley, our conception of free action is taken to involve origination as well as voluntariness. Austin, if he was no philosopher to enter into familiar sorts of philosophical speculation, must none the less be taken as going some way in the direction of Aristotle and his successors. It might be said, speculatively, that he is inclined to accept a certain end, the conclusion that we share a single conception inconsistent with determinism, and is unwilling to embrace the available means of attempting to explain it, ideas of the self, powers, faculties, and so on. Moore's reflections, via his conception of freedom, also take him close enough to the conclusion that we have a belief as to what morality requires by way of freedom, and that this freedom would be unaffected by the truth of determinism. Of Austin, it would be rash to say that he would actually conclude, via his different conception of freedom, that morality would be affected by determinism. His innovation in philosophy was accompanied by a great caution.

To revert to my principal point, it is that Moore and Austin take us to have a single conception as to freedom, which conception issues in an ordinary usage, and which conception settles what has been the dominant philosophical question of freedom and determinism, that of the consequences of determinism. Their view is shared by very many 'analytic' philosophers who have subsequently carried on the controversy about 'He could have done otherwise' and the like. The proposition is to be confirmed by a look at a run of issues of most general philosophical journals, or at collections of essays on determinism and freedom. (Berofsky, 1966; Honderich, 1973; Watson, 1982)

As is made explicit in almost all of the views considered so far, they have had to do with the consequences of determinism for morality. Does determinism affect or not affect that particular freedom in mental acts and bodily actions which is a condition of morality? They are views, as just remarked, which enter into the third and dominant part of the philosophical controversy about our freedom.

Another and wrongly differentiated part of the controversy, as we have seen, has concerned the Epicurean objection to determinism, having to do with determinism's alleged inconsistency with knowledge. It was our preliminary conclusion (6.7) that the objection, when seen clearly, is essentially an objection which has to do with determinism's consequences for that freedom in mental acts and bodily actions, whatever it is, which is necessary to knowledge, or, as might have been added, to knowledge and related things. Thus, it was said, the Epicurean tradition *is* part and parcel of the dominant part of the philosophical controversy about freedom. This, it was said, has not been seen by the parties to the dominant dispute. There are very rare exceptions to the generalization. That they are unusual in *this*, in their greater awareness of the breadth of the consequences of determinism, does not make them unusual in their Incompatibilism and Compatibilism.

The neo-scholastic philosophers Boyle, Grisez and Tollefsen (1976) maintain that it is an undeniable proposition that one must have a certain freedom, and no less than it, if one is rationally to affirm anything. They explicitly take this to be a proposition necessarily agreed by all of us. The freedom is one tied to an indeterminist picture of our existence, a freedom of origination. In their characterization, I have such a freedom in making a choice in this actual world if it is the case that there is a possible world such that in it I do not make the choice but everything else is the same in that possible world except for the consequences of my choice. (Cf. van Inwagen, 1974, 1983.)

More fully, rational affirmation is narrowly understood as affirmation which is *not* the sure result of necessary truth or direct experience. It is, rather, affirmation in a circumstance where there are two inconsistent possibilities as to truth, and the best that can be done is to determine which is the more reasonable. Rational affirmation is subject to a certain norm, different from the norms which govern affirmation in a circumstance of necessary truth or immediate experience. If I am rationally to affirm anything, this norm must be operative or in force in my case. That is a possibility only if I have the given freedom of choice. Since this freedom of choice is inconsistent with determinism, there is the conclusion that determinism has as a consequence the impossibility of rational affirmation. In particular, it has the consequence that it is impossible rationally to affirm determinism.

Grunbaum (1953, 1971) also has the distinction of seeing the Epicurean objection rightly as an objection having to do with freedom of decision and action rather than, say, the relation of determinism to truth or its consistency with a supposed conceptual connection of a reason and a conclusion. His view, however, is entirely different. Of the Epicurean tradition he writes:

> . . . it is first pointed out rightly that determinism implies a causal determination of its own acceptance by its defenders. Then it is further maintained, however, that since the determinist could not, by his own theory, help accepting determinism under the given conditions, he can have no confidence in its truth. Thus it is asserted that the determinist's acceptance of his own doctrine was forced upon him. I submit that this inference involves a radical fallacy. The proponent of this argument is gratuitously invoking the view that if our beliefs have causes, these causes *force* the beliefs in question upon us, against our better judgement, as it were. Nothing could be further from the truth; this argument is another case of confusing *causation* with *compulsion*. (1971)

These two views explicitly or implicitly ascribe to us a certain conception of freedom, different in each case, and maintain or imply that if we are not confused, we shall see that it is just what we believe to be necessary to rational affirmation or confidence in truth. In the one case the freedom in question is inconsistent with determinism, in the other case consistent. These two views pertaining to knowledge or the like display an evident congruity with those pertaining to morality—those of Aristotle, Kant, Bramhall, Bradley, and Austin on the one hand, and Hobbes, Hume, Mill, and Moore on the other.

Finally, there is a recent Compatibilist doctrine already noticed (p. 392) in another connection. (Dennett, 1984) It takes a somewhat different wider view of the consequences of determinism. It also has the distinction of illustrating or being coloured by, if not perceiving or asserting, something of the broad fact considered in Chapter 7—the fact of the two families of attitudes and the two responses. Certainly there are references to the question of what *wants* of ours are, and what wants are not, affected by a determinism, and also other consonant remarks. (Dennett, 1984, pp. 18, 171) In this respect, if not others, it can perhaps be regarded as marking a kind of transition to the view to which we shall come. Still, what we are mainly offered, indeed more or less officially and explicitly offered, *is* a Compatibilism, although a more uncertain one than has been supposed by critics. (G. Strawson, 1985; D. Locke, 1986; Watson, 1986)

For example, as a successor to the Compatibilist claim that causation is not compulsion or constraint, we have it that what is necessary, in considering determinism, is to escape the confusion of taking causation to consist in *control*. 'What we must do . . . is perform a long overdue bit of "ordinary language" philosophy, to see what we actually have in mind when we yearn for control and fear its loss.' We must find 'our everyday notion'. (Dennett, 1984, pp. 59, 51) We then see that the fact that I am caused to act as I do is not the proposition that I do not control my action. There is the different point, secondly, that the proposition that I am caused to act as I do is not the proposition that my actions are the product of any simple or 'Sphexish' causation. (p. 392) There is the point, thirdly, that we must see what people are actually thinking when they say someone could have done otherwise in a particular situation. (1984, p. 136) Finally, as already implied, the problem of determinism is fundamentally a matter of confusions, and may be an artefact of the methods philosophers have typically used to study it. (1984, p. 17) These have importantly to do with the metaphors of the Invisible Jailer and the like. (p. 392)

8.2 COMPATIBILISM AND INCOMPATIBILISM DEFINED

As remarked and as we have seen, and as would in any case be expected, the Incompatibilist party of philosophers is certainly not monolithic. Still, it is evident that almost all or at least very many members of the party, including many contemporary members we have not considered, share certain fundamental propositions and a certain attitude—an enumeration of which things gives us a definition of Incompatibilism. We can set out the definition or enumeration of commitments in such a way as to serve the ends of our inquiry, and in particular to make clear not only what distinguishes Incompatibilists from Compatibilists, but also what *unites* the two parties. What unites them, from the point of view of our inquiry, is more fundamental. Incompatibilists maintain the following things.

(1) Virtually all men have a single, settled conception of the initiation of action, which conception is written into their language. It is a conception that pertains to moral approval and disapproval, and praise and blame, and perhaps all of morality.

(2) The conception in question is not only of the voluntariness or willingness of actions, in a sense conveyed in our survey of the controversy about the consequences of determinism, but also of their origination, in a sense conveyed in the survey and also earlier in this inquiry. (Ch. 3) Each of voluntariness and origination enters into this conception of the initiation of actions.

(3) All men have a belief of some kind as to the connection between the initiation of action and moral approval and disapproval. This belief is not to be confused with another they may also have, as to what the facts actually are with respect to the initiation of action—importantly, whether we do in fact originate actions. Nor, of course, is the belief simply the conception of initiation which they have. Rather, precisely, it is a belief as to what the facts *must be* with respect to the initiation of action *if* moral disapproval and the like are in place. It is a belief that

only voluntariness plus origination constitute a sufficient condition, in so far as the initiation of action is concerned, of moral disapproval and the like.

(4) It follows that we must accept that there can be no moral responsibility if we never do originate our actions, which we never do if a determinism is true. Our belief in moral responsibility must in reason collapse if we become convinced of a determinism. This effect on moral responsibility, of course, is a matter of the logical incompatibility of origination and determinism.

(5) The principal problems faced by anyone concerned with determinism and freedom are of intellectual or theoretical kinds. They are problems of analysis, definition, clarity, consistency, other logical relations, truth, and the like. There is the problem, more generally, of seeing clearly the fundamental propositions: that we have a certain single conception pertaining to initiation, written into our language, and that we believe only actions which satisfy it are such as to allow for moral responsibility and what depends on it, and perhaps such other things as personal feelings. This requires that we become clear about origination, and clear about voluntariness. We need to reflect adequately on our own thinking, acquire a full perception of our shared conception and belief, examine the language which expresses them, and so on. The problem is importantly or even primarily a linguistic one. A clear awareness of the inconsistency of our conception with determinism is to be secured, perhaps by the provision of a proof—yet another proof. None of this has to do with attitude or feeling.

(6) It is in virtue of their power of origination, if they have it, that men have a certain adequate stature. An adequate stature depends on their not being subject to causation, or rather an effective determinism, and hence being distinct in kind from other existing things, including other living things. This, unlike its five predecessors, is an attitude rather than a proposition.

It is worth noting that the Incompatibilist party of philosophers is not defined as asserting the existence of origination or the falsehood of determinism. Most of them do so—and thus pass beyond items (1) to (6) to one of the indeterminist pictures of the mind considered earlier. However, Incompatibilists also include philosophers who are uncertain of the truth of an indeterminist picture. Austin might be included here, however tentatively. Finally, as well as including those who embrace an indeterminist picture of the mind, and those who are agnostic on that point, the Incompatibilist party includes some very few determinists who have conceived of the ground of moral responsibility and so on as necessarily including origination. They are, in the description of William James (1909), the Hard Determinists.

The second party of philosophers, the Compatibilists as defined, share the following commitments.

(1) Like their opponents, they assume and almost always assert that virtually all men share one settled conception of the initiation of action, one conception pertaining to moral disapproval and perhaps a little more. The conception is entrenched in language.

(2′) The conception is of an action that is no more than voluntary. It has in it nothing about origination.

(3′) All men somehow believe voluntariness by itself is sufficient, in so far as the initiation of action is concerned, to moral disapproval and so on.

(4′) It follows that a determinism, being compatible with voluntariness, carries no threat to moral responsibility. Our ascriptions of moral responsibility would be unchanged by our coming to believe a determinism.

(5′) The problems of determinism and freedom are, as Incompatibilists also believe, intellectual problems. There is the problem for Compatibilists of leading the opposing party of philosophers to take a clear and unconfused view of ordinary belief and language as to the conditions of responsibility, and to see their consistency with determinism. Merely philosophical habit and doctrine is to be escaped, and the plain fact of consistency is to be insisted upon. There is also the enterprise, taken as very necessary, of leading Incompatibilists to take up higher standards of clarity and contentfulness with respect to conceptions. These higher standards exclude indeterminist theories of the mind.

(6′) Causal connection is or may be what can be called the commanding fact of all that exists. It is or may be even more than, to use Hume's description of something else, 'the cement of the universe'. (1938 (1740), p. 32) Determinism is or may be a great fact, an imperative of inquiry, such that all else must be brought into line with it.

The Compatibilists as defined, as just implied, need not be committed to the truth of a determinism. Some of them, following Hobbes, Hume and Mill, are so committed, and hence fall into James's category of Soft Determinists. Moore is otherwise, a Compatibilist who is officially agnostic about determinism. He has very many companions among contemporary philosophers. If they do not accept determinism, they have not been persuaded either—to take one central point—that Quantum Mechanics falsifies it. Perhaps there are Compatibilists, thirdly, who add a denial of determinism to the above commitments. Their view is that an indeterminism, while true, is not needed for freedom. They would of course have to depart from Hume's idea that determinism is essential to moral responsibility. (Cf. G. Strawson, 1986.)

8.3 COMPATIBILISM AND INCOMPATIBILISM: BOTH FALSE

This inquiry into the consequences of determinism was begun with a consideration of, in fact an eliciting of, two conflicting families of attitudes which we share, attitudes to the initiation of action. (Ch. 7) They do or may give rise to two conflicting responses to determinism, dismay and intransigence. Time enough was taken to establish the truth of the empirical claim about two families of attitudes between which we move, which is fundamental. There is only one effective way of doing so, the one which was adopted. It is to prompt someone's consideration of his or her own attitudes, at bottom by actually evoking them. Doing so must be as much a matter of a kind of philosophical persuasion as anything else, or, to speak differently, a matter of invitation and reminder rather than argument in a more circumscribed sense.

It would be close to futile, or at the very least unpromising, to try by a certain means to establish the proposition that we all have the two kinds of attitudes—the means of a kind of merely linguistic argument, one which does not in effect make an active appeal to the attitudes of interlocutors. This is so since all of the system of terms having to do with the initiation of action—starting with 'freedom', and including even terms which are more apt for the expression of one attitude than another—in fact has a character akin to systematic ambiguity. The entire system is ambiguous as between different attitudes. To *stick with the words*, in a certain sense, is to fail to get in touch with the reality, the reality of the consequences of determinism.

I allow that there are problems raised by this reflection, having to do with what might be called the adequacy and inadequacy of language, but not such problems as must detain us. It is notable that Incompatibilists and Compatibilists, in their different purpose of attempting to establish the existence of a single conception, have been most successful when they have in fact been engaging in something like the evocation of attitude, however differently they have conceived their endeavours. This is the case with the Compatibilist essay which first widened consideration of the consequences of determinism. (P. F. Strawson, 1962)

As remarked already, I have no need to embrace or defend what does in fact seem to me the exceptionless truth that each of us can enter into both families of attitudes, and is capable of both responses—in fact, that in such circumstances of reflection and feeling as those imagined earlier, having to do with the scheming husband and so on, each of us *does* move between the attitudes and responses. It may be true, on the contrary, that some individuals are *immured* in one family of attitudes and one response, very likely intransigence rather than dismay, perhaps as a result of the kind of philosophical commitment which can at least obscure feeling. Those who are immured in this way—I certainly do not speak of the general run of philosophers inclined either to Compatibilism or Incompatibilism—are very few in number indeed. There cannot be many of us whose situation is one that may perhaps be suggested by the declaration with which the Compatibilist philosopher Moritz Schlick opens his chapter, 'When Is a Man Responsible?'

> With hesitation and reluctance I prepare to add this chapter . . . For in it I must speak of a matter which, even at present, is thought to be a fundamental ethical question, but which got into ethics and has become a much discussed problem only because of a misunderstanding. This is the so-called problem of the freedom of the will. Moreover, this pseudo-problem has long since been settled by the efforts of certain sensible persons; and, above all, the state of affairs just described has been often disclosed—with exceptional clarity by Hume. Hence it is really one of the greatest scandals of philosophy that again and again so much paper and printer's ink is devoted to this matter, to say nothing of the expenditure of thought, which could have been applied to more important problems (assuming that it would have sufficed for these). Thus I should truly be ashamed to write a chapter on 'freedom'. In the chapter heading the word 'responsible' indicates what concerns ethics, and designates the point at which misunderstanding arises. Therefore the concept of responsibility constitutes our theme, and if in the process of its clarification I also must speak of the concept of freedom I shall, of course, say only what others have already said better; consoling myself with the thought that in this way alone can anything be done to put an end at last to that scandal. (1956, pp. 143–4)

Perhaps Schlick's passion indicates that he would have been unmoved by any attempt whatever to have him disclose in himself any trace of a disposition to dismay or the attitude from which it springs. Perhaps, for example, he had no inclination whatever to an unfixed personal future, and no retributive attitudes of the kind described earlier. (pp. 430 ff.) I doubt it, but if so, no matter. There is no need for a universal truth, no more than there is a need for universal truths about experience in connection with other problems of philosophy that have to do with contingent matters.

If what has been maintained about our families of attitudes and the responses of dismay and intransigence is true, then all of the propositions espoused by Compatibilists and Incompatibilists are false. What has been maintained about the attitudes, and dismay and intransigence, to my mind, is indisputable. Each of the five propositions of each party then *is* false. As for the sixth commitment (pp. 473, 474) of each of the two parties, about human stature and the sway of causation, it may in a way depend on the false propositions. That is, to take the Incompatibilist inclination that origination is required if men and women are to be of a proper stature, it must at least be touched by accepting, if one has to, that we do not have a common belief that only the inclusion of origination can give a place to responsibility and the like.

Let us see all this in detail, and strengthen the fundamental argument in several ways, by working through the items in question, beginning with the first proposition shared by the two parties.

It is, to repeat, that virtually all men have a single conception of the initiation of action in so far as moral responsibility and the like are concerned. This conception provides the single definition of the word 'free', and the related single definition of many other terms and usages. This has, although rarely, been in a way questioned. (Ayer, 1964, p. 6) In fact it is demonstrably false. It has been demonstrated false by our reflections on our own shared attitudes. We do not have a single conception of initiation in so far as moral responsibility and the like are concerned. We do not have *any* definition of freedom in the relevant sense: where such a definition is the one correct specification of a single thing, fact, or phenomenon. What is true, rather, is that we attend to two different sets of actual or possible properties of the initiation of actions. The fact is reflected by dictionary definitions of 'free' and the like. They move between our two attitudes—all of the definitions do so, under defensible understandings. 'Free' and related terms, as remarked earlier, are thus in a way systematically ambiguous.

A fortiori, to turn to the second Incompatibilist contention (p. 472), it is not the case that we have, as a single conception of initiation in so far as responsibility and the like are concerned, a conception of voluntariness joined with origination. We do have such a conception but we also have another, which has nothing of origination in it. None of us is debarred from the feeling that malignant desires in an action, or want of principle or uncaring disdain in an action, when there is voluntariness, are sufficient for moral disapproval. Admittedly, we can also be taken aback, in our disapproval, by the consideration that the action in question was the necessitated outcome of necessitated antecedents. That is not to say that we cannot and do not in turn escape this constraint on our feelings and desires. We are not immured in the conception which enters into one attitude.

An anticipation of a kind of this central claim was noticed earlier.

(p. 471) The other one of which I know, which deserves more attention than it can have at a late stage of the writing of this book, is contained within a strong doctrine of freedom otherwise wholly different from our own. (G. Strawson, 1986) This doctrine, which also contains the idea considered earlier (p.177) that determinism does not need reflection since indeterminist views of the mind are futile or fatuous, does accept there is something called 'the ordinary strong sense' of the word 'free', and is in a way resolutely Compatibilist. Still, it speaks of us as being 'in some respects *natural compatibilists* in our thought about freedom, and . . . in other respects *natural incompatibilists*.' (G. Strawson, p. 19, cf. Ch. 2, p. 47, note 30)

It will have been noted of the third Incompatibilist contention (p. 472) that it is in a way vague. It is that all men have a belief *of some kind*. They have a belief *of some kind* that only voluntariness plus origination constitute a necessary and sufficient condition, in so far as the initiation of action is concerned, of moral disapproval and the like. This vagueness reproduces the actuality of the Incompatibilist tradition, as a look back at the quoted passages will confirm. What is suggested, but not explicitly asserted, is that we possess it as a literal truth, or something approximate to an ordinary truth, that something is a sufficient condition for something else. We have a belief in that fundamental sense.

The facts are otherwise. What we have is an attitude, which, to describe it one way, is a *taking* of voluntariness with origination together as sufficient for moral disapproval and the like. It is not the only attitude we have, but that is not the present point. The present point is that we do not have anything like a belief in a standard sense: *we do not have something that is true or false*.

Will it be suspected that too much distinction is being made, or too much weight being put on a distinction, between beliefs and attitudes? In the course of our inquiry so far into our attitudes and responses, relatively little of a general kind has been said of the nature of attitudes. It was taken that an attitude is typically an evaluative thought of something, feelingful and bound up with desire. (p. 382) It was noticed that we may speak alternatively of the same facts as (i) taking an attitude of moral disapproval with respect to an action, which attitude includes a conception of the action's initiation, and, as a moment ago, (ii) taking an attitude of disapproval on account of something separate, the conception of the action's initiation. (p. 449) Nothing important hangs on it, but let us here proceed in the second way.

What is it to take it that an action of a certain kind, of a certain initiation, is in a certain respect a sufficient condition of moral disapproval, of holding the agent responsible for it? Certainly this could not be a factual belief to the effect that the given facts—say facts having to do with voluntariness—are part of a *causal* or other *nomic* condition or requirement for moral responsibility. (Ch. 1) To suppose so would be a bizarre confusion, for a number of reasons. One is that the connection between the given facts and moral disapproval is not an explanatory but evidently a justificatory one. The facts provide a *ground*, not part of a causal circumstance.

Can it be, then, that in taking certain facts to be a sufficient condition of moral disapproval, in so far as the initiation of action is concerned, we take it that certain premisses of fact are a *logically* sufficient condition for that disapproval? That is, can it be that we take it that a certain entailment holds? Obviously not. If no complete account, as remarked, has been given of what it is to disapprove of a man in a moral way, it has been said that disapproval involves certain feelings and desires. Feelings of repugnance and desires for his discomfiture were mentioned. (p. 432) These feelings and desires are evidently to be understood in terms of what can be called evaluations and prescriptions. The desires, that is, can be rendered in some such way as this: 'Let it happen that he is discomfited or otherwise distressed'. There can then be no possibility that factual propositions about voluntariness can, so to speak, entail moral disapproval.

To come to the heart of the matter, and hence to show that what we have is properly described as an attitude, which thing is not a belief in the standard sense, or akin to one, it is clear that to say that we take certain facts as in a certain respect sufficient for moral disapproval can only be to say that we *regard* certain facts as providing reasons for certain feelings, desires and the like. What is in question here is in the ordinary sense not a belief at all, but rather something that is like moral and other evaluative judgements in not having a truth-value in the ordinary sense. If it can be called a belief, it is not a belief with a truth-value. Believing that certain facts are the sufficient condition of moral disapproval is precisely like, to take a random example or two, *believing that the existence of distress is a reason for a plan to end it*, or *believing that the length of time it takes to become a doctor is a reason for higher pay*. To have such a belief is not to have something which is true or false, but to have something which itself is properly described, as it has been, as an attitude or feeling.

Return for a moment to the scheming husband, who gives a large gift to his son in order to deprive his wife of her fair share both of the family money and the affections of the son. I may indeed take certain facts about the husband's action to be sufficient for my disapproval of him. But that can only be to say that I am disposed or inclined to take certain facts as reasons for feelings, desires, and related actions. I may take it that my desire that he be discomfited or distressed, and the judgement I may utter to that effect, rest on the good reason that his action was both voluntary and originated. At bottom what is in question is not a belief with a truth-value, but rather a ground-level commitment to the effect that certain facts support or justify feelings, desires, and so on.

If they are to be brought into distant reach of truth, what our Incompatibilists—and Compatibilists—must be converted into maintaining, although they certainly do not see this, is that virtually all men are alike in having a single attitude, in being of a single feeling. Incompatibilists and Compatibilists must be converted into exponents of the view that each of us is inclined to regard a certain set of facts, and these alone, as being good reasons for moral disapproval and related things. No doubt an overlooking of this necessary understanding of what is being maintained, or an avoidance of the question, has been of help to both parties of philosophers in their advocacy. It is easier to get the idea accepted that all men are of *one mind* about something—*each* man is of *one mind* about something—if the nature of the thing is left unexamined, and hence it is not made plain that what is in question is not a belief with a truth-value. The truth of a non-attitudinal belief—the facts which make it true—can be expected to produce in each of us a convergence upon it. The case is different with whatever gives rise to the *attitudinal* belief that something is a good reason for something else.

The true nature of the connection between the given founding-beliefs or founding-ideas and what they support, as we now understand it, reinforces what has been maintained, that each of us can take two conflicting attitudes, and make the response of dismay *and* the conflicting response of intransigence in connection with determinism. There can be no great surprise in the claim that we can entertain opposed 'beliefs' of the given kind, one which requires origination and voluntariness as a pre-condition for various things, and one which requires only voluntariness. It *would* be surprising if they were beliefs with truth-values, since the given claim would then be a claim as to the existence of a standard paradox, or something like one, and that would be an improbable claim. The nature of the given connection between founding-beliefs and what they support will be of significance in the argument to come, issuing in a conclusion about the choice between dismay, intransigence, and a third thing.

The fourth proposition and a further distinction of Incompatibilism is that moral responsibility was never a fact if determinism is true, since it excludes origination. Further, if a determinism comes to have general acceptance we can no longer persist in holding people responsible. To believe this, of course, is necessarily also to believe more, about more things than moral responsibility. While almost all Incompatibilists, as we know, have been overly concerned with the narrow effect on moral responsibility, and have given no attention to life-hopes and so on, some few of them—very few—have had a proper wider view. Some have been mentioned. (p. 470) The wider view does not save them from the present objection. Isaiah Berlin also takes a wider view, indeed a grand one. If we begin to take a determinism seriously, he writes, then

> . . . the changes in the whole of our language, our moral terminology, our attitudes toward one another, our views of history, of society, and of everything else will be too profound to be even adumbrated. The concepts of praise and blame, innocence and guilt, and individual responsibility . . . are but a small element in the structure, which would collapse or disappear. If social and psychological determinism were established as an accepted truth, our world would be transformed more radically than was the teleological world of the classical and middle ages by the triumphs of mechanistic principles or those of natural selection. Our words—our modes of speech and thought—would be transformed in literally unimaginable ways; the notions of choice, of responsibility, of freedom, are so deeply embedded in our outlook that our new life, as creatures in a world genuinely lacking in these concepts, can, I should maintain, be conceived by us only with the greatest difficulty. (1969, p. 113)

It cannot be that this wider view, perhaps a touch more apocalyptic than is required, or the usual lesser proposition about moral responsibility, is correct. *Whatever* is to be drawn from the premiss that we have a single belief to the effect that both voluntariness and origination are required for something—just moral responsibility, or that and a great deal more—is drawn from a false premiss.

Given the possibility of the attitude having to do only with voluntariness, it is wholly false to suppose that a general acceptance of determinism must leave a void where moral responsibility had been. A future counterpart of the scheming husband, who goes about his wretched business in the era of determinism, will in the voluntariness of his action be open to a certain moral disdain. It has been implied already that I do not take it that our choice in the matter of responses to determinism is between only dismay and intransigence. However, the very possibility of what might be called the voluntarist attitude to the initiation of action is quite enough to falsify the claim that determinism must be regarded as destructive of moral responsibility and what depends on it.

What has been established, about attitudes and responses, conflicts as sharply with the fifth Incompatibilist commitment, that the principal questions about the consequences of determinism are questions of definition, clarity, logical relations, truth, and the like. Certainly the question of the truth of determinism is just that, a question of truth. However, that is precisely not the case with the fundamental question of the sufficient condition, with respect to the initiation of action, of responsibility, and like things. The issue of the sufficient condition of moral approval and disapproval is not one that is to be settled by *discovery*. Neither truth in the ordinary sense—contingent truth—nor logic provides us with a resolution of that issue. The resolution of that issue, which will be attempted in the following chapter, will have a character akin to that of *decision*. This follows from what has been said already, about the nature of our attitudes as to the necessity of certain facts to responsibility and so on.

The sixth commitment of Incompatibilism, an inclination or motivation rather than a proposition, is that men and women have a certain stature or standing only if they possess the power of origination. There is no determinate conception of an acceptable stature shared by all Incompatibilists. It can be said that for Kant it is the stature of entering into what is of ultimate value, the moral law, through originated actions. For some Incompatibilists, indeed many, it is a stature which has to do with a life true to the great reality asserted by religion, a reality which includes eternity. For others it is the lesser stature of simply being distinct in kind from the rest of what we experience, all of which is subject to causation. Not to escape this necessity, in Bradley's view, is to fall very low indeed.

For the moment, what can be said is that this sixth item in Incompatibilism is open to the reply that not all conceivable standings for men and women are dependent on their being originators. Needless to say, the general idea of an acceptable standing is vague. However, the intransigent response to determinism, a response which does not involve origination, can give some place to conceptions of nobility of spirit, gentleness, courage, and other great virtues and excellences which have always been among the objects of our moral and personal feelings. The idea that human lives which possess these excellences as conceived could none the less be absolutely without stature or standing may be taken as something like eccentric.

To come to the propositions of the Compatibilist Party (pp. 474 f.), we have already considered and rejected the first, which is shared with their adversaries, to the effect that virtually all men have a single common conception of the initiation of action in so far as moral disapproval and the like are concerned.

As to the second Compatibilist proposition, that this single conception has to do only with voluntariness, its falsehood follows from that of the first proposition. All of us find in ourselves a way of thinking which does not all accord with the given proposition. There can be no doubt that we are made to pause in our disdain of a man by the thought that his action was the necessitated consequence of a causal sequence whose initial items were either external to him, or else internal to him at the time of his birth. (pp. 431 ff.) It does not help to reflect that some items in the sequence were desires and intention for the action, and hence that the desires and intention were necessary to the action. It remains true that *no action but the action he performed could have been performed*. That is here to say, of course, no more than what has been said already: that the action was necessitated in a certain way. The use of the telling abbreviation, however, is entirely in place. Furthermore, if we are made to pause in our disdain by the given thought, and to turn to another conception, it seems as true to say that we would be made actually to halt if the thought was taken up by us unreservedly as a truth. We resist taking it up unreservedly because we see and are opposed to its consequence. (pp. 435 ff.)

We come to the vague third proposition of Compatibilism, that all men somehow believe that it is only voluntariness that is sufficient, in so far as the initiation of action is concerned, for moral disapproval and so on. Will it perhaps be objected that Compatibilists can be taken, despite all that has been said here, as having really been concerned with attitudes? The objection, of course, bears some similarity to the one considered in connection with the third proposition of Incompatibilists. (p. 478) The objection here, to repeat, is that Compatibilists have in fact not supposed or maintained that we have anything like a belief in the fundamental sense about the initiation of action. The objection runs afoul of a good deal, and cannot be taken seriously.

For one thing, this understanding of the Compatibilist enterprise would leave its degree of success wholly unexplained. Who would have been persuaded by the proposition here supposed to have been advanced, that *we have but one attitude* to the initiation of actions? *That* proposition, given what we know of the general possibility and fact of divergent attitudes in ourselves, would have been singularly unpersuasive. Secondly, this understanding of the Compatibilist enterprise would make almost incomprehensible its burden of complaint about want of clarity, confusion and so on. Finally, and most simply, to put it no higher, it is not what the language of Compatibilists implies or suggests. '. . . according to this proper, and generally received meaning of the word, a FREEMAN, is he that . . . .' (p. 456) 'For what is meant by liberty . . . ? . . . we can only mean. . . .' (p. 459) ' . . . what I mean is simply after all that I could, *if* I had chosen . . .'

(p. 467) I accept that such claims are about a single conception of action-initiation we are supposed to have. They are not claims as to a belief about a sufficiency of that conception for something else. They do suggest or at the very least consort with such claims, partly because the conception and such claims are to some extent conceptually attached.

Have Compatibilists—and of course Incompatibilists—not been clear in their own minds as to what they were asserting, or must be taken as having been asserting? Have they not in their own minds distinguished between a claim about a standard belief and a claim about an attitude? No doubt this can be said of some or many of them. Perhaps few or even none of them have explicitly raised the question as to whether their subject-matter is a belief on our part or an attitude. None of this disturbs the proposition just defended, that what they have unreflectively assumed, and written, has to do with a belief.

That, however, is not what is most important in this connection. What is most important is that the main contention of the Compatibilists *requires* a standard belief rather than an attitude. Their main contention is to the effect that a certain sentence about a sufficient condition with respect to moral responsibility, a sentence about only voluntariness, is *correct*, and another sentence about the only sufficient condition, partly about origination, is *incorrect* or *mistaken*. Anything at all like the intended correctness and incorrectness are properties only of beliefs in the ordinary sense, not attitudes and the like. The usage 'correct attitude' does indeed exist, in political circles for example, but certainly does not denote something which has correctness in the sense that Compatibilists—and Incompatibilists—claim it in connection with their subject-matter.

The fact of our pairs of attitudes, together with the response of dismay, stands solidly in the way of the fourth Compatibilist idea, that a determinism does not touch the matter of moral responsibility, and that an acceptance of determinism would leave things exactly as they always have been. That cannot be true.

There is a further large reflection prompted by this supposedly anodyne idea about determinism's harmlessness, but a reflection that pertains as much to Kant and the other Incompatibilists as it does to Hume and his Compatibilist companions. It is a reflection which, to speak of my own thinking about the consequences of determinism, has been a particularly powerful one. It has contributed greatly to my confidence in what has been said of two families of attitudes and the two responses.

Hume and his companions, in order to sustain the idea that determinism does not touch moral responsibility, must offer some account of why the contrary has so persistently been supposed, no doubt more persistently than the other view. A suitable explanation of what is taken to be error is needed to deal with the anticipated and no doubt proper rejoinder that the idea in question owes its persistence to truth. The Compatibilists, to use a term lately introduced, do provide such an error theory. (Mackie, 1977) We have seen a good deal of it. To think about it is surely to conclude that it does not fall far short of being amazing. It reduces to this: what explains the persistence of the idea that determinism does affect moral responsibility is four or more centuries of *confusion*.

The reply must be that four or more centuries is a long time for a persistently discussed idea to be sustained by confusion. It is a stunningly long time if the idea in question, to remember Hume's words, and use them against him, has to do with 'any subject of common life and experience' (p. 459), and has been a matter not merely of persistent discussion but of acute controversy. It seems undeniable that one must find something better than confusion and indeed a kind of widespread weak-mindedness to explain the resilience of the idea that determinism affects responsibility. It seems undeniable that the only adequate explanation must undercut Compatibilism. *It is the existence of an entrenched attitude, widely or universally shared.* The attitude is to the effect that origination is needed for responsibility.

The same kind of reflection is entirely as pertinent to the counterpart error theory of the Incompatibilists. It is to the effect that it is unreflectiveness, or shallowness in reflection, or weak empiricism, or word-play that explains the persistence of the view that voluntariness does give an adequate foundation for moral disapproval and the like. That too is an extraordinary claim when it is taken as intended, and as it must be intended—as having to do with some sort of intellectual failing. Indeed, some might say that we have no need to mention four or more centuries of examination and cross-examination in order to cast doubt upon it. That is, to mention only Hume, it might be said that it offends against sense to consign him to the company of the unreflective, the reflectively shallow, the myopically empirical—or, God help us, *the muddled*. (Popper, 1982b, p. xix) Again, there must be a better explanation of the persistence of the view in question, and there is one which is nothing so congenial to Kant and his fellows. *It is the existence of an entrenched attitude, widely or universally shared.* The attitude is to the effect that voluntariness suffices for moral disapproval and the like. (Cf. p. 495.)

All this is as relevant to the fifth commitment of Compatibilists, shared with Incompatibilists, to the effect that the problems we are considering are of standard intellectual kinds, and that our failings in considering them are intellectual failings. In particular, according to Compatibilists, there is the problem of understanding and keeping before the mind the distinction between causation and compulsion, the fact that causation does not entail a want of voluntariness. (p. 471) The truth that causation, being what it is, does not entail any want of voluntariness has been indefatigably asserted and elaborated by Compatibilists. So too have they been persistent in judging the obscurity of origination, and, to repeat a description, its 'obscure and panicky metaphysics'. (P. F. Strawson, 1968 (1962), p. 96) Certainly the issue of determinism and freedom raises intellectual problems. As remarked already, however, one principal question—what has been taken as *the* principal question—is not of this kind.

The sixth feature of Compatibilism, of the order of an attitude itself, is its subservience to determinism. Compatibilism is inclined to take determinism as rightly imperious. What I have in mind is not the conviction that determinism is true. Indeed, the attitude in question, or some close variant of it, can coexist with an uncertainty about the truth of determinism. Rather, the attitude is one which involves an acceptance of determinism as being, or as likely to be, *the very principle of reflection about all of reality*. What follows from this is a resistance to the perceiving of any recalcitrant fact. The main fact here is our common attitude having to do with origination. Compatibilists, notably those who are determinists, give the appearance of being incapable of registering or at any rate coming to grips with this fact. Their resistance to it will not make it go away.

It needs to be allowed that there is more speculation in ascribing this sixth feature to Compatibilism than in the case of the others. Also, what has been said of this feature is vague. It does seem clear, however, that there is at the basis of Compatibilism some counterpart of the Incompatibilist attachment to an aspiration with respect to the stature of mankind. There is a kind of commitment which contributes as effectively to what one is tempted to call one-eyed philosophy. The description of this Compatibilist commitment might be supplemented by a consideration of intellectual temperaments, these having to do with more things than determinism. As in the case of Incompatibilism and human stature, more will be said in what follows of relevance to Compatibilism and its subservience to determinism.

That concludes my main consideration of the long battle over the consequences of determinism between the two orthodox schools of thought. Before going forward with other matters, it will be worth while glancing back over the argument so far about the consequences of determinism.

Each of us can focus on either of two conflicting sets of propositions, ideas, or images about actions. One set of these things has to do with voluntariness or willingness—in one of several summary definitions, they have to do with action issuing from embraced desires. We can take these propositions as the only essential ones entering into life-hopes, personal feelings, knowledge, and moral matters. If we do this, we may make the intransigent response to determinism, that it does not matter. On the other hand, we can focus on a larger set of propositions and the like about actions. They have to do with both voluntariness and origination. We can take it that only all of these considerations together provide good reasons for life-hopes and so on. If we do this, we may make a different response to determinism—dismay. Neither facts nor logic by themselves, then, force us into either of the two responses. They are in part a matter of the mentioned focusing, the attitudes. What is true, as a matter of logic, is that a determinism is consistent with the first set of considerations and inconsistent with the second. It follows, if determinism is a fact, that the first set of considerations is undisturbed and the second is false.

This view of our situation, to repeat, was advanced before anything else partly in order to frustrate the philosophically conditioned reflexes of Compatibilism and Incompatibilism. If it is not all that is to be said of our situation, it constitutes a refutation of both those traditional views. Our situation is not one of a single conception as to the initiation of action, and anything like a true or false belief as to what is sufficient for moral approval and disapproval, or, to take the properly wide perspective, for that and the other mentioned things. The fundamental mistake of both Compatibilists and Incompatibilists has been to seize on one attitude with respect to the initiation of action, convert it into something else—a single conception of initiation and something presented as a belief—and to ignore another attitude quite as real. To reject the two traditional views is to leave the way clear for something else, to which we now turn. If the two traditional views were the only possible responses to the problem of determinism, that would not reduce it to 'a dead problem'. (Earman, 1986, p. 235) Given that they are not the only possible responses, it is very far from being such.

9

Affirmation

9.1 THE TRUE PROBLEM OF CONSEQUENCES

The true problem of the consequences of determinism, as will have become clear, is the problem of settling on a satisfactory response to it, involving our two families of attitudes. It is a problem of dealing with our feelings, feelings challenged by determinism and about things that are important to us—our futures, other people, and so on. It is most importantly a problem of desire. Life-hopes *are* a species of desire; desires enter into appreciative and resentful personal feelings; we desire to be able to have certain of the attitudes, notably a confidence in knowledge. Although we have yet to look at this, the problem also involves what we do in our societies as the result of our desires. It is not a problem to be settled by the discovery of truth, either truths of logic or truths of fact. The problem is partly raised by the seeming truth of determinism, but certainly not settled by it.

Our situation, as we have seen, is that each of us has or is prone to have two conflicting families of attitudes with respect to things important to us, and that the seeming truth of determinism gives rise in the first instance to two conflicting responses in us. If this is a situation that falsifies Compatibilism and Incompatibilism, it is also an unsatisfactory situation, or more than that, for a number of reasons. What it calls for, if it can be had, is a change in our feelings. Perhaps it is not too much to say that it calls for a change in our feelings about the nature of our existence.

To succeed in affecting a satisfactory change, if this can be done, will be to resolve the problem of the consequences of determinism. Nothing else will resolve it. This is so since, to repeat, the problem *is* that of settling on a satisfactory response to determinism. It can be said with absolute certainty that the problem is *not* a matter of extracting a correct definition of 'free' or whatever from ordinary language, or of contriving yet another in the long futile sequence of proofs and persuasions, however arresting or meticulous (Kenny, 1975; van Inwagen, 1983), each to the effect that we can properly mean only one thing by our relevant words. Nor is it a matter of establishing that we have some single 'belief' as to sufficiency, for moral responsibility and so on, of a free action somehow defined. The endeavour to move towards such a change in our feelings which will resolve the consequence-problem is our present business, the main business of the last part of this book.

One of the discomforts of the situation in which we find ourselves is the simple existence of our response of dismay, taken by itself. It is no satisfactory thing to have or to contemplate having life-hopes threatened, feelings about others and oneself put in question, confidence of knowledge undermined. Each is such a thing as to make us try to avert our attention from it. The discomfort, further, is not a problem which should exist only in one's study, or in one's personal emotional life, as distinct from the world outside, a dismay of philosophical reflection rather than ordinary or practical life. We *act* on certain of our desires, notably certain of our retributive desires. Hence there must be at least an unhappiness having to do with the morality of fundamental social institutions, practices, and habits. (Ch. 10)

Given an acceptance of the seeming truth of determinism, there is also another kind of discomfort involved in dismay itself: that of being aware of being adversely affected in a way that is unreasonable. This is the unsatisfactoriness of being troubled by desires for what, given the truth of determinism, cannot be had—origination and what follows from it. This cannot but strike one, or weigh on one, as in fact irrational. The irrationality, put one way, is that of having desires of which one accepts that they cannot be satisfied and hence are not a means to one's well-being.

A further and fundamental discomfort of the situation is of course the inconsistency it involves. We feel that only voluntariness with origination is enough to sustain hopes satisfactorily, and so on, and also that voluntariness by itself can do so. (This inconsistency has nothing to do, obviously, with the inconsistency claimed by Incompatibilists, between 'freedom' and determinism.) As for the two responses, we feel both that our hopes are dashed and that they can persist. It needs to be allowed that since no attitude and no response is a truth-valued belief, this inconsistency does not in itself involve us in falsehood. What is the case, rather, to speak in one way, is that we are involved in inconsistent *judgements*. If my inclination to take voluntariness alone as sufficient for life-hopes is not a matter of truth or logic, it does none the less issue in what it is proper to call a judgement, and such judgements stand in logical relations with other such judgements and indeed other things. My inclination issues in the judgement that the fact of voluntary action *in itself* is an adequate reason for life-hopes. If this is not true or false, not a proposition in the usual philosophical sense of that term, it none the less can be said to stand in contradiction with the judgement that only the facts of voluntariness *and* origination together are an adequate reason for life-hopes.

There is the same kind of contradiction with ordinary imperatives, which also lack truth-values. One account of such contradiction, although not an explanatory or fundamental one, is that two judgements contradict one another if someone's having the idea that both can be affirmed together is a reason for saying that he fails to understand them, that he does not grasp the meanings of the terms involved. (Cf. Hare, 1952, Ch. 2) A more fundamental account of such contradiction, although we need not pursue it, will have to do with the contradiction, in the most fundamental sense, of certain propositions related to the judgements.

The fact that the contradiction in which we are involved does not in itself commit us to falsehood does not much lessen the unsatisfactoriness of our situation in this respect. Being trapped in inconsistent judgements is as unsatisfactory as being prone to issuing inconsistent orders, or, more precisely, being *subject to* inconsistent orders. There is what can be called the instability of the situation: we cannot rest easy, but are pulled in two ways with respect to what is, in one sense, the same subject-matter. We are, for example, pulled in two ways with respect to the subject-matter of our futures.

Finally, if dismay in itself and the inconsistency are nothing agreeable, intransigence in itself, despite what has been said of its recommendations (pp. 399, 406), can hardly be satisfactory. If I can persist in intransigence for a time, I cannot by fiat become *unaware* of the different attitudes I am then rejecting—above all the desires in them—which issue in dismay. I cannot simply forever *subtract* from my thinking—about my own life and the lives of others—thoughts or images as to the origination of action. Intransigence takes *effort*, and not congenial effort. It takes the effort required to maintain a pretence, that I am only what I am attempting to be, someone concerned only with voluntariness. This is a point not about instability or movement of feeling, but of the unsatisfactoriness of *one* feeling while one maintains it.

If none of this goes against what was said earlier of our two families of attitudes and two responses, that they are natural to us (pp. 392, 398), it is indeed clear that our course must be somehow to escape or alter the situation of attitudes and responses that has been described. It is by doing so that we shall resolve the problem of the consequences of determinism, since it arises from the unsatisfactory situation of those attitudes and responses.

Will it be thought that the enterprise of somehow making an escape from or altering our present situation is somehow misconceived? Is there some fundamentality about a particular attitude or response which stands in the way of the enterprise? Is there a fundamentality which, if it does not make the enterprise misconceived, at least must *direct* or *govern* it? Is there, in particular, such a fundamentality about the particular attitude which involves both voluntariness and origination? Some may think so. If the view we now have of our present situation has not been advanced before, philosophers have said things which can be so altered as to be applicable to it, and tend in the given direction. Let us look at three.

(i) It was said earlier, in summary of one conception of what it is for an action to be voluntary, that it is one done out of embraced desires. (7.3) But, it may be said, it cannot be that one acted *because one desired* or *wanted to*—in any sense at all—unless in a certain sense one *could have acted otherwise*. The idea here, perhaps, is that acting out of desire is acting out of preference for one rather than another possible situation. What this comes to, it may be claimed, is that it cannot be that one acted because one wanted to unless there was an alternative: the action was *not* the upshot of certain causal connections. It cannot be, in fact, that one acted because one wanted to unless the action was an originated one. This could have been considered earlier, as an argument to the effect that we *cannot* take voluntariness alone as a ground for life-hopes, as it has been maintained we do, but we *must* and therefore do instead take voluntariness together with origination. Not to reopen what has been settled, but to stick to the present stage of our reflections, we can consider whether our present project must take into account that voluntariness does involve origination for the given reason. The argument is suggested by, but not the same as, one offered in a kind of defence of an indeterminist picture of the mind. (Kenny, 1975, pp. 142–3; 1978, p. 26)

The initial premiss for this line of reflection may have a persuasive ring. In fact, we can accept that it is true that if one acted as one did *because one wanted to*, then in some sense one *could have acted otherwise*. But that does not necessarily get us to origination. It is entirely persuasive to say that the report that a man did buy a punnet of strawberries *because he wanted to* comes to this: his wanting to buy a punnet of strawberries was explanatory of his action, or a part of the explanation of his action. Slightly more fully, his desire was an ineliminable part of the explanation of his action. In that case, *if he had not had the desire, he would have done other than he did*. But that latter conditional proposition is easily enough expressed by way of the capacious 'could'. It *can* be what we mean by saying that he could have done other than he did. What would have made the difference, if he had done otherwise, would have been a fact about him, the absence of his desire.

Such an understanding of 'could have done otherwise' was of course noted earlier, in connection with Compatibilism. (p. 467) Here, what is being maintained is not that it is or may be the only possible correct understanding, but that it is a possible understanding which deals with the given problem. In a sentence, doing something because one wants to may entail the existence of alternatives, but these may be a matter not of origination but only voluntariness.

(ii) A second consideration which may be thought to commit us to the ground of voluntariness together with origination, or to establish its fundamentality, has to do with what is called reflectiveness. It is maintained that reflective persons operate with the conception of voluntariness with origination, and unreflective persons operate with the ground of voluntariness alone. A reflective person is one who has an awareness of determinism, as a result of acquaintance with philosophy, science or religion. (Cf. C. A. Campbell, 1951.) As will be apparent from what has been said already, this is false in its claim that many people—the unreflective as defined—lack ideas of the sort that issue in the indeterminist pictures of the mind, and so have ideas of voluntariness alone. It is not necessary, despite determinism's logical relation to origination, that our having ideas of origination requires our having an awareness of determinism, that our ideas of origination are somehow owed to an awareness of doctrines of determinism.

(iii) Differently, but again about something called reflectiveness, it may be said that each of us has reflective and unreflective moments in connection with moral responsibility in particular. When we are reflective we use the conception of moral responsibility which includes both elements, and when we are unreflective we use the conception which includes only voluntariness. This different reflectiveness is essentially a matter of emotional calm together with beliefs about the causal ancestry of a person's action, perhaps beliefs of a psychoanalytical kind having to do with experiences as an infant or child. Our unreflectiveness here is bound up with violent emotions, such as hate, anger, or indignation, typically having to do with an action of which we have been the victim. (Cf. Edwards, 1961; Hospers, 1961.)

This is perhaps *more* arguable than the distinction between kinds of people, and evidently stands in some distant relation to the view about shared attitudes to which we have come. That is not to say, however, that these considerations establish that the wider conception of moral responsibility is somehow superior to the narrower. If there is some connection between strong feelings about an action and the narrower conception, it is certainly not true that we focus on voluntariness by itself as a ground for moral responsibility only if we are pushed by such feelings. We can focus on voluntariness as a result, for example, of calm thoughts about the function of morality. (pp. 438 ff.) In any case, if it is granted that there is some closer connection between strong feelings about actions and the narrower conception, that would not establish anything significant by way of superiority. It would not make it possible to draw the conclusion that the narrower conception is *false* and the wider *true*, certainly, and nothing else is suggested.

Unhindered, then, by any idea to the effect that one of the two attitudes, or one of two responses, is somehow fundamental, we can set about reflecting on the possibility of bringing about a change in our situation, which is to say a change in our feelings. What particular course shall we take? We can make a response which is best described as an attempt. It is, by way of a first very general description, *the attempt to free ourselves from the attitudes which carry thoughts inconsistent with determinism, free ourselves by various means, including reflecting on our attitudes having to do only with what is consistent with determinism, which is voluntariness.*

To succeed in some degree in freeing ourselves from attitudes of the first kind would be, to some degree, to avoid the discomfort of dismay. It would also be, to some degree, to avoid the irrationality of desiring what we cannot have. It would, further, alleviate our condition of contradiction, our condition like the condition of being subject to inconsistent commands. To make this attempt to free ourselves from attitudes which carry thoughts or images of origination is not to make the response of intransigence, and hence we may escape its unsatisfactoriness as well. This is so for the following reason.

The defining feature of intransigence is the assertion, so to speak, that determinism affects nothing—that it is true or can be true without anything changing. (p. 399) To this end it encapsulates a kind of dismissal or denial of attitudes involving origination. The present response is in part a meta-response, greatly more so than is intransigence—or dismay. (pp. 406, 391) The present response, crucially, allows the existence of those attitudes and attempts to deal with them. It is not an attempt to disavow them or to do anything like pretend they do not exist. It involves our accepting a certain loss, centrally a certain frustration of desires we do have. It involves our accepting that if we have desires having to do only with voluntariness, we do also have desires which rest in part on ideas of origination. We have the attitude to others that their actions are both voluntary and originated, and we also have this attitude to ourselves about our own actions. (Cf. p. 501.)

Our third response to determinism, more particularly, will accept that we do indeed have life-hopes of a certain kind, which we must attempt to eschew, as well as hopes of another kind in which we can persist. My hope, say, that I shall overcome my various deficiencies and so achieve a distinction in my profession, in so far as the hope has to do with origination, rests on falsehood and I must seek to eschew it. My related hope, different in that it has to do only with voluntariness, may be taken to rest on a considerable truth or probability about voluntariness, and to be in no need of being eschewed. Reflection on the latter hope is one thing that may aid me in giving up the former.

Likewise, with respect to personal feelings, our response to determinism may be to accept that we do have feelings about others which carry ideas as to originated actions, and that we must seek to assuage these feelings, since the ideas are false—and the response will also include the fact that feelings about others which only carry beliefs as to voluntariness are often in place, since often the beliefs are generally true. So with the matter of knowledge, and with holding others and oneself morally responsible, and crediting others and oneself with moral responsibility. So too with the related moral matters. The attempted response to determinism in these connections will be one which accepts that we do have two kinds of feelings in this regard, one of which involves falsehood and needs to be eschewed, and one of which generally involves truth and so can persist.

The attempted response is better if still wholly generally and abstractly described as *the endeavour to give up certain attitudes, fundamentally the endeavour to accept the defeat of certain desires, by way of reflecting on the situation in which our success would put us*. It is *the endeavour to accommodate ourselves to what we can truly possess, mainly by seeing its value*. The two aspects of the endeavour, seeing the value of what we can have, and giving up with respect to what we cannot have, are of course intimately connected. The endeavour, as will be understood, is to arrive at attitudes beyond the pair involved in Compatibilism and Incompatibilism. (pp. 473, 474)

There is no supposing that we can satisfy ourselves about determinism and freedom without engaging in this endeavour. That is has not been much attempted in recent centuries within the dominant tradition of Western philosophy, is close to the explanation of the persistence of the problem of the consequences of determinism. That explanation, as anticipated in what was said of the 'error theories' of Compatibilists and Incompatibilists (p. 485), *is* the fact of our shared conflicting inclinations with respect to life-hopes, personal feelings, knowledge, and morality. We are pulled first in one direction, then in another. If we enter into the traditional Compatibilist and Incompatibilist misconception of our situation, a conception that mistakenly intellectualizes it, we are pulled back and forth between Compatibilism and Incompatibilism. Our attitude to the initiation of action which involves only voluntariness inclines us to Compatibilism, and our attitude which also involves origination inclines us to Incompatibilism.

The problem of the consequences of determinism has not persisted because of confusion, and it has not persisted either because the logical upshot of determinism is unclear—its consistency or inconsistency with other things. Compatibilists and Incompatibilists have maintained the former; many others have been inclined to the latter. The problem, as I hope it is a recommendation of this book to show, has persisted for the reason that determinism is unsatisfactory to the corpus of our desires—no matter which set of desires we concentrate upon—and not much attempt, and certainly no successful attempt, has recently been made to assuage that dissatisfaction.

To seek to do better—to seek to come to a certain judgement of our circumstance, and hence to seek to accommodate our desires to what we possess—is to try to arrive at a large part of what can properly be called a philosophy of life. A philosophy of life, in this understanding, is not rightly called an attempt to give meaning to life or to establish its meaning. It consists in a broad attitude to life, an attitude or group of attitudes which, most importantly, provides as much sustenance or support as can be had within the constraints of truth. The attitude or attitudes sustain or support one in the face of the defeat of our desires of various kinds, including those pertaining to the initiation of action which enter into life-hopes, personal feelings, and so on. The attitude or attitudes, further, encompass fundamental valuations, and, if the terms are not too diminishing, fundamental policies and strategies. The responses of dismay and intransigence, if either could be persisted in to the exclusion of the other, could also be advocacies of parts of philosophies of life, in the case of dismay something other than a satisfactory one.

Attempts to provide philosophies of life or parts of them have not been common in the dominant tradition of Western philosophy. It is the tradition which can be quickly identified as having Aristotle, Descartes, Hume, and Kant as exemplars, and as giving a priority to roughly that kind of argumentation which has been the stuff of our inquiry so far into determinism. Indeed, within the recent part of that tradition, there has been something like a disdain for philosophies of life. This may owe something to the obscurity, excess, idiosyncrasy, and in general the intellectual weakness of views of our existence offered by many who do not attempt to satisfy the usual standards of the dominant tradition—for whatever reason.

It is plain enough, however, that philosophies of life in the relevant sense have been of great importance in the work of some of the great philosophers within the tradition. Plato, Aristotle, Spinoza, Kant, and Russell have in different ways recommended philosophies of life. So too have able contemporary philosophers. (Klemke, 1981; Edwards, 1967a) My reason for attempting to arrive at a part of one, however, has nothing to do with any inclination to carry on what is mainly a philosophical practice of the past, and nothing to do with any inclination to be free of the usual standards of the dominant tradition. My reason, as has been explained, has to do with the nature of the final problem of this book. *Nothing but* the relevant part of a philosophy of life can complete our inquiry with any chance of success.

The attempt to come to such a thing can get some help from at least a part of what has been a subsidiary philosophical tradition, a variegated one of some depth and strength. It has Spinoza in it, and in a way Kant, Rousseau, and Hegel. Three other groups of members of it are Cicero, Marcus Aurelius and others of the Stoics; Duns Scotus, Luther and Maritain; and Comte, Marx, Engels, and Bakunin. Some would add Freud. The tradition is said to concern freedom. A full account of it has characterized it as concerning 'the acquired freedom of self-perfection'. (Adler, 1958)

It has to do, in fact, with the renunciation of some of our desires, and the embracing of what is right, good, true, natural, rational, wise, righteous, necessary, or historical. Despite what has been maintained earlier about the ambiguity of 'free' and about attitudes (p. 475), Bentham was perhaps not wrong to abuse the tradition, to abuse above all those who have said only *right* acts or the like are 'free' acts. 'They pervert language; they refuse to employ the word *liberty* in its common acceptation; they speak in a tongue peculiar to themselves.' (1950 (1840), p. 94; cf. Nozick, 1981, pp. 326 ff.) The tradition of thought in question cannot really be said to have to do with any freedom distinct from voluntariness and origination. It will be of some use none the less.

What has been said so far has been no more than programmatic. Let us now set about the attempt to resolve the problem of the consequences of determinism.

9.2 LIFE-HOPES

The least tractable issue is that of life-hopes. We *do*, as we have seen, have hopes of the kind which rest on and contain images or ideas of a self and of its activity in escaping environment and character, in securing for us an open, alterable, or unfixed future. (p. 385) These hopes, depending on and carrying these particular images or ideas of the initiation of actions, are in conflict with the theory of determinism. The hypotheses of the theory make our future actions the effects of certain causal sequences, by definition unbroken, whose initial items are bodily items at the time of our birth and environmental items then and after. There can be no doubt about the inconsistency. Given the theory of determinism, my feeling that my life so far has not gone so well as I hoped cannot be dealt with by means of the thought that I have a certain radical capability of achievement in the future.

Nothing can be contrived, consistent with determinism, that will actually *preserve* the life-hopes which get their identity partly from the given images or ideas—the germs of doctrines of origination in the indeterminist theories of the mind. It is possible, perhaps, to be attracted for a time to Kant's determination to have it both ways, to have a determinism and also to have what is inconsistent with it. In the end, his 'two worlds' enterprise (p. 462) is yet less satisfactory with life-hopes than with moral responsibility. Moral responsibility may seem to offer the possibility of *etherealization*, so to speak, but my life-hopes, very definitely, are *of this world*, the world of which we are taking determinism to be true. Nor is there any chance of going even any significant way in the Kantian direction. We shall not actually save our hopes of the given kind, by regarding them as involving *useful fictions*. (Cf. Vaihinger, 1924.)

The question must be not whether we can preserve these hopes consistently with a determinism, but rather, in part, whether we can go towards *abating* or *diminishing* them. The attempt to do so will involve various things: consideration of the content of these hopes, several possible compensations of determinism, a number of reassurances, and finally a certain prospect having to do with belief in determinism.

9.2.1 Conceptual Thinness

Our earlier inquiry into indeterminist theories of the mind, and the widespread philosophical scepticism about those theories, suggests a first possibility. The hopes with which we are concerned rest in part on what we can now call images of origination. For an action to be originated is (i) for it not to be a matter of certain nomic connections, such as those asserted in our theory of determinism, and (ii) for it to be somehow in the power of the agent. The first or negative part of this conception is clear. In one respect it has to do with an intention we form, the forward-looking intention to perform an action. It is the idea that this intention is not a certain effect, not a product of a certain kind of causal sequence.

The second or positive part of the conception includes the image or idea of a self, and some image or idea of its relation to the intention. (Cf. 3.5–3.8.) An attempt may be made to characterize the relation by way, say, of notions of teleology. Such notions, to recall, have to do with an earlier thing's being explained or determined by a later. They are unsatisfactory. Roughly the same is to be said, as we also know, of other attempts to give effective content to talk of the creative or productive relationship between the self and its intentions and the like. It may be said that in so far as the positive part of the conception of origination is concerned, we are left with a referring term that does not do much referring—*the self*—and a collection of verbs of doubtful content. The self *has a power* with respect to intentions and the like; it *controls* decision-making; it *acts*; it *gives rise to* choice; it *causes* things, but not in the standard way, such that its so-called effects are ordinary effects. The referring term and the verbs, it may be said, pass through the mind without leaving a great deal of trace. It can be argued that as they are used they are sounds to which not much by way of actual conceptions is attached.

The upshot is that it approaches being true that the relevant content of the given life-hopes can be articulated only negatively. To have a life-hope which involves thoughts of origination is to have a hope whose content having to do with the initiation of action is not much more than a proposition denying certain causal connections, connections expressed by certain conditional statements. My hope, perhaps, is to write another novel, and I take it that the necessary means to this end is a campaign of reflection and invention, which campaign will depend on my keeping various of my inclinations in control. My hope, being of the kind we are considering, is in its clear content for little more than the non-existence of certain causal connections. The little more is expressed by an ineffectual designator and the various verbs. What they reduce to, it seems, is a power-I-know-not-what.

To come towards the main point, can we make use of the thought that less is taken from us by the denial of such hopes than might have been supposed? My denied hope is less contentful than it might have been, far less contentful than the free-speaking and free-writing defenders of origination have pretended. They have often assumed that we all have some full and determinate conception of origination, but we have not. Can we then not set out to take the defeat of our hope as a lesser thing? To come to the main point, can we not with reason address a certain exhortation to ourselves? Let us, we can tell ourselves, seek to care less about being deprived of what is near to a mystery, care less about the absence of what is near to inexpressible. Let us try to escape being troubled by being denied what is conveyed by only a shuffle of elusive verbs with an evanescent subject.

The exhortation, to my mind, is not greatly effective, for four reasons.

The first is that it rests on a judgement or idea as to contentfulness, and such a judgement or idea is no very secure foundation. It did not turn out to be true, certainly, despite the considerable contributions of Logical Positivism to philosophy, that that movement succeeded in formulating a principle of contentfulness. (Ayer, 1936) It is all very well to speak of a shuffle of verbs with an evanescent subject, but they are not *nonsense*—actually without sense. Used in the elusive way, they are intimately connected with discourse of several kinds, which connections may themselves be taken as giving them significance. What is most important at this point is not the consideration of certain relevant doctrines in the philosophy of language, having to do with contextual meaning and the like. What is most important is that we *do accept* that the verbs and the like do bring into view something of importance to us.

A second reason why the given exhortation is less than persuasive is a general truth to the effect that our feelings do not depend so much on content as might be expected, or perhaps hoped. The example of a nameless or shapeless fear is relevant. There seems little reason to think that feelings which in some respect have little content are likely to abate on that ground alone. It is at least sometimes possible to argue in precisely the opposite direction.

Thirdly, the exhortation presupposes a certain possibility of managing our emotions, of managing hopes of the given kind. Certainly we can in some ways and to some extent control our emotions—if this were not possible, the whole enterprise on which we are now engaged in connection with the consequences of determinism would be pointless. We can, for example, sometimes forcibly turn our thoughts from a dark subject, and so alter our feelings. But can we by a rather intellectual consideration—about want of contentfulness—move far towards giving up something as fundamental to our lives as hopes of the given kind?

That is unlikely. There is a common intellectual temperament which easily withdraws from or even revolts against the conceptually unsatisfactory when the subject-matter does not touch on our lives so closely as the present one. The subject-matter of the nature of numbers will do as an example. There is also an intellectual temperament, not common, which withdraws or revolts even when the subject-matter *does* touch on our lives as closely as the present one. This latter temperament, perhaps exemplified by some of the Compatibilist philosophers, is not one which is anything like widely distributed among men and women. It is not widely distributed among philosophers, or, as can definitely be added, scientists.

There is a fourth reason why there is little use in the exhortation to care less about being deprived of what is near to a mystery, near to inexpressible. We have it that hopes involving origination have to do (i) with a denial of certain nomic connections and (ii) with a peculiar power on our part. It is the second and positive element that is obscure, and whose obscurity we are contemplating as an aid to abating the hopes in question. However, the second element, the conceptually slight element, is absolutely essential. It is this remainder that makes the difference between mere chance—a denial of determinism—and something else, which is what we want. (3.4) The second element is what is to save us from being victims of mere randomness. What we are being exhorted to do, then, is to free ourselves from certain hopes by way of denigration, if that is not too strong a word, of exactly something that is essential to them. It is at least as fundamental to them as the other element.

There is nothing irrational in the exhortation we have been considering—to try to eschew certain hopes by reflecting that they are in a certain respect minimally contentful, that they are not *for* something clear. It is hard to see that it can carry us far forward. It cannot be of much significance in the attempt to arrive at a satisfactory philosophy of life in so far as life-hopes are concerned, and in so far as the other matters having to do with the initiation of action are concerned, personal feelings and so on.

Before turning to a consideration of a second strategy, there is a general point that calls out for emphasis, although it has been in view already, and can be introduced by a certain similarity.

We are looking into possibilities with respect to the characterization of a certain response to determinism. Or, as it is as proper to say, we seeking to *make* this response. The first possibility we have been considering is an exhortation having to do with a conceptual thinness of essential ideas which enter into life-hopes inconsistent with determinism. Any reader with our past journey of inquiry in mind must have the true thought that there exists a certain similarity between the proposal we have been considering and something which was specified as part of the response of intransigence. A part of that response, it was said, is resistance to the unsatisfactory image of origination. (p. 399) Here in particular, therefore, there arises the question of the distinctness of the response with which we are now concerned and the response of intransigence. *Are they distinct?*

One part of the answer has already been given: that the response of intransigence is by definition a rejection or denial of certain attitudes, as distinct from an attempt to deal with them. (p. 493) The forceful frame of mind which is intransigence can make more use of the charge of conceptual thinness than the frame of mind we are now considering. A second and connected thing to be said is that our present enterprise is the self-conscious search for a *settled* state, an escape from a kind of oscillation between inconsistent things.

A third thing to be said in answer, which takes us to the general point that needs emphasis, is that we must not fall into the error or pretence that the matters with which we are concerned are so distinguished in the reality of our lives as they have been on the page. Distinguishable responses are not unrelated entities, anything like unique. The truth, rather, is that there are likenesses and continuities between what we distinguish. The response to determinism we are considering is *not* wholly different from intransigence. It is quite as true, however, that it is not wholly different either from dismay. We are in this part of our inquiry concerned with a spectrum of feelings. It is not a subject-matter that in itself is made up of wholly contrasting phenomena or entities. (Cf. pp. 399, 493.)

9.2.2 Escape from Chance

Comte serves as an example of a considerable number of philosophers, many of them traditional Compatibilists, who see in determinism the compensation that it rescues us from mere chance. (cf. Foot, 1957; Hobart, 1966) Comte takes it that the alternative to a world in which we are subject to determinism would be one in which my actions and the actions of others are owed to chance, which is to say owed to nothing. He also has something else in mind, having to do with the seemingly unlikely subject of politics. If a determinism 'may seem to chain us to external necessities', it has the recommendation of 'the elimination of the element of caprice, ever favourable to the worst instincts', the recommendation that it saves us from 'arbitrary will', 'arbitrary personal dictation'. (1875–7, Vol. iv, p. 194, Vol. i, p. 296, Vol. iii, p. 294, 1885, p. 435)

The main thing to be said of any such idea about chance alone is of course that we do not suppose that the alternative to a determinism is no more than chance, a denial of nomic connection. If a determinism saves us from being subject to chance, it equally deprives us of that which involves the absence of nomic connection but also involves control or agency. It deprives us, that is, of a self's origination. If our aim is to accommodate ourselves to a determinism's defeat of our life-hopes in so far as they rest on the idea of origination, we shall not much succeed in it by reflecting that determinism also defeats the idea of mere chance. We are not much compensated for the loss of a good by the reflection that what goes with it is something undesirable but never much contemplated by us.

With respect to Comte in particular, as distinct from other philosophers who have seen determinism as preserving us from chance, it is evident that he brings together an absence of nomic connection with such other things as wilful political authority or mindless tyranny. He thus also provides us with an example of confusions—there have been other kinds of them—that do not help. As is plain, what he speaks of is as consistent with a determinism: it cannot be that a determinism, in compensation for the loss of origination, gives us political freedom and rational societies. It does have certain political implications, but not necessarily these. (Ch. 10)

More generally, if one has in mind chance by itself, it is something less than clear that we would take *all* forms of it to be undesirable. That is, if we forget entirely about origination, there remains the fact that determinism in excluding *all* chance is not necessarily reassuring. Certainly we do desire the power of origination, clearly, and we cannot conceivably desire a world of chaos, one with very little or no order. It can be supposed, however, that that having subtracted origination from our world, determinism does us a separate hurt in ruling out what would be desirable, a world of *limited real uncertainty*. The idea is a vague one, but there seems the possibility that the fixity of a determined future is not preferable to the conceivable alternative of a future without origination but with limited kinds and degrees of want of necessity. (Cf. p. 65)

It must seem, therefore, that there is nothing to be gained from contemplating the idea that a determinism excludes chance. There is little more to be gained from related ideas that determinism gives us an intelligible world or that it provides us with a sense of security. (Snyder, 1972; Trusted, 1984, p. 49; Pagels, 1983, pp. 18–19) Nor does there seem to be anything like relevant truth in something which may be taken to be opposed to a fundamental proposition of our own inquiry, Sartre's famous theme to the effect that what we want is an escape from freedom, freedom to which we feel condemned and which issues in our 'bad faith'. (p. 466; M. Warnock, 1965, Ch. 2; Danto, 1975) There is a very different idea of determinism as compensatory, which calls for more consideration. It has a long history.

9.2.3 Nature

We have been concerned with the fact, as it can be described, that in certain of our life-hopes we desire a certain disconnection. We desire a disconnection from our own past lives and also from nature, where the latter is conceived as the non-human world. To have the power of origination would be to escape or to rise over nature. In a way, incidentally, it would be to have an ability to escape or rise over that part of the world which is constituted by one's own body. By contrast, it is true of all decent determinisms, like the theory of which we know, that they put us into connection with nature.

Each of the three hypotheses of our theory contributes to this. Our conscious lives consist in effects of certain causal sequences whose initial conditions are constituted not only by our bodily endowment at birth but by our environments then and thereafter. The same sequences are in the causal histories of our actions. There is the further proposition of our determinism, necessary to its completion, that mental events are in the connections with simultaneous neural events specified in the Union Theory. (3.3)

The compensatory idea must come to mind that a determinism in clearly asserting *a close and unproblematic connection with nature*, can be conceived as a source not only of frustration but of reassurance or even elevation. The unproblematic closeness is that of nomic connection. This particular idea about reassurance or elevation, tied to a determinism, is of course related to others. The taking of oneself as somehow but more loosely connected to nature, perhaps with nature more vaguely conceived than as just the non-human world, is something essayed and celebrated in a good deal of literature and in some religion, most familiarly in the literature of Romanticism. The particular recommendation of a determinism, if it is a recommendation, will have to do with its assertion of a close and unproblematic relation between ourselves and the natural universe.

If a determinism is true, it may be said, I cease to be a trivial existence. I cease to be what by contrast with the universe is a momentary and insignificant thing, the antithesis of anything of grandeur. I can, through my perceived *membership* in nature—my relation to it is properly so described, rather than as an external relation—escape the mereness of myself. I escape an isolation from the natural world. In place of triviality and isolation, I can identify with the greatest of realities. There is the satisfactory possibility, further, of having a certain view of my own species and its history, a view which brings it together with other species and so rises over a petty anthropocentrism. Determinism, therefore, may be claimed to be far from being 'the hideous hypothesis' (Hampshire, 1951, p. 179) and in fact the source of a deep satisfaction.

This internal connection with nature asserted by determinism, further, does not make me nature's slave. I can see, with Bakunin, that in my existence there 'is indeed no slavery at all, inasmuch as every kind of slavery presupposes two beings existing side by side and one of them subject to the other. Man being a part of nature and not outside of it cannot be its slave.' (1895, Vol. iii, pp. 213–4; cf. 1953) There is another related thought. It is that rather than being the creature of what must be regarded, in the context of the universe, as my footling series of desires, I go forward in the greatest of processes.

If this several-sided perception is a means to giving up my hopes, it may be said, it is also a compensation, or rather more than a compensation, for doing so. Furthermore, it may be said, if I have the sustenance of this several-sided truth, there remains a still larger truth, of greater sustenance. It is not only that I have what follows from clear internal connection with the natural universe, with the latter taken as the great *fact*. My membership, rather, is in that alone which is of *value*. My membership is in that alone which matters, and which matters supremely. Both the several-sided perception and the larger proposition about valùe have behind them a considerable tradition. Let us glance at four diverse parts of it, the thoughts of Zeno of Citium, Spinoza, Shelley, and Hegel, Marx and Engels.

Zeno of Citium, in the early part of the third century BC, conceived of nature, of which he took us to be part, as material and deterministic, but also as suffused with Spirit or the Divine Fire. He took us to have the possibility, in our awareness of determinism, of what can be described as a life in harmony with nature. This is also the life of virtue. Diogenes Laertius, speaking of the end or goal of man, records that 'Zeno was the first . . . to designate as the end "life in agreement with nature" (or living agreeably to nature), which is the same as the virtuous life, virtue being the goal towards which nature guides us'. (1925 (c.225), p. 195) To achieve this virtuous harmony or agreement with nature, and hence with the divine, is to have a life such that health, possessions, success, and indeed all the objects of ordinary life-hopes become of no account. I not merely abate but am *freed from* my hopes. It was Epicurus who supposed that the good man could be happy while being racked, but the idea is also in accord with Zeno's teaching.

According to Spinoza in his *Ethics* we are parts in a certain sense of the sum and system of the single substance that there is, God or Nature, and wholly subject to a determinism which is somehow a matter of logical necessity rather than the natural necessity of nomic connection. Through a certain kind of increasing knowledge of the single system, we can to some extent escape our passions, one of them being hope, which Spinoza defines as 'unsteady joy arising from the image of a future or past thing about whose issue we are in doubt'. (1910 (1678), p. 123) In particular, we escape our passions through knowledge of the necessity which governs all things. One of Spinoza's propositions is this: 'In so far as the mind understands all things as necessary, so far has it greater power over the passions or suffers less from them.' He adds: 'For we see that sorrow for the loss of anything good is diminished if the person who has lost it considers that it could not by any possibility have been preserved.' (1949 (1678), p. 258)

Spinoza also supposes that if our knowledge of the necessities within God or Nature increases, we achieve 'the highest possible peace of mind', 'the highest joy', which appears to be identical with what he famously called 'the intellectual love of God'. (1949 (1678), pp. 273 ff.) From a determinism, then, there comes the prospect of enjoying the most fulfilled of ways of life. It is, certainly, the life of virtue. Increase in the knowledge of the necessity of all things, and its culmination in the intellectual love of god, is entirely bound up with the life of virtue. (1949 (1678), p. 205) Determinism thus offers the overwhelming compensation—the word, of course, does not catch the great value in question—of identification with the divine and hence with the most fundamental goodness.

If it is in general true that Romanticism celebrates a lesser and more obscure connection between ourselves and nature than is asserted by a determinism, and perhaps by Spinoza, there is the exception of Shelley. *Queen Mab*, the poem of his youth which to some extent he later disavowed, explicitly celebrates a doctrine of determinism, which doctrine is expounded in his notes to the poem. The impersonal divinity of nature, as he is very certain, is not to be confused with the inferior manlike God of Christianity. What is apostrophized, further, is not the universe in its substantiality, however divine, but necessary connection itself.

> Spirit of nature! all-sufficing power,
> Necessity! thou mother of the world!
> Unlike the God of human error, thou
> Requir'st no prayers or praises; the caprice
> Of man's weak will belongs no more to thee
> Than do the changeful passions of his breast
> To thy unvarying harmony; the slave
> Whose horrible lusts spread misery o'er the world,
> And the good man, who lifts, with virtuous pride,
> His being in the sight of happiness,
> That springs from his own works; the poison-tree,

Beneath whose shade all life is withered up,
And the fair oak, whose leafy dome affords
A temple where the vows of happy love
Are registered, are equal in thy sight:
No love, no hate thou cherishest; revenge
And favouritism, and worst desire of fame
Thou know'st not: all that the wide world contains
Are but thy passive instruments, and thou
Regard'st them all with an impartial eye,
Whose joy or pain thy nature cannot feel,
Because thou hast not human sense,
Because thou are not human mind.

To move on quickly to Hegel, Marx and Engels, there is the theme prefigured in Spinoza that what is called man's freedom consists wholly or partly in knowledge of necessity. In the case of Hegel, the necessity is run together with the Rational, the Absolute. 'We are free in recognizing it [the Absolute] as law, and following it as the substance of our being.' (1944 (1840), p. 283) Engels gives credit to Hegel for seeing that 'freedom is the appreciation of necessity' and that 'necessity is blind only in so far as it is not understood'. Freedom, for Engels, 'does not consist in the dream of independence of natural laws, but in the knowledge of these laws, and in the possibility this gives of systematically making them work towards definite ends.' (1978 (1934), pp. 128–9)

There is in Engels the familiar idea that knowledge of necessities has an instrumental value in the advancement of human history: through knowing natural laws, whether of the non-human world or our own natures or society, we are enabled to achieve certain ends. It is of course a truism that knowledge of the working of natural laws is of great value to us. It increases voluntariness, in that it decreases obstacles to the realization of desire. However, as the quoted passages indicate, there is also a quite different idea in Engels. The freedom in which knowledge of necessity issues is evidently neither voluntariness nor origination, but a certain state of mind and feeling. This freedom, so called, consists in an *acceptance* of our existence. It consists in something approximate to a *tranquillity*, and indeed related to Spinoza's intellectual love of God, which, of course, is also properly named intellectual love of Nature. This state of mind and feeling is owed to what which a determinism can be said to assert, man's membership of nature. If nature is not made divine or god like, by Engels, it inevitably retains a grandeur. There is the same compensation of determinism offered to us by Marx, although it is not the first of his conceptions of freedom.

Of what strength are such assurances as the four at which we have glanced, those offered by Zeno, Spinoza, Shelley, and Hegel, Engels, and Marx? Is the compensation of membership in nature, in one of these forms or another—perhaps Russell's (1917b) or Santayana's (1930)—something which can assuage the defeat of our life-hopes in so far as they are informed by images of origination? Can it be that a determinism in this way gives us more than or as much as it takes away, and so enables us to come to tolerate the loss of what it takes away? I suspect, despite the existence of a venerable tradition, that readers will not have been reassured.

What determinism itself gives us is no more than a nomic connection with the non-human universe. Assuredly a determinism in itself does not include any conception of the non-human universe as spiritual, divine, or godlike, let alone any deification of necessity itself in Shelley's way. However, the attempt can be made, as it is by Zeno, Spinoza, Shelley, and many others, somehow to identify the universe with something personal, spiritual, or divine, or to ascribe to it such a character. Nothing in a determinism precludes such an attempt. There is an evident consonance, in fact, between religious ideas of omnipotence and omniscience and the idea of perfect nomic connection. It is a consonance exemplified by Calvinism and predestination.

However, it is evident that ascribing a spiritual, divine or godlike character to the universe, at least in the late twentieth century, is what might unkindly be called a minority inclination. Whatever has been true in the past, very few of us are now given to anything like Spinoza's vision, or to such approximations to it as forms of pantheism or panpsychism—philosophical defenders though they have. (Sprigge, 1983) The different mysticisms of Zeno and the youthful Shelley are yet more remote from the attitudes of almost all of us. Nor are many of us likely, by whatever sort of exertion, to change our attitudes—to bring ourselves to feel that the universe is spiritual or personal. Those who have such a feeling, it seems safe to say, have acquired it by some other means than philosophical or religious argument or instruction. If such aids have sometimes been involved in the genesis of such a feeling in individuals, that genesis has had greatly more to do with something which cannot be acquired through argument or instruction, or decided upon, or summoned. It is a temperament or susceptibility of a certain kind, one which is uncommon. In its direction on to the external world, it is distinct from any ordinary religious inclination.

That is not to say that a distantly related thing is uncommon. Almost all of us, in conceiving of the physical universe, have a conception of its grandeur. Awe is inevitable. It is explicitly expressed, often enough, in scientific descriptions. Here what is in question is nothing so rare as the mentioned temperament or susceptibility having to do with spirituality, but rather a common disposition of very nearly all men. Is it possible that the disposition to awe of nature can be put to use? Can a determinism's connecting of our existence with the grandeur of nature serve us in the endeavour of seeking to give up that character of our life-hopes which has to do with origination?

It is hard to think that we can have much success by this means. One reason is partly that the enterprise cannot be much different in effect from certain related ones mentioned at the beginning, say the enterprises in Romanticism of more loosely identifying with nature, which related enterprises do not include a clear idea of a close connection between ourselves and the non-human world. That is, we are unlikely to get much more sustenance, in whatever sort of frustration or melancholy, by contemplating nomic connection rather than some lesser relationship—and contemplatings of some lesser relationship have surely not often been significantly efficacious in dealing with the large challenges of life. (It might indeed be argued, as in the case of the vagueness of the idea of origination, that we are *more* moved by the unspecific (p. 499), and yet not moved much.) It is one thing to write odes and quite another to deal effectively with such experience as Samuel Johnson's in his recurring contemplation of death. (Boswell, 1934–64 (1791)) To suppose that men as they have been, and as they now are, can very often put aside such darkness, when it comes upon them, by the means of identification with the universe, is surely to confuse poetic or philosophical reverie with life itself. It is indeed hard to think that determinist identifications with nature, even if they do offer *more* hope, can do much to assuage the defeat of life-hopes having to do with origination. Still, at least one of the ideas which turn up in identifications with nature can be separated out and needs more consideration. (9.2.8)

9.2.4 Failure

A different possible compensation of a determinism is more intimately connected with it than are thoughts of nature. It has little tradition behind it, and can be set out quickly. To have a life-hope, to recall, is either to hope that one will achieve a certain thing, or else that some state of affairs will come to obtain. At least typically, with respect to the second kind of hope, one believes that the desired state of affairs will only come about through one's own doing, and so in both cases the realization of the hope *depends on oneself*, on one's own endeavours. Life-hopes therefore carry with them the possibility of personal failure. A hope is indeed to be characterized as an unsteady joy or like feeling about something of whose occurrence one is uncertain. (p. 505) At least typically the aspect of doubt in my hopes, as Spinoza might have added, is doubt as to whether I will succeed or fail in my endeavours. Will I win or lose, rise to the challenge or not, measure up or not, secure what I want or instead secure disappointment, embarrassment, grief, or disgrace? The contemplation of failure is no light thing. The experience of failure is likely to be a darker one.

Consider life-hopes of the two kinds in so far as they carry ideas as to origination. The prospect of failure and the actuality of failure are here a prospect and actuality having to do with origination. More particularly, our subject is the individual power of origination which each of us is inclined to attribute to himself or herself, and the possibility of its being insufficient to some challenge or project, or the fact of its having been insufficient.

My supposed power of origination is one which I obscurely identify with *myself*. That very claim is in itself obscure, but it is a claim which seems impossible to deny. The prospect or experience of failure, then, in so far as it has to do with origination, is absolutely the prospect or experience of *my failure*. This particular prospect and experience of failure are different from others which can be distinguished, those which involve a view of myself as fundamentally a corpus of dispositions. My experience of failure, if I put my failure down only to a less rational desire having carried the day over a more rational, brings with it a lesser self-accusation, a lesser diminishment or disgrace, than if I put my failure down to myself in the other sense, the very centre of my existence. The 'I have failed' which carries no idea about origination, and hence no failure in it, also carries less hurt.

Indeed, some will want to say more, by way of a certain question. *Is there* an 'I have failed' which carries no idea about origination? Can we, as we are, have something properly describable as an experience of failure which does not have in it the self-accusation which attaches to an idea of origination? The thought that a temptation has triumphed, or that a less rational desire has carried the day over a more rational, may seem to be more a dismayed report on what has happened than a self-accusation. Despite what was assumed above, some will say, it seems not really to be the thought that I have failed. This seems to me an understandable inclination, but incorrect. More will be said of it in due course in connection with the issue of determinism and personal feelings, on which we are now verging. It *is* correct, however, and the point that is now important, that the sharpest sense of failure is one which has to do with origination. It might be mitigated by reducing our sense of ourselves in a way that has been admirably expounded and indeed recommended. (Parfit, 1984, Chs. 10–12) but we have not yet done that.

All of this, as must be allowed, is impressionistic. The subject-matter of the self is one of which it is impossible to get a clear view, and it is as impossible that our conceptions for dealing with it can be very satisfactory. To speak of something that *is* possible, more might have been done to give a more specific account of many things that have been bundled into the two categories with which we are concerned, anticipation of failure and experience of failure. A certain conclusion can nevertheless be attempted.

If a determinism deprives us of certain life-hopes, all our life-hopes of a certain character, it can also be said to offer us *some* compensation. It saves us a certain apprehension and sharp experience of failure, a peculiarly sharp apprehension and experience of failure. It offers us, in place of certain fears of failure and self-accusations of having failed, a certain composure or tranquillity. This, in its calm, is related to such feelings as might be produced by a successful identification, if that were possible, with ongoing nature. However, the given composure or tranquillity is a quite different thing in having a quite different ground. Its ground has to do with me, and the lack of a conceivable power that I might have had, and not with the external world.

It is hardly possible to discuss our present subject without feeling uneasy, as I have before now, about falling into a kind of autobiographical reflection lacking a general relevance. Perhaps something of this kind is inevitable, since the enterprise we have in hand is in a way personal. It is an enterprise, by its very nature, which should not issue readily in confident generalizations. I propose, however, that the consideration about failure offers to all of us a *greater* aid than anything else considered so far. We can go some way towards abating certain of our life-hopes by persistence in the thought a determinism can preserve us from the anticipation and experience of a certain kind of failure. We can go some way in the given direction, which is not to say that we turn a corner, that a determinism in so far as it touches on life-hopes becomes a tolerable thing.

9.2.5 Unbounded Hopes

Let us, to consider the possibility of a certain reassurance, look back to our very first reflections on life-hopes. We took it that very often they are hopes that we shall achieve certain things, and, when they are not, they are likely to be hopes for states of affairs that will in fact depend on what we do. It is possible and natural, as we saw, to have in mind in either case life-hopes which depend on the idea of originated actions. I thus have the idea of a series of future moments of challenge in my life. At each such moment, according to this idea, I will be subject to two desires, one of them being a desire which will not serve my life-hope, no doubt a desire for immediate gratification. The outcome at each such moment will *not* simply be the result of the strength of this desire and that of the opposed desire to achieve my life-hope. It will not be the case, according to this idea, that what I do will be the effect of only these pulls. Rather, given both these desires, and all else that is true of me, I shall have a kind of possibility of two courses of action, the one serving my life-hope and the other providing only immediate gratification. (7.2)

On the other hand, as we saw, it is possible and natural to think that life-hopes do not depend on an image or belief about originated actions, but on the belief that our actions will flow, not from reluctant desires we shall have, but from our embraced desires. Our future circumstances of action will be enabling rather than frustrating circumstances. Thus a man's life-hopes may be taken to depend on what was called the way of his world, on the ascendancy in him of desires that do serve the realization of his hopes, on his independence of others, including their threats, and on the absence of such bodily constraints as illness or disability. His life-hope, say, to come to have some distinction in his profession, may be taken by me as depending on such things. (7.3)

Given a determinism, it is hopes of the latter kind that we can have, and the question now is their worth. A first thing to be said of them is that they are in a certain sense *no more limited than* life-hopes of the former kind, involving origination. Let us have in mind what can be labelled as the *object* of a hope, an action of achievement, with nothing said of the initiation of the action, or a certain state of affairs. For any life-hope involving origination, there can be a life-hope with the same object involving only embraced desires. Any object of a life-hope involving origination—professional distinction, to be an actress, the achievement of a certain peace—can be the object of a life-hope involving only embraced desires. This is of great importance.

If I contemplate my future first with origination in mind and then without it, it makes sense to say that whatever possible goods turn up in the first contemplation can also turn up in the second. It is not as if some *sector* of goods is closed off by a determinism. Any fantasy of grandeur, any vision of success, any picture of personal security, any thought of personal relationships or of the avoidance of disappointment or disaster—any of these can as readily be the stuff of the second kind of hope as the first. In a clear sense, hope can be as unbounded in the second way as in the first.

It may be objected that hopes are not only a matter of what have been called their objects. That is true enough, as we know. Hopes can be for actions expressly conceived as originated. They can be for states of affairs taken to depend on actions conceived as originated. If such ideas of origination are thin ones, as we have seen, that does not greatly matter. There is the fact of which we know, that something of value is subtracted from a hope by subtracting ideas of origination from it. It remains a truth, despite the objection, that such subtraction deprives us of no object of hope, and it can be said that objects of hopes are the great constituents of hopes. Any summation of objects, say Byron's in *Don Juan* (II, clxxix)—

> Glory, the grape, love, gold, in these are sunk
> The hopes of all men, and of every nation

—any summation of objects, however shallow or profound, is as true of a deterministic human world as it is of a human world of origination. It can be added, certainly, with respect to the constituent of hopes having to do with the initiation of action, that to take it as only voluntariness does allow us certain thoughts to the effect that what is to come is not inevitable, is within our control, and does contain opportunities. (Dennett, 1984)

9.2.6 Uncertainty

There is something else that is virtually as much a part of hopes which depend on beliefs that one's future actions will flow from one's own embraced desires. There is evidently a difference, already noticed, between hope and something else, the anticipation or expectation of a happy circumstance with the *certainty*, in so far as we can have it, that it will occur. There is a diffrence between hoping that I will get an award of merit and *knowing* or *fully* expecting that I will. Is it possible to feel that hope, when it is full and strong, is the better thing, a better species of happiness? One can take it, surely, that a life wholly informed by *certainty* as to the having of future goods would be a lesser life than one informed by good hopes, or at any rate one which was partly a matter of hopes rather than wholly undoubted expectations. No doubt this line of thought needs examination and more qualification, but is hard to resist the idea that a part of the value to us of our hopes is owned precisely to the uncertainty of our actually getting what we hope for.

To come to the essential point, there is virtually as much room for uncertainty—a desirable degee of uncertainty—with respect to the second kind of hopes. If my hope for fame rests on the belief that my future actions will indeed flow from my embraced desires, there is virtually as much room for uncertainty in this case as with hope when it involves origination.

Assuredly a hope involving only voluntariness is one which is consistent with the truth of a determinism, a fixed future. It thus is consistent with the logical possibility of a man's certainty as to what will happen. (6.1) There is not the same logical or conceptual possibility of certainty, based on nomic connection, with hopes that bring in origination. However, this makes little difference. Whatever logical possibility of certainty may attend the former hopes, there is uncertainty enough about them. It is very far from true that my world becomes in fact a predictable place by way of the truth of a determinism. I can be reassured that the flavour that is given to life by the possibility that good or great things will happen is not taken away by a determinism.

9.2.7 Credit

Perhaps there is also a reassurance that is as much a matter of the category of personal feelings as of life-hopes. Among my personal feelings towards myself, as remarked earlier, I may have the satisfaction of giving myself credit for something. (p. 403) A part of my hopes may be the anticipation of such credit. A larger part may be the anticipation of having the good judgements and feelings of others. What credit can I anticipate, from myself and others, by way of hopes which do not involve origination?

It needs to be allowed, consistently with what was said above of the tie between origination and failure, that there is a kind of credit that I cannot anticipate. I have no chance of the credit that could be mine if I had a power of origination. But that is certainly not the only credit conceivable. To say but a word here, I can in my future actions have the credit of rational action, action in accordance with desires that serve my fundamental end. Or I can have the credit of strong character, or sensitivity, or judgement, or decency, or of rising over mere conventionality. The list continues. Life-hopes having nothing to do with origination, therefore, can be hopes to do what will secure me the good feelings and judgements of myself and others.

9.2.8 Acceptance of Determinism

It was concluded earlier that the theory of determinism of this book is very strongly supported. It was said, further, that it is a theory of which it is possible to be convinced, as I myself am, although with the reservation that this conviction goes beyond the sum of evidence for and against. (p. 374) It was subsequently remarked, in the course of an argument having to do with moral responsibility and dismay, that almost all of us, at least in one great tradition of culture, do not in fact believe a determinism. (p. 435) It was added that we do not disbelieve it either.

To say more of this majority, it includes some or perhaps many who accept that all the evidence and argument on offer tells on balance in favour of a determinism, and yet do not believe it. To be in this position is not both to take it to be true, *simpliciter*, and yet not believe it, which is logically impossible. Rather, perhaps, it is to think that all the known evidence and argument supports it on balance, and also that contrary evidence or argument may turn up. The agnostic majority, secondly, also includes some or many who are inclined to think that all the known evidence and argument on balance supports determinism, but entertain a suspicion that there is error in that evidence and argument. The suggestion of contradiction in this position, surely, is a problem that might be dealt with if philosophers were to turn their hand to providing a perspicuous philosophy of the propositional attitudes, in part a differentiating account of kinds of assent to propositions. The agnostic majority, finally, must contain some or many people who are somehow still closer to dissent than the two groups mentioned.

It is possible to feel that the explanation of the existence of the agnostic majority is owed to rootedness of images of origination in our lives, where that rootedness is not to be identified with truth or conceptual necessity. It is not too much to say, as it seems to me, that in our reflection on determinism we are apt to be prisoners of culture. We are bound by our inherited ways of feeling. These get expression in our language and are given a reality in many fundamental practices both individual and social. That is not to say, to repeat, that we have *reasons* for thinking determinism false. We do not have rational grounds of opposition.

There is the possibility that further reflection on determinism, including reflection on the non-rational sources of opposition to it, will issue in a more settled and unqualified belief that it is true. *It is such belief that is certain to be more effective than anything we have so far considered in enabling us to abate life-hopes having to do with origination.* To approach the same conclusion by a slightly different path, there is a distinction between hopes and wishes. One can wish for the impossible, it seems, but not hope for it. At least a majority of us do not take origination to be impossible, whatever our attitude to evidence and argument. We hope for it. We shall move effectively towards ceasing to hope that it exists, and come merely to wish for it, when we come closer to a settled and unqulified belief in the truth of determinism and the impossibility of origination. We can move towards towards such a state by an awareness of what has been impeding it.

To express the matter from the point of view of one who *is* persuaded by the evidence and argument for a determinism, the best strategy for dealing with one's recalcitrant life-hopes is further persuasion of the truth of determinism. Spinoza was considered above in connection with the possible compensation having to do with membership in nature. As was noted, however, he also maintains that it is increasing knowledge of the necessity which governs all things which enables us to escape our passions. (p. 505) One can object to his general claim that sorrow over the loss of a good is diminished by the perception that the having of it was impossible. It does not seem that one can object to the proposition that certain of our hopes, if they can be abated somewhat by the consideration having to do with failure, can only be cured by a final acceptance of determinism. That, to repeat, is importantly a matter of rising over our history.

Of course, one cannot *choose* or *decide* to believe anything (p. 370), and so there is no possibility of abating our hopes by choosing or deciding to believe a determinism. The most that one can do is to consider that one's want of final assent to a determinism is owed not to evidence or argument against it, but to one's being a prisoner of inherited desires. To begin to do this—to suspect oneself—is to do something which may have effect. In this there is the greatest likelihood we have so far discovered of abating one's hopes having to do with origination.

What has just been said may be misconceived, and so seem curious or worse. It *would* be bizarre to attempt to prove the truth of a determinism by way of the idea that a desire gives rise to opposition to it. It would be perfectly proper to reply to any such attempt that the fact that determinism is unsatisfactory to a rooted desire is consistent with there also being good reason for not accepting it. What has been under consideration, however, has not been an irrelevant and necessarily failing attempt to prove or confirm determinism. The idea, rather, has been to try to ameliorate a certain effect which a theory has on life-hopes, rather than to confirm the theory. The idea has been to make a certain recommendation.

The recommendation comes to this: *let us try to escape certain hopes by coming fully to believe what is in fact true, a determinism*. It is the recommendation that we come to *accept* what can independently be seen to be a likely truth, by escaping a certain cultural inheritance, and so abate our life-hopes in so far as they involve origination.

9.2.9 *Affirmation*

It is at this point in this inquiry into the consequences of determinism that we come to its principal conclusion, although not a full awareness of the breadth of that conclusion. The latter cannot be had until we have looked again at the matters of personal feelings, knowledge, morality, and certain social institutions and practices, all of which are touched by a determinism, and have also glanced at things of great value to us which are not touched by it at all.

We have been considering a third response with respect to determinism and our life-hopes, a response different from those of dismay and intransigence. It is *the response of affirmation*. As anticipated (p. 493), it has several parts.

First, it accepts that a determinism does affect our life-hopes. It defeats those which have a character which they cannot have if determinism is true. This must be at least a deep disappointment. It is not as if we can submit ourselves to the commanding fact of causation—which submission is the attitude included in Compatibilism as defined. (p. 474) If there is assuagement to be had from certain things, and relief to be had in the future from one of them—a fuller or deeper belief in determinism—it remains a truth that we cannot escape being affected in our life-hopes by a determinism. We are far from the mistake of a Compatibilist who considers life-hopes and asserts that nothing changes with respect to them as a consequence of determinism. Nor is our response the response of intransigence.

The response of affirmation in a second part asserts that a determinism does not destroy life-hopes. We can in a fundamental way persist in them. It is entirely a mistake, the mistake of an Incompatibilist who considers life-hopes, to think that a determinism must end them. Nor is our response the response of dismay. We take it that if our life-hopes must have one character rather than another, that does not come close to saying that a determinism ends them.

There is a third part of the response of affirmation, separable from the propositions that a determinism at least touches our life-hopes but does not destroy them. The response is importantly that our life-hopes, conceived as they can be, remain at least sustaining things, life-sustaining things. They have at least that large value for us. This is owed to the facts we now know. They can be unbounded in a clear sense. The goods which are their objects are made no less good by a determinism. They can have what distinguishes them from foreknowledge of a good, the feature of uncertainty, which is a recommendation of them. They can be hopes for honour, standing, and in general all the forms of credit to which we so much aspire.

Thus, if a determinism touches my life, taken as consisting in a line of hopes, it none the less may leave my life a great deal more than merely tolerable in this respect. My life, in so far as it consists in a line of hopes, does not become dark. The line of my hopes supports me, and it may enlarge, delight, and enrich me. My life, as a line of hopes, can be a life of full anticipation and fulfilment. My hopes can enter importantly into a philosophy of life which is not one of resignation or defeat, such as to burden me, but rather can be one of celebration.

Will someone ask at this point for more precision, perhaps for a definition of what it is for something to be life-sustaining, or of what it is for a life to be one of full anticipation and fulfilment? There is nothing unreasonable about the request so long as it is not accompanied by unreasonableness about what can be done to satisfy it. It would be unreasonable, for example, to think that the conception of that which is life-sustaining is wholly a matter of discovery of facts. It is not, but is in good part a matter of decision or resolution. It would also be unreasonable, as always, to demand a kind of precision which the subject-matter does not allow.

Let us finish here by having clear the relation of affirmation to Compatibilism and Incompatibilism. Affirmation differs wholly from both in that it recognizes the existence of two attitudes where Compatibilism and Incompatibilism assert a single conception and a single connection with moral responsibility and the like. Affirmation does involve reliance on a single attitude, having to do only with voluntariness, which of course is related to the single conception of initiation which Compatibilists assign to us. Affirmation also has to do with the other attitude, pertaining also to origination, related to the single conception which Incompatibilists assign to us. It is not much more like Compatibilism than Incompatibilism. (Cf. p. 480.)

9.3 PERSONAL FEELINGS

What is to be attempted with personal feelings, as with life-hopes, is the response of affirmation. Personal feelings were divided into the appreciative and the resentful or accusatory kinds, the former being reactions to having the good feelings or judgements of others, the latter being reactions to having the bad feelings or judgements of others. Both kinds have large, fixed places in our lives, and the former enrich our lives. Resentful feelings were illustrated by the example of my bitterness about an adversary who makes trouble between myself and my friend and patron, thereby depriving me of support in something important to me. The appreciative feelings have gratitude at their centre, and might in fact be regarded as consisting in forms of gratitude.

Suppose, to elaborate the example just mentioned, that a woman comes to see my adversary's action as I do, and understands that I stand to lose by it. Perhaps out of affection owed to a past connection, and not because she has anything to gain, she goes out of her way to try to put matters right, and does so. I may feel that it was not merely that she had nothing to gain by her intervention, that for good reason she found it disagreeable to mix in the business. My gratitude to her is considerable.

In what does gratitude consist? To feel grateful is in part to have certain ideas or the like about the initiation of another's action, an action which one values as having benefited or been intended to benefit oneself, and in another part to have a desire or inclination to benefit one's benefactor. One wants to repay or at least recognize a good turn. More can be added, in line with the general conception of the attitudes or feelings noticed earlier. (p. 449)

My gratitude to my benefactor, then, includes ideas or the like about her initiation of the action which put matters right with my friend and patron. What are my ideas as to the initiation of the action? Evidently they must include beliefs as to voluntariness. There would be no possibility of my feeling gratitude to her if I took her action to have been the result of an effective threat or to have had some related inception. In short, it needs to have been an action which flowed from her embraced desires.

It may be that I also have an image or idea of origination in connection with her action. I may regard the putting-right of my connection with my patron as other than something that was inevitable, an event fixed by the past. A very natural expression of my gratitude to my benefactor, my saying that she need not have done what she did, is naturally taken as expressing not only my belief as to her voluntariness in her action but also its not having been the upshot of certain nomic connections. The image in my mind's eye is of my benefactor out of her own originative volition settling on doing the thing, resolving to seize an opportunity, or pressing on with her defence of me. As noticed already, such an image or thought of origination, in so far as the clear and the literal is concerned, reduces to not much more than a denial of determinism. Still it is of no small importance to my feeling.

If the relations between the components of gratitude are unclear, it seems at least arguable that the ideas or whatever about initiation have a certain consequence for other components. The *character* of the evaluation, and of the inclination or desire, owe something significant to the given ideas or whatever. If that claim is abstract and general, it seems none the less to contain a truth. What we come to, then, is that gratitude, in so far as it involves images or ideas as to an originated action, as it certainly may, is in other parts different from what it can be without those images or ideas. One is tempted to say that in so far as gratitude carries images or the like as to origination, it is more profound, perhaps what can be described as a more moving emotion. It is, to recall another component of an attitude, also different in feeling in the narrow sense.

It may well be, then, that there is more than one way in which an effective determinism conflicts with gratitude of a certain kind. Certainly it conflicts in being inconsistent with the image of origination which may be part of our gratitude. However, in the case of this gratitude, it seems to make sense to say that it conflicts with the character of other elements of the feeling. The same is to be said of the resentful personal feelings. Here a determinism fights with feelings which we do not esteem in anything like the way that we esteem their appreciative counterparts, partly as a consequence of the fact that the resentful feelings have in them desires to return a hurt. (p. 401) However, there can be no denying that the resentful feelings are deep-rooted. Here, if a determinism is not a challenge to what we value, it is a challenge to what we may feel we cannot succeed in giving up.

Given all this, we must accept that a determinism cannot leave our personal feelings untouched. How can we be helped to eschew the personal feelings which are inconsistent with determinism? What was said of identifying with nature in connection with life-hopes (p. 503), which enterprise was anyway judged unpromising, can hardly be transferred to the personal feelings. The reflection that determinism excludes a certain sort of failure is at least germane to life-hopes (p. 508), but hardly germane to giving up the relevant kinds of appreciation and resentment. There is a possibility, obviously, of reflecting on the impossibility of a certain failure in connection with *reflective* personal feelings, feelings about oneself (p. 403), but they are of lesser importance than other personal feelings. Gratitude to oneself, or resentment directed to oneself, are surely rare phenomena. What is to be said is much the same as what was said of giving credit to oneself. (p. 513)

A stronger idea that must come to mind in connection with the personal feelings is the idea that there would be no loss to our lives, but rather a gain, in withdrawing from *resentful* feelings in so far as they contain ideas of origination. One side of this several-sided idea of gain is that certain feelings of hatred and the like involve false accusations, which is to say false propositions to the effect that a person not only intended out of embraced desires, but also originated, a hurtful, damaging, or vicious action. There is thus a general unfairness in our feelings, the unfairness of being disposed to vengeful action on the basis of falsehood. (There is also the unfairness of *action itself*, of course, which will be attended to in Chapter 10.) We escape the falsehood and the resulting unfairness by eschewing the given feelings.

Another side to the idea of gain does not have to do with the possibility of our coming to have a better moral standing, but the possibility of serving self-interest by escaping the relevant part of the experience of hatred and the like. Such feelings are not a satisfaction to us. They are not a satisfaction in themselves, and there is also the fact that they must often be frustrated, sometimes for reasons of prudence. The situation is greatly different, certainly, with the appreciative feelings.

It seems to me, although there is again the danger of falling into the merely autobiographical, that there is some hope of making the acceptance of a determinism more tolerable by such a means. If it is said, in opposition to this, that being deprived of appreciative feelings of a certain character could not be compensated for by way of a withdrawal from like resentful feelings, there is a certain rejoinder. It is that there is also falsehood in the given appreciative feelings, if a determinism is true, and an unfairness. Here, the point is not so much that those to whom we feel grateful are unfairly benefited, but that those to whom we do not feel grateful are unfairly deprived. If origination is a fiction, true of no one, then to distinguish in our feelings between persons in terms of origination is essentially unfair. However, this rejoinder can at best give us no more than was claimed, the idea that a determinism in so far as it affects the personal feelings may be made *more tolerable* by reflection on the escape it offers from an aspect of the accusatory feelings.

One more fundamental strategy for accommodating ourselves to the necessary alteration of our personal feelings must be the attempt to accept determinism. (9.2.8) Another such strategy, of which more can usefully be said, has to do with those personal feelings that remain to us, those based on voluntariness, most importantly the appreciative ones. The situation *is* in this respect like that with life-hopes. (p. 510) We can persist in a full spectrum of appreciative personal feelings, and they are not to be underestimated. They are based on and informed by things which are of great value to us: the love of others for us, their loyalty, perseverance in forwarding our lives, humanity, tolerance in dealing with us, and also their admiration, esteem, and approval with respect to our natures and our endeavours.

A determinism does not wholly subtract any of these things from our existence. A determinism does alter a perception of mine of my sister's tenderness for me. It seems to make sense to say, none the less, and may be true, that in my remaining perception of her she is not made *less tender* towards me. The possible degrees or shades of tenderness would not be greater in a human world containing origination than in a human world without it. So with, say, the sagacious admiration of someone for me, if I happen to earn it. He is, it seems, not *less admiring* in virtue of the truth of a determinism. There is in a sense no level of admiration that is possible if determinism is false, but not possible if it is true, no level of praise that is possible in the first case but not in the second. To return to the earlier example as elaborated, my adversary's act of unscrupulousness does not come from lesser or different motives if a determinism is true, and the goodwill of the woman who helps me out is not in itself reduced by the fact of a determinism. There is obscurity and uncertainty here, but surely also a fact.

It is a fact which is part of what can tempt one to embrace intransigence, to declare that a determinism in no way affects our feelings for others. What we are attempting now is something else, more secure and in general more satisfactory. It includes accepting, despite what has just been said, that there *is* a kind of personal feelings to be given up. It also involves the assurance that we can persist in personal feelings of another kind. It involves, further, the assurance that these remain a great source of satisfaction to us. An unoriginated act of love is an act of love, an unoriginated act of magnanimity is an act of magnanimity, an unoriginated true assessment of our worth is a true assessment of our worth.

9.4 KNOWLEDGE

The response of dismay, with respect to knowledge, is that a determinism undermines or constricts it, condemns us to an agnosticism or scepticism. The response derives from the attitude that for us to have what are good reasons for a claim to knowledge, they need to be owed to mental acts and ordinary actions which are not only voluntary but also originated. For facts or propositions to be good reasons they must have in their background originated acts and actions. That is, there must have been a certain possibility of discovering whatever might have unseated these facts and propositions as good reasons. It needs to have been the case, as was remarked, that we have explored reality rather than been guided on one tour of it.

The attitude need not include the idea that my belief now, that the paper I am writing on is white, is owed to a nearly simultaneous originated mental act or action. It *is* arguable that my belief, involuntary as it is, is owed to an act of attention, and that without such an act I would have a kind of awareness that falls short of belief, but let us pass by that matter. I do require with any belief, if I take the attitude in question, that voluntary and originated acts and actions lay in the further background of the belief, in the growth of my knowledge and of a conceptual scheme. The attitude is stronger in the case of knowledge-claims not having to do with immediate experience, where we take acts of judgement as very much part of the recent background of the knowledge-claims.

The conflicting response of intransigence is that a determinism does not at all trouble our existence as knowers. No significant change need occur in my view of my own beliefs, my confidence in them, if I come to accept a determinism. The response rests on the attitude that facts or propositions, if they are good reasons for a claim, need be owed only to acts and actions that are voluntary. More fully, I take the attitude that for reasons to be good ones, what is required is only that they are wholly owed to my desires for information, as distinct from reluctant desires, and furthermore that my mentioned desires for information were themselves partly owed to acts and actions which themselves issued from desires for information.

The consequence of determinism for knowledge is different in detail from its other consequences, as they are different among themselves. This difference includes one having to do with the identities of the agents in question. My life-hope of either kind that I shall achieve some rank or standing is a feeling whose content includes beliefs or the like pertaining to *my* future actions, beliefs that they will be both voluntary and originated, or simply voluntary. With respect to personal feelings of either kind, except the reflexive ones, they have to do with initiation of *others'* actions towards me—say the action of the woman who remonstrates on my behalf. My moral disapproval of the husband facing divorce also has to do with his action, although not towards me. There is a related fact about the rightness of particular actions, such as his, and a different related fact about particular judgements as to moral standing over time, judgements on such specific individuals as the husband, but not judgements with respect to a single action. (p. 421) As for general judgements or principles as to good men and women, and right actions, they carry presuppositions about the initiation of acts and actions generally, assigned to no persons in particular.

In the case of my beliefs or claims to knowledge, to return to them, what I take to be good reasons are conceived by me as facts or propositions owed to the acts and actions related to those reasons, *my* acts and actions but also those of *indefinitely many other persons*, those in the history, so to speak, of my beliefs or knowledge-claims and also my conceptual scheme.

It may be that the indefiniteness of the class of relevant acts and actions has something to do with a more important proposition about the distinctiveness of the consequence of determinism for knowledge. The proposition is that despite what has been said, the consequence of determinism for knowledge is a lesser consequence than others, and hence that there is less obstacle to the response of affirmation. Other consequences of determinism, it might be added in support of this, have to do with my life where it has a pressing emotional content—consider life-hopes and resentment—but the consequence for knowledge has to do with what touches me less, my reflective or philosophical life. It is a consequence, more particularly, tied up with reflection on the nature of reality, with metaphysics. The consequence is most importantly one for those of a certain philosophical turn of mind, and for them when they are subject to it or indulging it.

Still, there is at least aspiration and a kind of desire in such reflection. The affirmative response accepts that an aspiration is defeated, that our access to reality is in a way limited. We have a thin but significant conception of what may exist and be forever beyond our reach. A parallel can be seen. We take it without hesitation that our knowledge is different from, and greater than, that of lesser species. It seems we must accept the thought that our access to reality stands to another conceivable one as knowledge of lesser species stands to ours. The consideration that underlies the former difference, a determinism, is of course different from the consideration that underlies the latter. Our position, therefore, is related to those to which we are assigned by Plato, Kant, Spinoza, and other philosophers who bar us from *a reality beyond*. It is my own feeling that the determinist argument for a conclusion of this kind may well give us more pause than arguments of a more metaphysical or wholly metaphysical kind, including those phenomenological ones which regard our immediate experience as a kind of screen between us and reality.

Can we assuage the defeat of an aspiration by a certain reflection? We considered, as a means of dealing with the defeat of certain of our life-hopes, the thinness of our ideas or images of origination. (2.1) The agem, it was judged, could not be very successful. There is the ...ne conceptual thinness in connection with knowledge, of course, but also another, noticed a moment ago in passing and also earlier in connection with the response of intransigence. (p. 419 f.). There is a particular sparseness about my conception of what I might now possess, a wider or deeper knowledge, if the history of my present beliefs had been a matter of acts and actions that were originated as well as voluntary. I can have no *grip* on the possible *reality beyond* from which I am excluded. The objects of my hopes are clear, as are the actions I resent or am thankful for, but that which origination might have provided with respect to knowledge is beyond my sight. Is renunciation therefore easier here? It is possible to think so.

Can the compensation having to do with membership in nature be specially shaped or adapted so as to work better with knowledge than with life-hopes? The contrary may seem more likely. Is there then some small compensation to be had having to do with the impossibility of a certain kind of intellectual failure? Perhaps. What may be more important is another consideration of which we know something, that determinism excludes chance. (2.2) This has to do with the clear fact that a world of causal and nomic order offers a possibility of knowledge that is absolutely denied by a chaotic world, a world of widespread chance. There is the difficulty, however, that what we are contemplating, as an alternative to a world of determinism, is one which is nomic except for the origination which would, if it existed, serve the end of knowledge.

The response of affirmation must rely not so much on what has been called assuagement or compensation but on a recognition of what knowledge we possess despite determinism. What we have is the present state, the continuing growth, and the large future of knowledge —conceived as the product of no more than desires for information. (It may also owe something significant, no doubt, to what is distantly related to origination, which is causally determined but in a sense 'random' searching. (Dennett, 1984, pp. 66 f.) If our good reasons lack a certain character which we may, in one mood, aspire to have for them, that is not near to allowing that our reasons have no recommendation. After all, truth and evidence, or anyway ordinary truth and evidence, are defined *within* the existence we have. Further, no particular reason, and no particular set of reasons, is thrown into question. As we need abandon no particular object of hope on the premiss of determinism, so we need abandon no particular belief on that premiss, other than any which is in the ordinary way inconsistent with it.

Finally, and relatedly, we need not on balance succumb to a dejection. One large point here must be that we need not suppose that the arrays of our desires and inclinations, including our life-hopes, are devalued by the true consequence of a determinism for knowledge. We need not suppose, that is, that our desires and inclinations have no decent foundation in our knowledge as it is. Yet more fundamentally, perhaps, we need not suppose that great satisfactions of what we can properly continue to call knowledge, satisfactions in knowledge itself, are denied to us. It is not as if philosophy and neuroscience, to mention only two bodies of knowledge, become anything like dreams. We need not suppose we are in anything like the situation of illusion to which Plato among others assigns us.

9.5 MORALITY

The burden of what was said earlier about moral responsibility is that to hold a man morally responsible for an action, to disapprove of him morally for it, is to have one or the other of two complexes of feeling about him. (7.7) This simplification ignores variations of several kinds, but catches hold of what is fundamental. The response of affirmation has to do with both of the things distinguished.

To speak again of the husband preparing for his coming divorce, the first complex of feelings includes as one of its elements a repugnance for his desires and intention to cheat his wife of their son's affections and of a fair share of the monies of the marriage. He himself is also repugnant as the tolerator and indeed the forwarder of the desires and intention. The first complex of feeling also includes the element of retributive desires. We are disposed to act against him, to subject him to some degree of distress, perhaps so little as the discomfiture produced by an expression of our disapproval, perhaps more. Our retributive desires are in fact attached to a third element of our disapproval, images or ideas of origination—his origination of his intention and action. There is no logical connection linking our retributive desires to our ideas of origination, but it is nevertheless true, as we now are, that the desires do depend on the given ideas. If we reflect on holding the man responsible in the way in question, we will take our ideas of origination to be a necessary part of the reason we can offer for our desires. The truth of a determinism, then, conflicts with these retributive desires, which we must seek to escape.

There is a like complex of feelings in crediting someone with responsibility, approving of him morally for an action. To do this may not only to be attracted to his desires, rather than repelled by them, and hence attracted to rather than repelled by him, but also to be disposed to benefit him for a particular reason, at least by offering him the satisfaction of one's expressed moral approval, one's commendation. These desires are as linked to ideas of origination. They too are vulnerable to the truth of a determinism, and we must seek to deal with them.

We may have an inclination to resist the latter uncongenial conclusion, about moral approval. This is to be explained, it seems, by what is taken to be a lesser need to justify desires to benefit as against desires to distress. The inclination can hardly be defended. It needs to be noted that what is under consideration is not at all a further element in the first complex of feelings, which is desires to direct future behaviour, desires to serve the ends of morality. What is under consideration is the question of whether a discriminably different desire to benefit someone by at least commendation for an action can persist if the action, the intention in it, and so on, are perceived as effects of causal sequences of the kind specified in the theory of determinism. If one clearly distinguishes the given desires to benefit as we now have them from other things, it seems that they are in place only if there was a certain possibility of the agent's having acted in other than in the creditable way he did. The desires to benefit are in place only if he possessed a power of origination.

Retributive desires are deep-rooted in us, and desires to benefit are both deep-rooted and more satisfactory to us. The response of affirmation with moral disapproval and approval, to repeat, is in part the attempt to give up both categories of desire. Here, as with the resentful personal feelings, we may have the help of an awareness of the unfairness of such desires. We can look to an escape from a certain moral guilt. We can, with respect to desires to benefit, escape the unfairness of rewarding some, and thus depriving others, on the ground of a fiction, that all possess a power of origination.

Can we, despite much that has been said, indeed despite what has been a refrain of this inquiry, contemplate a certain possibility, certain very different strategies of affirmation? We have it that the connection between our images or ideas of origination and our desires to benefit and desires to distress is a contingent one. To repeat, it is not as if there were a logical connection between the ground for the desires, so to speak, and the desires themselves. The situation, rather, is that we *regard* a certain conception of the initiation of action as a reason for the desires. We are disposed to this. This is the way we are. (p. 479) The fact of contingency can give rise to certain questions.

Why should we found retributive and the related desires on origination? Why should we desire the distress or benefit of a man on the ground of his having originated a malevolent or otherwise wrongful action, rather than on the ground alone of his having acted out of malevolent or otherwise wrongful desires, desires which he embraces? If it can be said to be reasonable on either, does it not seem as reasonable to do so on the latter ground—the ground of voluntariness or willingness—as the former? There is a second and different question. Why should we not attempt the radical course of eschewing desiring-to-distress but persisting in desiring to benefit?

Consider desiring to benefit a woman on the grounds alone that she is a moral exemplar, which is to say someone whose commitments, desires, inclinations, and so on give rise to our overwhelming admiration, and are wholly voluntary. There would be no *mistake* in the impulse to reward such a person. Would there be unfairness in it? Would it be unfair in that the existence of the desires of the moral exemplar are as fixed by antecedents as the desires of a moral monster? It might be said in reply, of course, that the charge of unfairness would have to be based on the supposition that the kinds of benefit in question had to be *earned*, in such a way as to bring in origination. But why should we be overborne by this thought?

There is, of course, the suggestion of inconsistency in the idea of attaching desires to benefit to voluntariness but not desires to distress. As elsewhere, it is possible to consider escapes from inconsistency. It might be said, for a start, that distress and satisfaction are different in such a way that desires for the satisfaction of some persons carry no implication about desires for the distress of other persons. That origination is a reason for doing one thing does not commit us to its being a reason for doing another, given that the two conceivable actions are different as well as alike.

Both prospects—attaching both kinds of desires to voluntariness, or only desires to benefit—are attractive. But to contemplate them, surely, is to fall into a kind of utopian thinking. I cannot with any optimism *resolve* to change what is my nature. Perhaps the most that can be said of the two prospects has to do with a possibility noticed in connection with life-hopes. We may come, or we are likely to come, eventually, to a full and true belief in determinism. It is that which will be a cure for the problem of the consequences of determinism. (p. 513) Along with a final acceptance of determinism might come a more real possibility of alteration in our dispositions having to do with desires to benefit and desires to distress.

We have so far been considering moral approval and disapproval when they consist in one complex of feelings, a complex which includes images or ideas of origination. The other part of the response of affirmation with our present subject has to do with moral approval and disapproval when they consist in a second and different complex of feelings. This complex is the same as the first in desires and feelings except that it lacks the retributive desires. What we have here is in no way inconsistent with determinism. What we have here rests only on beliefs as to the voluntariness of actions. We are in no way prohibited by a determinism from approving and disapproving of others and ourselves where what is in question is no more than actions which flowed from embraced desires.

What we have is of value to us. We may persist in approving of actions which are truly describable as fair, human, tolerant, or otherwise commendable. We are not prohibited from disapproving of those whose actions flow from unjust, prejudiced, vicious, or otherwise unacceptable desires. In a clear sense a morally splendid act becomes no less morally splendid for not being originated. In a clear sense, having to do with voluntariness, it is an act of self-denial, or kindly perception, or courage. In this sense, likewise, a monstrous act remains monstrous, despite not being originated. We can persist, further, in prevention and encouragement—in general, in desires to affect the future. We can seek to promote non-maleficence, beneficence, fairness, and truth. (p. 440) There is the large fact that the truth of a determinism does not conflict at all with the purpose of morality. In this alone we have a thing of great value. It is perhaps not too much to say that determinism would be a calamity or somehow unthinkable if it conflicted with the very purpose of morality. It does not do that.

What we have, in sum, is that a determinism forces us to give something up in connection with moral responsibility and leaves us with much. It leaves us with certain immediate reactions to the desires and intentions of others. It leaves us with the possibility of acting in accordance with the purpose of morality. It leaves us with the possibility of which much is made, as we have noticed in passing, in the tradition of thought having to do with 'the freedom of acquired self-perfection'. (p. 496) The situation is the same, of course, with respect to what we have not had in mind, moral responsibility and oneself. If I cannot in a certain sense be a moral agent, given a determinism, that carries with it an escape from a kind of guilt and self-torment. I do none the less have the possibility of regarding myself morally. My desires and intentions, in my own view of them, are things which can morally sadden or gladden me. They can be seen by me as obstructing or furthering the purpose of morality.

So much for the response of affirmation with respect to the bearing of a determinism on moral responsibility. Let us now quickly consider the remainder of morality, the connected issues of right actions and good agents. The first is the question of what actions are right, its general answer being a moral principle or a set of moral principles. The second question is that of the nature or standing of the good man or woman, and its general answer will be in terms of certain dispositions. An answer to either question, as we saw, also provides an answer to the other, and in fact the two questions can be expressed in this one way: How ought the world to be, in so far as we can affect it? (p. 423)

Right actions and good men or women, it was maintained, presuppose judgements about responsibility. Only an action for which an agent can be held responsible or credited with responsibility can be right or wrong. This is so since morality has to do with the world in so far as we can affect it, in so far as our actions are within our control. As for the judgement as to standing that someone is, say, morally sensitive, this is a judgement as to a disposition of the person. The disposition is one to perform certain right actions. In the absence of responsibility, there would be no right actions and hence no such dispositions. There would then be no men or women who were good in any moral way.

According to one of our conceptions, a responsible action is one that is both originated and voluntary. The other conception of responsibility is such that a responsible action is one which is no more than voluntary. It is evident that the conceptions of right actions and good men and women cannot be said actually to *depend on* the first conception of responsibility. It is not as if right actions and persons of moral standing cannot exist without origination, and hence do not exist if a determinism is true. Right actions can be taken as being within the class of actions that are controlled in the sense of being voluntary and no more than that. Good men and women can be taken as those who are disposed to perform right actions of this kind of initiation.

It is at least arguable, however, that the permissible conceptions of right action and good men and women—the conceptions consistent with a determinism—have a cast or character different from the conceptions that are related to the moral responsibility tied up with origination. The fact is a counterpart of facts having to do with life-hopes, personal feelings, knowledge, and moral responsibility itself. There is little doubt that we all have a sense of right actions and moral standing which is akin to that which moved Kant to tie morality to a moral self, at bottom to origination.

The response of affirmation, here as before, can involve a consideration of compensations, including an escape from a certain guilt and sense of moral failure. The response of affirmation must rest, however, on perceiving the value of what we have, the remaining possibility of right actions and of good men and women. It is not as if a determinism subtracted fair or just decisions from the world, or the possibility of celebrating morally great persons. A determinism does not consign me to a world from which moral grandeur has been removed. It has in it, to mention one principal category of moral grandeur, those men who have kept in sight their knowledge of the wretchedness of the lives of others, and have given over their lives to political struggles whose goal was to end or reduce that wretchedness.

There are two principles, or rather a possible principle and a principle, both mentioned earlier, that are affected by determinism. One is the idea, of a Kantian kind, that part of what makes an action right is that it was originated. (p. 437) That is, it is a right-making feature of an action, on a par with fair consequences or the like, that the action was originated. This is distinct from the idea, already rejected, that it is a logically necessary condition of an action's being right or wrong that it be a responsible one in the sense of being both originated and voluntary. The principle, if it can count as such, is obscure. It conflicts with a determinism but its loss, it must seem, is tolerable.

The other loss, although perhaps for an unexpected reason, is what can be named the Principle of Desert, if the name does not suggest something too determinate. (p. 437) It is, roughly, that every man is to get what he deserves. We are to act in such a way that this happens. What this comes to, in one part, is that each man is to be morally commended when this is deserved, and blamed when that is deserved. What the principle actually has to do with has long been a matter of mystery and dispute, but it must appear to be bound up with origination. It will be considered in the last section of this inquiry—into the consequences of determinism for fundamental institutions and the like in our societies, notably the institution of punishment.

9.6 THE POSSIBILITY OF AFFIRMATION

The response of affirmation, it was said earlier, is a part of a philosophy of life. By one possible reckoning, or in terms of one sense of things, it can better be described as entering into one large part. It enters into that part of a philosophy of life which has to do with a sector of our lives which is bound up with attitudes to action and its initiation. The response, a response to determinism, is that despite a certain defeat, the defeat of expectations which the history of our culture forces upon us, we can persist, at least satisfactorily, on the basis of one conception of the initiation of actions. We can persist in kinds of hopes, personal feelings and so on, resting on actions taken as no more than voluntary.

The response of affirmation, to repeat, does not in itself constitute all of an attitude to this sector of our lives bound up in the given way with action and its initiation. Two other things required, to fill out this part of a philosophy of life, are one's own appraisal or choice of goods, and a morality. (Cf. Rawls, 1972.) One's choice of a set of goods can be seen as an evaluation of, and selection from, the range of possible objects of life-hopes. As we know, a morality when brought into clarity and order is a principle or set of principles which provides a conception of right action, of the moral worth of persons, and of the grounds of approval and disapproval with respect to particular actions. A morality brings with it, implicitly or explicitly, a politics.

It is one thing to come to see and feel, in general, that voluntariness by itself provides at least an adequate foundation for an outlook on our existence in so far as the initiation of action is concerned. It is another to build on that foundation, It can be built ,upon in diverse ways. I begin, so to speak, with the commitment that voluntariness is in a way sufficient in so far as the initiation of action is concerned, and then add some or other conception of the good to be pursued for myself and one of various possible moralities. Not much constraint is put upon me by the foundation. One constraint that is put upon us has to do with the Principle of Desert, whose consideration has been postponed, and not much else.

A philosophy of life, to mention one other part, cannot be complete without an attitude to the ending of life. A determinism has very little logical consequence with respect to death. There is nothing in it inconsistent with doctrines and aspirations having to do with immortality, and there have of course been religious and theological determinisms. (Adler, 1958) On the other hand, there is a particular consonance between an acceptance of determinism and an acceptance of mortality. To accept the first is surely to be more able to deal with the second. It may be that the tight connection between determinism and nature is more sustaining as a part of one's attitude to death than it is as a part of one's attitude to life. The contemplation of oneself as a part of ongoing nature seems to be of more sustenance in connection with the end of one's life than in the acceptance of adversity within it.

I have not attempted to recommend a completed philosophy of life, but only a constituent of the part most concerned with action and its initiation. It is the acceptance of this constituent of many possible philosophies of life that is recommended in the response of affirmation, and which, it can be believed, will end the history of the problem of the consequences of determinism.

Are there good reasons for resisting the recommendation and doubting the prediction? My own explicit answer, certainly presupposed in what has been said already, can be approached by way of an essay mentioned several time already. (P. F. Strawson, 1962; cf. 1985,

Ch. 2; G. Strawson, 1986, Ch. 5) It is supposed in this essay that Compatibilists or some of them take a determinism to have the consequence, which they may recommend to us, that we must eschew absolutely the personal and moral feelings, and limit ourselves to taking *the objective attitude* to everyone. The objective attitude, as described earlier (p. 446), is a disengagement from ordinary connection with other persons, an attitude of a detached kind. It is the attitude that we now ordinarily have only to individuals who are grossly abnormal, those whom we treat, control, guard ourselves against, detain for their own good, and so on. It is in reaction to this rebarbative proposal assigned to some Compatibilists, it is supposed, that other philosophers have produced the obscure metaphysics of origination, ideas thought to be needed in order to rescue us from the threat of being imprisoned in the objective attitude.

My interest, however, is not in the speculation that Compatibilists have proposed that a determinism has such a consequence, and that it be accepted, but rather something else: something related to the questions of (i) whether it is actually possible that we should ever come to take up the objective attitude in a general way, as a consequence of accepting a determinism, and (ii) whether, if it is possible, it would be rational to do so. The questions suggest related questions. Is it the case that a different proposed alteration in our lives—the alteration that is proposed in the response of *affirmation*—is possible? If it is, would it be rational to make *it*?

It needs to be emphasized that the alteration proposed is *not* one which would limit us to the disengagement or detachment of the objective attitude. There is, admittedly, an analogue to the objective attitude in one part of what has been said of affirmation and the moral attitudes. It was said that several categories of feeling can enter into the approval and disapproval which is consistent with a determinism, and that one of these categories is of desires to direct or influence future behaviour, desires to gain the ends of morality. There is a partial analogue to the objective attitude in these particular desires, but no more than that. (A similar partial analogue was noticed earlier between the objective attitude and the attitude which enters into the response of intransigence in connection with morality. (p. 447)) To be disposed to attempt to influence the behaviour of others in the given way is *not* necessarily to detach them from the human community in the way that we now detach grossly abnormal individuals from that community. The desires in question have *all* men as their objects, and are consistent with accepting that we ourselves are suitable objects of such desires on the part of others. Related remarks can be made in connection with affirmation and the personal feelings.

That is one significant distinction between advocacy of affirmation and advocacy of the objective attitude. There is another. The alteration to our lives proposed in the response of affirmation—an alteration which subtracts ideas of origination and their concomitants—allows for a further category of feeling in connection with morality. We can persist in certain responses to the desires and intentions of others, and hence to them. There is no obstacle to my abhorrence of the desires and intention of the treacherous husband foreseeing his divorce, or, more important, to my abhorrence of him, a man whose personality and character are consistent with these desires and intentions, and support them. There is no obstacle to the related but very different feelings about those who behave well or better. Related remarks pertain to affirmation and the personal feelings.

A third distinction, although one not in view until now, is that affirmation in no way involves taking others as non-rational, as not open or less than fully open to rational considerations. Taking the objective attitude, it seems, does centrally involve regarding others as beyond the reach of ordinary argument and persuasion.

Affirmation, then, is far from recommending a retreat into the objective attitude. It recommends no such bloodlessly managerial an attitude to others. Still, it may certainly be thought to raise the same questions. Is it possible that we should come to live, in so far as moral responsibility and the personal feelings are concerned, in the way prescribed in the response of affirmation? If it is possible that we do live in this way, would it be rational to do so?

It may well be that there can be no real question of our retreating into the objective attitude with respect to all people. It may well be, as is said, that if it is not absolutely inconceivable, it is 'practically inconceivable' that we go over to a general objectivity. It is indeed close to certain that 'the human commitment to participation in ordinary inter-personal relations is . . . too thoroughgoing and deeply rooted for us to take seriously the thought that a general theoretical conviction might so change the world that, in it, there were no longer any such things as inter-personal relationships as we normally understand them. . . .' (P. F. Strawson, 1962, p. 197) It is close to certain that our natural human commitment to ordinary inter-personal attitudes may indeed be 'part of the general framework of human life, not something that can come up for review as particular cases can come up for review within this general framework.' It may be true, further, in connection with rationality, that 'if we could imagine what we cannot have, viz. a choice in this matter, then we could choose rationally only in the light of an assessment of the gains and losses to human life, its enrichment or impoverishment; and the truth or falsity of a general thesis of determinism would not bear on the rationality of *this* choice'. (1962, p. 198; cf. Ayer, 1980; Bennett, 1980; G. Strawson, 1986, Ch. 5)

What is *not* true, to consider affirmation rather than objectivity, is that there is no practical possibility of our making the response of affirmation, and living in accordance with it. To do so, admittedly, will involve an acceptance of the defeat of a large aspiration which gives a character to our moral and personal feelings as they are, and also to our life-hopes and our confidence in our knowledge. It is also to be admitted that no argument or book will produce a general acceptance of affirmation. It would be innocent to suppose that the alteration of our outlook in question could be affected by what, given the rootedness and pervasiveness of that outlook, is so slight a thing. That is not to say that the cumulative and no doubt very indirect effect of philosophical argument and the advance of the relevant sciences, to mention the principal considerations, will forever fail to produce a general acceptance of affirmation, and consistent ways of life. It is not to say, putting aside the matter of a future general acceptance, that the response of affirmation and what follows from it are not within the capabilities of individuals now. While it is a mistake to minimize what is involved in this renunciation, it is equally a mistake to maximize it in certain ways.

The idea of 'the general framework of human life', something not open to review or reconsideration, is certainly of use in thinking of the possibility of affirmation. There are facts about our existence—say the fact of desire itself, the fact that we are desiring creatures—which for whatever kind of reason cannot be supposed missing from anything that could count as human life. Such facts are in an obscure but very strong way essential. By contrast, ideas of origination and the considerable concomitants of these ideas cannot be regarded as such a fact about our existence. Compare the fact that we have and act on ideas of origination with the fact of our being desiring creatures, and the facts in past epochs of pantheistic ideas, or religious hopes and fears, or ideas of an individual's God-given place in the hierarchy of a society and indeed a universe. It seems clear enough that the fact having to do with origination is more to be put with the later three facts and not the fact of desire. It would be at least audacious to look back into the past and by some conceptual stipulation tie the emergence of 'recognizably human life' to that time at which ideas of origination emerged. The enterprise, surely inevitably, would depend on an arbitrary conception of human life.

Ideas of origination and what goes with them are a part of human history, however long a part, as distinct from some sort of condition of that history. It is tempting but it is surely mistaken to take ideas of origination and the structure of life which goes with them, however developed, extensive, and substantial that structure, as some foundation of our existence that cannot be thought away, some primordial *given* which always was, is, and will be. The emergence of ideas of origination and associated facts was a long historical event, an event with an explanation. The persistence of these ideas has also been such an event, with an explanation. It would be strange to suppose that there could not possibly occur conditions issuing in the decline of ideas of origination and their consequences. A general decline of desire, or, perhaps, a change to a universal objectivity of attitude, are things such that it is difficult or impossible to think of causal circumstances for them. This is not the case with ideas of origination and their consequences.

The second question has to do not with possibility but with rationality. Might it be other than rational to make the response of affirmation and to try to live in accordance with it? Might it be that an assessment of the gains and losses to human life, its enrichment or impoverishment, leads us to reject affirmation? Certainly it involves accepting what is reasonably regarded as a loss, a kind of impoverishment. There is succour in ideas of origination, most notably in connection with our life-hopes.

The question is that of the rationality of affirmation as against something else in the situation where determinism has been accepted, but it will be as well to notice, first, the absence of alternatives with respect to such an acceptance. One truth that bears on this matter, a truth which has entered our inquiry before now, is that belief in the fundamental sense is involuntary, that we cannot choose between believing and not believing, decide between thinking of a proposition that it is true and that it is false. We can of course choose to act as if such-and-such were true, or to accept such-and-such as an hypothesis. The fact of the involuntariness of belief in the fundamental sense, it seems, is not much put in doubt by the most relevant phenomenon which can be adduced against it, that of self-deception. The latter is best regarded not as successful lying to oneself, which might be taken to involve choosing to believe, as well as the logical impossibility of believing and not believing, but as choosing to remain in a state of ignorance, typically by taking care not to come to have the experience or evidence that will settle an issue.

No question can then arise of choosing to believe whether such a determinism as the one that has been expounded is true. No question can arise, further, of choosing whether such a determinism is inconsistent with ideas of origination. With respect to these two matters, then, we do not have alternatives, and so no question of rationality of a certain fundamental kind can arise. Such a question of rationality, of choosing the best course in terms of gains and losses, can arise only if there are two possible actions between which one can choose. The rationality in question, differently described, is of course the rationality of efficiency, as it is sometimes called—choosing effective and economical means to an end.

One way in which an issue of rationality as efficiency might be thought to arise, consistently with what has been said, is by way of the possibility of self-deception where that is not doing what is logically impossible, but choosing to remain in a certain ignorance. That is, most importantly, there can be the question of whether it would serve our human lives if we were to take care to avoid that which may settle the question of the truth of a determinism. It may be contentious to describe such a course as one of self-deception. Perhaps, with the aid of certain philosophical assumptions, it could be given a better name. Certainly such a course would offend against typical commitments to truth. There is, however, a more telling consideration. Most of us are such that it is not within our powers deliberately to take such a course. It is not the ignobility of it, but rather the impossibility or great difficulty of it, that is decisive. If the explanation of this fact is hard to come by, as it seems to be, it seems none the less a fact.

To come to the main question, that of what it is or would be rational to do given a full and clear acceptance of determinism, it is of course important to have the right alternatives in mind. The choice we have, given what has been claimed in this inquiry into the consequences of determinism, is between affirmation on the one hand, and, on the other, the discomforts of our situation as described in the first part of our inquiry—the discomforts of being settled in two inconsistent attitudes to the initiation of action, and the two inconsistent responses of dismay and intransigence. The alternative to affirmation is not something contemplated earlier, in connection with morality. (p. 526) We do not have the possibility of attaching to voluntariness by itself the feelings which were previously attached to voluntariness and origination.

Certainly the situation of inconsistency and conflict, although it includes dismay, cannot be regarded as so intolerable as would be a retreat into the objective attitude. But it is far from agreeable. It is this disagreeable situation which is to be weighed up against affirmation, or which will in the future be weighed up against affirmation. The choice concerns not only matters of which we know—life-hopes, personal feelings and so on—but also social matters to which we are coming. (Ch. 10) It will be clear already, however, that affirmation arguably can be taken as the more rational course, the course to be chosen in the light of an assessment of gains and losses for our lives.

9.7 RECAPITULATION

The argument about the consequences of determinism has been a long one, and can usefully be brought into a succinct form. The following paragraphs are derived from sections of the previous two chapters and the present one.

7.2 All our life-hopes involve thoughts to the effect that we somehow initiate our future actions. Some involve not only beliefs as to voluntariness or willingness but also an idea, or what is more an image, of our originating our future actions. To think of life-hopes of this kind, and their manifest inconsistency with determinism, and to accept the likely truth of determinism, is to fall into dismay. We are deprived of the hopes.

7.3 We also have life-hopes involving only beliefs as to voluntariness —that we will act not from reluctant desires and intentions, but from embraced desires and intentions, that we will act in enabling circumstances rather than frustrating ones. These circumstances have to do with at least the way of my world, the absence of self-frustration, independence of others, and absence of bodily constraint. Thinking of hopes of this kind, and noting the clear consistency of a determinism with them, may issue in intransigence. These life-hopes are not at all significantly threatened by determinism.

7.4 We have appreciative and also resentful feelings about others, owed to their actions deriving from good or bad feelings and judgements about us. Both sorts of personal feelings involve assumptions somehow to the effect that others could do otherwise than they do. It is natural in one way of thinking and feeling to take the assumptions to amount to this: others act with knowledge, without internal constraint, in character, and in line with personality, not out of abnormality, not because of constraint by others. This second one of a set of fundamentally like conceptions of voluntary action, wholly consistent with determinism, may lead us to make the response of intransigence with respect to personal feelings. However, we also have other personal feelings, having a certain person-directed character and including an assumption as to a power or control of their actions by others. The assumption is inconsistent with determinism and may lead to dismay.

7.5 We accept that our claims to knowledge derive in part from beliefs and assumptions as to our mental acts and our ordinary actions, by which we come to have evidence and the like. We may take it that originated acts and actions are necessary, and, taking them as ruled out by a determinism, suffer a want of confidence in our beliefs, a dismay having to do with the possibility of a further reality. Inevitably, however, we can have a different kind of confidence, owed only to an assumption as to voluntariness, the possibility of our satisfying our desires for information. Hence intransigence about knowledge. These are facts which the Epicurean tradition of objection to determinism has greatly misconstrued.

7.6 One fundamental question in morality is that of how the world ought to be in so far as we can affect it. However, it allows us to concentrate either on the nature of good men and women, or the nature of right actions. The other fundamental question is that of moral approval and disapproval of agents for particular actions, the responsibility they must have for their actions. An action's being right, and a person's having a good moral standing, presuppose that we do somehow have responsibility for our actions. Hence determinism's effect on all of morality can be considered by way of its effect on moral responsibility.

7.7 What feelings enter into our moral disapproval of the vicious husband and father anticipating his divorce? We may have tendencies to act against him, retributive desires for at least his discomfiture. These desires, by a kind of direct reflection (pp. 434) can be seen to be vulnerable to a determinism. The result may be dismay. However, reflection on the purpose of morality brings into view a kind of moral disapproval, and approval, which rest not on an image of origination but only certain beliefs as to voluntariness. There is no conflict between them and determinism. Intransigence with respect to determinism and morality is as possible and natural as dismay.

8.1 There are two traditional views of the challenge of determinism, Compatibilism and Incompatibilism. Considering them throws into greater definition the fact that each of us has two families of attitudes, including two sorts of life-hopes and so on, and may respond to determinism with at least dismay and intransigence. The two traditional views also demand consideration as the principal alternatives to the correct resolution of the problem of the consequences of determinism.

8.2 Compatibilist philosophers ascribe to us a single conception of the initiation of action, and a kind of belief as to the sufficiency of this initiation in so far as moral approval and disapproval are concerned. The conception is that of a voluntary action, and hence a determinism is taken to affect moral responsibility not at all. Incompatibilists also ascribe to us a single conception of the initiation of action, which includes origination, and a belief as to its role. They take it that the truth of determinism would destroy moral responsibility. Both philosophical parties take the problem of the consequences of determinism to be of an intellectual or theoretical kind, to which can be added that Compatibilists are in a way overwhelmed by the great fact of causation generally, and Incompatibilists are greatly desirous of our having a certain stature, of elevating us.

8.3 Our two families of attitudes, and the two responses, establish the falsehood of both Compatibilism and Incompatibilism. We do not have a single conception of the initiation of action, or a single belief as to the role of such a conception. Our circumstance is not either that a determinism leaves moral approval and disapproval untouched, or that it destroys it. To suppose that it destroys it, as Incompatibilists do, is to ignore our attitudes which may issue in intransigence. To suppose that a determinism leaves moral approval and disapproval untouched, as Compatibilists do, is to ignore our attitudes which issue in dismay. Compatibilism and Incompatibilism are as mistaken in other respects, not least in offering what are very nearly absurd explanations of the persistence of the problem of the consequences of determinism.

9.1 The true problem of the consequences of determinism is to escape the unsatisfactory situation in which we find ourselves, prone to two inconsistent families of attitudes, and two inconsistent responses. It is fundamentally a problem of dealing with desires. In trying to make this escape, we are not restrained by some fundamentality of origination as against voluntariness. Our endeavour must be to accept the defeat of certain desires, by reflecting, in part, on the satisfaction of others. It is an endeavour which enters into arriving at a philosophy of life.

9.2 In so far as origination is a fiction, life-hopes which we have are affected, and the damage cannot be assuaged by the reflections that ideas of origination are faint ones, or that a determinism saves us from chance. There is little solace in the fact that determinism gives us a particular membership in nature. There is more in the escape from failure which it allows. There is also the fact that our life-hopes in a deterministic world are no more bounded in their objects than life-hopes would be in a world of origination. If these hopes also have other recommendations, a final acceptance of our situation will depend on *full belief* in a determinism. We may respond to determinism, nevertheless, in so far as our life-hopes are concerned, with affirmation rather than dismay or intransigence. This includes the endeavour to accept what must be accepted, by several means, and also the recognition that our life-hopes can be life-sustaining things. They can enter into a celebratory philosophy of life.

9.3 A determinism conflicts with personal feelings of the kind that involve an image of origination, and an acceptance of this is included in the response of affirmation. The renunciation, particularly of the appreciative personal feelings of this kind, is made more tolerable by a related escape from the resentful ones. The response of affirmation also includes an assertion of the great value of the personal feelings as they can exist in a deterministic world. To make the response is to keep one's balance, which balance allows for a recognition of the great worth of an existence enriched by facts of personal relationship.

9.4 We can be said to be barred by determinism from knowledge of a possible reality. Thus there is a truth distantly related to propositions of Plato, Spinoza, and others. That is not to say that our lot must be a kind of unhappy agnosticism. Affirmation, as elsewhere, gives a place to both considerations.

9.5 Moral approval and disapproval, since they may rest partly on origination, are affected by a determinism. Our specifically retributive desires are affected. There is more consolation here than with life-hopes, however, and perhaps more than with personal feelings. The moral responsibility untouched by determinism is of a large significance. For one thing, each of us has a moral standing. There are corollaries having to do with right action, and good men and women.

9.6 The response of affirmation enters importantly into a number of possible philosophies of life. It may be asked if it is possible for us really to make the response, since it involves a significant change in our lives. It may be asked if it would be rational to make the change. In fact the change is possible, and the question about whether it would be rational can be answered in the affirmative.

These answers effectively give the main ideas of a resolution of the problem of the consequences of determinism, the problem which has most exercised philosophers. What remains is a consideration, for which we now have some guiding principles, of certain fundamental social and political facts.

10

Punishment, Society, Politics

10.1 INSTITUTIONS, PRACTICES, HABITS

There is unavoidable greater complexity with respect to each of this set of subjects to which we now turn. More differentiations need to be made with each of these subjects than with life-hopes, personal feelings, knowledge, moral responsibility, right actions, and the moral standings of persons.

This is so for three reasons, the first being that it would not be at all effective to proceed in a general way, to think of any of the institutions, practices or habits with which we shall be concerned as one thing, a single type of thing. It would not be at all effective to ask what the consequence of determinism is for this type, since, given our interests, it has importantly different instances or tokens. To speak of punishment, there now exist discriminably different institutions of punishment by the state. There have been others in the past, and perhaps there are more to come. What we have is a range of possible institutions, several of them now actual. To try to settle on a satisfactory response in the matter of determinism and punishment is to do something which involves a choice or choices from this range of possible institutions.

The second reason for complexity has to do with the fact that punishment, like the other institutions, practices and habits, is a matter of action and the result of action, not of attitude taken more or less in itself. It has thus been the subject of moral and other reflection and argument in a way that life-hopes and so on have not. The given range of possible institutions is inevitably subject to moral judgement. We necessarily value one or some, defend it or them, and do not value or defend others. To try to settle on a satisfactory response in the matter of determinism and punishment, then, is to do something which is complicated by moral commitments and arguments. One result is that we may be able to give up something inconsistent with determinism by way of becoming convinced of what we take to be true independently of determinism, that it lacks a moral defence. More important we may have the support, in limiting ourselves to something, that it is by our lights in accordance with fundamental moral principle. Here then there are particular aids, if that is not too small a word, to affirmation.

To be more specific, we can and indeed must conceive of punishment in ways which make for moral disagreement: as a matter of a man's getting what he deserves, as the fulfilment of an agreement, as something whose aim and recommendation is Utilitarian, as an institution having a Utilitarian character but also features having to do with desert, as an institution serving fairness somehow conceived, and so on. That is not to say, of course, that we can conceive of our *actual* institutions of punishment in our present societies in all of these ways, but that we can conceive of different possible institutions, one or several of them identical or more or less identical with one or more of our actual institutions. Each of these possible institutions will be described by and justified by one particular *theory of punishment*, as such things are called.

There is no real counterpart to these facts with, say, life-hopes. Life-hopes, at any rate, have not been in such a way distinguished into kinds, kinds discriminated by means of general moral principles and necessarily also a matter of possible moral disagreement. Personal feelings have not been and could not in such a way be the subject of differentiating moral theories. So with confidence in knowledge. Nor is the situation in fact greatly different with the moral matters we have been considering, although to be sure we might have looked into the aid to affirmation provided by moral conviction. It was possible and effective, certainly, to proceed in the general way we did with moral approval and disapproval, right actions, and moral standing.

Our project, then, first with respect to punishment, must be to consider from a certain point of view—although out of a concern for our present state of affairs, for actual institutions—each of a number of theories and associated possible institutions. What, we need to ask, is determinism's consequence for it? This consequence will be fixed by the fact of what is required by each theory or possible institution in terms of the initiation of action. Above all, does it depend on the assumption that men and women are not only voluntary in their actions but also originate them? Is it the case, with a possible institution of this kind, inconsistent with determinism, that it is anyway morally indefensible, that determinism's inconsistency with it is no loss? Is it the case that a different institution, taken independently of determinism, *is* morally defensible? Is it the case with this institution, or this with another, that it can properly be the subject of, or enter into a response of, affirmation? Where do our actual institutions fall in this range of possibilities? If necessarily we must consider possible institutions, it is our actual institutions which must be our main concern.

So—the enterprise is complicated for the reason that punishment is not effectively conceived as a single type of thing, and the further reason that we are in an area of moral dispute. It is also complicated for a third reason, related to the second. If we have been urged by a long line of moral advocates to conceive of punishment in various ways, and to act on the conceptions, it is far from clear what some of this advocacy comes to. What *is* the nature of the advocated possible and perhaps actual institution in question?

It is far from clear, above all, what it is to conceive of punishment as *deserved*, what it is to conceive of the possible or actual institution of punishment whose nature is fixed wholly or partly by desert. For this to be unclear, of course, is for the moral recommendation of the institution, if any, to be unclear. Talk of desert, using the word itself or a variety of equivalent locutions—the locutions have become more popular than the tainted word itself—has always been pervasive on the part of judges and other officers of punishment by the state. It remains so. That is not to say that the talk is clear. There are similar difficulties about punishment taken as a matter of the fulfilment of agreement, punishment taken as a matter of fairness, and so on.

Our inquiry into determinism's consequences for punishment, then, and for other institutions and the like, must have large parts different from our inquiry into the previous consequences considered. We need to clarify certain lines of moral advocacy, thereby coming to see what possible institution or practice is being recommended, and have some view of its worth. The endeavour is a large one, and cannot be organized and carried forward with as much confidence as those which have been completed. It will involve quickness, generality, and selectiveness in dealing with large subjects which themselves call for, and have often had, close and extended attention.

What is true of punishment, to repeat, is as true of the other six institutions and the like to be given less consideration. Each is in fact best regarded as a set of possible institutions or whatever. Let us quickly survey the six, however, in terms of what are most important to us, their instantiations in our contemporary societies. These instantiations are like our present institution of punishment in raising moral questions. In particular, they seem to have to do significantly with desert.

Some say that punishment has no counterpart institution of *reward*, an institution of the same social fundamentality. That is at least misleading, but not because of the relatively unimportant fact that specific rewards are offered for information in connection with particular criminal offences, relatively few of them, or because of other special inducements to actions of one kind and another. Rather, it is misleading since a society can properly be said *generally* or *systematically* to reward law-abiders, law-abiders in general, for their obedience to law. They are rewarded in that they are left to themselves, in peace, which is to say not subjected to the attentions of the criminal law. Certainly we can and do say that they deserve this, whatever argument or consideration may be conveyed by that, and whatever else may be offered in justification of this reward.

It can be said, thirdly, and often is, that our actual institution or practice of different *incomes* for different jobs, positions, and ranks is justified by desert, or has desert as a part of its justification. A managing director or a doctor deserves more money than a plumber or salesgirl, a more cunning policeman more money than a less. Those who think that the great disparities of income, and the great disparities in things of real value attached to them, are not defensible, or think that the scale is in one part mistaken or in several parts mistaken, may still suppose that some or even many disparities can be defended at least in part by some consideration of desert. Alternatively, advocates and critics of income-distributions may depend on other propositions, perhaps having to do with agreement.

Fourthly, partly for reason of the connection of income with *wealth and property*, limited though the connection is, it is not uncommon partly to defend our distribution of wealth and property by means of such talk of desert. It may also be defended as deserved by the idea that some of those who have wealth and property have used it for the common good, or have been patrons within a community, or simply by the idea that they have protected or preserved it. It is said of those who do not have it, or many or some of them, that they do not deserve to have it, or indeed that they deserve not to have it.

A fifth practice or institution, again in a somewhat extended sense of those words, has to do with positions of *relative power* of several kinds. If we put aside democratic decision-making, in a society as a whole and in greater or lesser parts of it, where we are at least in some distant approximation to being on a level, and also put aside much of private life, it is perhaps reasonable to say that the rest of our existence falls under relations, systems, or hierarchies of power. We have positions in systems of seniority, chains of command, or authority and the like. It is often suggested, in one way or another, that we deserve our positions of relative power. It is also suggested, for example, that they have a Utilitarian justification or something like it.

So too, sixthly, with what can be distinguished from power, if uncertainly, which is *rank, authority, standing, respect*, and the like.

Here there are precedences and offices of various kinds, public awards and honours, and also grades in examinations and prizes in competitions and games. There are also impeachments, dismissals, cashierings, discharges, failings, and disqualifications. These too are said to be deserved.

There remains a seventh item, related to what concerned us earlier in this inquiry, moral disapproval and approval, holding people responsible for things and crediting them with responsibility. Moral approval and disapproval in this sense, as we know, consist in feelings. They do not consist in actions of *commendation or condemnation, praise or blame.* (pp. 379, 421) These latter things, it can seem, are of the stuff of much of our social life, a seventh and last practice or habit to be considered. It includes the actions of praise and blame, often directed to social groups, in which officers of a society and other public persons engage. Moral condemnation by judges is distinct from the punishment they impose. There are also politicians, industrialists, economists, churchmen, leader-writers, and other overt or covert praisers and blamers. There are also the various forms of praise and blame in private or more private life. In all practices and habits of praise and blame, it is common to speak of them as wholly or partly explained and justified by desert.

It is clear then, as philosophers have needed to be reminded (Feinberg, 1970), that desert has to do with a good deal more than punishment—or punishment and reward, and praise and blame. Indeed the list of relevant things might be extended further than we have, to such items as practices of compensation and reparation for injury or loss, and habits of gift-giving and denial as a consequence of such personal feelings as gratitude and resentment. There are also such social provisions as old-age pensions. Let us go no further, however. It is safe to say that much of the nature of our societies is owed to the seven institutions, practices, and the like. Each of them, and hence our societies, are in *some* degree shaped and defended by considerations of desert, whatever those may come to.

To come to a principal if anticipated point, it is at least natural to think that some claims that someone deserves something for an action, which I shall call *desert-claims*, and practices that are affected by such claims, depend on taking the action to have been not only voluntary but also originated. On the other hand, the idea that there is this dependency has often been denied. It has been denied, although in the course of their misunderstanding of the problem of the consequences of determinism, by some Compatibilists, who take morality, including a principle of desert within it, to require only voluntariness. Mill's words quoted earlier might be taken to suggest that desert is independent of origination. (p. 463) Also, and as important, various analyses of desert-claims—including several at which we shall look—seem not to require the assumption of origination.

That it would be rash to draw conclusions quickly about the issue of whether desert-claims and practices into which they enter do depend on ideas of origination is also indicated by other facts. It is not only *we* who are said to deserve things, but also other animals, and also such entities as problems, theories, proposals, and works of art. Some problems deserve close attention, some hypotheses deserve reconsideration, some bills of legislation deserve to be passed, and a portrait or a still life by Monet or Coldstream deserves admiration. Nor is it evident that these latter sentiments can be reduced to claims about the desert of some person or persons. In any case it seems not absurd to say that a piece of natural beauty, perhaps a sunset on the Alberta prairies near Lethbridge, deserves looking at or even contemplation. Given that things other than persons can be deserving, it cannot be certain that when persons are taken as deserving, this is so in virtue of a proposition about them which is threatened by determinism.

If it is the case that some claims of desert do depend on taking actions to be both voluntary and originated, and that these claims do affect or shape our actual institutions, then we must, as with other things of which we know, attempt the response of affirmation. We must escape what will be our situation here as elsewhere, dismay and intransigence. Affirmation here, unlike affirmation with most of our previous subjects, may have somewhat more to do with the seeming necessity or unavoidability of things—above all punishment—than the warmth of our feelings for them. The situation is rather like that with *one* previous subject, the resentful personal feelings, which do not improve our lives but are entrenched in them.

The nature or size of the enterprise of affirmation in connection with the seven social institutions, practices, and the like will depend importantly, as already implied, on what view we have, or come to have, of the worth or defensibility of arguments having to do with desert, taken independently of determinism. The nature and size of the enterprise of affirmation will also depend as much on something else, the *extent* to which the seven things are affected by desert-claims. Certainly they are *significantly* affected, but that is not to say a great deal. It *may* be that they are greatly more affected by other feelings, claims, principles, or whatever. To speak differently, are we or are we not to draw a general conclusion to the effect that the truth of determinism, given its relation to desert, has great consequences for our societies as they are, that great changes are called for? That has been declared by philosophers opposed to or sceptical about determinism. They have sometimes supposed, as it seems, that our societies as we know them could barely survive a general acceptance of determinism.

The principal subject-matter of this chapter is in fact determinism and punishment. More particularly, as explained, it is determinism's effect on theories of punishment and their associated possible institutions, above all the theories associated with our actual institutions. Our conclusions about determinism and punishment, when we have them, will be a guide to what will get very much less attention, the other mentioned institutions and the like. If punishment is of great importance itself, and the subject of ongoing controversy, we can by considering it also come to have much of the means of dealing with the other institutions and the like, and also some of the means to a final speculation, about determinism and politics.

Desert-claims in themselves and the problem of their analysis, as already indicated, cannot be near to all of our subject. It is evident, to repeat, that desert-claims do somehow affect our present social institutions, that some institutional facts do derive from them. How large these facts are will depend on the extent to which our existing institutions derive from quite different claims or propositions. We must ask to what extent they do. There is also another question about these different claims—separate from the question of their relative effect. It is the question of whether any of these claims not having to do with desert are none the less affected by determinism. One category of them in particular, already noticed in passing, is both like and unlike desert-claims, and needs to be thought about in this connection.

They can be called *agreement-claims*, and are to the effect that someone has agreed to do or to accept a thing. They can be offered as reasons for his doing or accepting the thing. Agreement-claims are like desert-claims in that they enter into a defence or justification of something which can seem to rest wholly upon a past action, in the case of agreement-claims the past action of making an agreement, giving consent, or the like. Agreement-claims are unlike desert-claims, to mention one way, in that they do not essentially involve any moral or evaluative appraisal of the past action—that of agreeing. Whether I think you ought or ought not to have signed the lease making me the tenant of your cottage, and indeed whether or not I have any moral view at all about it, I may think that the fact that you did sign it is of importance. You ought to give me the keys. The distinction between desert-claims and agreement-claims issues in the linguistic fact that I cannot naturally say you deserve to give me the keys because you signed the lease, let alone deserve to do so *for* signing the lease. I may allow, without prejudice to my argument, that there is no way in which you do deserve to give me the keys.

We shall at first follow a certain procedure in dealing with what, as will now be evident, is a large and not easily manageable subject. Our sequence of questions, to be followed with a number of propositions about punishment, will be roughly as follows.

(1) How are we to understand or analyse a given agreement-claim or desert-claim about punishment, and what reason or argument for the institution does it in fact involve?

(2) Is the latter really a significant or prima facie reason or argument for punishment?

(3) If so, is it affected by a determinism? That is, does it involve origination?

(4) If the reason is affected by a determinism, is it none the less the case that other justificatory propositions untouched by a determinism are of greater importance in logically sustaining the proposed institution of punishment, perhaps our actual one?

The latter questions in this sequence arise only if certain answers are given to those before. It is mainly at the end of each sequence—if we reach it—that we shall consider the response of affirmation.

10.2 PUNISHMENT AND AGREEMENT-CLAIMS

There is a certain advantage in beginning not with desert but with the considerable tradition to the effect that all offenders have somehow agreed to their punishment, or even that they want or choose it for themselves. The penalties which we now impose on offenders in our actual systems of punishment, it is said, can be justified, partly or wholly, by certain defensible agreement-claims. Let us look at two strong and recent examples of the tradition. Certainly they have recommendations lacked by their predecessors.

The first begins from the nature of consent in ordinary agreements or contracts in civil law, perhaps the agreement one makes by getting into a taxi-cab and giving an address to the driver. (Nino, 1983) That example illustrates that an act of consent need not be in a certain sense explicit. No taxi-driver expects to hear from the back seat 'I consent to pay the specified fare if you will take me to Ritson Road.' Indeed we can with a little trouble imagine a taxi case, in line with the case of a nod at an auction, or handing the ice-cream man a coin and pointing to the chocolate, where no words at all are spoken. What is required for consent in law, so long as two other things are true, is a somehow free action *of any type*. In particular, it need not be an action of speaking or writing. One of the other two things required, if it is proper to separate it from the requirement that the action be somehow free, is that the agent in question knows of consequences of a certain kind of his action—for example, that the taxi-driver will have a legal right to the fare. The other, it is said, is that the relevant law is somehow morally defensible.

It is important that for the passenger to have consented, certain other things are *not* required. It does not matter if he believes he can avoid paying, intends not to pay, or does not want to pay, either at the time he consents or when he gets to his destination. None of these things affects the fact that he has in law consented. Nor, it is said, does any of these things affect what importantly follows from his consent, more particularly from his consent to undertake a legal obligation, which is that others acquire a moral right to enforce the obligation he has acquired.

Now consider an offender. His very action of breaking the law—pulling the trigger, or diverting the funds, or cheating the client—can be regarded as his giving a certain consent. This has nothing to do with any prior social contract which some philosophers deem him to have made, simply as a member of the society. His offence *is* his consent. What he consents to is said to be the giving up of his legal immunity to punishment. This is an immunity which we all possess if we do not offend, a matter of our individual legal rights.

His consent, to be more exact, is said to consist in his acting with the knowledge that loss of his legal immunity to punishment is a legal consequence of his action. He acts with this knowledge, which is to say nothing of his moral culpability or moral blameworthiness. The *Consensual Theory*, to give it a name, has nothing to do with moral culpability or blameworthiness. (Nino, 1983, p. 293) The general idea of consent of this kind is only that one is said to consent to all of what one knows to be the somehow necessary consequences of one's action. Thus the offender is not said to consent to his actual punishment, which itself is not a necessary consequence of his action. It may not happen.

As in the case of an ordinary legal agreement, it does not matter if the offender believes he will not be punished, intends to avoid it to the best of his ability, and wants not to be punished. He has nevertheless consented to give up his legal immunity to punishment, and it follows from this that we acquire a moral right to enforce the agreement he has made. We need not exercise our right, as with any right, but we may also choose to do so, for the reason that we suppose his punishment will have the effect of preventing future offences, perhaps by him.

What justifies us in making use of him to this end, in short, is not that he deserves it, but that he has in the given way consented. It is part of the view we are considering, as it is of others, that the preventive reason by itself would not justify punishment.

The Consensual Theory has the remarkable recommendation, not had by other theories in its tradition, of struggling to give a specific account of agreement-claims with respect to our actual institution of punishment and other possible ones. Does it, however, to pass on the the second question to be asked, give us a significant or prima facie-reason for punishment?

There are several grounds for saying no. One has to do with what we need to keep clearly in mind, which is that *no* claim has been made that the offender consents to his *punishment*. One reason the Consensual Theory does not claim this, foreshadowed above, is that if it did, it could provide no reason for punishment whatever where offenders *believe*, as some certainly do, that they will not be punished. A man cannot conceivably be said to consent in the given way to something he does not believe will happen or be the case. Thus the essential claim of the theory must indeed be the weaker one, that he consents only to the loss of his immunity to punishment. Despite what has been said so far, what exactly does this come to? What are the details of this consent? That is uncertain, partly because his loss of immunity is repeatedly said to be a 'normative' consequence or a 'legal normative consequence'. (Nino, 1983, pp. 296–7) Perhaps his conpenting to lose his immunity comes to something so strong as this, which is favourable to the Consensual Theory, that he acts with the knowledge that the law is such that if he is apprehended, and if the authorities make no mistakes, he will not be regarded as having an immunity to punishment.

We might reflect a good deal on this matter but we need not. Evidently whatever he consents to is wholly consistent with his precisely *not* consenting, in any sense whatever, to *his punishment*. He will not foresee *it* as a necessary consequence of his offence. Further he will not consent to it in another more ordinary and fundamental sense, difficult to specify clearly. Here, very roughly, to consent to a thing is not merely to see it as a necessary consequence of one's act of consent. Here, to consent is to be more willing and decisive, or at any rate more desirous or happy, with respect to the very thing itself. I consent to pay in this ordinary sense, as well as the other, when I point to the chocolate ice-cream. In this ordinary sense, to repeat, the offender does not consent to his punishment. On the contrary, he is against it, attempts to escape it, and so on. With respect to his punishment, he in both senses and indeed every sense *dissents*.

The general idea of the Consensual Theory, and all theories of punishment having to do with agreement-claims, is that in punishing a man we have the defence that somehow we are acting in accordance with *his* past or present will, volition, autonomy, self-determination, end, or desires. In the Consensual Theory, it is said in particular that we are treating him as an end. This is understood with commendable clarity as our acting *in accordance with his ends*. But if we take into account both his consent of a kind to lose his immunity, whatever the details of that consent, and his dissent in every sense with respect to his punishment, it is at least arguable that in punishing him we are acting against what are his ends. We are certainly acting against his predominant ends. We can be said, as well, to be acting against his predominant will, volition, self-determination, desires, or whatever.

It is to take a partial view of the situation, to say the least, to focus upon his consent in a secondary sense to what is less significant to him, and to overlook his dissent in every sense from what is greatly more significant to him. He consents in a secondary sense to some necessary condition of his punishment, and dissents in every sense from his punishment itself. It is false, or as near to it as makes no difference, to say that punishment among other things 'is the product of the will of the person who suffers it'. (Nino, 1884, p. 297)

Should we suppose that the consent which consists in the offender's acting with knowledge of a certain consequence is none the less somehow decisive because, after all, it is the kind of consent which is entrenched in the ordinary law of contract? Should we suppose, for this reason, that the objection that has just been made must somehow be ill-judged? One might well resist the assumption that the kind of consent which is decisive in contracts is also decisive with punishment. There is another consideration, however. It bears not only on the question of whether an offender's consent to lose his immunity is somehow decisive, but also on the prior and basic idea of the Consensual Theory, that there is a close analogy between that consent which is given in an offence and, on the other hand, ordinary consent in the law of contract.

What we have with respect to punishment is an action done with knowledge of some legal consequence, but also an action done, at least typically, in the perfectly apparent desire, intention, and determination to avoid punishment. We are given every evidence of this. What is the proper analogue of this action in terms of the law of contract? It is too quickly assumed that the proper analogue is a standard or ordinary contract. On reflection, it must instead be a curious case which has the feature among others that someone successfully conveys from the beginning that he has no intention of doing or accepting a certain thing.

Consider the remarkable case where I get into the taxi cab, give an address, and then beyond doubt really do persuade the driver that I intend not to pay. For whatever reason, none the less, he takes me where I want to go. Have I consented to pay in the relevant sense, and do I have a legal obligation to pay when we get there? The answer is surely no. Or, certainly, it cannot be that the answer is a simple yes. If *this* curious case is the analogue to the offender's consent, then the offender's consent is *not* the kind which enters into ordinary or standard contracts.

More generally, what we are given in the Consensual Theory is a certain account of what it is to consent in connection with an ordinary contract. This does not include anything about intending to keep to the contract. That is held not to be necessary—in anticipation of what is going to be said about consent in punishment. But what if we find an accurate analogue of the offence? What if it is such that the consent-condition is satisfied *and*, in the judgement of all parties, the person in question successfully convinced the other person that he intended not to keep the contract? It may be that there can be no such case, that there is incoherence in the speculation. But surely *any* tolerable response to the question must put into question the line of argument that an offender's action is somehow decisive consent because it has an analogue of consent within the law of contract. There is room for more argument (Nino, 1987), but it remains difficult to accept the idea of the Consensual Theory that there is an effective analogy between punishment and ordinary contracts.

There is a further ground for saying the Consensual Theory does not give a prima-facie or significant moral reason for punishment. Suppose a man sees the possibility of an action profitable to him but greatly harmful to others, one which will ruin their lives. It is by any moral test an action of viciousness. He sees too what is also true, that it is not illegal, although it is indubitably of the moral gravity of illegal acts. How it has happened that it is not illegal—there are various possibilities—does not need detailing. The Consensual Theory of punishment does not apply here at all. In performing the awful act, he cannot be said to have consented to lose his legal immunity to punishment. That is precisely not a foreseeable consequence. Here, then, the theory gives no reason whatever for punishment.

This is unfortunate for the theory, for a particular reason. The theory assumes, as all such theories do, that there is *some* moral reason for punishment in an offender's past action, taken by itself. The assumption has been hard to resist. The theory, so to speak, recommends its reason for punishment partly on the ground that there is wide agreement that there does exist a reason of the given sort. However, *its* reason is *not* of the given sort. This is so since anyone at all inclined to think there is such a reason will take it that it somehow *does apply* to the imagined case. But the analysis in terms of consent has the consequence that in this case there is *no* backward-looking reason for punishment. There might be attempted the reply, among others, that the imagined case is not one of an *offender*, and hence not one where *punishment* is in question. To that kind of reply there is a rejoinder. (Honderich, 1976, pp. 62 f.)

Let us not pursue the Consensual Theory further, however. If the objections to it have not been fully developed, and certain complexities and difficulties glossed over, it is difficult to avoid the conclusion that although the theory gives a novel account of agreement-claims in connection with punishment, it can hardly be taken as providing a significant or prima-facie reason or argument for it. It purports to apply to our actual system of punishment, but it cannot be said to give us a morally defensible institution.

Is the reason for punishment offered by the Consensual Theory consistent with determinism? Does the reason necessarily bring in origination? That third question is apposite since the reason requires, as remarked, a somehow *free* action, a free action of consenting. (p. 548) We shall not consider the question, or the further one, to which an answer has been implied, of the relative weight of other propositions pertaining to the defence and nature of the institution. Thus we do not come to the matter of the response of affirmation. That matter does not arise. It is not the case, for example, that we here have a kind of justification on which we might be able to depend if we are forced by determinism to give up another kind.

To turn now to a second and large view of agreement-claims in connection with punishment, it is one clearly suggested but not developed in an impressive and greatly successful political philosophy, a theory of justice. (Rawls, 1972) This second view of agreement-claims is best approached by way of something it includes, the Hypothetical Contract Argument about justice. If we take time to get both things clear, it will not take much more to come to a judgement on what is important to us.

The Hypothetical Contract Argument is in limited analogy with a certain simple method of reflection, whatever its nature and worth. That is the method whereby in an attempt to solve a moral or other problem, I ask what view would be taken of it by a person or persons with certain attributes. Suppose my problem is about divorce. I may suspect that my personal history prejudices or clouds my judgement. It will be useful, I suppose, to ask what view would be taken by a different person of another history. He or she—the less prejudiced person—does not need to be an existing person, but can be a matter of my own imagining. The less prejudiced person, real or imagined, is of course not identical with me.

Suppose the less prejudiced person in fact or in my reflections takes the view that the proper ending for marriages of a certain kind is an equal division of property between husband and wife. This may now seem right to me. Notice, however, whatever is to be said of that opinion, that there obviously is no possibility of a certain line of thought. Partly since the less prejudiced person is not me, there is no possibility of a certain *agreement-claim* pertaining to me. I cannot later be said to have *agreed* to a 50/50 division, or *consented* to it, or made any kind of *contract*, in virtue of the fact that the less prejudiced person took that view, or was imagined as taking that view. None of this is the case although it is irrelevantly true that I can be said to *agree with* the less prejudiced person.

So much for the analogy. The Hypothetical Contract Argument, in sum, is that imaginable persons or persons in an imaginable situation would make a certain contract or agreement as to the society they are in process of establishing. It is a hypothetical contract in the sense that it is not one that ever was or ever will be made, in actuality, but one that would be made if certain imaginable conditions obtained.

One of the conditions is that each of the contractors in what is called the Original Position is equally free to propose principles for the governance of the society to come. A second condition is that each is rational. That is to say that he will choose principles which are effective means to securing his end, which is the possession of certain goods, and that he is not envious—he will not worsen his own position, in absolute terms, in order to reduce the difference between himself and someone better off. Thirdly, each contractor is self-interested, concerned for himself. Finally and importantly, the contractors have a good deal of *general* knowledge or belief about people and societies, but each is entirely in ignorance as to his—or her—own individual future in the society to come. Each is in ignorance as to his or her own future position and personal characteristics. This ignorance is complete, and so extends to each person's future wealth, power, standing, sex, race, physical well-being, intelligence, desires, interests, even moral beliefs. The imagined situation, then, is one which absolutely precludes any contractor's being influenced by the contingencies of personal good fortune and natural advantage which in ordinary life guide or distort an individual's choices as to principles and policies.

Taking into account all the enumerated conditions, the situation is a *fair* situation for making an agreement. The principles of distribution that are agreed upon will be owed to a fair agreement: an agreement made under conditions of fairness, a recommendatory situation. This is what we want. True principles of justice are those that would issue from an agreement arrived at in such a situation of fairness. We see justice rightly when, for what a much-used abbreviation is worth, we see *justice as fairness*.

According to the Hypothetical Contract Argument, we can reason that the imagined persons would in fact reject certain alternatives and agree on what are perhaps best described as three fundamental principles of justice to govern the institutions of their coming society. The principles have to do with (i) certain traditional liberties, (ii) equal opportunity, and (iii) socio-economic inequalities.

The argument as a whole is of course directed to us. It seeks to persuade us that given principles are correct for our actual societies. Despite the imagined agreement, and as with the less prejudiced person and the question of divorce, the argument as we have it has nothing to do with agreement-claims bearing on us. Certainly we can speculate, about our imagined contractors, that *they* will be subject to certain agreement-claims in their coming society which we imagine. That is nothing to the present point. We may or may not be persuaded to act in accordance with the three principles, to support or struggle for actual social institutions in our actual societies which are in line with the three principles. That is not to say that any of *us* is subject to any agreement-claim having to do with the Original Position. None of us is or was a person in the Original Position.

A little attention is given, in this theory of justice, to certain other arguments analagous to the one just sketched. That is, we can imagine our contractors, having settled in the first stage on the fundamental principles for their society, moving into a second, third, and fourth agreement-making positions or situations. (Rawls, 1972, pp. 195–201) The second is called a constitutional convention, the third the legislative stage, and the fourth a stage in which rules are applied to particular cases by judges and administrators. The four-stage sequence involves, among other things, an increase in individual knowledge—knowledge of the attributes and positions of particular persons. In the fourth stage it is complete. In each of the three later stages of this process there arises the matter of the institution of punishment. Indeed the institution is literally a part of the fourth stage.

Clearly we are to suppose that our imagined persons decide on an institution of punishment, although little is said of it. Further, given their situations, that decision may be argued to have a certain recommendation. Just as the moral view about divorce of the less prejudiced person can be taken to be an argument for my conduct, and the choice of the original contractors an argument in favour of certain fundamental principles for our societies, so their choices in subsequent stages can be taken as arguments for our guidance with respect to particular actual institutions, including punishment.

What is important now is only the same point as before. We can, to repeat, imagine certain persons, or persons in certain situations, who would opt for an institution of punishment. That they would do so, given their imagined attributes or situation, may lead us to think that the institution has a certain justification. What is also true is that *none of us*, by engaging in just this method of inquiry or argument, becomes subject to any agreement-claim whatever.

So much for the Hypothetical Contract Argument in general and specifically with respect to punishment. It is not in itself our present concern, but I have given time to it partly since it is essential to distinguish it, and whatever strengths or virtues it may have, from a larger and yet more speculative argument into which it enters, and with which it is run together. The larger argument, which *does* have to do with a supposed agreement by us, is referred to as 'the Kantian interpretation of justice as fairness' or 'the Kantian interpretation of the Original Position'. (Rawls, 1972, pp. 251–7) It might better be named the Kantian use of the Hypothetical Contract Argument, or, as I shall name it, for reasons which will be clear in a moment, *True-Nature Consensualism*.

It is proposed or implied in the paragraphs below. In them, as needs to be noted, the persons referred to are not the imagined contractors but we ourselves, at one point we ourselves imagined as being in the Original Position. The reasoning referred to in the third paragraph is the reasoning that the persons in the Original Position would choose the three principles of justice having to do with liberty, equal opportunity, and socio-economic inequalities.

. . . there is a Kantian interpretation of the [Hypothetical Contract] conception of justice . . . . This interpretation is based upon Kant's notion of autonomy.

. . . he begins with the idea that moral principles are the object of rational choice. They define the moral law that men can rationally will to govern their conduct in an ethical commonwealth. Moral philosophy becomes the study of the conception and outcome of a suitably defined rational decision. . . .

. . . Assuming, then, that the reasoning in favour of principles of justice is correct, we can say that when persons act on these principles they are acting in accordance with principles they would choose as rational and independent persons in an original position of equality. The principles of their actions do not depend upon social or natural contingencies, nor do they reflect the bias of the particulars of their plan of life or the aspirations that motivate them. By acting from these principles persons express their nature as free and equal rational beings subject to the general conditions of human life. For to express one's nature as a being of a particular kind is to act on the principles that would be chosen if this nature were to be the decisive determining element. (Rawls, 1972, pp. 252–3)

If this is not wholly clear, and perhaps gives less than the most effective statement possible of True-Nature Consensualism, as to my mind is the case, an explicit argument can certainly be derived from it. It is, to repeat, a larger argument which includes the Hypothetical Contract Argument within it.

(i) The fundamental premiss is that each of us can be and should be autonomous. That is, we should be free makers of rational choices. This is our shared *true nature*. It is not to be confused with our merely individual properties, either our personal attributes or our inclinations owed to our positions in society, both of which are contingencies.

(ii) It is this true nature which is the only source of true morality, including correct principles of justice and justified institutions, including punishment.

(iii) If we wish to distinguish these principles and institutions, we can do so first by fixing on our true nature, and then noting its deliverances. More particularly, we can imagine certain persons, those in the Original Position and subsequent positions. In the Original Position and to some degree subsequently, none is distracted by his or her particularity. At least in the Original Position the contractors are, so to speak, true abstractions of ourselves, imagined beings who have only our true natures, or rather are forced to proceed as if so.

(iv) Given that they would agree on certain principles of justice and in particular on an institution of punishment, what attitude are we to take to the principles and the institution? We must conclude *more* than that the principles and institutions are supported by the Hypothetical Contract Argument in itself. Given, so to speak, that we in our true natures are identical with the contractors, we must conclude that punishment, to speak of it in particular, is something to which *we* can be said to have assented and to which we do now assent. If we are also offenders, and so resist it in our own cases, we nonetheless in our true natures or autonomy consent to it. *The offender agrees to his punishment.* In punishing him, we do his own rational bidding. In the words of another proponent of True-Nature Consensualism, we *respect* him. 'Respecting a man's autonomy, at least on one view, is not respecting what he now happens, however uncritically, to desire; rather it is respecting what he desires (or would desire) as a rational man.' (Murphy, 1973, p. 229)

It is possible, despite the great attention paid to the Hypothetical

Contract Argument by political philosophers, to remain unimpressed by it (Honderich, 1975), and hence by this larger sequence of argument, True-Nature Consensualism, which contains it. Many objections do not pertain directly to our present concern, which is the given doctrine about agreement-claims advanced in justification of punishment. Let us here try to consider the larger argument.

Its first premiss is that we have a true nature, as free makers of rational choices, and the second is that this is the only source of true morality, and in particular any correct moral view about punishment. The initial and indecisive description of our true nature—we are free makers of rational choices—is first given some content by contrasting that nature with our individual properties, our personal attributes, and our inclinations owed to social position. Our true nature is further specified, in the third premiss, as that part of ourselves which would issue in moral choices if we were in total ignorance of our individual properties in the given sense. That is not and cannot be the end of the specification. Our true natures, since they are brought into view by what is said of the imagined contractors, *also* involve our having certain general knowledge, or general belief, about persons and society. Such knowledge or belief is obviously wholly necessary to what follows.

On the assumption that our true natures, so called, have been adequately specified for the purposes of the argument, together with the assumption that our true natures produce true morality, we can now ask in particular what their upshot is in connection with punishment. To put the question differently, as we are advised, we can ask what the imagined persons would decide. It can indeed be supposed that if they would in fact decide on the three principles of justice, this decision itself implies support for an institution of state punishment. It can be supposed, further, on the same assumption as to their initial decision about principles, that they would explicitly establish such an institution in their subsequent constitutional and legislative stages.

Indeed, as in effect is allowed by the proponent of the view we are considering, it can be *guaranteed* that the contractors would decide on the three principles and on punishment. We can do this by adjusting our conception of them, in fact by adding to it. To be plain. we can be more specific about the general beliefs which we have given them. We can add one to the effect that the lack of an institution of punishment in almost any society would be disastrous. That it can be a guaranteed conclusion that imagined contractors would opt for punishment is an instance of a wider possibility. It is in fact possible to produce a proposition, about any moral or political view whatever, from Apartheid to Zen Buddhism, to the effect that a certain imaginable assembly would agree on it. It is one of the fundamental contentions we are considering, of course, that only the Original Position is an imagined assembly that should lead us to accept what it decides on.

We are here in the neighbourhood of the general objections to the Hypothetical Contract Argument mentioned above, not pertaining precisely to our present concern. There are reasons for doubting the worth or the relative worth of all arguments based on hypothetical contracts. However, let us stick to the matter of agreement-claims about actual persons and punishment. If we do so, we can in fact also ignore something else, the mysterious second premiss about right morality. That I have agreed to something does not depend on, and is independent of, the thing's actually being morally right. True-Nature Consensualism can in fact be reduced to this argument:

(a) Imaginable persons would opt for or approve of an institution of state punishment bearing on themselves.

(b) Those persons are in a sense identical with our shared true natures.

(c) Hence all of us in our true natures opt for the institution of punishment.

(d) The offenders among us therefore agree to their punishments.

Two questions arise, and the answer to each of them, to my mind, defeats the argument. Do all *offenders* have the given true nature? That is, in part, do all of them have the necessary general beliefs—do they have just the general beliefs that must be assigned to the hypothetical contractors? That is the first question.

No reason is given for thinking its answer is yes, and in fact the answer must be no. Admittedly it is arguable that most or all offenders are inclined to have beliefs about *some* conceivable society, with punishment in it, such that they would opt for punishment in that society. In that society it would without exception be essential. Consistent with this is the fact that almost all offenders, like very many non-offenders, are inclined to have beliefs about their own actual society, and the punishments in it, such that they would *not* opt for all the punishments that are imposed. Nothing is more common on the part of offenders than general beliefs about their societies to the effect that some members are allowed by law to have agreeably large shares of what is going, while others, including the offenders themselves, must break the law to have the hope of anything like such a share. They precisely lack a belief which they must have if their true natures, so called, are as True-Nature Consensualism requires. Since at least many offenders lack beliefs which they would need to have in order for the given agreement-claim to be put upon them, True-Nature Consensualism fails.

It fails, I think, even if that objection is waived. The second question about the doctrine can falsely presuppose that all offenders *do* have precisely the true nature which the doctrine assigns to them. They have exactly the properties, including the general beliefs, which can be assigned to the imagined contractors so as to guarantee their support for punishment by the state. On the assumption that all offenders have the specified true nature, does it follow that they consent to their punishments? Do they agree to them?

It is impossible to say so. My agreements with other people are not agreements between *certain of my properties* and other people. They are not agreements between certain of my beliefs, commitments, inclinations, etc., and other people. No doubt I may be subjected to a good deal of argument and pressure to the effect that *really* I do accept something, perhaps to make a donation to the National Union of Mineworkers, or that in my true self I do agree with doing so. This may be true, and moreover I can agree that it is true. That is, it may be true that certain grounds which I accept, or certain feelings which I have, or certain other facts about me, should incline me to make the donation. They may indeed be grounds, feelings, or facts that are somehow central or fundamental to the person I am. I can agree with all this. It does not follow that I have agreed to, or do agree to, give the donation. The situation, rather, is that there are others of my beliefs and feelings, or other facts about me, that tell against it. Hence, I do not agree or consent, whether or not that can be described as going against my better judgement. I do not perform an act of agreeing, in words or on paper, or in any other way. I do not do a thing of which I know certain necessary consequences. Nor do I agree or consent in the other more ordinary and fundamental sense mentioned in connection with the the theory considered earlier, the Consensual Theory. (p. 550) That is, it is not true that I consent in the sense, difficult to specify briefly, where I am more willing and decisive and also more desirous with respect to precisely the thing consented to.

True-Nature Consensualism is in fact a large elaboration of an ordinary response to an offender. It is that if he were not facing a penalty himself, he would agree that punishment is necessary, justified or whatever. That may well be true, and it is conceivable that something of interest can be discovered to follow from it. What is not true is that he has agreed or does agree to his penalty.

It was anticipated earlier (p. 548) that a sequence of questions arises about agreement-claims and desert-claims with respect to punishment. The very first question was that of the understanding of such claims, and the reason or argument for the institution which they involve. True-Nature Consensualism does not establish that our actual or indeed any possible institution of punishment does involve agreement-claims, or that these claims give rise to a significant or prima-facie reason for punishment. Therefore we need not go forward with questions about the effect of determinism on such a reason and the relative weight of such a reason. We shall not here consider the response of affirmation with punishment.

Our general conclusion is that neither the Consensual Theory nor True-Nature Consensualism establishes that agreement-claims are important with respect to punishment. Many related theories about offenders' agreeing to their punishments are as unpersuasive. This is so, certainly, of the classical theory of this kind, owed to Hegel (1942 (1840)) and discussed elsewhere. (Honderich, 1976, pp. 45–8) If agreement-claims are not of importance in the justification and nature of punishment, however, they may be so with other social institutions, practices and habits. In fact they are. We shall consider the question in due course, but let us now turn to desert-claims in connection with punishment.

10.3 PUNISHMENT AND DESERT-CLAIMS—TRADITIONAL VIEWS

Desert-claims are to the effect that someone deserves something for something else. In connection with punishment, they are to the effect that someone deserves a particular penalty, or something bound up with a particular penalty, for a particular offence, or something involved in a particular offence. As in the case of all desert-claims, at least as standardly made, these are somehow to the effect that there exists a certain *relation,* which relation serves as or enters into a reason or justification for something, the thing said to be deserved. We shall take sufficient time—which is no little time—to understand what in general is involved in desert-claims before looking at the particular theories of punishment in which they issue or to which they contribute.

What, more clearly, is the thing for which an offender deserves something? The still inexplicit but correct answer must be a culpable action, which, by way of initial description, is an action somehow open to moral disapproval. In the ordinary course of things, this will also be an illegal action, but it seems plain enough, despite a common view to the contrary, that it is not its character as illegal which is essential to its being deserving of something. I may of course suppose that many penalties are justified for reason of being deserved for actions that are among other things illegal. But I shall also refuse to accept that all penalties visited on persons in connection with illegal actions have been deserved, or that certain conceivable actions, if made illegal, would deserve any penalty whatever. I shall refuse to accept this of actions for which I am convinced there is moral justification or no need of moral justification.

That is not to claim that we suppose all actions rightly punished are morally culpable actions, or that all actions rightly punished deserve to be punished. There are offences prohibited under statutes of strict liability, and hence, in effect, actions defined specifically without reference to intention in the ordinary way or to any negligence or failing. (Hall, 1960; G. Williams, 1961) They can be, despite suspicions about many offenders, actions whose intentions and degree of care do not in any way call for moral disapproval. Strict-liability offences include the unintentional selling of bad food under certain circumstances, certain motoring offences, some actions in contempt of court, the possession of drugs under certain circumstances, certain financial transactions, and a great many other actions. A large part of the given rationale of strict liability is that it is difficult or impossible to prove intention or negligence with respect to certain actions, and yet essential to secure the prevention of more of them, to reduce their incidence. It is commonly allowed in the law that at least some penalties under strict-liability statutes, however defensible the penalties, are not deserved.

A culpable action, to repeat, can be described as one for which we feel moral disapproval, hold the agent responsible. Somewhat better, a culpable action is one for which we feel moral disapproval and which we take to be sufficiently grave that it ought not merely to be blamed or condemned but ought to be illegal and therefore open to punishment. It may well be that our moral disapproval is in part or importantly owed to the belief that the agent knew the action to be illegal, but that is not to say that our relevant view of the action, in so far as desert is concerned, is that it was illegal. Nor is the fact that it is the culpability which is important to desert, and nothing but the culpability, affected by the fact, if it is one, that we take the act to be culpable partly because illegal.

To recall what was said earlier (7.7) of moral disapproval, it involves repugnance for an agent's desires and intention, repugnance typically owed to our awareness of the consequences of his action, of which we also take him also to be aware. We may also have desires for his distress in some degree, at least for his discomfiture. These may be at least as fundamental to our state of feeling. We may also have desires pertaining to the future and prevention. We want to prevent like actions. These and other elements of moral disapproval are raised in us by two properties of the action, one of them being its wrongfulness. The other is the agent's initiation of it or, as we can say, his responsibility for it. This is the fact that it was a matter of his voluntariness, or his voluntariness and origination. A culpable action in itself, then, is one of wrongfulness and responsibility, the wrongfulness such as to call for legal prohibition of the action. This familiar truth can be usefully elaborated. (Fletcher, 1978; Nozick, 1981)

What is the other term of the relation involved in desert-claims about punishment? The standard answer is a penalty, but the answer requires clarification. Does a rapist get what he deserves in the barely conceivable case where his gaol sentence is what he wants above all, something which causes him no distress? It may be correct, it seems, no doubt depending on *why* he wants it, to say no. Does he get a penalty? It is certainly possible to say yes. What emerges from this, evidently, is that if a penalty is given as one term of the desert-relation, it must be a penalty in the sense of a distress itself, rather than a penalty in the more standard sense of an alteration in his circumstances, which alteration is specified independently of its relation to his desires. A penalty in the latter sense, further, is an alteration which typically but not necessarily gives rise to a penalty in the former sense. What can be said to be deserved, then, if we are more careful than to refer to an end by its typical means, is a penalty in the non-standard or less standard sense of a distress.

If a culpable action and a distress are the terms of the desert-relation in the case of punishment, what is the relation? It is this fundamental question of analysis which has received many conflicting answers. Let us look at four of them, in several cases very quickly. Others have been considered elsewhere. (Honderich, 1984b) All of them, in terms of a certain contrast, can be described as of *traditional* kinds. (p. 571)

The most common and least reflective answer, the *Uninformative Answer* as it might be called, is to the effect that there is some factual relationship of *equivalence* between an action of a certain culpability (c) and a certain distress (d) of a penalty. It is in an ordinary sense true or false that a particular culpability is in this relation of equivalence with a particular distress—as it is true or false that I am of the same height as you. The factual relation is also described as one such that d with respect to c is *proportional, corresponding, fitting, commensurate, reciprocal, merited, owed,* or *retributive*. Or, d is *according to* c.

In the absence of an analysis or account of this factual relation, as distinct from still more synonyms as unenlightening, this answer to the general question of the nature of the desert-relation is useless. It is a useless answer which very few pieces of jurisprudential writing (N. Lacey, 1988) have the great merit of avoiding. Certainly there is no commensurability in the literal sense of that term between culpability and distress taken by themselves, and hence no such possibility of a literally equivalent or non-equivalent penalty in an ordinary sense. If common units can be contrived for the direct measurement of culpability and distress, no one has yet done the job.

A second and better answer to the general question of the content of desert-claims, and a possible analysis of the one just glanced at, is that *c* and *d* are so related that persons of somehow standard *preferences* would be indifferent between being the victim of the culpable act and having *d* imposed upon them—indifferent, say, between being assaulted in a certain way and having the distress typically caused by a certain gaol sentence. (A. H. Goldman, 1979) There are certain standard difficulties in this line of thought, the *Indifference Answer*, having to do with theory about preferences, but let us ignore them. Let us accept, that is, that we do here have an adequate *content* given to desert-claims with respect to punishment. What is the value of a desert-claim, so construed, when given as a reason for the distress? Do we, to pass on to the second question (p. 548), have a significant or prima-facie reason for punishment?

It seems plain that we have no reason at all. To speak generally, suppose I do so act as to satisfy or frustrate another person's desires, and it is possible for someone to act so as to satisfy or frustrate my desires. Suppose further, and crucially, that in terms of the preferences of standard persons, there would be indifference between having the satisfaction or frustration of the person I affected, and having the satisfaction or frustration which I might experience. Does this equivalence in itself give any reason whatever for someone to act in the given way, so as to satisfy or frustrate my desires?

Who can think so? I please the waiter to a certain extent by a generous tip. There is, we can suppose, standard indifference between having his experience and having my experience of getting to the shop in time to buy batteries for my bicycle lamp. The fact of indifference is no reason why someone should secure that I get to the shop on time. Yet, *ex hypothesi*, there obtains precisely the stipulated equivalence. There can be no more force, evidently, where the indifference involves two undesired experiences. Indeed, if there *were* a recommendation in indifference-equivalence in itself, then there would be a recommendation in causing a certain experience to Green even though he had absolutely nothing to do with the causing of an earlier experience of Brown's.

The objection has gone unnoticed for a particular reason. Here and elsewhere, proposals as to the content of desert-claims, and arguments that if they are so understood they constitute reasons for imposing distress, are bound up with other quite different considerations about punishment. The doctrine of punishment on offer, as in the present case, may also contain the recommendation of it that it is preventive of offences, something regarded as essential to justification, or the definitional constraint that punishment is the work of an authority. It is plainly essential that these surrounding propositions do not distract attention from what is in question. That is a proposal as to the sense and argumentative worth of desert-claims, and nothing else.

The third answer to the question of what relation holds between a culpable action and a distress, when the latter is said to be deserved for the former, is a more long-running and a puzzling one. It requires more attention. The *Intrinsic-Good Answer*, as it is given, is just that the relation is such that *the suffering of the guilty* is intrinsically good. (L. H. Davies, 1972) In the terms we have been using, what we have is that the relation is such that it is intrinsically good that distress somehow be experienced by anyone who has performed a culpable action.

This is to be distinguished from the proposition that it is intrinsically good that such a person have a penalty in the standard sense imposed on him—where that is to say that it is intrinsically good that we do to him what typically causes distress, perhaps gaol him. The latter proposition—a fourth answer to our question, to be noticed only in passing—cannot usefully be given as a justification of what we do, since it itself *is* just to the effect that there is a justification of what we do. The *Petitio Answer*, as we can call it, produces only the useless argument that something has a justification because it has just that justification. The argument begs the question, by using the conclusion as its premiss, and is of course an instance of the fallacy of *petitio principii*.

The third and different answer, to repeat, is that there is intrinsic goodness in the state of affairs which consists in the distress of the culpable. That proposition offers the formal possibility of an argument, since it is not identical with some proposition which it can be taken to support, about our embracing the means to the good end, acting so as to bring about the distress. What we have, rather, is that the distress of the culpable does not have extrinsic or instrumental value, as a means to something else which is of value, but rather is something which is good in itself. We see or feel that the relation that holds between a man's culpability and his distress is that the two constitute an intrinsic good.

What this must come to or entail, in part, is that it is *better* that a man's culpable action be followed or accompanied by his distress than not be so followed or accompanied. There is, as just suggested, the further entailment that there exists *some* reason for our acting so as to bring about the given intrinsic good. This reason need not consist in a moral obligation to cause distress to the culpable man, in its being wrong not to do so, let alone an obligation which overcomes all other obligations not to act—perhaps the conflicting obligation precisely not to cause distress. It may be, rather, that what is taken to follow from the possibility of securing the intrinsic good is merely the absence of a certain moral obligation on us not to act—its not being wrong to act. Further, the initial claim as to intrinsic goodness, however construed, is presumably not merely in logical connection with some such judgement or judgements about the rightness of actions, but also, to remember our earlier account of morality (7.6), in logical connection with our judgements on agents as to their general standing and also our moral approval or disapproval of particular actions. Something follows, from your having so acted as to earn a kind of moral disapproval, about my somehow earning approval or disapproval for actions bearing on your subsequent distress or the lack of it.

Must the view as to the intrinsic good of the suffering of the guilty in fact involve the postulation of not one intrinsic good but rather an untold number of them? This is implied by the thought that a proponent of the view surely cannot suppose, with respect to a certain degree or sort of culpability, that *any* subsequent distress to the agent would make for an intrinsic good. That is, in line with the supposition that a petty offender does not deserve a long gaol sentence, it is natural to think that someone who gives the Intrinsic-Good Answer to our question cannot suppose that a *petty* culpability together with the distress of twenty years in gaol make for an intrinsic good. Rather, the possible intrinsic good will consist in the given petty culpability and some lesser distress. The Intrinsic-Good Answer, it seems, will therefore be that for each of the many degrees and sorts of culpability, there is a particular intrinsic good.

Is it possible to maintain against this that really there is but one intrinsic good, a general intrinsic good consisting in culpability and *some distress or other*, and that we are to use different considerations —perhaps the degree of need for prevention of a type of offence—to fix what particular distress ought to be imposed on an offender? That would be singularly uncompelling. It is *desert* that is being analysed by way of the idea of intrinsic good, and it is precisely desert that is used to justify particular penalties rather than others. The Intrinsic-Good Answer, then, must indeed involve the postulating of very many intrinsic goods.

This multitude of intrinsic goods is not reassuring, but entirely consonant with a larger fact of relevance. It is that there appears to be no logical or other effective barrier to the postulating of intrinsic goods in any area. In the history of moral philosophy, and in ordinary reflection, there are two categories of intrinsic goods. One of these, to say the very least, has immediate appeal. This is the category, to describe it very generally, which has to do with the satisfaction of desire, both amounts and distributions of satisfaction. Traditionally what has been spoken of is happiness. Perhaps it is happiness above all, whether conceived as satisfaction or differently, that has most often been taken to be a good in itself. The second category, which can be defined only negatively, contains intrinsic goods which are not a matter of satisfaction and its distribution. It has contained the intrinsic goods of promise-keeping, truth, truth-telling, beauty, certain intentions, a good will, personal autonomy in Kant's sense or a like one, and, as we know, the distress of the culpable. All of these, if they *are* in the category in question rather than the other one, are conceived wholly independently of satisfaction and its distribution. There appears to be no logical or other effective barrier to adding, say, the intrinsic good of persons of long family lineage having twice the income of others, or, for that matter, the intrinsic good of the colour mauve, or of Irish-Norwegians being arranged in straight lines.

Since there seems to be no logical or other barrier to the affirming of any intrinsic good, we cannot refute the claim that the distress of the culpable is such a good. (Cf. Kenny, 1978, p. 73.) At the same time, the plethora of affirmable intrinsic goods of the second category seems to make each of them doubtful, and in fact all of them have been disputed. The contrast between intrinsic goods of the two categories is, in this very relevant respect, a sharp one.

To express it in one way, consider a person who denies that the fulfilment of desire is an intrinsic good. She feels there is no reason, in the frustation itself of the sick or the enslaved, to end that frustration. Her position is not, as is often true, that she sees reason in the frustration for ending it, but also believes that other reasons of whatever kind require its continuation. Rather she sees in the distress of the victim *no* reason of whatever weight for ending that distress. Such a person is less than human, and in danger of being assigned to some medical category. If she really does see or feel no difference between pressing the Distress Button or the Happiness Button in a case we can imagine—she knows *only* that pressing one will cause distress, and pressing the other will cause happiness—we shall, not to put too fine a point on it, regard her as insane. We have no such attitude to someone who denies the intrinsic good of, say, beauty which gives no satisfaction whatever, or a good-will in itself— or the suffering of the guilty.

All of this sceptical reflection, while persuasive, must be admitted to be inconclusive. It seems to remain a kind of possibility to maintain that the suffering of the guilty *is* an intrinsic good, as much so as the intrinsic goods of satisfaction or happiness, and that those who fail to see or feel this are somehow morally insensitive, and that the proponents of most other intrinsic goods in the second category are also such, and so on. There is the same possibility, if no more, of affirming more elaborate intrinsic goods of the same kind, including those which are bound up with kinds of knowledge on the part of the person punished and also with his reformation. (Nozick, 1981, pp. 363–97) Perhaps this possibility can somehow be closed off, but the attempt to do so, by way of a good deal of moral philosophy, will not be made here.

However, it is one thing to allow that there seems no strict refutation of the claim that the suffering of the guilty is good in itself, and another thing to allow that the desert-relation which we have been considering can in fact be construed as consisting in such a good.

In the first place, it seems settled, as much so as such facts ever are, that to claim a man deserves a certain distress for his culpable action, and somehow to use the claim to defend imposing the distress on him, is to depend on some *claim of fact*. It is to claim and to rely on something which is in the ordinary sense true or false. In defending ten years of imprisonment for a man by saying that he deserves the distress of it for a killing, people do not mean to argue merely that ten years is right since that conclusion follows from some other moral conviction, however logically independent, which thing *also* lacks a truth-value. This would be the conviction as to an intrinsic good. They would not accept that they are offering no reason of fact whatever. No doubt to say that he deserves ten years is ordinarily to imply that it is right that he have it, but to take that implication as the total content of what is said can safely be said to be simply mistaken.

There is a connected but more conclusive consideration, which applies as decisively to the other answers given so far to the question of the desert-relation—the Uninformative Answer, the Indifference Answer, and the Petitio Answer. As long as men have been punished, their punishments have been defended or justified, wholly or partly, as deserved. The defence has been given of very different institutions of punishment. It would be no less than bizarre if this tradition of desert-claims, defined and entrenched in law, had in it only an *obscure, uncertain, or yet weaker* argument. There may be no *finally effective* argument in the tradition—that question remains open—but it cannot be that the tradition has in it only an obscure or uncertain argument. (p. 572)

What ws said before against the Intrinsic-Good Answer can perhaps be regarded as less than decisive. What is beyond question is that what we get from the Intrinsic-Good Answer at best gives us no more than an uncertain argument for punishment. We can then draw the conclusion that we have not arrived at the correct answer to the question of the nature of the desert-relation in connection with punishment. We need not consider the question of how the traditional answers we have stand to origination and hence to determinism. The correct answer will take us further forward, to various possible institutions of punishment, some of which do claim attention, and may enter into a response or indeed *responses* of affirmation with respect to punishment.

10.4 PUNISHMENT AND DESERT-CLAIMS—ANOTHER VIEW

Let us make another start, by way of a certain realism. Let us remember facts which are basic to at least much talk of desert in connection with punishment—and also remember, so to speak, their felt nature for us. Some of us are the victims of others, of their culpable offences. We are vilified and slandered, or terrified in the street and robbed, or maimed by the negligence or wilfulness of others. We are treated offensively and attacked for doing our jobs, or the colour of our skins, or for intervening to help others. We have our houses or rooms broken into and our things taken, are defrauded out of our property, or are led into vicious financial arrangements. We are victimized by employers, or assaulted and lied about by policemen. We suffer the ordeals of threat, harassment, extortion, and rape. Our children are ill-treated, molested, or corrupted. We have deformed children because some drug companies have too clear an eye to their profits. Our husbands and wives are tortured, or murdered, for gain or for political purposes.

When such things happen, we want the offender to *get what he deserves*. We may say just that, or something very like it—perhaps that we want him to get his *just deserts* or his *rightful deserts*. We also say, evidently with the same end in view, that we *want justice*, or to *have justice done*, or to *have the law take its course*, or to *have things put right*. We may say, more openly, but to the same end, that we want him *not to get away with it*, to *pay* or to *pay his debt*, to *get his punishment*, *get what is owing to him* or *coming to him*. We may say plainly that we want *satisfaction*.

Certainly this want, expressed in these various desert-claims and still more forcibly in more common speech, may not be our only want, or all that moves us. We may also demand compensation or restitution. We may be anxious about the prevention of more such injuries to ourselves and others. We may perhaps have some general concern for the public good. We may have kinds of understanding of those who have injured us. We may have an uncomfortable sense of circumstances of social injustice which lie in the background of the offences against us. Few or none of us, however, if we can have it at a tolerable or even bearable cost, and without danger, fail to have the desire which is variously expressed. It evidently is a desire which is in some way focused upon the culpable agent who injured us, and somehow also has much to do with ourselves.

Is what we want most clearly expressed when we say one of the mentioned things, that we *want justice*? As might be thought from that expression of our desire, is what we want just that the criminal law as it stands, in the judgement of its officers, and with the care or diligence they in fact bring to bear, be carried forward with respect to the offender? Do we, that is, want the fact of *legality*? This is often supposed, but cannot be right, since we may not get what we want when exactly what is legal happens. We do not get what we want when we disagree with the law as it stands or the course it takes, properly in its own terms but unsatisfactorily to us. This is the case when we take the law to be soft, unfair, or worse. On the other hand, we may get just what we want when we suppose the law to have been departed from in our favour, its officers to have exceeded their legal roles or rights. Here we take the law or its operation to have been improved upon, but not, so to speak, by having been made more legal.

It is clear from such considerations that when we are satisfied, and the law has been carried forward properly in its own terms, what we have got is not the satisfaction of legality, the satisfaction simply that the law has been observed. In fact the end in question, when separated from other things, is an elusive and abstract one, unlikely to move many of us. If we are moved by it, and do get the satisfaction in question when the law is carried forward, that is not the different satisfaction we get which is expressed in the various ways as focused on the culpable doer of injury but having much to do with us.

Is what we want most perspicuously expressed when we say we want our injurer to get his *rightful desert*, or we want to *have things put right*? That is, is what we want that the right thing be done, the *morally right thing* by our lights? Certainly we say and believe that, and we believe it with full confidence and without self-deception. But that is not to exclude another fact. Consider two persons. One is the victim of an offence, a man whose life has been wrecked, or has just been injured or suffered a loss about which he cares. The other is a person in no particular connection with the victim, and who does not identify with him for some particular reason—but a person of whom we can also truly say that he wants the right thing to be done. He thinks that the offender, like all such offenders, ought to be punished, for whatever reason or reasons.

The victim may indeed be described as wanting the right thing done. But his state of desire, if the words do fit it, is none the less not the state of desire of the disinterested person. We rightly hear a good deal more in the words of the victim when *he* says he wants the right thing done. We hear self-interest. We hear it, partly, because unlike the disinterested person, he says other things, or we can readily think of him saying other things, some of those noticed above. The most revealing is that he wants *satisfaction*.

It seems clear enough, despite several things that obscure the matter, that desert-claims about punishment are properly understood in terms of a certain fact. It is a fact from which we try to avert our eyes, which jurisprudents are keen to exclude from their subjects, and defenders of what might be called 'the moral realm' find intolerable. It is that when we want justice, or the man who has injured us wrongly to get what is owing to him, or satisfaction, what we want is that *he suffer a distress, of some degree or kind*. We have a desire for his distress, aimed at that itself. It is a desire that will be satisfied only through the belief that the person who has wrongly injured us has been subjected to an effecive reprimand, a disgrace, deprivation, hurt, injury, or worse, and, no doubt, with a knowledge of why. We thus have a desire that, like any other, can be less than satisfied, or more than satisfied, or satisfied.

To come to the central proposition, it is that the indubitable if unwelcome fact of what can be named our *grievance-desires* is what is fundamental to at least many claims of desert in connection with typical institutions of punishment. The relation between an offender's culpability and a certain distress when the latter is taken as deserved for the former—the relation so unsuccessfully characterized in various face-saving ways—is essentially that the distress satisfies a desire to which the offender has given rise by his offence. It does no less and no more than satisfy it. The deserved penalty is in fact the satisfying penalty, or any rate that is the principal fact about it in so far as argument for the penalty is concerned.

This is a conclusion with some history behind it, despite its conflict with what have rightly been called the traditional analyses of desert. It has the inexplicit support, which is worth something, of many of those whose reflections on the law and punishment are most in touch with it—judges, policemen, and some legal theorists. Few of them are slow to say, whatever else they say of the law, that it stands in a certain relation to revenge, in the sense of satisfaction obtained by the repayment of injuries. The classical statement of the view is that of the Victorian judge James Fitzjames Stephen.

The benefits which criminal law produces are twofold. In the first place, it prevents crime by terror; in the second place, it regulates, sanctions, and provides a legitimate satisfaction for the passion of revenge. I shall not insist on the importance of this second advantage, but shall content myself with referring those who deny it is one to the works of the two greatest of English moralists, each of whom was the champion of one of the two great schools of thought upon that subject—Butler and Bentham. The criminal law stands to the passion of revenge in much the same relation as marriage to the sexual appetite.

Of these two advantages, the first—the prevention of crime by terror—must, from the nature of the case, be co-extensive with the criminal law. The second—the pleasure of revenge—is obtained in those cases only in which the acts forbidden by the law excite feelings of moral indignation. (Stephen, 1863, pp. 98–9; cf. Lotze, 1885, p. 98)

Mill is of a similar view, as the following two passages indicate, although he brings it together less than clearly with a good deal else.

. . . the two essential ingredients in the sentiment of Justice are, the desire to punish a person who has done harm, and the knowledge or belief that there is some definite individual or individuals to whom harm has been done. (1972c (1859), p. 47)

The sentiment of justice, in that one of its elements which consists of the desire to punish, is thus, I conceive, the natural feeling of retaliation or vengeance, rendered by intellect and sympathy applicable to those injuries, that is, to those hurts, which wound us through, or in common with society at large. This sentiment, in itself, has nothing moral in it; what is moral is, the exclusive subordination of it to the social sympathies, so as to wait on and obey their call. (1972c (1859), p. 48)

More needs to be said in explanation and qualification of the conclusion that at least many desert-claims in connection with typical institutions of punishment are somehow to be understood in terms of the fact of grievance-desires and their satisfaction. Certainly the conclusion is a long way from philosophical thought about the intrinsic good of the suffering of the guilty, declarations about wholly elusive or morally ineffectual factual relations involving only culpability and distress, and the like. It is no popular conclusion. Let me first enumerate reasons for it.

(i) It does indeed go against sense to suppose that desert-claims in connection with punishment do not provide or involve any clear and firm reason for it. That desert-claims are empty of relevant content has in fact been supposed by many philosophers, but it must go against sense to suppose that men have not seen and got some clear gain in the repetition, over millennia, of claims that others deserve penalties. As remarked above, particularly in connection with the Intrinsic-Good Answer to the question of the desert-relation, the tradition of desert-claims cannot have been directed to an obscure or uncertain good. (p. 568) The account just given of desert-claims, the *Grievance Theory*, discovers in them an argument of the most fundamental kind, that something satisfies an existing desire, a persisting and strong desire. It is an argument, although more will be said of the matter, that depends on an unproblematical instrinsic good of the first category. No other analysis of desert in punishment passes this crucial test. (Cf. Kenny, 1978, Ch. 4; Mackie, 1982.)

(ii) As already implied, it is clear that the analysis issues directly from the most entrenched and explicit ways of talking of desert, and is at least consonant with all of them, including theoretical and institutional usages and constructions. The entrenched and explicit usages are that the deserved penalty for the offender is the debt he is to be made to pay, what is owed by him, what is owed to him, what is coming to him, what gives satisfaction. The theoretical or institutional usages are that the deserved penalty is the equivalent, corresponding, proportional, commensurate, or reciprocal penalty.

With respect to these latter expressions, it is in fact unthinkable that in so far as they involve a reason for punishment, they have to do only with a bare equality taken for itself, like the equality of two things in length or weight. Rather, they have to do with a transaction, in which one party provides and one party receives something, which thing is in some or other relation of equality with something else. We do not put people into gaols with the aim of increasing the number of bare equalities in the universe. That is an aim, to have its pointlessness clear, which might be had by a dotty passenger who puts effort into trying to make all his bus journeys as many minutes long as there are passengers on the bus.

(iii) The Grievance Theory does of course give a factual reason for punishment, as theories about intrinsic goods do not. (p. 568) Whatever the final argumentative worth of the proposition that a penalty satisfies a grievance-desire, it is true or false in the ordinary sense, however much a matter of judgement. As with any other propositions about desire, incidentally, there are behavioural criteria as to truth or falsehood.

(iv) We do have a clear relation with respect to a man's culpability and a certain distress. The distress is so related to the culpability that it satisfies the desire deemed to be owed to that culpability. The distress is unlike any other in that it alone satisfies the grievance-desire deemed to have been brought into being.

(v) To come to something not so far anticipated, it can reasonably be asked of any account of desert and punishment that the account issues from, or at least fits, a defensible account of things more fundamental, out of which punishment mainly arises. These things are the resentful personal feelings, and morality, and in particular our moral disapproval of persons for particular actions. One kind of the resentful feelings, as was remarked in passing, have in them desires to return a hurt. So too with one kind of moral disapproval. (p. 433) With respect to the vicious husband, our feelings in part are that he should have a return for his action, that he should at least have the pain of moral criticism or condemnation.

To speak summarily, typical institutions of punishment are, in part, our resentment and moral disapproval carried into action on our behalf by others, not just the action of speaking out. Both kinds of resentment and disapproval enter into these institutions. The given account of desert in punishment, based on grievance-desires, is an account, in fact *the* account, which we can rightly expect to have, given the nature of one kind of resentful feelings and moral disapproval, and the fact that punishment is importantly a kind of product of them. We can be confident that we have a right account of a part of punishment since it is an account that fits sources of it. It might be replied that desert conceived in the way of the Intrinsic-Good Answer reflects the presence of certain judgements of intrinsic good in moral disapproval. Similar remarks might be made about other answers to the question of the desert-relation. These replies would not connect punishment with the very nature of the accusatory feelings and moral disapproval, in particular certain desires which are fundamental to them, whatever the content of the feelings in terms of this or that value, principle, or commitment.

As remarked, further explanation and qualification is needed of the general conclusion that when a certain distress is somehow said to be deserved for a certain culpability, the fact of importance is that the distress will satisfy what is deemed to be the grievance-desire created by the offence, and the argument for the distress is that the distress will satisfy the desire. This explanation and qualification of the Grievance Theory will be given by way of considering five objections to it.

(i) Some care has been taken, as may have been noted, to frame the theory in a limited way. It has been said that *the proper understanding* of many desert-claims about punishment is in terms of grievance-desires (p. 571); that they and their satisfaction are *fundamental to* desert-claims about punishment (p. 571); that the reason *provided by or involved in* many desert-claims is that penalties satisfy grievance-desires (p. 572), and so on. It has not been claimed, then, that when we say in this way that a certain penalty is deserved for a certain offence, *what we standardly mean* is that the penalty will satisfy the caused grievance-desire. Any objection to that effect is misdirected. It is perhaps arguable that some of the entrenched locutions, about the offender paying, or giving satisfaction, *can* have their meaning properly characterized in terms of grievance-desires, but I intend no general claim to that effect. It is perhaps proper to say that desert-claims about punishment, whenever made, do carry an implication about the satisfaction of grievance, but let us leave alone the parts of linguistics and the philosophy of language which are concerned with such matters.

It can be supposed without damage to the general conclusion that has been drawn that the desert-claims in question are properly to be understood as essentially vague, as suggesting some factual connection or other between offence and penalty, such that the connection justifies the penalty. It would not damage the conclusion if, as is unlikely, these desert-claims were claims as to indifference scales (p. 564), or if they were claims to the effect that a penalty was in accordance with precedent or a penalty-system—which view was passed over in our survey. Nor would it damage our conclusion if, as may be more likely, they were non-factual claims as to the intrinsic good—the fittingness—of the suffering of the guilty.

This is so since our conclusion is not essentially a piece of linguistic analysis but an answer to this question: What is the argument for punishment that is suggested by and involved in certain desert-claims about it? Does some doctrine of the philosophy of language resist the idea that what is *said* may not convey the essential argument of an enterprise? If that is so, then, to my mind, so much the worse for the doctrine. What is true here, of course, may not be true elsewhere. Elsewhere, as we shall see, desert-claims may call for a similar linguistic analysis but not involve the given argument.

(ii) It may be objected against the conclusion, simply enough, that it is in error to suppose that the satisfaction of grievance is what is offered in defence of punishment by any desert-claims: 'Nobody . . . would say that a judge ought to sentence because the people outside in the street are baying for blood, and that's what it amounts to.' (N. Walker, 1983, p. 15) If the conclusion did amount to that, it would of course be absurd. It does not.

A first thing to be kept in mind here as well as elsewhere is that the conclusion does not include any proposition as to the relative weight of desert-claims or the given argument in connection with punishment, as against, say, considerations of prevention. We have not come to that matter. A second is that there is no doubt that a good deal of the law and punishment has little or nothing to do with desert, and is not defended by reference to it. Stephen makes such a point in the passage quoted above (p. 572), and it has more application to legal systems since his time. The conclusion that has been drawn here has to do with desert-claims where they do figure in punishment, and not with the extent to which they figure.

The objection, to repeat, is that the conclusion reduces criminal justice or the argument from desert to the mistaken proposition that a judge is or ought to be governed by the mob in the street. In reply to one thing that might be intended by the objection, something needs to be allowed. It is not proposed that the given desert-claims involve the argument that penalties satisfy *whatever* desires for satisfaction happen to exist in particular cases, for whatever reason. To restrict ourselves to the actual victims of offences, either persons harmed or persons close to them, who are of prime importance, it is not proposed that judges, in so far as they are influenced by matters of desert, are in fact attempting to gauge the actual desire for satisfaction of these individual victims, whether or not unreasonable, peculiarly vindictive, or inflamed. For a number of clear reasons, despite the fact of *some* attention paid to individuals, it is rather the case that a penalty system is to be seen as reflecting what can be called standard grievances.

One thing that determines a standard grievance is certainly a conception of a person of ordinary or average responses of feeling. Again, a standard grievance is conceived in relation to a particular culpability. It is not as if such a grievance were understood as a desire arising out of, say, popular hatred for some group of people, or a popular fear of some type of offender. There is also a consideration of another kind which touches on the idea of a standard grievance. It is evident that the institution of punishment is very subject to the principle of formal justice, that of treating like cases alike. With respect to penalties, this operates in much more than the obvious way: requiring a like distress for offenders of like culpability. The most relevant at the moment is its operation with injuries and grievance-desires. A like injury is to be taken as giving rise to a like grievance-desire.

What has been said so far, in answer to the simple objection, may well bring to mind certain other objections which will be treated below. To finish with the simple objection itself, it might be improved into this: Punishment by the state, in so far as it is governed or influenced by desert or by anything else, is rule-governed, a matter of principle, consistency, precedent, objectivity, specific resistances to kinds of partiality and feeling; this punishment, further, is independent in different ways of society generally, of parts of it, and of victims themselves; therefore it is mistaken to regard desert-claims in punishment as a matter of grievance-satisfactions for victims.

The premisses as they stand and a good deal more about the nature of the law, all of it in fact fundamental to Stephen's view (p. 572), can be granted. The conclusion nevertheless does not follow. The claim that a man deserves ten years for the rape is indeed a claim made within an institution, a claim shaped by and subject to a considerable number of kinds of constraint. This complexity, which has been little more than indicated, does not begin to defeat the proposition that desert-claims and arguments from desert are to be made sense of in terms of a ruling idea: that the distress of a man's penalty satisfies a certain desire. The objection is no better than conceivable ones about other institutions, say that of old-age pensions, whose aims are also in various ways defined and constrained, and whose categories are necessarily standard or general.

(iii) It may be objected very differently, that the given account of desert-claims and arguments from desert cannot be right because of the existence of a certain very settled custom. Theories about the justification of punishment, or different justifying considerations within theories, divide without much important remainder into two groups: those that attempt to justify it somehow by reference to agreement and desert, and those that attempt to justify it by reference to the prevention of future offences. The former group, by settled custom, are conceived of and spoken of as finding an argument for punishment in only the past offence, or only in a relation between it and a punishment, which relation does not bring in consequences of the punishment. They are backward-looking. By contrast, the second group of theories or considerations are forward-looking or consequentialist. The account that has been given here of desert-claims goes against this customary view. Desert-claims have been understood as having to do fundamentally with a forward-looking or consequentialist argument. It is that a penalty is defensible or has a recommendation because it has the consequence of satisfying desires.

One admirable philosopher who made this objection did unwittingly weaken it, and in effect point to its proper rejoinder, by also arguing something else. This is that traditional attempts to clarify and defend an argument from desert in connection with punishment have been failures. '. . . a retributive principle of punishment cannot be explained or developed within a reasonable system of moral thought. . . .' '. . . attempts to make sense of the principle of positive retributivism, as an independent principle with immediate moral authority, have signally failed.' (Mackie, 1982, pp. 3, 6) He went on to offer an *explanation* of the fact of desert-claims with respect to punishment, a persuasive explanation derived from biology and in particular evolutionary theory, but to persist in the position that they can involve no significant *argument*.

My answer to the objection can be anticipated, resting as it does on an argument already given. It is not enough to grant that there exists a causal explanation of the fact of desert-claims. What needs also to be granted is that the tradition of them and of punishment, entrenched over time, cannot be without effective rational content. (pp. 568, 572) It cannot be that in this continuing tradition men have not aimed at and got something of value and capable of clear enunciation. By way of brisk summary, offenders have always been seen as having debts of some kind to pay, and have been forced to pay them—*how could it be that no one received anything*?

One can reasonably have a certain readiness to be impressed by customary philosophical reflection, and in particular to be impressed by the customary philosophical idea that desert is not forward-looking or consequentialist. However, if a struggle over centuries to produce an argument of the supposedly right character has failed, and if an argument of a different but fitting character can be found, it is more reasonable to take that argument to be the one which has informed and does inform talk of desert. By an argument of a fitting character I mean, in part, one that is in accord with desert-claims and only with them.

(iv) Will someone say that the argument taken as fundamental to desert in punishment does not really fit because, at bottom, it defends penalties as giving satisfaction, and has little to do with equivalence or whatever between offence and penalty? Certainly, as has been granted, these desert-claims *are* claims as to some sort of equivalence. The objection needs only a brief reply, in two parts.

It is of course true that penalties are defended in the argument by the fact that they give satisfaction. But the clear idea of equivalence involved in the argument is bound up with or integral to that fact. The idea, to repeat, is that the penalty is equivalent to the offence in that it does not do less or more than satisfy the relevant grievance, but rather just satisfies it. That the penalty must not do less than satisfy it is simply entailed by by the requirement of satisfaction. If the penalty were to do more than satisfy the grievance, the additional distress would be pointless—it would lack exactly the defence advanced by the argument for the distress of an equivalent penalty in the given case. The further distress would in fact be indefensible in terms of the argument.

(v) If the given argument is what is fundamental to desert-claims about punishment, then punishment is indeed like revenge. That is not to say that it *is* revenge, since, for one thing, punishment by proper definition is the work of an authority rather than the free-lance activity of the person harmed or those close to him. There are other differences. (Nozick, 1981, pp. 366–70) However, if the given argument is right, punishment is like revenge in involving desires for the distress of others. This conclusion cannot be said to be obviously false, to be a consequence of the Grievance Theory which in fact refutes it. Punishment, as already noticed, *is* often enough likened to revenge. (p. 572) What an objector to the theory must do is identify a respect in which the theory makes punishment like revenge, and show that punishment does not in fact have the supposed feature. Perhaps the theory gives to punishment a moral shortcoming had by revenge, and in fact punishment does not have that shortcoming. There is such a line of thought, as follows.

It may be held that desires for the distress of others, of all desires, are most evidently those which we ought not to have. They may be described as indecent, vicious, or degraded—such as to make revenge what it is, which is abhorrent. Therefore they are desires whose satisfaction should be absolutely ignored in considering the worth of any action. That an action gives rise to grievance-satisfaction is no recommendation whatever of it. Surely any such satisfaction should even count *against* the action. Thus it cannot be that the argument from desert, as it has been presented, is even a tolerable one. It cannot be that any institution can even in part be defended by saying that it satisfies such desires. On the assumption that there is an argument of some force pertaining to desert and punishment, this cannot be it. We have several times assumed in this inquiry, not that there must be a conclusive reason for punishment having to do fundamentally with a man's past action, but that there must be a clear and a significant or prima-facie reason, something of value which is capable of explicit enunciation. The objection now before us is that we have no such reason in the consideration having to do with grievance-desires. The consideration is clear, but, it is unique in its absolute want of value or significance, indeed in being abhorrent.

The objection must not be confused with something else to which we are coming. That is, what we have is not the inevitable proposition that while the satisfaction of grievance-desires is a significant and firm reason for punishment, it is outweighed by other things. Rather, the objection is that it is indecent or whatever to desire someone else's distress as an end in itself, and hence that any satisfaction of grievance-desires produced by a punishment counts as *no* reason for it, or even a reason against it.

This has nothing to do with opposition to taking the ordinary means to that satisfaction, the actual causing of distress to the person in question. If it were possible to satisfy a man's grievance-desire by a deception, such that no distress at all was caused to the person he has in mind, we would none the less have no reason to satisfy his desire and would perhaps have a reason to frustrate it. We ought perhaps to prevent him having the false belief that the person was or would be in distress, and hence prevent him having his grievance satisfied. This would not be on account of anything about truth and falsehood, but on account of the baseness of the satisfaction. Nor, incidentally, does any of this have to do with the possibility of certain undesirable consequences of the satisfying of grievance-desires, say the man's getting some abnormal or heightened sexual passion for the causing of suffering. It has to do with the nature of ordinary grievance-desires and grievance-satisfaction.

What is to be said of this? Certainly, when we are not ourselves the victims of the culpable actions of others, we find something pertaining to the fact of grievance-desires unattractive or worse. We may, when we are such victims, be less than happy with our having such desires ourselves. We may do a good job of concealing them from ourselves, at least some of the time, by diverting our attention from them to more elevated considerations, perhaps the supposed need to impose a punishment in order to prevent more offences. This disinclination is in the background of what can be called the *concealing* character of certain of the ordinary and entrenched ways of talking of desert, as when we say we want things put right. (p. 570) It is also in the background of the concealing theoretical and institutional usages mentioned earlier, whereby a penalty is said to be equivalent, corresponding, proportional, commensurate, or reciprocal. (p. 563)

The fact of disinclination, however, is not at all sufficient to establish the moral conclusion that grievance-satisfaction is to be ignored in considering the worth of an action or practice, let alone the stronger view that it counts against it. For one thing, it is possible to give an explanation of this disinclination which is consistent with the grievance-satisfaction itself having a positive value. The explanation of course has to do, in part, with our very naturally associating a man's grievance-desire with what is indubitably bad in itself, which is the proposed suffering he has in mind. Another part of the explanation has to do with the seemingly undeniable proposition that the world would be a better place if it were very different, perhaps unrecognizably different, such that our grievance-desires were replaced by desires with other effects, desires for the well-being of others.

A further and more conclusive reply to the objection is that in fact no clear reason has been given for the proposition that the possibility of grievance-satisfaction is to be ignored in considering the rightness of actions, let alone that it counts against rightness. We do not get a reason, as it seems to me, in the moralist's rhetoric about the indecent, vicious, base, or other nature of the desires and the satisfaction in question. What we get, surely, if we get something precisely to the point, is no more than an anticipation or a version of the conclusion for which we are looking for a reason. Further, while it is indeed possible to be in a frame of mind where condescension is possible, another frame of mind is also possible, and it certainly has a reasonableness about it. Is is, for what the point is worth, the frame of mind of judges, or very many judges, as they go about their work.

Consider again the extraordinary but simplifying case where it is possible to satisfy a man's grievance-desire by deception. He has been savagely treated, perhaps maimed. *What* reason do we have, aside from the obvious irrelevant reason of truthfulness, for thinking that whatever is to be said against, there is *nothing* to be said for giving the victim content in place of discontent, peace in place of bitter passion, and escape from a kind of wretchedness? To my mind, any reason that exists has never been brought into clarity. It is open to anyone to declare, of course, that grievance-satisfaction is simply intrinsically bad, as it is open to anyone to declare that the suffering of the guilty is intrinsically good. Few of us, however, are likely to be reassured or persuaded by these fiats. We want a *reason*, surely, for abandoning what seems to be the reasonableness of replacing discontent by content, bitterness by peace, and so on.

It can be allowed that we might think better of the maimed victim if he had escaped or risen over his grievance. But even if we would think better of him if he were quite different, that does not entail that when we consider him as he is, we have a reason for leaving him in his self-mortifying passion. We might think better of a man in some quite different situation if he were willing to forgo an absolutely indubitable moral right which he has to something, but it does not follow at all that we are not obliged to respect his right. That he might become Christlike does not dissolve our obligations to him as he is. Nor, in fact, might our obligations be dissolved if he did become Christlike, since his renunciation would certainly leave intact many of our obligations to him.

It was remarked earlier in connection with the idea that the suffering of the guilty is an intrinsic good, that in the history of moral philosophy two categories of intrinsic goods can be distinguished. (p. 567) The first has to do with quantities and distributions of satisfaction—very roughly, well-being and fairness. The second has to do with the other postulated intrinsic goods. The goods of the first category, despite the fact of great disagreement about their proper characterization and ordering, are in some way beyond any serious question. In no substantial inquiry do we stop to ask whether pain or frustration considered in itself might be good, or pleasure or satisfaction in itself bad. Nor do we stop to ask, of the situation of which we know only that there is enough food for two equally starving people, whether we ought to give all to one and none to the other, or most to one and little to the other. By contrast the goods of the second category, although some of them certainly appeal to us, are open to doubt. All, I as remarked earlier, have in fact been doubted.

No reason has been given, and in my view none exists, for excluding grievance-satisfactions from the first category. In themselves—and we are wholly concerned with them themselves, rather than any costs, concomitants, or effects of them—they are satisfactions like any others. They are open to a range of descriptions inseparable from the descriptions we apply to other satisfactions. For example, they may be said to consist in, or to give, contentment. Like other satisfactions, they are struggled for, and, like other satisfactions, their denials or frustrations can be a torment and an affliction. They have just the properties or character which give satisfactions generally the place which they have in our lives.

Like very nearly any goods, there can most certainly be countervailing reasons against having them or providing them. By way of a random example of another equally vulnerable good, there is perhaps no moral view which can take a hold on us which does not allow *any* conceivable circumstance in which there would be countervailing reasons against the good of preserving an innocent life. That grievance-satisfactions can for good reasons be denied is one truth. That they do, despite that, provide a clear and firm reason for punishment, is another truth. It cannot successfully be objected that an argument based on grievance-desires cannot be the argument from desert since an argument from grievance-desires is insignificant. It is not.

10.5 THE THEORIES OF PUNISHMENT

We now have an understanding of the clear and firm argument for punishment, the argument based on grievance, which is involved in desert-claims about it. We need now to deal with two further questions, as anticipated earlier. (p. 548) The first is the bearing of determinism on the argument from desert. The second is that of the relative weight of the argument, as against other justificatory propositions about punishment, and hence the relative importance of certain features or a certain character, having to do with desert, of our actual institution of punishment and also possible institutions.

There is the possibility, among others, that determinism does affect the argument from grievance—in fact we already have grounds for thinking so—but that the argument is not of great relative weight. In particular, the argument is not of great importance within justifications attempted of our actual institutions of punishment.

The answer to the first question is in accordance with findings in earlier stages of this inquiry. One kind of our personal feelings of the resentful kind, as noted again recently (p. 401), have within them desires to return the hurt done to us by those whose actions give evidence of bad feelings towards us or bad judgements on us. Moral disapproval of one kind, to turn to it, also has within it at least an inclination to the discomfiture or worse of persons whom we take to have acted wrongfully. (p. 433) It was said of the latter desires, and implied of the former, that they could not survive our actually coming to believe that persons lack the power to originate different actions than those they perform. (pp. 435, 407) Our desires against others on account of their actions could not hold up if we came to believe of them that *they could not have done otherwise*, in the specific sense of the words having to do in part with origination. One argument for the vulnerability of these desires was that in certain circumstances we do now resist any doubt or denial of origination, taking it as a threat to our resentment and disapproval. (p. 435)

The feelings had by victims of offenders, expressed in desert-claims, are perhaps best regarded as being within the class of the given kind of resentful personal feelings, rather than the like feelings of moral disapproval, although an argument could be attempted for assigning them to the latter class. The feelings had by victims of offenders are the strongest of the given resentful feelings, involving enlarged and dominant retributive desire, such desires as rightly have the name of grievance-desires. (p. 571)

The true argument from desert, to repeat, is to the effect that a penalty will satisfy grievance-desires. The desires, and hence the argument, are vulnerable to a determinism. The argument, if determinism is taken as true, depends on desires that will not survive if determinism comes to be accepted. However, in so far as determinism is not generally accepted, and grievance-desires persist, the argument keeps its basis. Admittedly, as we know, it cannot be said that grievance-desires owed to an action do logically presuppose the proposition that the action could have been otherwise in the sense just mentioned. It cannot be that the expression of the desire, perhaps a prescription, *entails* the proposition that the agent was not only voluntary in his action but might have originated a different action. That does not matter. Just as we are so constituted to take the fact of suffering as *a* reason against an action, so we are constituted to take grievance-desires as requiring the reason of an originated action. That our grievance-desires depend on our ideas of actions as originated is indeed a brute fact. (p. 479)

The remaining and culminating issue is that of the relative importance of desert or grievance with respect to a number of possible theories of punishment and possible institutions of punishment into which it enters, above all our actual institutions. Among the theories are a certain family of theories commonly presented as justifying our actual institutions. This family, *Mixed Theories* as they will be called, can be set out most effectively by contrasting them with others. In specifying these other theories, *Pure Retribution Theories, Utilitarian Theories*, and *Fairness Theories*, we shall in passing deal with the further question of the bearing of determinism on them and hence on the associated possible institutions of punishment. Some of the latter were perhaps actual institutions in the past, and some may be actual in the future.

(i) *Pure Retribution Theories* have been curiously persistent despite the fact that it is very arguable that the possible institution of punishment they propose, if it was actual once, has ceased to be. Certainly it is not the institution found in societies with which most of us are familiar. Pure Retribution Theories depend or purport to depend entirely on considerations of desert in justifying punishment. All else is excluded. Some, like Kant's, are to the effect that we are not merely justified in punishing offenders purely on grounds of their deserts, but are obliged to do so.

> Even if a Civil Society resolved to dissolve itself with the consent of all its members—as might be supposed in the case of a People inhabiting an island resolving to separate and scatter themselves throughout the whole world—the last Murderer lying in prison ought to be executed before the resolution was carried out. This ought to be done in order that everyone may realize the desert of his deeds. . . .' (Kant, 1887 (1797), p. 198)

There has been speculation on Kant's own understanding of desert-claims, and hence the particular argument from desert on which he depends, and the questions remain open. Let us proceed on the basis of the conclusion we have, that there is but one clear and firm argument from desert in connection with punishment, the argument based on grievance-desires. Pure Retribution Theories, conceived in terms of this argument, are no more defensible than they have generally been taken to be when otherwise conceived. It is morally unthinkable that we can disregard all else about punishment, above all what can be said against it, or that we can suppose that the satisfaction of grievance by itself outweighs all of what can be said against it. It is unthinkable that we are obliged to engage in punishment solely because it satisfies grievance. Our present actual practice of punishment is certainly not owed to this. The theory would be destroyed by a general acceptance of the truth of determinism, since such an acceptance would dissipate the desires which are the theory's subject-matter. This, at least from the moral point of view just expressed, is of little matter.

Pure Retribution Theories have often been taken, rightly or wrongly, to have a certain virtue despite their indefensibility. They are said to respect the claims of individuals. This has to do with the fact that they do not merely seek to justify punishments, but also to prohibit and limit them. They in certain ways prohibit what can be called, if improperly, punishment of the innocent, and they limit the punishment of the guilty. That is, only those who deserve it may be punished, and no one is to be punished more than he deserves. These features, whatever their worth—we shall look at the question later in another connection (p. 588)—are to some extent preserved when Pure Retribution Theories are conceived in terms of grievance-satisfaction. So conceived, they are to the effect that no punishment is justified which does not satisfy grievance-desires—whether punishment of non-offenders or offenders. Non-offenders, since they do not give rise to grievance, are at least in general not to be punished, and offenders are not to be punished more than satisfies grievance. A general acceptance of determinism, issuing in the decay of grievance-desires, would of course affect these prohibitions. In the absence of grievance-desires, the given distinction between offenders and non-offenders, that the former do and the latter do not give rise to them, would no longer exist. Similarly, in the general absence of grievance-desires, the given limit on punishment would no longer exist.

Pure Retribution Theories conceived in terms of grievance-satisfaction are of course different in character from such theories conceived traditionally in terms of some other idea or image of desert. To the objection that to conceive them in our different way is too resolute, or even cavalier or pointless, since it is not to consider such theories as they have been understood and advanced, there is the reply that will be anticipated. As our inquiry into the desert-relation indicates, it is only by conceiving them in terms of grievance-satisfactions that we do have something firm and significant to consider. It has generally been assumed that determinism conflicts with Pure Retribution Theories traditionally conceived. The uncertain grounds for this assumption (p. 545) can continue to go unconsidered. It is a merit of the interpretation of desert in terms of grievance-satisfaction that it makes clear why desert in connection with punishment is affected by determinism.

(ii) *Utilitarian Theories* of the justification of punishment have typically contained dismissals of talk of desert, and been to the effect that punishment is justified when it is economically preventive of offences. The prevention may be the result of incapacitation, as when a man is prevented from many offences by being in gaol, or deterrence in several forms, or the creation and reinforcement of habitual, unreflective obedience to law. There are of course many difficulties here, of a factual kind. The theories in question, to state them more generally, are to the effect that punishment is justified when, in whatever way, it produces a greater total balance of satisfaction or a lesser total balance of dissatisfaction than any alternative to punishment. What is dismissed by Utilitarian Theories, in so far as desert is concerned, has been traditional argument which makes use of desert-claims but does *not* have to do with the satisfaction of grievance. Utilitarians have been rightly disparaging about such items as the instrinsic good of the suffering of the guilty.

The argument from desert in justification of punishment, as it has been conceived here, is the argument from grievance. In fact it could and in consistency should enter into any Utilitarian justification for imposing punishment, for the reason that grievance-satisfactions are among satisfactions produced by punishment. Bentham's general enumeration of the kinds of pleasures and pains in fact includes 'the pleasures of malevolence', those 'resulting from the view of any pain supposed to be suffered by the beings who may become the objects of malevolence'. (1970 (1789), p. 44))

It is not the case, however, that a recognizably Utilitarian theory could in consistency use considerations of desert, understood in terms of grievance-satisfactions, to prohibit the punishment of the innocent and the over-punishment of the guilty. A theory could not consistently defend punishment by consequences of satisfaction having to do with prevention, and prohibit it, despite *any* such consequences, on grounds of absence of grievance-satisfactions. If satisfactions having to do with prevention were of a certain magnitude, it would be inconsistent to prohibit some particular punishment of an offender on the ground that it did more than satisfy existing grievance-desires.

Utilitarian Theories of punishment have in fact long been objected to on the ground that they somehow ignore the claims of individuals. It is a consequence of the theories, it has been claimed, that the innocent may or must be treated as offenders in conceivable circumstances where this would be called for by a comparative judgement of satisfaction and dissatisfaction. Further, it is a consequence of Utilitarian Theories that offenders may or must be punished excessively. Both objections have been advanced on the basis of some or other principle of desert.

However, it is very far from being the case that the only way of ignoring the claims of individuals is by ignoring considerations of desert, however conceived. The claims of individuals are ignored when we treat them, as we say, *unfairly or unequally*. Utilitarianism can be argued to conflict not with Retributive Justice but with Distributive Justice. Utilitarian Theories of punishment as we have defined them, which are approximate to Bentham's, and hence do not themselves incorporate elements of Distributive Justice, can be argued to be untenable for reason of that conflict.

These theories, to the effect that punishment is justified when it produces a greater total balance of satisfaction or a lesser total balance of dissatisfaction than any alternative practice, face no obstacle based on determinism. This is so since the theories so understood, if some of them rightly take grievance-satisfaction into account, do not *depend* on the existence of grievance-satisfaction. In so far as determinism is concerned, then, the situation with Utilitarian Theories may be argued to be in an important way like that with Pure Retribution Theories. Although the retribution theories necessarily are vulnerable to determinism and the Utilitarian Theories as naturally understood are not, neither fact, it may be argued, is of much importance. This is so since theories of both kinds are open to objection not having to do with determinism—in the case of retribution theories, overwhelming objection. The retribution theories fail without help from determinism, so to speak, and the Utilitarian Theories are not to be saved by their consistency with it. Let us not attempt to draw a firm conclusion about the worth of Utilitarian theories. More can be said in defence of some of them than is often supposed.

There is another conclusion that *is* to be drawn. Utilitarian theories, if they are not judged to be intolerable, offer the possibility of *a* response of affirmation with respect to punishment, the first possibility we have encountered in these reflections. It is a possibility for someone who supposes that the Utilitarian theories *can* be saved from the mentioned objections. *He* can seek to free himself from commitment to whatever is inconsistent with determinism by relying on a Utilitarian institution of punishment. This is one of several possible responses of affirmation that will be noticed. As will have been anticipated, the essential features of the response of affirmation are consistent with different moral views.

(iii) *Mixed Theories* of the justification of punishment make up a majority of contemporary theories. They are at bottom amalgamations of what are taken to be the virtues of the theories of the first two categories, perhaps with further additions. Pure Retribution Theories, as noticed, do have the feature of placing a certain limit on what can be done to individuals—the limit of their desert, somehow conceived. Utilitarian Theories do have the virtue of recommending punishment on the basis of an indubitable good, at bottom the prevention of offences. Mixed Theories are fundamentally to the effect that punishments are justified only when they are both preventive and deserved. These theories thus escape what appears to be the pointlessness of theories of the first category when desert is conceived not as we have, but traditionally, and also escape the bizarre concentration on satisfactions of one category, which is the principal fact about Pure Retribution Theories when conceived as we have, in terms of grievance. Mixed Theories are also supposed to escape the general objection to Utilitarian Theories having to do with the claims of individuals.

There are very many theories of this general kind, the work of philosophers, jurisprudents, less theoretical legal writers, and others. They differ very considerably in emphasis and detail, and many of them, at least to a philosophical eye, are sadly inexplicit, above all about desert. It will be best not to attempt any further unitary summary of them, but, at the risk of too much complication, instead provide models of three varieties of them.

(iiia) Mixed Theories on the first model require of any justified punishment that (1) it be no more than deserved. This first requirement can be expressed, as typically it is in jurisprudential and legal writing, by some choice from the plethora of desert-locutions noticed earlier. (p. 563) The punishment must be no more than 'proportional', 'commensurate', or whatever. If we give satisfactory content to the requirement, it is that a punishment must not do more than satisfy grievance, which allows either that it does satisfy it or does less than that. (2) It is also required on these theories that a justified punishment must have a certain preventive effect. That is to say, precisely, that it prevents what are in fact fixed as criminal offences in the society. That as it stands may be regarded as insufficient, since it sanctions punishments whose preventive effect is trivial. It is then to be added (3) that the punishment must be economical: prevent more distress than it causes. It may be added, further, that (4) if the punishment is one of a number which can be judged to have the same preventive effect, it is the least severe of them.

It is possible to object to such a justification of punishment. To be brief, and despite what has been supposed by many about desert, such a justification may be argued to issue in an unacceptable denial of the claims of individuals. There is in fact no guarantee in the four requirements, it can be said, that individuals *will* be *fairly* treated, no guarantee that their treatment will be according to an acceptable principle of Distributive Justice. To consider the first of the four requirements, it is possible that a penalty will not do more than satisfy grievance, and yet be in a sense unfair. That it does to some extent satisfy grievance does not ensure its fairness. Nor, it may be said, does any of the other three requirements ensure its fairness.

There is also a wholly different difficulty. As already remarked, developed legal systems punish those who cannot be said to deserve it. (p. 562) These are offenders convicted under statutes of strict liability, or statutes interpreted as requiring only strict liability. There is also the fact of exemplary punishments, imposed when it is believed that there is a particular need of deterrent examples. Both matters are to some extent controversial, but it appears necessary to allow exceptions to the first requirement of the justification we are considering, that all punishments be deserved. It is unclear what lesser thing is to be put in place of the absolute requirement.

What determinism here affects, then, as of course it does, is not an indubitable justification of punishment. As for the effect, a general acceptance of determinism and hence the decay of grievance-desires would make pointless the first requirement of the justification. In the absence of grievance-desires, there could be no distinction between punishments that do more than satisfy them, and punishments that either satisfy them or do less than that.

(iiib) A second model of Mixed Theories of punishment differs from the first only in its preventive part. It does not require that punishment prevents what in fact are in part fixed as criminal offences in the society. Rather, punishment must prevent actions which are such that they give rise to such grievance-desires as can be satisfied only by distress of the extent which is caused by punishment. It is not a conceptual or other necessity that the class of actions which theories on the first model seek to prevent will be identical with the like class of actions in the second model. There are the same possible objections to this second mixed justification of punishment, the main one having to do with Distributive Justice, and the same situation with respect to determinism. In fact this second mixed justification, given its greater reliance on desert, would be more undermined than the first by an acceptance of determinism.

(iiic) A third model of Mixed Theories, in a way harsher, does not require of any justified punishment only that it be *no more than* deserved. It specifies, rather, that any such punishment be no more and *also no less than* deserved. In clear terms, what is required is that a justified punishment does no more and no less than satisfy grievance. There are added certain requirements as to prevention. These cannot simply be to the effect that there be economical prevention, since a fully deserved punishment may not be economical, and an economical punishment may be less than deserved. The difficulty is often overlooked or ignored in Mixed Theories, but we shall not consider it further. Mixed Theories on this third model are also open to the objection having to do with the claims of individuals and Distributive Justice. The consequence of an acceptance of determinism is also of the same kind as with the previous two models.

It is evidently the Mixed Theories which can be taken as providing a kind of general characterization of punishment by the state as we have it in our societies. That punishment is, so to speak, an effect of mixed theories. It is a difficult matter of judgement which of the three models of Mixed Theories or a related one is closest to our practice. It may be that our practice in one part or feature is a kind of realization of one of these models, in another part or feature a realization of another. (Hart, 1968) Whatever is to be said of these secondary questions, it is evidently one or more of the mixed justifications which can be taken as telling us of our practice. (N. Lacey, 1988)

Let us not try to draw a firm conclusion about the defensibility of Mixed Theories. What is rather to be concluded is that they stand in some considerable conflict with determinism, given their content having to do with desert. That is to say, since one or more of them does give us our actual institutions of punishment, that our actual institutions are in some considerable conflict with determinism. Change is required. That change, a response of affirmation, is the keeping of what can be kept, and the giving up of what must be given up. This can be conceived not as necessarily involving a commitment to some different theory and institution of punishment, the Utilitarian or another, but as involving a valuing of and a choice from the elements of Mixed Theories. Their desert-element, conceived as we have conceived it, must be abandoned. Other elements, most importantly the element of prevention, can be maintained. More can be said than that, since Mixed Theories may have more to them than has been indicated in the quick summary given. In particular, although this complication has so far been suppressed, they may involve matters of desert so far unconsidered, desert where it is *not* vulnerable to determinism. (p. 595)

(iv) Finally, and briefly, there is a further category of theories of punishment, different from the Utilitarian but like them in facing no serious threat from determinism. The category, which has but few announced members, although more are implied by certain political philosophies and ideologies, consists in theories which rest on some principle of Distributive Justice. They are the *Fairness Theories*. One is the theory which I myself take to provide the only defensible justification of an institution of punishment. It is, in essence, that an institution of punishment is justified if it is in accordance with the Principle of Equality. That principle requires of any social institution that it have a certain function, which is to say that it contributes to making well off those who are badly off. The well off and the badly off are defined in terms of satisfaction or the lack of it with respect to certain fundamental desires. (10.7; Honderich, 1981c, 1983a, 1984c) There can be no objection to the Principle of Equality taking account, although necessarily it will be slight account, of existing grievance-desires. It is in this respect vulnerable to determinism. However, like the Principle of Utility, the Principle of Equality does not depend on the existence of grievance-desires, and can of course be stated without reference to them. Like Utilitarianism, although the fact was not mentioned before, it has the consequence, on consequentialist grounds, that we should try to diminish them. Shall we ever have a practice of punishment in accordance with this justifying theory? That can be no more than a matter of less than confident hope. The given theory, evidently, can enter into a response of affirmation different from those so far noticed. As in the case of the other responses, it will involve the support, so to speak, of what can be taken as a fundamental principle of morality.

10.6 OTHER SOCIAL FACTS

It was remarked at the beginning of this discussion of the social and political consequences of determinism that our wide use of talk of desert, across many contexts, does not entail that we are advancing a single argument, or a single kind of argument, whenever anyone or anything is said to deserve something. Talk of what is deserved is in this respect like talk of what is just or fair, which is also diverse in intention and content. Thus it cannot be supposed that since much talk of desert in connection with punishment has to do with a certain argument about grievance, so too does *all* talk of desert with respect to persons, let alone other animals and inanimate things. Still, to turn now to the first of the remaining social institutions, practices, and habits listed earlier (p. 543), it is to be expected that there will at least be a similarity, having to do with determinism, between punishment and *moral blame and praise, moral condemnation and commendation.*

The similarity between punishment and blame, in so far as determinism is concerned, is indeed that desert-claims typically made about both are vulnerable to an acceptance of determinism. More particularly, just as our grievance-desires which are fundamental to desert-claims about a man's punishment cannot survive our coming to believe that his action was not originated, so there cannot be survival of certain related desires or inclinations somehow involved in typical claims that a man deserves blame or condemnation. That is not to say, as we have seen, that punishment must come to an end. There is more to be said for punishment than that it gives grievance-satisfaction, indeed much more. To see clearly how matters stand with blame, let us recall certain essential distinctions already made or implied.

Moral disapproval, the source of blame, consists in two complexes of elements. Each complex is properly described as an attitude or feeling. One complex carries a belief or idea as to an agent's origination of a wrongful action, and our own related retributive desires to act against him. Moral disapproval when it is different has in it, in so far as ideas as to the initiation of action are concerned, only beliefs as to voluntariness, and does not include retributive desires. It does have in it, as the first complex also does, feelings of repugnance, and forward-looking desires as to the prevention of further actions of the kind in question, and perhaps restitution.

Our *blaming or condemning* is an activity, usually linguistic, not an attitude. It is evidently to be characterized, however, by reference to moral disapproval. To reprove, upbraid, or accuse a man or a group may or may not be to convey or presuppose ideas or beliefs about his origination of wrongful action, and may or may not be to convey or imply retributive desires. Thus blaming may or may not be an analogue to punishment in the latter's backward-looking character.

Thirdly, there is our *desert-claim that a man deserves blame or condemnation.* The claim, perhaps 'He deserves a reprimand', has to do with the effect of the activity of blaming him. We have in mind some kind or degree of distress. The desert-claim also has to do with his initiation of his wrongful action. It may have to do in part with his presumed origination of his wrongful action, or only with his voluntariness. There is also the matter of the relation or connection which is claimed to hold between the wrongful action and the distress. This requires attention.

As with punishment (p. 575), it is unlikely that a single relation is always asserted. Sometimes what is asserted is no more than the fact that by some precedent or rule, such wrongful actions get such blame. This is analagous to speaking of an offender's penalty as deserved, and meaning that it is the penalty according to a certain penalty-system in the law as it stands. Sometimes what is asserted is the moral judgement that it is fitting or right that such a wrongful action bring such pain or discomfiture on the agent. That is, the action and the distress form an intrinsic good. This is no claim of fact, but it can serve as a kind of premiss for the conclusion that someone was right to speak as he did. Such a point was noticed with the Intrinsic-Good Answer to the question of the nature of the desert-relation with punishment. (p. 568) Sometimes, finally, what is asserted *may* be partly that the blameworthy agent's pain or discomfiture satisfies desires on the part of the victim or victims of the blameworthy action, desires for just that pain or discomfiture.

Be all that as it may, it can hardly be doubted that our desert-claims about blame, even when they do not convey it, may involve desires or inclinations which we ourselves have for the pain or discomfiture of those we take to have behaved badly, wrongly, treacherously, or viciously. That this is true of our claims that blame is deserved cannot be surprising, given the fact that moral disapproval, so to speak, is a source of or enters into such desert-claims.

The contention that our desert-claims about blame may involve such desires on our part must not be confused with a related one, which is unacceptable. What was said of punishment and desert-claims, centrally, was that such claims *involve a single clear and firm argument* having to do with the satisfaction of the grievance-desires. Blame is different. It cannot be said of blame that in so far as we speak of it as deserved, there is involved a significant argument for it based on the satisfaction of retributive desires of either victims or ourselves. With punishment, there *is* the argument based on grievance, but it can hardly be said that there is a counterpart argument for blame. At any rate, to say so would be to engage in overstatement. Consider my claim that the moral condemnation by someone of the husband anticipating his divorce was deserved. It would at least be unpersuasive to say that the desert-claim involves something properly called the argument that the condemnation satisfies certain desires of either his wife or myself.

There are clear explanations of this discontinuity. One is that we can engage in blaming persons, and very often do, without their knowing it and hence without their being adversely affected by our activity. We also blame the dead. In neither case is there any possibility of satisfaction of the relevant kind for anyone. Punishment is different. Secondly, it is the case that desires for others' distress in connection with punishment give rise to demands or claims in a way that such desires in connection with blame do not, since punishable actions are more serious than actions which merit only blame. A third difference is that blame gives less satisfaction than punishment. A fourth, although not a simple one, is that the distress or frustration of persons blamed is less than the distress or frustration of persons punished. With blame, therefore, on the whole, there is less need for justification.

These latter reflections adequately explain the absence, with blame said to be deserved, of an argument analagous to the argument from grievance. They also serve, incidentally, to defeat a certain belated sceptical thought about our earlier conclusion about deserved punishment and the fundamentality to it of the argument from grievance. The sceptical thought is that that conclusion is put in doubt by the want of analogy with what is related to it, blame.

As for the main question before us now, the answer already suggested is not affected by what we have. That is, the absence of an argument analagous to the argument from grievance, and the reflections which explain the absence, do not put in doubt what was said about the consequence of determinism for moral blame: one kind of moral blame *is* vulnerable to an acceptance of determinism. This is so, to sum up, partly because the claim that blame is deserved itself may involve retributive desires or inclinations on the part of the claimer, which desires or inclinations rest on an idea of origination. Further, such desert-claims may involve assertions having to do with related retributive desires on the part of victims. However, another kind of blame is not vulnerable to determinism. It has to do only with voluntariness. It may involve, among other things, forcing a self-awareness upon a wrong doer, an awareness of his moral standing or rather the lack of it. This blame is a kind of analogue to institutions of punishment that are untouched by determinism.

So—an acceptance of determinism would not leave blame as it is, or leave us with nothing worth the name. This is the response of affirmation. The circumstance is essentially the same with moral praise or commendation. Consider, in this regard, a woman attracted to a comfortable position in her social world. Her life is lived among people prone to be embarrassed or worse by activities of a certain moral or political character. She none the less carries forward such a campaign, of which we approve, at some cost to herself. We take praise of her to be deserved.

Our doing so may rest on the idea that she could have done otherwise—that what she did was a matter of her own origination—and our praise does involve our desire or inclination to benefit her on this account. To attempt to deny this pair of facts, as some philosophers have, is simply to escape for a moment, usually in the pursuit of theory, from an awareness of entirely common feelings intimately connected with an idea as common, the origination idea. To come to see that origination is a fiction is to be unable to persist in the given desire. We may well feel attracted to her and thus inclined in a certain way to favour her over others. We may take ourselves to have reason to reward her, having to do with the future. But these are matters distinct from the feelings that get their nature from the belief that an agent in a fundamental sense could have done a lesser thing, but did a greater. Like blame, then, praise is necessarily affected but not demolished by an acceptance of determinism.

That is not the end of the story about blame and praise. Let us notice a large complication which so far has been suppressed, but also has relevance to punishment and blame. It is relevant too to the social institutions and the like still to be surveyed.

Punishment, moral blame, and moral praise, as we have seen, have grounds which are affected and grounds which are not affected by determinism. These grounds give rise to parts of the nature or features of different possible institutions and practices. The grounds and the natures or features which are affected have been spoken of in terms of desert. As for the grounds and features which are not affected, it has been said that they have to do not with origination of actions, but with their voluntariness. This is so with the preventive or like features of each of punishment, blame, and praise, and also their feature of involving feelings such as those of repugnance which are not intimately connected with origination.

In each of punishment, blame, and praise, certainly, talk of desert *does* most importantly have to do with origination, although to different degrees, and this fact has been reflected in our inquiry. However, as already emphasized, desert-locutions are put to very varied uses. To come to the main point, the large complication, desert-locutions *can* in fact be used to speak of the grounds and the nature or features of moral blame and praise, and perhaps punishment, which are *not* vulnerable to determinism.

Return to the worldly woman who nevertheless carries forward the estimable moral or political campaign. It is entirely natural to say that she deserves praise for her character, or for her good and firm intentions. That may be to say that she deserves praise on the ground of estimable actions, which, considered from the point of view of their initiation, are taken only as voluntary. Consider the vicious husband anticipating his divorce. It is, despite all that was said above, entirely natural to say that he deserves blame for his despicable intentions, and to conceive of them as no more than voluntary. All that is required here, it seems, for him to deserve blame, is for him to have acted as he did out of his own intentions, rather than for him to have been somehow ignorant of what he was doing, or to have acted in some other way involuntarily. (Cf. Parfit, 1984, pp. 323–6.)

There are uncertainties and obscurities here, but the main point is clear. Not all talk of our desert, of persons' deserving things, is welded to the idea of origination and what goes with it. If more reason for this conclusion is needed, some can be had from the fact already noticed, that pictures, other artefacts, and also natural objects, can be said to deserve attention, approval. It would be odd if this were so, but admirable human properties unconnected with origination were excluded from discourse in terms of desert.

None of this affects the conclusions that have been drawn about punishment, blame, or praise. These conclusions can as well be stated in a way which takes into account what has belatedly been allowed, the fact of the different uses of desert-locutions. Let us divide these uses into two categories, those which take origination as a condition of what is said to be deserved, and those which do not. We can label the first and main category of uses as having to do with origination-desert. Our conclusions are then essentially as follows. Each of punishment, blame and praise depends in part on, and has a nature or features in part owed to, considerations of origination-desert. Each stands in the same way to other considerations, including other considerations of desert. In so far as each depends on and is affected by origination-desert, it is vulnerable to determinism. In so far as each depends on and is affected by other considerations, including other considerations of desert, it is not vulnerable to determinism. Affirmation consists in part in eschewing origination-desert and embracing the other considerations.

To proceed now with the remaining five social institutions and the like, having to do with (ii) reward for obedience to law, (iii) income, (iv) wealth, (v) power, and (vi) kinds of rank or standing, the first of these is different from the rest. This is so partly because the reward in question, a life untroubled by policemen, judges, and gaolers, is in a way the same for all law-abiders. In the case of income and the other things, there is great variation in what is received or possessed. Also, reward in the given sense is also different in that the ground for getting it is in a way single and simple: obedience to law. Both points need qualification but let us not pause.

If we ask our principal question, whether an acceptance of determinism would somehow affect the given ground for reward, a certain answer may be attempted—no. This is the case, it may be said, since obedience to law is not necessarily something dependent on origination. In particular, if the reward of an untroubled life is taken as deserved by those who keep the law, the desert-claim is not to be taken as vulnerable to determinism. The desert-claim is to be characterized, perhaps, in terms of an idea not so far mentioned, that of compensation. We do commonly speak of compensation as deserved in many contexts. To say law-abiders deserve an untroubled life in the given sense is to say, or mainly to say, that such a life is a kind of compensation for the obedience to law, and more precisely for the restraint or self-denial of keeping to the law. The self-denying in question is to be understood, roughly, not as denying oneself goods in an ordinary sense, but as denying oneself all but the legal means of coming to possess or enjoy them. The self-denial consists in denying to oneself short-cuts to things we all desire. (Honderich, 1984b) To think of the self-denial as deserving the given reward, it may be said, is not to think of something affected by determinism. The self-denial is a matter of voluntariness.

Things may not be so simple, as can perhaps be shown by a piece of imagining. Imagine we were to come to believe in what can be called partial determinism. That is, there are two kinds of persons. Some of us are subject to determinism, and some of us have the capability of origination. In particular, the law-abiders contain some of the determined and some of the originative. Do all of the law-abiders deserve the same? To return to something like strict liability (pp. 562, 589), suppose further that we believe there are grounds for infringing ordinary expectations of some law-abiders, grounds having to do with the need for prevention. If we could, would we be inclined to choose from the class of the determined rather than from the class of the originative? It is possible to conjecture that we should take those who are law-abiding but subject to determinism as *less* deserving of the given reward than those who are law-abiding, but, in the requisite sense could have been otherwise.

Thus there is the thought, which might be supported in other ways, that our present situation, in which we do not believe in either partial or total determinism, is one where we take the general reward for obedience to law to rest in part on origination-desert. Let us move on to consideration of the remaining social facts, however, and draw a general conclusion at the end.

In the societies with which most of us are familiar, inequality in the distribution of *income* and what goes with it is great. Some idea of it can be had from the facts that the lowest tenth of families in terms of income in Britain and the United States receive something less than 3 per cent of the total income after tax, and the highest tenth receives upwards of 23 per cent. (Honderich, 1984b, p. 192) The justifications offered of the spread of incomes may be divided into three classes, those which have to do with desert of whatever kind, those which have to do with agreement-claims, and those which have to do with claimed economic and other effects of income-inequalities. (Cf. Dick, 1975.)

Of justifications of the third class, to begin with it, the most familiar has to do with incentive. The possibility or the actuality of earning more, it is said, gives rise to a greater social total of material and other goods. This, it may be said, is to the benefit of all, including those who earn less or nothing. It is possible to believe, as certainly I do myself, that the argument is greatly weaker than it is taken to be, and consists largely in unreflective habit and perhaps self-deception. It involves ignoring the conceivability and indeed the actuality of productive societies not organized by the income-incentive principle, whatever else is to be said of them. It is true, however, that in many societies some form of the income-incentive argument is accepted by many members. The argument, taken as having whatever weight, can of course be detached from Utilitarianism and also from doctrines of Conservatism, to which it is sometimes connected. It can figure, and does, in social and political philosophies whose principles of distribution are neither Conservative nor Utilitarian.

The second class of justifications, having to do with agreement-claims, is evidently of importance. It is with income above all, perhaps, as against other social institutions and the like, that agreement-claims are significant and of weight. Employers and employees agree that a certain salary or wage is to be paid for certain work. Here, despite considerable difficulties about degrees of voluntariness of agreements, and difficulties of other kinds, there is no need for the kind of speculativeness and unpromising struggle that goes into the idea of which we know, that offenders agree to their punishments, or the idea that we all somehow contract to abide by social arrangements.

The remaining class of income justifications in itself illustrates the diversity of talk of desert. A man is said to deserve higher pay or other financial benefits on one or more of many grounds: ability, skill or talent; effort, energy, or industriousness; productivity or profit-making; qualifications and the time it took to get them; a burden of responsibility or the danger or disagreeableness of a job; contribution to general economic well-being; popularity or simply demand for services. He may be said to deserve higher pay not for present attributes of these kinds but for his past history of effort, profit-making, or whatever. With respect to each of these attributes, he may be compared both to others in his own line of life, say the civil service, and to others in other lines of life.

Let us postpone consideration of the character and implications of these agreement-claims and desert-claims, and turn now to *wealth*. It is yet more unequally distributed than income, and the effect of this distribution on society is greater than that of income. As with income, a general idea of the distribution of wealth in many societies can be had from the shares of total personal wealth of the bottom and top deciles of families in terms of such wealth in Britain and the United States. The bottom decile of families has less than 1 per cent. The top tenth has roughly 80 per cent in Britain and 60 per cent in the United States. (Honderich, 1984b, p. 193) The attempted justifications of the distribution perhaps fall into roughly the same three classes as with income, partly for the reason of the connection between income and wealth. A part of a man's personal wealth may be saved or invested income, earned wealth.

The class of attempted justifications in terms of certain general *economic and social effects* includes the familiar idea that wealth concentrations are essential to provide capital for investment and economic advance. The claim in itself is false, as the existence of socialist societies demonstrates. Samples of other different arguments of this class are the argument from incentive, the argument that concentration of wealth allows for the patronage essential to art and the preservation of certain traditions of culture, and the argument, if it can be taken seriously, that such concentrations give a power to some which is important in the preservation of personal liberties generally. At least the first two arguments, whatever their worth, have a considerable acceptance. Both, perhaps like most arguments for wealth-concentration, can be offered in defence of inherited as well as earned wealth.

The second class of arguments offered by or on behalf of the wealthy is that their situation is the outcome of certain *agreements*. So too can the situations of others be presented as outcomes of agreements. It can be maintained, with earned wealth, that it is the outcome of the explicit agreements already mentioned in connection with income. The claim enters into a larger one covering inherited wealth as well. It is that the wealth distribution is the outcome of voluntary transfers between individuals. It is the outcome of exercises of defensible liberties. (Nozick, 1974b) The argument is at the very least open to question, essentially because it is incomplete. It is far from tolerable to suppose that *all* voluntary transactions between two individuals, whatever the effect on others, are anything like sacrosanct. Also, it is plainly insufficient to *declare* something to be a defensible liberty. Others can be as declarative. Substantial argument is required. Again, however, the argument from voluntariness and liberty has a considerable acceptance and a wide effect.

To come to the class of *desert-arguments*, what is said of deserved income evidently applies to the earned part of personal wealth. Desert-arguments surely have a kind of summation in one seemingly fundamental part of the most famous of justifications of private property, that of Locke. (1960 (1690)) A man's private property is partly justified by the fact of his having *mixed his labour* with something, perhaps a plot of land or raw materials. Too many inheritors of Locke's idea, for several reasons, do not attempt to explain why it is that mixing one's labour with something should give one a right to it or to some product. It is hard to resist the view that what is in question is fundamentally the claim that a man deserves something in virtue of his labour, effort, industriousness, or the like. Whatever the obscurity of talk of desert, that view of Locke is clearer and more effective than the idea, say, that a man has some sort of right to his labour and so comes to own the thing with which it has become inextricably mixed. (Cf. Becker, 1977.)

The last two of the array of social facts, more difficult to characterize briefly, are those of power and rank. With *power*, what is of most interest to the present inquiry are the many systems, hierarchies, and relations of relative power within societies: the devolution of authority in national and local political systems; pyramidal control in business, industry, the civil service and so on; chains of command in the military and the police; the organization of the judiciary; ordinary relations of overseeing and the like with respect to jobs. These are to be seen as effective decision-making systems, and the power of positions in them is perhaps well judged in terms of the number of people affected, according to the rules, by decisions taken at the positions.

In so far as these systems are true to a predominant theory of them, persons within them have the position, as it is said, to which they are suited. They have positions that suit their abilities, these being open to tests and judgements of various kinds. That is, the rationale is effectiveness in terms of the goal of the enterprise. None the less, judgements of desert are common, and often acted upon. Whatever the character and implications of such judgements, men and women are said to deserve to be made judges, drivers, editors, clerks. It is not merely that of two equally qualified candidates for a post, the more deserving in some sense may be appointed, but that a more deserving candidate may be appointed rather than a somewhat more able. Still, it is perhaps safe to say that in so far as relative power is distributed or possessed on grounds at all, as distinct from coming to someone through friends, social connection, and the like, it is effectiveness in terms of the goal of the endeavour that is paramount.

Positions of *rank, standing, respect, and the like*, by very rough definition, are desired positions which none the less are not positions of power. They do not, at any rate, involve defined relations of decision-making. Almost all of those which are hereditary, say aristocratic rank, are now but weakly defended by uncertain considerations. In the main these defences actually have more to do with what commonly goes with hereditary rank, which is to say wealth or power. Non-hereditary positions of rank and the like are various, but many are of the nature of awards or prizes. They are therefore very different from hereditary positions. Here there are memberships of orders of merit, prizes for books, tributes for long service. There are also more informal practices of respect. Certainly it can be said of all of these non-hereditary positions that they have the recommendation of encouraging achievement and excellence. Of all the seven social institutions, practices, and habits, however, it is the practice of according rank, standing, or respect to individuals that somehow is most governed by desert.

To look back and draw a general conclusion, it seems evident that reward for obedience to law, and the facts of income differences, wealth differences, power, and rank, all have grounds having to do with origination as well as grounds having only to do with voluntariness. Some of the grounds of the first kind are of origination-desert. Others, it can be argued, have to do with agreement-claims that do bring in origination. What we have is the conclusion that the given social institutions and the like are or will be *significantly affected* by determinism. Each of the institutions and the like makes *a* distinction between people having to do with origination, although a uniquely general one in the case of reward to law-abiders.

With the gradual acceptance of determinism, we shall have one less kind of reason for distinguishing as we do between people. The upshot, to the extent to which our existence is subject to rationality, a significant extent, must be a lesser difference between what people enjoy, possess and have imposed on them. That human life will change in this way does not involve the conclusion, as hardly needs saying, that it will change out of all recognition, or anything of the sort.

The proper response, less difficult to make with these public matters than with the more internal matters of life-hopes and the like, is again properly described as the response of affirmation. It is that change is forced upon us, not destruction, and that the change is better than tolerable. It may be aided by being perceived, as it can be, as having quite independent moral recommendations.

10.7 POLITICS

Let us finally speculate about something related, determinism and politics. What are, or will be, the political consequences of determinism, its consequences for the principal ongoing political traditions? These traditions include attitudes to and judgements on the social institutions and the like just considered, but are open to general characterization. Fundamentally they give encompassing if often veiled answers to the central question of political philosophy, that of the principle or principles on which our societies ought to be organized. The question, at bottom, is about the distribution of the things we all desire. The traditions, which make up what can properly be called the spectrum of politics, despite occasional and certainly misguided scepticism about that idea, are those of the Left, Centre, and Right. More particularly, let us get a view, although necessarily an impressionistic one, of the Left, Liberalism, and Conservativism.

The Left in politics has in it diverse principles, movements, parties, and national states, including opposed ones. It includes social democracy and democratic socialism, various ideologies of an egalitarian kind, the Eastern Communist states, other somehow Marxist states, some national liberation movements, and also labour or workers' movements, terrorist organizations, and the weak tradition of anarchism. Despite the range of ends and means exemplified, it is possible to bring the Left under general description, by way of certain pervasive features, and so to distinguish it from the rest of the political spectrum.

(i) It is traditional and correct to take the Left as committed to various ideas and ideals of *equality*, and correct to see it as having, in some places, partly realized some of them. Social democracy and democratic socialism, the latter giving more place to socialism, defend what is called social justice. 'We need to persuade men and women who are themselves reasonably well off that they have a duty to forgo some of the advantages they would otherwise enjoy . . .' (Jenkins, 1972) Many national movements are opposed to racial inequality and to the great imbalance in conditions of life between the world's rich and poor societies. The Eastern Communist states have acted to a considerable extent on Marx's declaration, 'From each according to his ability, to each according to his needs'. Marxist and Trotskyist ideologies which concentrate on the fact of what is called exploitation are in part protests against the unequal use of people by people, and also resulting inequalities.

(ii) A second large feature of the tradition of the Left, as is illustrated by good accounts of it (Berki, 1973), has consisted in demands and struggles for *freedoms* of many kinds. Democratic socialists and some social democrats have in effect sought greater democratic rights than those supported in the tradition of Liberalism. That is not to say they have been more zealous about the individual's right to vote and related political rights, but that they have been more committed to extending the reach or extent of governmental power, and hence, it can be argued, the worth of individual political rights. Democratic socialists and social democrats, in their support of such rights, distinguish themselves from the other part of the Left, more influenced by Marx.

It has defended political freedoms of a different and in a way a lesser kind, those of citizens of the one-party state or proletarian democracy, thereby producing a great divide in the Left. This more Marxist part of the Left, however, has pursued and in large part secured what is called freedom from economic oppression and distortion within politics. It can further be argued, with some success, that a fundamental Marxist commitment is to an ideal of freedom that has to do with the full development of individuals. (Brenkert, 1979, 1983; cf. G. A. Cohen, 1978) Finally, with respect to freedom and the Left, anarchism has traditionally taken the impulse of personal autonomy to an extreme degree.

(iii) There can be no doubt that *fraternity and community* have a considerable importance among ideals of the Left. In place of each of us being moved by kinds of self-concern, we are to have societies where each identifies with others and in an effective way with his or her society as a whole. Rousseau requires, if obscurely, not only that 'each of us puts his person and all his power in common under the supreme direction of the general will', but that 'in our corporate capacity, we receive each member as an indivisible part of the whole'. (1973 (1762)) China and Cuba have in ways been realizations of this.

To come to several related and more particular doctrines and practices, (iv) the capitalist system has been and continues to be condemned or questioned by the Left, to a greater or lesser degree, for its presumed injustice, inefficiency, waste, and irrationality. (v) Policies of public ownership or social control of the means of production, or at any rate of greater governmental entry into the economy, have derived from this condemnation and also from the various ideas and ideals of equality and freedom. (vi) The great divide in the Left in the matter of government, with the more Marxist part opposed to the traditional democracy of the multi-party state, is of course accompanied by a divide on other political means. There has remained a Marxist acceptance, not merely formal, of the means of armed revolution.

This slight sketch of the Left in politics, as does a fuller picture (Honderich, 1982b), suggests what can be taken as its fundamental commitment. It is a commitment to effective means to the end of making well off those who are badly off. The conditions of being well off and badly off may be defined in terms of fundamental human desires: for a decent length of life; an enlargement of life dependent on more material goods than merely enough to sustain a decent length of life; freedom and power in larger and smaller contexts; respect and self-respect; personal and wider human relationships; the goods of culture. The effective means to the end, making well off those who are badly off, include many crucial practices of equality, of which 'One man, one vote' is one, but not of overwhelming importance. The principle of the Left, that we should adopt effective means to the end of making well off those who are badly off, is partly for reason of these crucial practices properly named the Principle of Equality, mentioned earlier in several connections. (pp. 420, 591); Honderich, 1981c)

The first of the defining features of the Left specified above, a concern with various ideas and ideals of equality, is of course in accord with this view of the fundamental commitment of this part of the political spectrum. The concern for freedoms is a concern both for a part of the end of the principle of Equality and also for effective means to that end. Related remarks are to be made about the remaining features.

It is necessary to allow, certainly, that the Principle of Equality is no more than the *fundamental* commitment of the Left. It is true that other commitments and impulses are evident in both its past and its present. It is most relevant to ask this question: do considerations of desert of whatever kind play a significant part? Certainly what seem to be voices of desert have often been heard. Socialists have commonly claimed that those who toil must have their just reward. It needs to be allowed, and also for the reason that retributive impulses are a part of all our natures, that ideas of desert have been significant in the tradition of the Left. That they are secondary to the fundamental commitment is beyond question.

To turn now to the Centre in politics, if one seeks to characterize it, as is reasonable, as being other than an assortment of compromises between Left and Right, the most promising tradition is that of *Liberalism*. It can be described, although clarification is needed, as having two principal and connected ideals, those of political freedom and individualism.

(i) Liberalism has been an opposition to constraint by the state. (Ryan, 1970, Ch. 13; Manning, 1976). In the eighteenth and nineteenth centuries, in its assertion of political freedom against monarchical, sectional, and other unrepresentative government, it advocated what can be called balanced democracy, such as gives a certain general political freedom. This advocacy of political freedom has subsequently issued in opposition to movements and ideas of the Left. Liberalism has been opposed not only to systems of proletarian democracy, which does not much distinguish it, but also to increasing egalitarianism in the traditional democracies. It has had a fear of the increasing democratic power of the working classes. It has often enough seemed to be somewhat more concerned with Left than Right as enemy of the political freedom it defends, which tendency may be a true reflection of its character.

(ii) As for individualism, liberals of the eighteenth and nineteenth centuries were to the fore in seeking and defending rights of individuals in the matters of economic life, religion, thought and discussion, and private life. Mill's *On Liberty* (1972c (1863)) is the *locus classicus* of much of this individualism. The essay's clear intention, in which it does not notably succeed, is to state a clear, single principle of individual liberty, to specify a large part of life where individuals are to be left to govern themselves, free of interference by both state and society. The main idea or at any rate declaration is that individuals are to be free of interference in their lives so long as they do not harm others, although the essential definition of harm is far from evident and has been a matter of persistent controversy. (Honderich, 1974, 1982c)

Already in *On Liberty* there is evident another Liberal proposition with respect to individual life, one which has been more prominent since. It has to do with the state's interference, as it may somewhat controversially be called, to support and help citizens. Mill not only prohibits state interference in the lives of individuals if they do not somehow harm others, but also prohibits interference in their lives to help them. By inference at least, he also prohibits interference in the lives of some individuals in order to help others. The first group are not to be made to contribute to the second. He is opposed to the end of what we now know as the Welfare State, and at any rate by implication to its essential means. We are all to be responsible for ourselves, self-supporting and self-reliant, and indeed are to be constrained to do this. Liberalism, like all political traditions, has had to make some accommodation with history, and does so at this point, but its impulses are in a line of descent from Mill's.

If these are perhaps the principal features of Liberalism, there are other related ones worth distinguishing. (iii) Unlike much of the Left, and like Conservatism, Liberalism has in various ways defended the institution of private property and related rights. (iv) Both political freedom and those freedoms which enter into individualism are to be secured by way of, and to be rooted in, the rule of law. There is in Liberalism a greater attachment to the rule of traditional law than is evident on the Left. (v) Political and social change is to be gradual and democratic in nature. There is in Liberalism, also, a persistent tendency to call for the scrutiny of social and political change in terms of what is called rationality, reasonableness, enlightenment, or judgement. Liberals have taken themselves to be well placed with respect to such virtues. (vi) Mill was in no way atypical in his very considerable faith in the possible development of individuals, partly through education. The hope for individuals within Liberalism has perhaps sometimes been a quite general one, but perhaps more often not. Some are capable of worthwhile advance, the mass of individuals not so capable.

As shown in a fine account of Liberalism, it is not much more uniform than other political traditions, and certainly shows national differences. (Cranston, 1967) As with the Left, however, let us attempt a judgement as to the fundamental commitment or commitments of Liberalism, initially by way of a reflection on its two principal features, having to do with freedom and individualism. The political freedom defended by Liberalism has necessarily been a matter not only of one vote for each person, and of like political rights, but of the scope of political decision-making, the effective power of the elected government. The political freedom defended by Liberalism therefore becomes clearer, as does the associated idea of balanced democracy, when one brings to mind Liberalism's other principal feature, individualism. Individualism of the given kind sets a limit, whether or not a defensible one, to whatever political freedom is advocated along with it. Indeed, this has been clearly enough indicated in the Liberal tradition. Sir Henry Maine provides but one early example of this in his rooted opposition to

> the omnipotent democratic state . . . which has at its absolute disposal everything which individual men value, their property, their persons, and their independence, . . . the state which may make laws for its subjects ordaining what they shall drink or eat, and in what way they shall spend their earnings . . . and which, if the effect on human motives is what it may be expected to be, may force us to labour in it when the older incentives to toil have disappeared. (Maine, 1886, p. 156)

It is difficult to resist the idea, since a large part of what the doctrine of individualism defends has to do with personal achievement—the same personal achievement not to be impeded by the given political freedom—that Liberalism includes a considerable commitment to ideas of desert. More particularly, what is it that justifies constrained individualism, the doctrine that people are to be left to fend for themselves? Is it in part that they are to have no more than their earned deserts? What supports the related right of others to keep what they have gained by their own efforts? It is clearer here that desert of some kind is essential to the explanation. Further, ideas of desert are consonant with the secondary features of Liberalism, and to the fore in several of them, notably those having to do with private property and the development of more gifted individuals.

The tenor of many Liberal pronouncements are of relevance to this contention. To hinder the farmer from sending his goods at all times to the most profitable market, according to Adam Smith, 'is evidently to sacrifice the ordinary laws of justice to an idea of public utility, to a sort of reasons of state: an action of legislative authority . . . which can be pardoned only in cases of most urgent necessity'. (1844 (1776), p. 354) Under socialism, according to Maine again, 'no man is to profit by his own strength, abilities, or industry, but is to minister to the wants of the weak, the stupid and the idle.' (1886, p. 158)

In *On Liberty*, Mill announces that his Liberalism rests wholly on the Principle of Utility, but adds what he does not explain, that he means 'utility in the largest sense, grounded on the permanent interests of man as a progressive being'. (1972c (1863), p. 74) His subsequent defence and celebration of individuality is so relentless as to obscure matters further. Still, it is to be allowed that Mill's Liberalism does stand in some connection with some principle of utility, perhaps one whose end is no more precise than the general good. This is true as well of Liberalism generally. It needs to be allowed, also, that Liberalism has been and is informed by certain ideas and ideals of equality. This particular collection of such things is not identical with the related collection with distinguishes the Left and is not such as to allow summary by way of the Principle of Equality. It is such as to distinguish Liberalism from Conservatism.

It is reasonable to allow, then, that Liberalism has fundamental commitments to other than ideas of desert. The commitment to desert somehow conceived, however, is indubitable, and indubitably greater than the commitment of the Left.

Conservatism is sometimes characterized as resistance to fundamental social and political change, but misguidedly. Even if it is conceived in a restricted way, so as not to include all of the Right, and in particular not to include Fascism, which did advocate and in ways produce great and terrible change, Conservatism has often enough not resisted change but sought to secure it, change in the direction of the past, a Golden Age. As good accounts of Conservatism illustrate, Edmund Burke has in this respect had successors. (O'Sullivan, 1976) However, a more fundamental objection to characterizing Conservatism as resistance to change—which is *a* truth about it—is that this is to fail to get hold of what is basic. In fact, if Conservatism were no more than general resistance to change, it would be no less than wholly irrational. General resistance to change itself, on the *sole* ground that it is change, like general support for change in itself, is indeed wholly irrational.

There is some diversity in Conservatism, but evidently not so much as in the Left, or in all of the Right. Still, it is certainly possible to see social and political ideals, convictions, and practices that Conservativism in its continuing history has favoured, and certain that it has opposed. Most of these features do not openly reveal its fundamental commitment or nature, and none reveals all of it. However, they do point to that commitment or nature.

(i) Conservatism as we now recognize it came into existence as a response to the French Revolution, and it is indubitably a part of its nature that it is anti-egalitarian. It has been against almost all ideas and movements that can reasonably be called egalitarian. It has defended and continues to defend actual élites, and has occasionally aspired to the creation of new ones. Coleridge in his advocacy of a new spiritual leadership for English society, a clerisy, provides an example. (1972 (1830))

(ii) A very great deal of Conservatism has involved some ideal of an organic, hierarchical, ordered, or balanced society. An organic society is one of spiritual unity, or unity of feeling. It depends on authoritative institutions, above all strong government, and on a general acceptance of a certain ideology, insufficiently described as an ideology of a traditional state, an orthodox or pure culture, and perhaps an orthodox church. (Cf. Scruton, 1980) All of this is most notable in German thought, including that of Fichte, Novalis, Muller, Treitschke, and, in his way, Hegel. It was Novalis who recommended that a citizen should pay his taxes to the state in the spirit in which a lover gives presents to his mistress. Carlyle provides an English variation in his vision of an industrial society united by a deep sense of community and of mutual responsibility between social classes. (1888 (1843))

(iii) A defence of private property, including private ownership of the means of production, and also a defence of related structures of law and power and systems of morality, have always been features of Conservatism. It has been more encompassing and unyielding in this regard than Liberalism.

(iv) As already remarked, political traditions do necessarily accommodate themselves to history. Conservatism has done so with democracy. It is none the less not by impulse or conviction democratic. It remains opposed to the further democratization of democracy, as might be secured, say, by constraints on the financing of political parties, and also to the further democratization of institutions and practices within democracy.

(v) Conservatism has been very committed to restricting certain roles of the state and government in society, to preserving in certain ways a separation between state and society. The English *laissez-faire* economists will come to mind, as will American Conservatives of this century, sometimes dedicated above all to the proposition of limited government in so far as business and welfare services are concerned. In connection with limitations of the first kind, it has been common to speak of the defence of individual liberties.

(vi) Conservatism has tended to a low or pessimistic view of human nature, such that no great improvement in human affairs is possible. What has been delivered to us by the past has stood the test of time and is superior to any new order that could be the product of our own limited rationality, knowledge, and our strong passions. We ought to eschew *theory*. One related line of thought is that we are of very limited fellow-feeling, such that the more able will use their abilities only if they are given certain greater rewards, at bottom economic rewards. A further related line of thought is that we are such that any tolerable society must be one where we are subjected to traditional contraints of an external kind. It is not conceivable that a tolerable society could count on fraternity, community, or anything other than law as we know it, police, judges, and prisons, in order to preserve itself.

(vii) It needs to be mentioned that the inequalities Conservatism defends or prescribes have sometimes been defended by the claim that they are to the benefit of all. In English politics the idea is perhaps owed to Disraeli (1835). It is the idea, put in plain form, that the rich must be as rich as they are so that the poor are not poorer. The claim is related to ideas of paternalism, social obligation, and stewardship.

(viii) It needs to be mentioned too that Conservatism, despite much philistinism, has in several of its forms had connections with the defence of cultural excellence and its advance. T. S. Eliot provides an example. (1939) It has sometimes condemned the vulgar materialism of commerce and industry.

Anti-egalitarianism, the first of these features, is indeed fundamental to Conservatism. However, this is no more than a negative side. We require an enlightening positive characterization of Conservatism in order to have a true grasp of it. This is implied by its defence of élites but not made explicit. Is an enlightening positive characterization to be found in the second feature, having to do with a kind of society?

Hardly, since there are in fact many possible societies, including societies of the Left, which could fall under the description of being organic, hierarchical, ordered, balanced, or unified. The additional idea of a traditional ideology calls out as much for clarification. It is some particular *nature or character* of our traditional ideology that is fundamental and needs to be made clear. Certainly Conservatism is not committed to *any* conceivable traditional ideology.

The third feature, having to do with private property, has often issued in the perception of Conservatism as being no more than the self-interest of the well-placed. There is no doubt that Conservatism *is* an ideology of self-interest, but it can hardly be distinguished by that. It is as true that movements of the Left have an impulse of self-interest. It is quite as persuasive, also, to regard Liberalism as being moved in part by the self-serving inclinations of a bourgeoisie or middle class. For our purposes the relevant question about the third feature of Conservatism, like the others, is that of what positive principle or commitment, of a fundamental kind, can be offered in justification of private property and its accompaniments.

Nor does the fourth feature of Conservatism, its disinclination to democracy, being of a negative character, take us far foward. It is of course in intimate connection with the anti-egalitarianism. Like anti-egalitarianism, the disinclination to democracy does point to a conclusion which is left inexplicit. So with the fifth feature, a limited role for government in certain prescribed areas. What is the recommendation of limiting the activity of governments with respect to business and the welfare services? It may be thought that things are made plainer by the proposition that Conservatism is devoted to and defends individual liberties. But what liberties, in terms of their basis, are in question? To approach the same point differently, a liberty is evidently a power or want of constraint, which thing is in some way justified. What is the justification with the particular liberties defended by Conservatism?

The remaining features are no more revealing. With respect to the sixth, what *is* the recommendation of the ordering of things that has stood the test of time? It is certain that some characterization is assumed by Conservatives. If there were none, then of course the unavoidable response would be to try to mend or improve human nature, or to try to alter circumstances so as to take better account of our limited rationality, knowledge, and fellow-feeling, and our strong passions.

With the seventh feature, we do indeed come to an explicit claim as to the commitment or nature of Conservatism. It is, we are told, the political tradition moved by the proposition that the inequalities it defends are the means to the betterment of all, without exception. The proposition is false, as is the claim that it is anything like the mainspring of Conservatism. Both consist in what of course is no distinguishing feature of Conservatism, which is to say false propaganda. It is the most rebuttable of propositions that the rich must be as they are in our societies so that the poor are not poorer. Many counterparts of the proposition, pertaining to earlier wealth-distributions, social orderings, and a good deal else, are now accepted by all as having been no more than fictions.

Finally, the connection of Conservatism with cultural excellence is by no means general. The defence of such excellences, where it does exist, is wholly subordinate to such other features as the defence of private property.

All or most of these reflections can be encapsulated in a single question. What is the fundamental commitment or nature of Conservatism, which is in accordance with the defence of élites and opposition to egalitarianism and democracy, which recommends societies somehow unified by the particular traditional ideology of which we all know, which issues in defences of private property, of the restriction of government in so far as individual initiative or responsibility is concerned, and of our societies such as they are, and which is at least consistent with views as to our fallen nature and perhaps a certain aestheticism?

The answer, to repeat, cannot be that Conservatism is resistance to change, a pure principle of traditionalism. It would not be a service to Conservatism, either, to take it as no more than some congeries of intimations, intuitions, and unreflective feelings, as is done by defenders, without the approval of others. (Oakeshott, 1962, p. 168; Kirk, 1982, pp. xi ff.) What cannot be said cannot be a defence, and cannot reasonably be taken to have informed a political tradition of great strength. Nor could Conservatism be characterized as no more than an acceptance of what has been produced by a history of transactions or transfers, with little more said of the transactions than that they were somehow voluntary. (Nozick, 1974b) Certainly, it cannot be that Conservatism is at bottom informed by a Utilitarian principle, as may be suggested by the seventh feature, the mistaken claim that inequalities serve the well-being of all. That is not to deny, as can be added, that it has involved a certain sense of obligation to some of those in need. That falls entirely short of a Utilitarian commitment, or a commitment to anything like the Principle of Equality.

The only answer to the question of the fundamental commitment or nature of Conservatism is that it consists in a body of desert-claims. It is not too much to say, despite the misleading suggestion as to unity and simplicity, that Conservatism must be taken as founded upon the Principle of Desert, that each of us is to have what he or she deserves.

It is this which is very nearly in view with its defence of élites and its opposition to what are in effect refusals to go by desert—egalitarianism and democracy. It is this commitment which accords with its defence of our societies as they are, in which desert plays so considerable a part. So with its defence of private property in particular, and of the limitation of government in such ways as to leave room for or to compel individual initiative. The answer is supported, as well, by considering feelings with respect to the seven social institutions, practices, and habits. What feelings are identified with Conservatism? There is little possibility of disagreement. Conservatism stands in peculiar connection with retribution in punishment, and with defences in terms of desert of the general practice of reward, our income and wealth distributions, and positions of power and standing.

It would be a large task, and not greatly rewarding, to attempt to analyse meanings with respect to the body of desert-claims which are fundamental to Conservatism. Some tolerable idea of this body of desert-claims, of course, can be had from the reflections on punishment and the other institutions and the like, and of course on the identifying features of Conservatism. Some of the desert-claims in question are best seen in terms of the argument from grievance, others in terms of an image of a factual relation of proportionality or whatever, others in terms of an intrinsic good or of compensation.

What we have in sum, then, is that the commitment of the Left is to the Principle of Equality, and the commitment of Conservatism to desert. Inevitably there are other things to be said of Conservatism, but it is not to be regarded as having several distinct commitments in the way of Liberalism.

Desert-claims, by our earlier distinction, divide into two categories: those which take the origination of an action as a condition of what is deserved, and those which do not. The first are directly vulnerable to an acceptance of determinism. Our conclusion here, since Conservatism does to some significant extent rest on origination-desert, more so than any other political tradition, must therefore be that it is peculiarly vulnerable to determinism. The gradual acceptance of determinism, to the considerable extent that political traditions are rational, will peculiarly affect this political tradition. It will do less to Liberalism, and little to the Left. I leave to Conservatives the matter of reflecting on a response of affirmation.

10.8 RETROSPECT OF THE BOOK

We have a clear conception of the connection between a causal circumstance and an effect, got from our experience of the natural world. On this conception there rests a philosophy of mind and action, free of ancient and modern mystery. It, or something not fundamentally different from it, is likely true. It makes our choices and decisions, and our actions, into certain necessitated events. They cannot then derive from what is named origination. To be inclined to accept this is to have a problem of feeling, which one can attempt to resolve by a certain affirmation. That response first involves the main category of possible consequences of determinism, which includes life-hopes, personal feelings, knowledge, moral responsibility, the rightness of actions and the moral standing of persons. These great things are affected by determinism, but persist, and our lives do not become dark, but remain open to celebration. This response can also be made with a second category of possible consequences of determinism, more of the order of necessities than great things. These have to do with our social lives. It was Schopenhauer's view, perhaps, that our existence is to be mourned, that we would decline the gift of life if we could anticipate its nature beforehand. (1883 (1818), 1962 (*c.*1840)) Nietzsche, in his way also a determinist, said differently, that we may affirm life. (1954 (*c.*1880), 1966 (*c.*1880)) It is Nietzsche with whom we can and must agree.

References

ABELSON, R. R. *et al.*, 1980, commentaries on Searle, 1980, *The Behavioural and Brain Sciences*.

ADDIS, L., 1984, 'Parallelism, Interactionism, and Causation', in P. A. French *et al.*, 1984.

ADLER, M. J., 1958, *The Idea of Freedom: A Dialectical Examination of the Conceptions of Freedom*. Garden City, Doubleday.

ALEXANDER, M. P. and ALBERT, M. I., 1983, 'The Anatomical Basis of Visual Agnosia', in Kertesz, 1983.

ALEXANDER, P. P., 1866, *Mill and Carlyle: An Examination of Mr. John Stuart Mill's Doctrine of Causation in Relation to Moral Freedom*. Edinburgh, Nimmo.

ALSTON, W., 1967, 'Emotion and Feeling', in Edwards, 1967.

—— 1976, 'Self-Warrant: A Neglected Form of Privileged Access', *American Philosophical Quarterly*.

ANDERSON, A. R. and BELNAP, N. D., 1962, 'The Pure Calculus of Entailment', *Journal of Symbolic Logic*.

ANSCOMBE, G. E. M., 1957, *Intention*. Oxford, Blackwell.

—— 1971, *Causality and Determination*. Inaugural Lecture. Cambridge University Press. Reprinted in Sosa, 1975.

—— 1972, 'The Causation of Action'. Address to the Institut International de Philosophie, Cambridge, 1972.

ARISTOTLE, 1915 (*c*.350 BC), *The Works of Aristotle: Ethica Eudemia*, trans. W. D. Ross. Oxford University Press.

—— 1953 (*c*.350 BC), *The Ethics of Aristotle: The Nicomachean Ethics Translated*, trans. J. A. K. Thomson. Harmondsworth, Penguin.

ARMSTRONG, D. M., 1968, *A Materialist Theory of the Mind*. London, Routledge and Kegan Paul.

—— 1978a, *Nominalism and Realism*. Cambridge University Press.

—— 1978b, *A Theory of Universals*. Cambridge University Press.

—— 1980, *The Nature of Mind*. Brighton, Harvester.

AUSTIN, J. L., 1950, 'Truth', *Supplementary Proceedings of the Aristotelian Society*. Reprinted in Austin, 1961.

—— 1956, 'Ifs and Cans', *Proceedings of the British Academy*. Reprinted in Austin, 1961.

—— 1961, *Philosophical Papers*, ed. J. O. Urmson and G. J. Warnock. Oxford, Clarendon.

—— 1961a, 'Unfair to Facts', in Austin, 1961.

AYER, A. J., 1936, *Language, Truth and Logic*. London, Gollancz.

—— 1940, *The Foundations of Empirical Knowledge*. London, Macmillan.

—— 1954, *Philosophical Essays*. London, Macmillan.

—— 1954a, 'Privacy', in Ayer, 1954.

AYER, A. J., 1954b, 'The Indentity of Indescernibles', in Ayer, 1954.
—— 1954c, 'Freedom and Necessity', in Ayer, 1954.
—— 1961. *The Foundations of Empirical Knowledge*. London, Macmillan.
—— 1963, *The Concept of a Person and Other Essays*. London, Macmillan.
—— 1963a, 'What Is a Law of Nature?', in Ayer, 1963.
—— 1964, *Man as a Subject for Science*. Auguste Comte Memorial Lecture. London, Athlone.
—— 1972, *Probability and Evidence*. London, Macmillan.
—— 1976, *The Central Questions of Philosophy*. Harmondsworth, Penguin.
—— 1980, *Hume*. Oxford University Press.
—— 1984, *Freedom and Morality and Other Essays*. Oxford University Press.
—— 1984a, 'On Causal Priority', in Ayer, 1984.
—— 1984b, 'Identity and Reference', in Ayer, 1984.
—— 1984c, 'Freedom and Morality', in Ayer, 1984.
AYERS, M. R., 1968, *The Refutation of Determinism*. London, Methuen.
BAKUNIN, M., 1895, *Œuvres*. Paris.
—— 1953, *The Political Philosophy of Bakunin: Scientific Anarchism*, ed. G. P. Maximoff. New York, Free Press.
BECKER, L. C., 1977, *Property Rights*. London, Routledge and Kegan Paul.
BELINFANTE, F. J., 1973, *A Survey of Hidden-Variable Theories*. Oxford, Pergamon.
BELL, J. S., 1966, 'On the Problem of Hidden Variables in Quantum Mechanics', *Reviews of Modern Physics*.
—— 1971, 'Introduction to the Hidden Variables Question', in B. d'Espagnat, ed., *Foundations of Quantum Mechanics*. New York, Academic Press.
—— 1984, 'Beables for Quantum Field Theory', CERN preprint, TH.4035/84.
BENACERAFF, P., 1973, 'Mathematical Truth', *Journal of Philosophy*.
BENENSON, F. C., 1984, *Probability, Objectivity and Evidence*. London, Routledge and Kegan Paul.
BENNETT, J., 1980, 'Accountability', in van Straaten, 1980.
BENTHAM, J., 1950 (1840), *The Theory of Legislation*, trans. R. Hildreth. London, Routledge and Kegan Paul.
—— 1970 (1789), *An Introduction to the Principles of Morals and Legislation*, ed. J. H. Burns and H. L. A. Hart. London, Athlone Press.
BERKI, R. N., 1973, *Socialism*. London, Dent.
BERLIN, I., 1969, *Four Essays on Liberty*. Oxford University Press.
BEROFSKY, B., 1966, ed., *Free Will and Determinism*. New York, Harper.
—— 1971, *Determinism*. Princeton University Press.
BINKLEY, B., BRONAUGH, R., MARRAS, A., 1971, eds., *Agent, Action and Reason*. Oxford, Blackwell.
BISHOP, J, 1981, 'Peacocke on Intentional Action', *Analysis*.
BLAKEMORE, C., 1977, *Mechanics of the Mind*. Cambridge University Press.
—— 1985, 'The Nature of Explanation in the Study of the Brain', in Coen, 1985.
BODEN, M., 1977, *Artificial Intelligence and Natural Man*. Brighton, Harvester.
BOHM, D., 1957, *Causality and Chance*. London, Routledge and Kegan Paul.

—— 1962, 'A Proposed Explanation of Quantum Theory in Terms of Hidden Variables at a Sub-Quantum-Mechanical Level', in S. Korner, ed., *Observation and Interpretation in the Philosophy of Physics*. New York, Dover.

—— 1980, *Wholeness and the Implicate Order*. London, Routledge and Kegan Paul.

BOHR, N., 1934, *Atomic Theory and the Description of Nature*. Cambridge University Press.

BOSWELL, J., 1934–64 (1791), *Boswell's Life of Johnson*, ed. G. Birkbeck Hill and L. F. Powell. Oxford, Clarendon.

BOYLE, J. M. JR., GRISEZ, G., TOLLEFSEN, O., 1976, *Free Choice: A Self-Referential Argument*. Notre Dame, University Press.

BRADLEY, F. H., 1927, *Ethical Studies*. Oxford University Press.

BRADLEY, R. and N. SWARTZ, 1983, *Possible Worlds: An Introduction to Logic and Its Philosophy*. Oxford, Blackwell.

BRAITHWAITE, R. B., 1953, *Scientific Explanation*. Cambridge University Press.

BRAMHALL, J., 1676, 'A Defence of True Liberty', in his *Works*. Dublin. Passages quoted in Hobbes, 1839–45a.

BRAND, M. N., 1980, 'Simultaneous Causation', in van Inwagen, 1980.

—— and D. WALTON, 1976, *Action Theory*. Dordrecht, Reidel.

—— 1980, 'Simultaneous Causation', in Van Inwagen, 1980.

BRANDT, R. and KIM, J., 1967, 'The Logic of the Identity Theory', *Journal of Philosophy*.

BREITMEYER, B. G., 1985, 'Problems With the Psychophysics of Intention', *Behavioural and Brain Sciences*.

BRENKERT, G., 1979, 'Freedom and Private Property in Marx', *Philosophy and Public Affairs*.

—— 1983, *Marx's Ethics of Freedom*. London, Routledge and Kegan Paul.

BRENTANO, F., 1973 (1874), 'The Distinction Between Mental and Physical Phenomena', in his *Psychology From An Empirical Standpoint*, ed. O. Kraus and Linda McAlister. London, Routledge and Kegan Paul.

BRIDGMAN, P. W., 1927, *The Logic of Modern Physics*. New York, Macmillan.

—— 1936, *The Nature of Physical Theory*. Princeton University Press.

BROAD, C. D., 1925, *The Mind and Its Place in Nature*. London, Routledge and Kegan Paul.

—— 1934, *Determinism, Indeterminism and Libertarianism*. Cambridge University Press.

BRODAL, A., 1981, *Neurological Anatomy*. 3rd edn. Oxford University Press.

BRODMAN, K., 1909, *Vergleichende Lokalisationslehre der GroBrhirnrinde*. Leipzig, Barth.

BROGLIE, L. de, 1953, *The Revolution in Physics*, trans. R. W. Niemeyer. New York, Noonday.

BROWN, H. I., 1987, *Rationality*. London, Routledge and Kegan Paul.

BUDD, M., 1985, *Music and the Emotions*. London, Routledge and Kegan Paul.

BUNGE, M., 1959, *Causality: The Place of the Causal Principle in Modern Sciences*. Cambridge, Mass., Harvard University Press.

—— 1980, *The Mind–Body Problem: A Psychobiological Approach*. Oxford, Pergamon.

BURNS, B. D., 1968, *The Uncertain Nervous System*. London, Arnold.
BUSER, P. and ROUGEUL-BUSER, A., 1978, *Cerebral Correlates of Conscious Experience*. Amsterdam, Elsevier.
BUTLER, R. J., 1962, ed., *Analytical Philosophy*. 1st series. Oxford, Blackwell.
BUTTER, C. M., 1968, *Neuropsychology: The Study of Brain and Behaviour*. Belmont, Brooks/Cole.
CALVIN. W. H. and OJEMANN, G. H., 1980, *Inside the Brain: Mapping the Cortex, Exploring the Neuron*. New York, Mentor.
CAMPBELL, C. A., 1951, 'Is Free Will a Pseudo-Problem?', *Mind*.
CAMPBELL, K., 1970, *Body and Mind*. Garden City, Doubleday.
—— 1981, 'The Metaphysics of Abstract Particulars', in French *et al.*, 1981.
CARGILE, J., 1975, 'Newcomb's Paradox', *British Journal for the Philosophy of Science*.
CARLSON, N. R., 1977, *Physiology of Behaviour*. 1st edn. Boston, Aleyn and Bacon.
—— 1985, *Physiology of Behaviour*. 3rd edn. Boston, Aleyn and Bacon.
CARLYLE, T., 1888 (1843), *Past and Present*. London, Routledge.
CARNAP, R., 1962, *Logical Foundations of Probability*. University of Chicago Press.
—— 1966, *Philosophical Foundations of Physics*. New York, Basic.
CARPENTER, R. H. S., 1984, *Neurophysiology*. London, Arnold.
CARTWRIGHT, N., 1979, 'Causal Laws and Effective Strategies', *Nous*.
CAWS, P., 1979, *Sartre*. London, Routledge and Kegan Paul.
CHANGEUX, J. P., 1983, *L'Homme neuronal*. Paris, Fayard.
CHIHARA, C. S., 1972, 'On Alleged Refutations of Mechanism Using Godel's Incompleteness Results', *Journal of Philosophy*.
CHISOLM, R. M., 1957, *Perceiving: A Philosophical Study*. Ithaca, Cornell University Press.
—— *1966, 'Freedom and Action', in Lehrer, 1966.*
—— *1971, 'On the Logic of Intentional Actions', in Binkley et al.*, 1971.
—— 1976, 'The Agent as Cause', in Brand and Walton, 1976.
CHOMSKY, N., 1971, review of B. F. Skinner, *Beyond Freedom and Dignity*, *New York Review of Books*.
CHURCHLAND, P. M., 1970, 'The Logical Character of Action Explanations', *Philosophical Review*.
—— 1981, 'Eliminative Materialism and the Propositional Attitudes', *Journal of Philosophy*.
—— 1984, *Matter and Consciousness*. Cambridge, Mass., MIT Press.
CHURCHLAND, P. S., 1981, 'Is Determinism Self-Refuting?', *Mind*.
—— 1986, *Neurophilosophy: Towards a Unified Science of the Mind/Brain*. Cambridge, Mass., MIT Press.
COEN, C. W., 1985 ed., *Functions of the Brain*. Oxford, Clarendon.
COHEN, G. A., 1978, *Karl Marx's Theory of History: A Defence*. Oxford University Press.
COHEN, L. J., 1980, 'The Problem of Natural Laws', in Mellor, 1980.
COLERIDGE, S., 1972 (1830), *On the Constitution of Church and State*. London, Dent.

COMTE, A., 1875–7, *System of Positive Polity*, trans. J. H. Bridges. London, Longman Green.

—— 1905, *Positive Philosophy*, trans. H. Martineau. New York, Calvin Blanchard.

COTMAN, C. W. and MCGAUGH, J. W., 1980, *Behavioural Neuroscience*. New York, Academic Press.

COWAN, J. L., 1969, 'Deliberation and Determinism', *American Philosophical Quarterly*.

CRANSTON, M., 1967, *Freedom: A New Analysis*. London, Longman.

CRICK, F. H. C., 1979, 'Thinking About the Brain', in Piel *et al.*, 1979.

CURTIS, B. A., 1972, *Introduction to Neuroscience*. Philadelphia, Saunders.

DACEY, R., SIMMONS, R. E., CURRY, D. J., and KENNELLY, J. W., 1977, 'A Cognitivist Solution to Newcomb's Problem', *American Philosophical Quarterly*.

DANTO, A. C., 1973, *Analytical Philosophy of Action*. Cambridge University Press.

—— 1975, *Jean-Paul Sartre*. New York, Viking.

—— 1985, 'Consciousness and Motor Control', *Behavioural and Brain Sciences*.

DAVIDSON, D., 1980, *Essays on Actions and Events*. Oxford, Clarendon.

—— 1980a, 'The Logical Form of Action Sentences', in Davidson, 1980.

—— 1980b, 'Causal Relations', in Davidson, 1980.

—— 1980c, 'The Individuation of Events', in Davidson, 1980.

—— 1980d, 'Events as Particulars', in Davidson, 1980.

—— 1980e, 'Mental Events', in Davidson, 1980.

—— 1980f, 'Intending', in Davidson, 1980.

—— 1980g, 'Freedom to Act', in Davidson, 1980.

—— 1980h, 'Actions, Reasons and Causes', in Davidson, 1980.

DAVIES, L. H., 1972, 'They Deserve to Suffer', *Analysis*.

DAVIES, P. C. W., 1979, *The Forces of Nature*. Cambridge University Press.

—— 1984, *Quantum Mechanics*. London, Routledge and Kegan Paul.

DAVIS, L., 1979, *Theory of Action*. Englewood Cliffs, Prentice-Hall.

DAVIS, W. A., 1983a, 'Weak and Strong Conditionals', *Pacific Philosophical Quarterly*.

—— 1983b, 'The Two Senses of Desire', *Philosophical Studies*.

—— 1984, 'A Causal Theory of Intending', *American Philosophical Quarterly*.

DAVIS, W. H., 1971, *The Freewill Question*. The Hague, Nijhoff.

DEEKE, L., SCHEID, P, KORNHUBER, H. H., 1969, 'Distribution of Readiness Potential, Pre-Motor Positivity and Motor Potential of the Human Cerebral Cortex Preceding Voluntary Finger Movements', *Experimental Brain Research*.

DELONG, M. R., 1971, 'Central Patterning of Movement', *Neuroscience Research Programme*, Bulletin 9.

DENNETT, D. C., 1972, review of J. R. Lucas, *The Freedom of the Will*, *Journal of Philosophy*.

—— 1979, *Brainstorms: Philosophical Essays on Mind and Psychology*. Hassocks, Harvester.

DENNETT, D. C., 1984, *Elbow Room: The Varieties of Free Will Worth Wanting*. Oxford, Clarendon.

—— and HOFSTADTER, D. R., 1982, eds., *The Mind's I: Fantasies and Reflections on Self and Soul*. New York, Basic.

DENYER, N., 1981, *Time, Action and Necessity: A Proof of Free Will*. London, Duckworth.

DEUTSCH, J. A., 1973, *The Physiological Basis of Memory*. New York, Academic Press.

DICK, J. C., 1975, 'How to Justify a Distribution of Earnings', *Philosophy and Public Affairs*.

DIMOND, S. J., 1980, *Neuropsychology*. London, Butterworth.

DIOGENES LAERTIUS, 1925 (*c*. AD 225), *Lives of Eminent Philosophers*, trans. R. D. Hicks. London, Heinemann.

DISRAELI, B., 1884 (1835), *Vindication of the English Constitution*. London, Field and Tuer.

DOWNING, P. B., 1958–9, 'Subjunctive Conditionals, Time-order and Causation', *Proceedings of the Aristotelian Society*.

—— 1959, 'Levels of Discourse', Ph.D. dissertation, University of London.

—— 1970, 'Are Causal Laws Purely General?', *Supplementary Proceedings of the Aristotelian Society*.

DUMMETT, M., 1954, 'Can an Effect Precede its Cause?', *Supplementary Proceedings of the Aristotelian Society*.

—— 1978, *Truth and Other Enigmas*. London, Duckworth.

DYKES, R. W., 1983, 'Parallel Processing of Somatosensory Information: A Theory', *Brain Research Reviews*.

EARMAN, J., 1986, *A Primer of Determinism*. Dordrecht, Reidel.

EASTERBROOK, J. A., 1978, *The Determinants of Free Choice*. New York, Academic Press.

ECCLES, J. C., 1953, *The Neurophysiological Basis of Mind*. Oxford, Clarendon.

—— 1970, *Facing Reality*. London, English Universities Press.

—— and K. R. POPPER, 1977, *The Self and Its Brain*. Berlin, Springer.

EDDINGTON, A. S., 1939, *The Philosophy of Physical Science*. Cambridge University Press.

EDWARDS, P., 1961, 'Hard and Soft Determinism', in Hook, 1961.

—— 1967, *The Encyclopedia of Philosophy*. New York, Macmillan and Free Press.

—— 1967a, 'The Meaning and Value of Life', in Edwards, 1967.

EINSTEIN, A., 1950, *Out of My Later Years*. New York, Philosophical Library.

—— 1953, 'The Fundaments of Theoretical Physics', in Feigl and Brodbeck, 1953.

ELIOT, T. S., 1939, *The Idea of a Christian Society*. London, Faber and Faber.

ENGELS, F., 1978 (1934), *Anti-Dühring*, trans. E. Burns. London, Lawrence and Wishart.

EPICTETUS, 1928 (*c*. AD 100) *Discourses*, trans. W. A. Oldfather. London, Heinemann.

EPICURUS, 1926 (*c*.300 BC), *The Extant Remains*, ed. C. Bailey. Oxford, Clarendon.

EVANS, D. A. and LANDSBERG, P. T., 1979, 'Freewill in a Mechanistic Universe?', *British Journal for the Philosophy of Science*.
FARRELL, B., 1950, 'Experience', *Mind*.
FEIGL, H., 1953, 'Notes on Causality', in Feigl and Brodbeck, 1953.
—— 1958, 'The "Mental" and the "Physical" ', in Feigl *et al.*, *Minnesota Studies in the Philosophy of Science*, Vol. 2. Minneapolis, University of Minnesota Press.
—— and BRODBECK, M., 1953, eds., *Readings in the Philosophy of Science*. New York, Appleton-Century-Crofts.
—— and SELLARS, W., 1949, eds., *Readings in Philosophical Analysis*. New York, Appleton-Century-Crofts.
FEINBERG, J., 'Justice and Personal Desert', in *Doing and Deserving*. Princeton University Press.
FEIRTAG, M. and NAUTE, W. J. H., 1979, 'The Organization of the Brain', in Piel *et al.*, 1979.
FEYERABEND, P., 1962, 'Comments on Bohm', in S. Korner, ed., *Observation and Interpretation in the Philosophy of Physics*. New York, Dover.
—— 1963, 'Materialism and the Mind–Body Problem', *Review of Metaphysics*.
FEYNMAN, R., 1965, *The Character of Physical Law*. Cambridge, Mass., MIT Press.
FICHTE, J. G., 1873, 'The Closed Commercial State', in *Fichte's Works*, ed. W. S. Smith. London, Trubner.
FINDLAY, J., 1942, 'Gödelian Sentences: A Non-Numerical Approach', *Mind*.
FLANAGAN, O. J., 1984, *The Science of the Mind*. Cambridge, Mass., MIT Press.
FLETCHER, G., 1978, *Rethinking Criminal Law*. Boston, Little, Brown.
FLEW, A., 1965, 'A Rational Animal', in Smythies, 1965.
FODOR, J., 1968, *Psychological Explanation*. New York, Random House.
—— 1979, *The Language of Thought*. Cambridge, Mass., Harvard University Press.
FOLEY, R., 1977, 'Deliberate Action', *Philosophical Review*.
FOOT, P., 1957, 'Free Will as Involving Determinism', *Philosophical Review*.
—— 1978, *Virtues and Vices*. Berkeley, University of California Press.
FRANKFURT, H., 1969, 'The Principle of Alternative Possibilities', *Journal of Philosophy*.
—— 1971, 'Freedom of the Will and the Concept of a Person', *Journal of Philosophy*.
—— 1975, 'Concepts of Free Action', *Supplementary Proceedings of the Aristotelian Society*.
—— 1978. 'The Problem of Action', *American Philosophical Quarterly*
FREGE, G. 1960 (1892), *Translations from the Philosophical Writings of Gottlob Frege*, ed. P. Geach and M. Black. Oxford, Blackwell.
FRENCH, P. A., UEHLING, T. E., and WETTSTEIN, H. K., 1979, eds., *Midwest Studies in Philosophy IV: Studies in Metaphysics*. Minneapolis, University of Minnesota Press.
—— 1981, eds., *Midwest Studies in Philosophy VI: The Foundations of Analytic Philosophy*. Minneapolis, University of Minnesota Press.

FRENCH, P. A., UEHLING, T. E., and WETTSTEIN, H. K., 1982, eds., *Midwest Studies in Philosophy VII: Social and Political Philosophy.* Minneapolis, University of Minnesota Press.
—— 1984, eds., *Midwest Studies in Philosophy IX: Causation and Causal Theories.* Minneapolis, University of Minnesota Press.
GALL, F. J. and SPURZHEIM, J., 1810, *Anatomie et physiologie du système nerveux en général et du cerveau en particulier.* Paris, Schoell.
GARDEN, W., 1983, *Modern Logic and Quantum Mechanics.* Bristol, Hilger.
GARDNER, M., 1973, 'Mathematical Games: Free Will Revisited, With a Mind-Bending Prediction Paradox by William Newcomb', *Scientific American.*
GAUTHIER, D., 1971, 'Comments', on Hare, 'Wanting: Some Pitfalls', in Binkley *et al.*, 1971.
GAZZANIGA, M. S., 1970, *The Bisected Brain.* New York, Appleton-Century-Crofts.
GEACH, P., 1957, *Mental Acts.* London, Routledge and Kegan Paul.
—— 1969, *God and the Soul.* London, Routledge and Kegan Paul.
GETTIER, E., 1963, 'Is Justified True Belief Knowledge?', *Analysis.*
GLOVER, J., 1970, *Responsibility.* London, Routledge and Kegan Paul.
—— 1987, *I.* Harmondsworth, Penguin.
GÖDEL, K., 1934, Lectures at Institute of Advanced Study, Princeton.
GOLDMAN, A. H., 1979, 'The Paradox of Punishment', *Philosophy and Public Affairs.*
GOLDMAN, A. I., 1967, 'A Causal Theory of Knowledge', *Journal of Philosophy.*
—— 1970, *A Theory of Human Action.* Englewood Cliffs, Prentice-Hall.
—— 1976, 'The Volitional Theory Revisited', in Brand and Walton, 1976.
GOODMAN, N., 1965, *Fact, Fiction and Forecast.* London, Routledge and Kegan Paul.
GREER, K., 1984, 'Physiology of Motor Control', in Smyth and Wing, 1984.
GREGORY, R. L., 1981, *Mind in Science.* London, Weidenfeld and Nicolson.
GRICE, H. P., William James Lectures at Harvard. Unpublished.
—— 1975, 'Logic and Conversation', in P. Cole and J. Morgan, eds., *Syntax and Semantics,* Vol. 3. New York, Academic Press.
—— and P. F. Strawson, 1956, 'In Defence of a Dogma', *Philosophical Review.*
GRUNBAUM, A., 1953, 'Causality and the Science of Human Behaviour', in Feigl and Brodbeck, 1953.
—— 1971, 'Free Will and Laws of Human Behaviour', *American Philosophical Quarterly.*
GUSTAFSON, G., 1973, 'A Critical Survey of the Reason vs. Causes Arguments in Recent Philosophy of Action', *Metaphilosophy.*
HALDANE, J. B. S., 1932, *The Inequality of Man.* London, Chatto and Windus.
—— 1954, 'I Repent an Error', *The Literary Guide.*
HALL, J., 1960, *General Principles of Criminal Law.* 2nd edn. New York, Bobbs-Merrill.
HAMPSHIRE, S., 1951, *Spinoza.* Harmondsworth, Penguin.
—— 1959, *Thought and Action.* London, Chatto and Windus.
—— 1965, *Freedom of the Individual.* London, Chatto and Windus.
—— 1966, 'The Uses of Speculation', *Encounter.*

—— 1972a, 'Freedom of Mind', in *Freedom of Mind and Other Essays*. Oxford, Clarendon.
—— 1972b, 'Spinoza and the Idea of Freedom', in *Freedom of Mind and Other Essays*. Oxford, Clarendon.
——, GARDINER, P. L., MURDOCH, I., PEARS, D. F., 1963, 'Freedom and Knowledge', in Pears, 1963.
HANNAY, A., 1979, 'Proximality as a Mark of the Mental', in G. Ryle, ed., *Contemporary Aspects of Philosophy*. London, Oriel.
—— 1988, 'Coping with Consciousness'. Forthcoming.
HANSON, N. R., 1961, 'The Gödel Theorem: An Informal Exposition', *Notre Dame Journal of Formal Logic*. —— 1967, 'Philosophical Implications of Quantum Mechanics', in Edwards, 1967.
HARE, R. M., 1952, *The Language of Morals*. Oxford, Clarendon.
—— 1971, 'Wanting: Some Pitfalls', in Binkley *et al.*, 1971.
—— 1981, *Moral Thinking*. Oxford University Press.
HARMAN, G., 1973, *Thought*. Princeton University Press.
—— 1976, 'Practical Reasoning', *Review of Metaphysics*.
HARRE, R. and MADDEN, E. H., 1975, *Causal Powers: A Theory of Natural Necessity*. Oxford, Blackwell.
HART, H. L. A. , 1968, *Punishment and Responsibility*. Oxford University Press.
—— and HONORÉ, A. M., 1959, *Causation in the Law*. Oxford, Clarendon.
HASSETT, J., 1978, *A Primer of Psychophysiology*. San Francisco, Freeman.
HEGEL, G. W. F., 1929 (1840), *Science of Logic*, trans. W. H. Johnston and L. G. Struthers. London, Allen and Unwin.
—— 1942 (1840), *Philosophy of Right*, trans. T. M. Knox. Oxford, Clarendon.
—— 1944 (1840), *The Philosophy of History*, ed. J. Sibree. New York, Wiley.
HEIDELBERGER, H., 1966, 'On Characterizing the Psychological', *Philosophy and Phenomenological Research*.
HEISENBERG, W., 1930, *The Physical Principles of the Quantum Theory*, trans. C. Eckart and F. C. Hoyt. New York, Dover.
HEMPEL, C. G., 1962, 'Deductive Nomological vs. Statistical Explanation', in H. Feigl and G. Maxwell, eds., *Minnesota Studies in the Philosophy of Science*, Vol. 3. Minneapolis, University of Minnesota Press.
—— 1965, *Aspects of Scientific Explanation*. New York, Free Press.
—— 1966, *Philosophy of Natural Science*. Englewood Cliffs, Prentice-Hall.
—— and OPPENHEIM, P., 1953, 'Studies in the Logic of Explanation', in Feigl and Brodbeck.
HENN, F. A. and NASRALLAH, H. A., 1982, eds., *Schizophrenia as a Brain Disease*. Oxford University Press.
HERRNSTEIN, R. J. and BORING, E. G., 1966, *A Source Book in the History of Psychology*. Cambridge, Mass., Harvard University Press.
HESSLOW, G., 1976, 'Two Notes on the Probabilistic Approach to Causation', *Philosophy of Science*.
—— 1981, 'Causality and Determinism', *Philosophy of Science*.
HINTZ, H. W., 1961, 'Some Further Reflections on Moral Responsibility', in Hook, 1961.

HOBART, R. E., 1966, 'Free Will as Involving Determinism and Inconceivable Without It', in Berofsky, 1966.
HOBBES, T., 1839–45 (*c*.1650), *Works*, ed. W. Molesworth. London, Bohn. —— 1839–45a (*c*.1650), *Of Liberty and Necessity*, 1839–45 (*c*.1650).
—— 1839–45b (*c*.1659), *Leviathan*, 1839–45 (*c*.1650).
HONDERICH, T., 1968, 'Truth', in *Studies in Logic Theory, American Philosophical Quarterly* monograph, ed. Nicholas Rescher.
—— 1969, *Punishment, The Supposed Justifications*. London, Hutchison. Revised edition, 1984b. Harmondsworth, Penguin.
—— 1970, 'A Conspectus of Determinism', *Supplementary Proceedings of the Aristotelian Society*.
—— 1973, ed., *Essays on Freedom of Action*. London, Routledge and Kegan Paul.
—— 1973a, 'One Determinism', in Honderich, 1973.
—— 1974, 'The Worth of J. S. Mill's *On Liberty*', *Political Studies*.
—— 1975, 'The Use of the Basic Proposition of a Theory of Justice', *Mind*.
—— 1980a, *Violence For Equality: Inquiries in Political Philosophy*. Harmondsworth, Penguin.
—— 1980b, critical discussion of Kenny, *Will, Freedom and Power*, and *Freewill and Responsibility*, *Mind*.
—— 1981a, 'Psychophysical Lawlike Connections and Their Problem', *Inquiry*.
—— 1981b, 'Nomological Dualism: Reply to Four Critics', *Inquiry*.
—— 1981c, 'The Problem of Well-being and the Principle of Equality', *Mind*.
—— 1982a, 'Causes and *If p, even if x, still q*', *Philosophy*.
—— 1982b, 'Determinism and Politics', in French *et al*., 1982.
—— 1982c, '*On Liberty* and Morality-Dependent Harms', *Political Studies*.
—— 1982d, 'The Argument for Anomalous Monism', *Analysis*.
—— 1982e, 'Against Teleological Historical Materialism', *Inquiry*.
—— 1983a, 'The Principle of Equality Defended', *Politics*.
—— 1983b, 'Anomalous Monism: Reply to Smith', *Analysis*.
—— 1984a, 'Smith and the Champion of Mauve', *Analysis*.
—— 1984b, See Honderich, 1969.
—— 1984c, 'The Principle of Equality: Reply to Nathan', *Mind*.
—— 1984d, 'Actions and Psychophysical Intimacy', *Inquiry*.
—— 1984e, ed., *Philosophy Through Its Past*. Harmondsworth, Penguin.
—— 1985, 'The Time of a Conscious Sensory Experience and Mind–Brain Theories', *Journal of Theoretical Biology*.
—— 1986, 'Mind, Brain and Time: Rejoinder to Libet', *Journal of Theoretical Biology*.
—— 1987, 'Causation: Rejoinder to Sanford', *Philosophy*.
—— 1989?, 'Causation: One Thing Just Happens After Another', in L. E. Hahn, ed., *The Philosophy of A. J. Ayer, The Library of Living Philosophers*. LaSalle, Open Court.
—— and BURNYEAT, M. F., 1979, eds., *Philosophy As It Is*. Harmondsworth, Allen Lane.
—— and BOTTOMS, A., EDMUND-DAVIES, E., FLOUD, J., GOSTIN, L., GUNN,

J., TAYLOR, L., WALKER, N., 1983, 'Symposium: Predicting Dangerousness', *Criminal Justice Ethics*.
HOOK, S., 1961, ed., *Determinism and Freedom in the Age of Modern Science*. New York, Collier.
HORNSBY, J., 1980, *Actions*. London, Routledge and Kegan Paul.
HOSPERS, J., 1956, *Introduction to Philosophical Analysis*. London, Routledge and Kegan Paul.
HOSPERS, J., 1961, 'What Means This Freedom?', in Hook, 1961.
HUBEL, D. H. and WIESEL, T. N., 1977, 'Functional Architecture of Macaque Monkey Visual Cortex', *Proceedings of The Royal Society*.
—— ——, 1979, 'Brain Mechanisms of Vision', in Piel *et al.*, 1779.
HUBY, P., 1967, 'The First Discovery of the Freewill Problem', *Philosophy*.
HUME, D., 1888 (1739), *A Treatise of Human Nature*, ed. L. A. Selby-Bigge. Oxford, Clarendon.
—— 1902 (1748), *An Enquiry Concerning Human Understanding*, ed. L. A. Selby-Bigge. Oxford, Clarendon.
—— 1938 (1740), *An Abstract of a Treatise of Human Nature*, ed. J. M. Keynes and P. Straffa. Cambridge University Press.
—— 1963 (1748), *An Enquiry Concerning Human Understanding*, ed. L. A. Selby-Bigge. Oxford, Clarendon.
HUXLEY, F. H., 1893, 'Animal Automatism', in *Collected Essays*. London, Macmillan.
JAMES, W., 1890, *The Principles of Psychology*. Cambridge, Mass., Harvard University Press.
—— 1909, 'The Dilemma of Determinism', in *The Will to Believe and Other Essays*. New York, Longman.
JAMMER, M., 1966, *The Conceptual Development of Quantum Mechanics*. New York, McGraw-Hill.
—— 1974, *The Philosophy of Quantum Mechanics*. New York, Wiley.
JENKINS, R., 1972, *What Matters Now*. London, Fontana.
JOHNSON, W. E., 1940, *Logic*. Cambridge University Press.
KANDEL, E. R. and SCHWARTZ, J. H., 1985, *Principles of Neural Science*. 2nd edn. New York, Elsevier.
KANT, I., 1887 (1779), *Philosophy of Law*, trans. W. Hastie. Edinburgh, Clark.
—— 1943 (1781), *Critique of Pure Reason*, trans. J. M. D. Meiklejohn. London, Bell.
—— 1949 (1788), *Critique of Practical Reason*, trans. L. W. Beck. University of Chicago Press.
—— 1950 (1781), *Critique of Pure Reason*, trans. N. Kemp-Smith. London, Macmillan.
—— 1959 (1785), *Foundations of the Metaphysics of Morals*, trans. L. B. White. Indianapolis, Liberal Arts.
KAVKA, G. S., 1980, 'What Is Newcomb's Problem About?', *American Philosophical Quarterly*.
KENNY, A., 1963, *Action, Emotion and Will*. London, Routledge and Kegan Paul.
—— 1975, *Will, Freedom and Power*. Oxford, Blackwell.

KENNY, A., 1978, *Freewill and Responsibility*. London, Routledge and Kegan Paul.
—— 1979, *Aristotle's Theory of the Will*. London, Duckworth.
—— LONGUET-HIGGINS, H. C., LUCAS, J. R., WADDINGTON, C. H., 1972, *The Nature of Mind*. Edinburgh University Press.
KERTESZ, A., 1983, ed., *Localization in Neurophysiology*. New York, Academic Press.
KEYNES, J. M., 1952, *A Treatise on Probability*. London, Macmillan.
KIM. J., 1971, 'Materialism and the Criteria of the Mental', *Synthese*.
—— 1972, 'Psychophysical Supervenience', *Philosophical Studies*.
—— 1973, 'Causation, Nomic Subsumption, and the Concept of Event', *Journal of Philosophy*.
—— 1979, 'Causality, Identity and Supervenience in the Mind–Body Problem', in French *et al.*, 1979.
—— 1981, 'Causes as Explanations: A Critique', *Theory and Decision*.
—— 1982, 'Psychophysical Supervenience', *Philosophical Studies*.
—— 1984, 'Epiphenomenal and Supervenient Causation', in French *et al.*, 1984.
—— and BRANDT, R., 1967, 'The Logic of the Identity Theory', *Journal of Philosophy*.
KIRK, R., 1982, *The Portable Conservative Reader*. New York, Penguin.
KLEMKE, E. D., 1981, ed., *The Meaning of Life*. Oxford University Press.
KOHLER, W., 1966, 'Psychophysical Isomorphism', in Herrnstein and Boring, 1966.
KORNHUBER, H. H. 1974, 'Cerebral Cortex, Cerebellum, and Basal Ganglia: An Introduction to Their Motor Functions', in Schmitt and Worden, 1974.
KRAUT, R., 1986, 'Feelings in Context', *Journal of Philosophy*.
KRIPKE, S., 1971, 'Identity and Necessity', in M. K. Munitz, ed., *Identity and Individuation*. New York University Press.
—— 1980, *Naming and Necessity*. Oxford, Blackwell.
KUFFLER, S. W., NICHOLLS, J. G., MARTIN, A. R., 1984, *From Neuron to Brain*. 2nd edn. Sunderland, Sinauer.
KUHN, T. S., 1962, *The Structure of Scientific Revolutions*. Chicago University Press.
LACEY, H. and JOSEPH, G., 1968, 'What the Gödel Formula Says', *Mind*.
LACEY, N., 1988, *Punishment and Political Principles*. London, Routledge and Kegan Paul.
LACKNER, J. and GARRETT, M., 1973, 'Resolving Ambiguity: Effects of Biasing Context in the Unattended Ear', *Cognition*.
LAMETTRIE, J. O. de, 1974 (1748), *Man a Machine*, La Salle, Open Court.
LANDSBERG, P. T. and EVANS, D. A., 1970, 'Freewill in a Mechanistic Universe?', *British Journal for the Philosophy of Science*.
LAPLACE, P. S. de, 1951 (1820), *A Philosophical Essay on Probability*, trans. F. W. Truscott and F. L. Emory. New York, Dover.
LASHLEY, K., 1950, 'In Search of the Engram', *Society of Experimental Biology*, Symposium 4.
LEHRER, K., 1966, ed., *Freedom and Determinism*. New York, Random House.
LEIGHTON, S. R., 1985, 'A New View of Emotion', *American Philosophical Quarterly*.

LEPORE, E., 1986, *Truth and Interpretation: Perspectives on the Philosophy of Donald Davidson*. Oxford, Blackwell.

LEVI, I., 1975, 'Newcomb's Many Problems', *Decision and Theory*.

LEWIS, D., 1966, 'An Argument for the Identity Theory', *Journal of Philosophy*.

—— 1972, 'Psychophysical and Theoretical Identifications', *Australasian Journal of Philosophy*.

—— 1973, *Counterfactuals*. Oxford, Blackwell.

—— 1975, 'Causation', in Sosa, 1975.

LIBET, B., 1978, 'Neuronal vs. Subjective Timing for a Conscious Sensory Experience', in Buser and Rougeul-Buser, 1978.

—— 1985a, 'Subjective Antedating of a Sensory Experience and Mind–Brain Theories: Reply to Honderich', *Journal of Theoretical Biology*.

—— 1985b, 'Unconscious Cerebral Initiative and the Role of Conscious Will in Voluntary Action', *Behavioural and Brain Sciences*.

—— , WRIGHT, E. W., FEINSTEIN, B, PEARL, D. K., 1979, 'Subjective Referral of the Timing for a Conscious Sensory Experience', *Brain*.

LICHTENBERG, A. J. and LIEBERMAN, M. A., 1983, *Regular and Stochastic Motions*. New York, Springer.

LIGHTHILL, J., 1986, 'The Recently Recognized Failure of Predictability in Newtonian Dynamics', *Proceedings of the Royal Society*.

LOCKE, D., 1978, 'How to Make a Newcomb Choice', *Analysis*.

—— 1986, review of Dennett, 1984, *Philosophical Books*.

LOCKE, J., 1960 (1690), *Two Treatises of Government*, ed. P. Laslett. Cambridge University Press.

LOTZE, H., 1885, *Outlines of Practical Philosophy*, trans. G. T. Ladd. Boston, Ginn.

LOUX, M. J., ed., 1970, *Universals and Particulars: Readings in Ontology*. Notre Dame University Press.

LUCAS, J. R., 1961, 'Minds, Machines and Godel', *Philosophy*.

—— 1962, 'Causation', in Butler, 1962.

—— 1967, 'Freedom and Prediction', *Supplementary Proceedings of the Aristotelian Society*.

—— 1970, *The Freedom of the Will*. Oxford, Clarendon.

LURIA, A. R., 1973, *The Working Brain: An Introduction to Neuropsychology*, trans. B. Haigh. Harmondsworth, Allen Lane.

LYCAN, W. G., 1969, 'On Intentionality and the Psychological', *American Philosophical Quarterly*.

—— 1982, 'Functionalism and Psychological Laws', *Southwestern Journal of Philosophy*.

LYONS, W., 1975, 'Determinism and Knowledge', *Analysis*.

—— 1980, *Emotion*. Cambridge University Press.

MACDONALD, C. and G. M., 1986, 'Mental Causes and Explanation of Action', *Philosophical Quarterly*. Reprinted in Stevenson, Squires and Haldane, 1986.

MCGAUGH, J. L. and COTMAN, C. W., 1980, *Behavioural Neuroscience*. New York, Academic Press.

MCGINN, C., 1982, *The Character of Mind*. Oxford University Press.

MACINTYRE, A., 1967, *A Short History of Ethics*. London, Routledge and Kegan Paul.

MACINTYRE, A., 1971, *Against the Self-Images of the Age*. London, Duckworth.
—— 1971a, 'The Antecedents of Actions', in MacIntyre, 1971.
—— 1971b, 'Psychoanalysis: The Future of An Illusion', in MacIntyre, 1971.
—— 1971c, 'Emotion, Behaviour and Belief', in MacIntyre, 1971.
—— 1972, ed., *Hegel: A Collection of Critical Essays*. Garden City, Doubleday.
—— 1984, 'Hegel: On Faces and Skulls', in MacIntyre, 1972, and Honderich, 1984e.
MACKAY, D. M., 1960, 'On the Logical Indeterminacy of a Free Choice', *Mind*.
—— 1964, 'Brain and Will', in G. N. A. Vesey, ed., *Body and Mind*. London, Allen and Unwin.
—— 1966, 'Cerebral Organization and the Conscious Control of Action', in J. C. Eccles, ed., *Brain and Conscious Experience*. Berlin, Springer-Verlag.
—— 1967, *Freedom of Action in a Mechanistic Universe*, Eddington Memorial Lecture. Cambridge University Press.
MACKENZIE, B. D., 1977, *Behaviourism and the Limits of Scientific Method*. London, Routledge and Kegan Paul.
MACKIE, J. L., 1965, 'Causes and Conditions', *American Philosophical Quarterly*.
—— 1973, *Truth, Probability, and Paradox*. Oxford, Clarendon.
—— 1973a, 'Conditionals', in Mackie, 1973.
—— 1973b, 'Concepts of Probability', in Mackie, 1973.
—— 1974, *The Cement of the Universe: A Study of Causation*. Oxford, Clarendon.
—— 1976, *Problems From Locke*. Oxford University Press.
—— 1977, *Ethics: Inventing Right and Wrong*. Harmondsworth, Penguin.
—— 1979, 'Mind, Brain and Causation', in French *et al.*, 1979.
—— 1980, 'The Transitivity of Counterfactuals and Causation', *Analysis*.
—— 1981, 'The Efficacy of Consciousness: Comments on Honderich's Paper', *Inquiry*.
——, 1982, 'Morality and the Retributive Emotions', *Criminal Justice Ethics*. Reprinted in 1985b.
—— 1985, *Logic and Knowledge*, ed. J. Mackie and P. Mackie. Oxford, Clarendon.
—— 1985a, 'Newcomb's Paradox and the Direction of Causation', in Mackie, 1985.
—— 1985b, *Persons and Values*, ed. J. Mackie and P. Mackie. Oxford, Clarendon.
MAINE, H., 1886, *Popular Government*. London, Murray.
MAINZ, F., 1955, *Foundations of Biology*. University of Chicago Press.
MALCOLM, N., 1968, 'The Inconceivability of Mechanism', *Philosophical Review*.
MANNING, D. J., 1976, *Liberalism*. London, Dent.
MARKS, L. E., 1974, *Sensory Processes: The New Psychophysics*. New York, Academic Press.
MARX, K., 1971 (1859), *A Contribution to the Critique of Political Economy*. London, Lawrence and Wishart.

MATEER, C. A., 1983, 'Localization of Language and Visuospatial Functions by Electrical Stimulation', in Kertesz, 1983.

MAXWELL, N., 1985, 'Are Probabilism and Special Relativity Incompatible?', *Philosophy of Science*.

MELCHERT, N., 1986, 'What's Wrong with Anomalous Monism?', *Journal of Philosophy*.

MELDEN, A. I., 1961, *Free Action*. London, Routledge and Kegan Paul.

MELLOR, D. H., 1971, *The Matter of Chance*. Cambridge University Press.

—— 1980, *Prospects for Pragmatism*. Cambridge University Press.

—— 1982, 'Chance and Degrees of Belief', in R. McLaughlin, ed., *What? Where? When? Why?* Dordrecht, Reidel.

—— 1986, 'Fixed Past, Unfixed Future', in B. Taylor, ed., *Michael Dummett: Contributions to Philosophy*. The Hague, Nijhoff.

MELZACK, R. AND WALL, P. D., 1982, *The Challenge of Pain*. Harmondsworth, Penguin.

MILL, J. S., 1924 (1873), *Autobiography*. New York, Columbia.

—— 1961 (1843), *A System of Logic*. London, Longman.

—— 1972a (*c*.1860), *Utilitarianism, Liberty and Representative Government*, ed. H. B. Acton. London, Dent.

—— 1972b (1863), *Utilitarianism*, in Mill, 1972a (*c*.1960).

—— 1972c (1859), *On Liberty*, in Mill, 1972a (*c*.1860)

—— 1979, *Collected Works of John Stuart Mill*, ed. J. M. Robson. University of Toronto Press and London, Routledge and Kegan Paul.

——1979a (1865), *Examination of Sir William Hamilton's Philosophy*. In 1979.

MOODY, E. A., 1967, 'William of Ockham', in Edwards, 1967.

MOORE, G. E., 1912, *Ethics*. London, Williams and Norgate.

MORGENBESSER, S. and WALSH, J., 1962, eds., *Free Will*. Englewood Cliffs, Prentice-Hall.

MOYER, K. E., 1980, *Neuroanatomy*. New York, Harper.

MULLER, A., 1955, excerpts in H. S. Reiss, ed., *The Political Thought of the German Romantics*. London, Macmillan.

MUNITZ, M., 1971, *Identity and Individuation*. New York University Press

MURPHY, J. G., 1971, 'Three Mistakes About Retributivism', *Analysis*.

—— 1973, 'Marxism and Retributivism', *Philosophy and Public Affairs*.

NAGEL, E., 1953, 'The Causal Character of Modern Physical Theory', in Feigl and Brodbeck, 1953.

—— 1979, *The Structure of Science: Problems in the Logic of Scientific Explanation*. London, Routledge and Kegan Paul.

—— and NEWMAN, J. R., 1958, *Gödel's Proof*. London, Routledge and Kegan Paul.

NAGEL, T., 1979, *Mortal Questions*. Cambridge University Press.

—— 1979a, 'What Is it Like to Be a Bat?', in Nagel, 1979.

—— 1986, *The View From Nowhere*. Oxford University Press.

NAUTA, W. J. H. and FEIRTAG, M., 1979, 'The Organization of the Brain', in Piel *et al*., 1979.

NEWTON-SMITH, W. H., 1981, *The Rationality of Science*. London, Routledge and Kegan Paul.

NIETZSCHE, F., 1954 (*c.*1880), *The Portable Nietzsche*, ed. W. Kaufman. New York, Viking.

—— 1966 (*c.*1880), *Basic Writings of Nietzsche*, ed. W. Kaufman. New York, Viking.

NINO, C., 1983, 'A Consensual Theory of Punishment', *Philosophy and Public Affairs*.

—— 1987, 'Consenting to be Punished: A Reply to Professor Honderich'. Forthcoming.

NOVALIS, 1891, *The Thought of Novalis*, ed. M. J. Hope. London, Stott.

NOWELL-SMITH, P. H., 1954, *Ethics*. Harmondsworth, Penguin.

NOZICK, R., 1969, 'Newcomb's Problem and Two Principles of Choice', in N. Rescher, ed., *Essays in Honour of Carl G. Hempel*. Dordrecht, Reidel.

—— 1974a, 'Reflections on Newcomb's Problem: A Prediction and Free-will Dilemma', *Scientific American*.

—— 1974b, *Anarchy, State and Utopia*. Oxford, Blackwell.

—— 1981, *Philosophical Explanations*. Oxford, Clarendon.

OAKESHOTT, M., 1962, *Rationalism in Politics and Other Essays*. London, Methuen.

OAKLEY, D. A., 1985, ed., *Brain and Mind*. London, Methuen.

OCHS, S., 1965, *Elements of Neurophysiology*. New York, Wiley.

O'CONNOR, D. J., 1971, *Free Will*. Garden City, Anchor.

O'HEAR, A.,1984, 'Reply to Glassen', *British Journal of the Philosophy of Science*.

—— 1985, review of Popper, 1982a, 1982b, *mind*.

O'KEEFE, J., 1985, 'Is Consciousness the Gateway to the Hippocampal Cognitive Map?—A Speculative Essay on the Neural Basis of Mind', in Oakley, 1985.

OLAFSON, F., 1967, 'Jean-Paul Sartre', in Edwards, 1967.

OLDS, J. and MILNER, P., 1954, 'Positive Reinforcement Produced By Electrical Stimulation of Septal Area and Other Regions of Rat Brain', *Journal of Comparative and Physiological Psychology*.

OPPENHEIM, P. and HEMPEL, C., 1953, 'The Logic of Explanation'. in Feigl and Brodbeck, 1953.

O'SHAUGHNESSY, B., 1980, *The Will: A Dual-Aspect Theory*. Cambridge University Press.

O'SULLIVAN, N., 1976, *Conservatism*. London, Dent.

PAGELS, H. R., 1983, *The Cosmic Code: Quantum Physics as the Language of Nature*. London, Michael Joseph.

PAPINEAU, D., 1985a, 'Probability and Causes', *Journal of Philosophy*.

—— 1985b, 'Causal Asymmetry', *British Journal of the Philosophy of Science*.

PARFIT, D., 1984, *Reasons and Persons*. Oxford, Clarendon.

PEACOCKE, C. 1979a, 'Deviant Causal Chains', in French *et al.*, 1979.

—— 1979b, *Holistic Explanation: Action, Space, Interpretation*. Oxford, Clarendon.

PEARS, D., 1963, ed., *Freedom and the Will*. London, Macmillan.

—— 1967, 'Are Reasons for Actions Causes?', in A. Stroll, ed., *Epistemology*. New York, Harper and Row.

—— 1968, 'Predicting and Deciding', in P. F. Strawson, 1968, and Pears, 1975b.
—— 1971a, 'Wanting: Some Pitfalls', in Binkley *et al.*, 1971 and Pears, 1975b.
—— 1971b, 'Two Problems About Reasons for Actions', in Binkley *et al.*, 1971.
—— 1972, *What Is Knowledge?* London, Allen and Unwin.
—— 1973, 'Rational Explanation of Actions and Psychological Determinism', in Honderich, 1973.
—— 1975a, 'The Appropriate Causation of Intentional Basic Actions', *Critica.*
—— 1975b, *Questions in the Philosophy of Mind.* London, Duckworth.
PENFIELD, W., 1967, *The Excitable Cortex in Conscious Man.* Liverpool University Press.
—— and RASMUSSEN, T., 1957, *The Cerebral Cortex in Man: A Clinical Study of Localization of Function.* New York, Macmillan.
PERRY, J., 1975, ed., *Personal Identity.* Berkeley, University of California Press.
PHELPS, M. E. and MAZZIOTTA, J. C., 1985, 'Positron Emission Tomography: Human Brain Function and Biochemistry', *Science.*
PIEL, G. *et al.*, 1979, eds., *The Brain.* San Francisco, Freeman.
POLLOCK, J. L., 1976, *Subjunctive Reasoning.* Dordrecht, Reidel.
POPPER, K. R., 1982a, *Quantum Theory and the Schism in Physics*, ed. W. W. Bartley III. London, Hutchinson.
—— 1982b, *The Open Universe*, ed. W. W. Bartley III. London, Hutchinson.
—— and ECCLES, J. C., 1977, *The Self and Its Brain.* Berlin, Springer.
POWERS, J., 1982, *Philosophy and the New Physics.* London, Methuen.
PRICHARD, H. A., 1949, *Moral Obligation.* Oxford, Clarendon.
PUTNAM, H. , 1965, 'A Philosopher Looks at Quantum Mechanics', in R. G. Colodny, ed., *Beyond the Edge of Certainty: Essays in Contemporary Science and Philosophy.* Englewood Cliffs, Prentice-Hall.
—— 1975, *Mind, Language and Reality.* Cambridge University Press.
—— 1975a, 'The Meaning of Meaning', in Putnam, 1975.
—— 1981, *Reason, Truth and History.* Cambridge University Press.
—— 1983, 'Probability and the Mental', in D. P. Chattopadhyaya, ed., *Humans, Meanings and Existences.* New Delhi, Macmillan. Also in LePore. 1986.
QUINE, W. V., 1953, 'Two Dogmas of Empiricism', in *From a Logical Point of View.* Cambridge, Mass., Harvard University Press.
—— 1960, *Word and Object.* Cambridge, Mass., MIT Press.
—— 1970, 'On the Reasons for Indeterminacy in Translation', *Journal of Philosophy.*
QUINTON, A., 1973, *The Nature of Things.* London, Routledge and Kegan Paul.
RAWLS, J., 1972, *A Theory of Justice.* Oxford, Clarendon.
REID, T., 1969 (1788), *Essays on the Active Powers of the Human Mind*, ed. Baruch Brody. Cambridge, Mass., MIT Press.
RESCHER, N., 1968, ed., *Studies in Logical Theory. American Philosophical Quarterly* monograph.
ROBINSON, D. N., 1973, *The Enlightened Machine.* Encino, Dickenson.
RORTY, R., 1970, 'In Defence of Eliminative Materialism', *Review of Metaphysics.*

ROSE, S., 1976, *The Conscious Brain*. Harmondsworth, Penguin.

ROSENBLUETH, A., 1970, *Mind and Brain: A Philosophy of Science*. Cambridge, Mass., MIT Press.

ROUSSEAU, J. J., 1973 (1762), *The Social Contract and Discourses*, trans. G. D. H. Cole. London, Dent.

ROUTTENBERG, A. and LINDY, J. 1965, 'Effects of Availability of Rewarding Septal and Hypothalamic Stimulation on Bar Pressing for Food Under Conditions of Deprivation', *Journal of Comparative and Physiological Psychology*.

RUCH, F. L. AND ZIMBARDO, P. G., 1971, *Psychology and Life*. Glenview, Scott Foresman.

RUSSELL, B., 1917, *Mysticism and Logic. London, Allen and Unwin.*

—— *1917a, 'On the Notion of Cause, with Applications to the Freewill Problem', in Russell, 1917.*

—— *1917b, 'A Free Man's Worship', in Russell, 1917.*

—— *and* WHITEHEAD, A. N., *1910, Principia Mathematica*. Cambridge University Press.

RYAN, A., 1970, *The Philosophy of John Stuart Mill*. London, Macmillan.

RYLE, G., 1949, *The Concept of Mind*. London, Hutchinson.

SANFORD, D., 1976, 'The Direction of Causation and the Direction of Conditionship', *Journal of Philosophy*.

—— 1985, 'Causal Multiplicity and Causal Dependence', *Philosophy*.

—— 1988, *'If P, then Q': Theories of Conditionals Past and Present*. London, Routledge and Kegan Paul.

SANTAYANA, G., 1930, *The Realm of Matter*. London, Constable.

SARTRE, J.-P., 1957 (1943), *Being and Nothingness*, trans. H. Barnes. London, Methuen.

SCHAFFER, J. A., 1967, 'The Mind–Body Problem', in Edwards, 1967.

—— 1968, *Philosophy of Mind*. Englewood Cliffs, Prentice Hall.

SCHLESINGER, G., 1974, 'The Unpredictability of Free Choices', *British Journal for the Philosophy of Science*.

—— 1976, 'An Important Necessary Difference Between People and Mindless Machines', *American Philosophical Quarterly*.

SCHLICK, M., 1956, 'When is a Man Responsible?', in his *Problems of Ethics*, trans. David Rynin. New York, Dover.

SCHMITT, F. O. and WORDEN, F. G., 1974, eds., *The Neurosciences Third Study Program*. Cambridge, Mass., MIT Press.

SCHOPENHAUER, A., 1883 (1818), *The World as Will and Idea*, trans. R. B. Haldane and J. Kemp. London, Routledge.

—— 1962 (c.1840), *The Will to Live—Selected Writings of Arthur Schopenhauer*, ed. R. Taylor. Garden City, Doubleday. SCHRODINGER, E., 1935, *Science and the Human Temperament*, trans. J. Murphy. London, Allen and Unwin.

SCRIVEN, M., 1964, critical notice of E. Nagel, *The Structure of Science*, *Review of Metaphysics*.

SCRUTON, R., 1980, *The Meaning of Conservatism*. Harmondsworth, Penguin.

SEARLE, J. R., 1980, 'Minds, Brains and Programmes', *The Behavioural and Brain Sciences*.
—— 1981, 'Analytic Philosophy and Mental Phenomena', in French *et al.*, 1981.
—— 1983, *Intentionality: An Essay in the Philosophy of Mind*. Cambridge University Press.
—— 1984, *Minds, Brains and Science*. Reith Lectures. London, BBC.
SELLARS, W., 1966, 'Fatalism and Determinism', in Lehrer, 1966.
—— 1976, 'Volitions Reaffirmed', in Brand and Walton, 1976.
SHEPHERD, G. M., 1983, *Neurobiology*. Oxford University Press.
SHOEMAKER, S., 1980, 'Causality and Properties', in van Inwagen, 1980.
SIDGWICK, H., 1966 (1874), *The Methods of Ethics*. London, Macmillan.
KYRMS, B., 1980, Causal Necessity: A Pragmatic Investigation of the Necessity of Laws. New Haven, Yale.
SLOTE, M., 1980, 'Understanding Free Will', *Journal of Philosophy*. *MART, J. J. C., 1959, 'Sensations and Brain Processes', Philosophical Review.*
—— 1963, *Philosophy and Scientific Realism*. London, Routledge and Kegan Paul.
—— and WILLIAMS, B, 1973, *Utilitarianism: For and Against*. Cambridge University Press.
SMITH, A., 1844 (1776), *An Inquiry Into the Nature and Causes of the Wealth of Nations*. London, Nelson.
SMITH, P, 1982, 'Bad News For Anamalous Monism?', *Analysis*.
—— 1984, 'Anomalous Monism and Epiphenomenalism: Reply to Honderich', *Analysis*.
SMYTH, M. M. and WING, A. M., 1984, *The Psychology of Human Movement*. London, Academic Press.
SMYTHIES, J. R., 1965, ed., *Brain and Mind: Modern Concepts of the Nature of Mind*. London, Routledge and Kegan Paul.
SNYDER, A. A., 1972, 'The Paradox of Determinism', *American Philosophical Quarterly*.
SOLOMON, R. C., 1986, 'Emotions, Feelings and Contexts', *Journal of Philosophy*.
SOMMERHOFF, G., 1950, *Analytical Biology*. Oxford University Press.
SORABJI, R., 1980, *Necessity, Cause and Blame: Perspectives on Aristotle's Theory*. London, Duckworth.
SOSA, E., 1975, ed., *Causation and Conditionals*. Oxford University Press.
SPERRY, R. W., 1952, 'Neurology and the Mind–Brain Problem', *American Scientist*.
—— 1968, 'Mental Unity Following Surgical Disconnection of the Cerebral Hemispheres', *The Harvey Lecture Series*.
SPICKER, S. F. and ENGELHARDT, H. T., 1976, *Philosophical Dimensions of the Neuro-Medical Sciences*. Dordrecht, Reidel.
SPINOZA, B., 1910 (1678), *Ethics*, trans. W. Hale White. Oxford University Press.
—— 1949 (1678), *Ethics*, ed. J. Gutmann. New York, Hafner. SPRIGGE, T. L. S.,

1971, 'Final Causes', *Supplementary Proceedings of the Aristotelian Society*.
—— 1982, 'The Importance of Subjectivity: An Inaugural Lecture', *Inquiry*.
—— 1983, *The Vindication of Absolute Idealism*. Edinburgh University Press.
SQUIRES, E., 1986, *The Mystery of the Quantum World*. Bristol, Hilger.
STALNAKER, R. C., 1975, 'A Theory of Conditionals', in Sosa, 1975.
STEIN, J. F., 1982, *An Introduction to Neurophysiology*. Oxford, Blackwell.
—— 1985, 'The Control of Movement', in Coen, 1985.
STEINER, M., 1986, 'Events and Causality', *Journal of Philosophy*.
STEPHEN, J. F., 1863, *A General View of The Criminal Law of England*. London, Macmillan.
STEVENSON, L., SQUIRES, R., HALDANE, J., 1986, eds., *Mind, Causation and Action*. Oxford, Blackwell.
STICH, S., 1978, 'Autonomous Psychology and the Belief-Desire Thesis', *Monist*.
—— 1981, 'On the Relation Between Occurrents and Contentful Mental States', *Inquiry*.
—— 1983, *From Folk Psychology to Cognitive Science: The Case Against Belief*. Cambridge, Mass., MIT Press.
STOUTLAND, F., 1980, 'Oblique Causation and Reasons for Action', *Synthese*.
STRAWSON G., 1985, review of Dennett, 1984, *Times Literary Supplement*.
—— 1986, *Freedom and Belief*. Oxford University Press.
STRAWSON, P. F., 1950, 'Truth', *Supplementary Proceedings of the Aristotelian Society*.
—— 1952, *Introduction to Logical Theory*. London, Methuen.
—— 1959, *Individuals: An Essay in Descriptive Metaphysics*. London, Methuen.
—— 1962, 'Freedom and Resentment', *Proceedings of the British Academy*. Reprinted in Strawson, 1968.
—— 1968, ed., *Studies in the Philosophy of Thought and Action*. Oxford University Press.
—— 1979, 'Universals', in French *et al.*, 1979.
—— 1985, *Skepticism and Naturalism: Some Varieties*. London, Methuen.
—— and GRICE, A. P., 1956, 'In Defence of a Dogma', *Philosophical Review*.
SUDBERY, A., 1986, *Quantum Mechanics and the Particles of Nature*. Cambridge University Press.
SUPPES, P., 1970, *A Probabilistic Theory of Causality*. Amsterdam, North Holland.
—— 1984, *Probabilistic Metaphysics*. Oxford, Blackwell.
SWAIN, M., 1978, 'A Counterfactual Analysis of Event Causation', *Philosophical Studies*.
—— 1980, 'Causation and Distinct Events', in Van Inwagen, 1980.
TARSKI, A., 1944, 'The Semantic Conception of Truth', *Philosophical and Phenomenological Review*. Reprinted in Feigl and Sellars, 1949.
TAYLOR, R., 1966, *Action and Purpose*. Englewood Cliffs, Prentice-Hall.
THALBERG, I. G., 1977, *Perception, Emotion and Action*. Oxford, Blackwell.
—— 1978, 'Mental Activity and Passivity', *Mind*.

THORP, J., 1980, *Freewill: A Defence Against Neurophysiological Determinism*. London, Routledge and Kegan Paul.
TREITSCHKE, H. VON, 1963, *Politics*, trans. B. Dugdale and T. de Bille. New York, Harcourt Brace.
TRUSTED, J., 1984, *Freewill and Responsibility*. Oxford University Press.
UTTAL, W. R., 1978, *The Psychobiology of Mind*. Hillsdale, Lawrence Erlbaum.
UVAROV, E. B., CHAPMAN, D. R., ISAACS, A., 1979, *The Penguin Dictionary of Science*. Harmondsworth, Penguin.
VAIHINGER, H., 1924, *The Philosophy of 'As If'*, trans. C. K. Ogden. New York, Harcourt Brace.
VALENTINE, E. R., 1982, *Conceptual Issues in Psychology*. London, Allen and Unwin.
VAN HEIJENOORT, J., 1967, 'Gödel's Theorem', in Edwards, 1967.
VAN INWAGEN, P., 1974, 'A Formal Approach to the Problem of Freewill and Determinism', *Theoria*.
—— 1975, 'The Incompatibility of Freewill and Determinism', *Philosophical Studies*.
—— 1980, ed., *Time and Cause: Essays Presented to Richard Taylor*. Dordrecht, Reidel.
—— 1983, *An Essay on Free Will*. Oxford, Clarendon.
VAN STRAATEN, Z., 1980, ed., *Philosophical Subjects: Essays Presented to P. F. Strawson*. Oxford, Clarendon.
VENDLER, Z, 1962, 'Effects, Results and Consequences', in Butler, 1962.
VON NEUMANN, J., 1955, *Mathematical Foundations of Quantum Theory*. Princeton University Press.
VON WRIGHT, G. H., 1971, *Explanation and Understanding*. Ithaca, Cornell.
WALKER, N., 1983, comments in Honderich *et al.*, 1983.
WALKER, R., 1978, *Kant*. London, Routledge and Kegan Paul.
WALL, P. D., 1985, 'Pain and No Pain', in Coen, 1985.
WALTON, D. and M. N. BRAND, 1976, *Action Theory*. Dordrecht, Reidel.
WARNOCK, G. J., 1961, ' "Every Event Has a cause" ', in A. Flew, *Logic and Language*. 2nd Series. Oxford, Blackwell.
—— 1971, *The Object of Morality*. London, Methuen.
WARNOCK, M., 1965, *The Philosophy of Sartre*. London, Hutchinson.
—— 1973, 'Freedom in the Early Philosophy of J.-P. Sartre', in Honderich, 1973.
WATSON, G., 1979/80, critical notice of Kenny, *Will, Freedom and Power*, *Journal of Philosophy*.
—— 1982, ed., *Free Will*. Oxford University Press.
—— 1986, review of Dennett, 1984, *Journal of Philosophy*.
WEATHERFORD, R., 1982, *Philosophical Foundations of Probability Theory*. London, Routledge and Kegan Paul.
WEIL, V. M., 1980, 'Neurophysiological Determinism and Human Action', *Mind*.
WERTHEIMER, M., 1966, 'Psychophysiological Isomorphism', in Herrnstein and Boring, 1966.

WESTERMARCK, E., 1932, *Ethical Relativity*. London, Kegan Paul, Trench, and Trubner.

WHITE, M. G., 1965, *Foundations of Historical Knowledge*. New York, Harper.

WICK, W., 1964, 'Truth's Debt to Freedom', *Mind*.

WIGGINS, D., 1970, 'Freedom, Knowledge, Belief and Causality', in *Knowledge and Necessity*. Royal Institute of Philosophy Lectures, 1968/9.

—— 1973, 'Towards a Reasonable Libertarianism', in Honderich, 1973.

—— 1980, *Sameness and Substance*. Oxford, Blackwell.

WILKES, K. V., 1978, *Physicalism*. London, Routledge and Kegan Paul.

—— 1980, 'Brain States', *British Journal for the Philosophy of Science*.

WILLIAMS, B., 1960, 'Man As Agent: On Stuart Hampshire's Recent Work', *Encounter*.

—— 1963, 'Postscript', in Pears, 1963.

—— 1973, *Problems of the Self*. Cambridge University Press.

—— 1978, *Descartes: The Project of Pure Inquiry*. Harmondsworth, Penguin.

—— 1981, *Moral Luck*. Cambridge University Press.

—— and SMART, J. J. C., 1973, *Utilitarianism: For and Against*. Cambridge University Press.

WILLIAMS, G., 1961, *Criminal Law, The General Part*. London, Stevens.

WILSON, E., 1980, *The Mental as Physical*. London, Routledge and Kegan Paul.

—— 1981, 'Psychophysical Relations', *Inquiry*.

WITTGENSTEIN, L., 1967, *Zettel*, trans. G. E. M. Anscombe and G. H. von Wright. Oxford, Blackwell.

—— 1968, *Philosophical Investigations*, trans. G. E. M. Anscombe. 2nd edn. Oxford, Blackwell.

—— 1980, *Remarks on the Philosophy of Psychology*, trans. G. E. M. Anscombe. Oxford, Blackwell.

WOODFIELD, A., 1976, *Teleology*. Cambridge University Press.

WORKMAN, R. W., 1959, 'Is Indeterminism Supported by Quantum Theory?', *Philosophy of Science*.

YOUNG, J. Z., 1964, *A Model of the Brain*. Oxford, Clarendon.

—— 1978, *Programs of the Brain*. Oxford University Press.

—— 1985, 'What's In a Brain?', in Coen, 1985.

—— 1987, *Philosophy and the Brain*. Oxford University Press.

ZEKI, S. M., 1980, 'The Representation of Colours in the Cerebral Cortex', *Nature*.

ZIMBARDO, P. G. and RUCH, F. L., 1971, *Psychology and Life*. Glenview, Scott Foresman.

Index

BOCA RATON PUBLIC LIBRARY, FLORIDA
3 3656 0586565 5